EAC Occasional Paper No. 5

Occasional Publication of the Aerial Archaeology Research Group No. 3

Remote Sensing for Archaeological Heritage Management

EAC Occasional Paper No. 5

Europae
Archaeologiae
Consilium

Occasional Publication of the Aerial Archaeology
Research Group No. 3

aarg
Aerial Archaeology
Research Group

Remote Sensing for Archaeological Heritage Management

Proceedings of the 11th EAC Heritage Management Symposium,
Reykjavík, Iceland, 25-27 March 2010

Edited by David C Cowley

ENGLISH HERITAGE

Royal
Commission on the
Ancient and
Historical
Monuments of
Scotland

FORNLEIFAVERND RÍKISINS
The Archaeological Heritage Agency of Iceland

MENNTA- OG MENNINGARMÁLARÁÐUNEYTIÐ

SWEDISH NATIONAL HERITAGE BOARD
RIKSANTIKVARIEÄMBETET

EAC Occasional Paper No. 5

Occasional Publication of the Aerial Archaeology Research Group No. 3

Remote Sensing for Archaeological Heritage Management

Edited by David C Cowley

Published by:
Europae Archaeologia Consilium (EAC), Association Internationale sans But Lucratif (AISBL), Siège social

Koning Albert II-laan 19	Avenue Roi Albert II 19
P.O. Box 10	Boîte 10
1210 Brussel	1210 Bruxelles
Belgium	Belgique

www.e-a-c.org

In association with:
Aerial Archaeology Research Group

ISBN 978-963-9911-20-8

Brought to publication by Archaeolingua, Hungary
Managing editor: Elizabeth Jerem
Copy editing by Dorottya Domanovszky
Layout and cover design by Gergely Hős
Printed by Aduprint Printing and Publishing Ltd, Hungary
Distribution by Archaeolingua, Hungary

Cover image:
Airborne Laser Scan (LiDAR) of a forested area before and after filtering
(St. Anna in der Wüste, Austria). © *Michael Doneus and Klaus Löcker, LBI-ARCHPRO, Vienna*

Contents

Foreword

Katalin Wollák, **President of Europae Archaeologiae Consilium**

This publication is based on the 11th EAC Heritage Management Symposium held in Reykjavik during March 2010. The symposium emphasised the significance of remote sensing in exploration, identification and documentation, and in monitoring archaeological heritage in the living landscape. This topic is vitally important to heritage management, and the presentations demonstrated to the heritage experts, and scientific researchers alike, that these scientific techniques significantly enrich our knowledge of the past. Additionally, I refer to a nearly twenty year old document, the Valletta Convention of the Council of Europe, which encapsulates the principles of archaeological heritage management. Certain elements of the Convention could be updated, but the part of Article 3 that emphasizes the importance of applying, wherever possible, non-destructive methods of investigation, is as current today as it ever was.

The practice of archaeological heritage management has changed significantly over the last three decades. Besides excavation as the most common diagnostic practice the role of the different preventive techniques, such as geophysical survey, remote sensing and laboratory analysis has been increasing gradually. The papers illustrated the ways in which these techniques broaden perspectives towards a landscape approach. Therefore in those countries where both research and heritage management concentrate on individual elements of heritage, the application of remote sensing methods and the interpretation of the results is a huge stimulus to achieving a broader understanding of the historic environment.

At the symposium, which was organised within the framework of the 11th Annual Meeting of Europae Archaeologiae Consilium, 18 speakers gave presentations from eight countries. It was a special pleasure for me to have speakers (and additional authors in the publication) from the post socialist countries since in these parts of Europe aerial reconnaissance could only really be applied in practice during the last twenty years. Among the new techniques presented, the most spectacular was LiDAR (Light Detection And Ranging), also known as Airborne Laser Scanning (ALS), which offers several opportunities for archaeology. The most effective application is the exploration and mapping of features beneath tree canopy, which thus provides a way to explore such areas at a broad scale that cannot be equalled by other ground-based techniques.

The round table discussion at the end of the symposium offered speakers and participants an opportunity to talk about experiences from different countries. This discussion helped the editor of this volume to commission further articles. The diverse papers in this publication, therefore, provide detailed insights into the application of remote sensing methods in different contexts, and demonstrate the ways in which these techniques can be used for archaeological prospection, for mapping, digital recording, archiving and modelling of archaeological features and previously unknown phenomena, as well as monitoring of scheduled ancient monuments, explored sites and the historic landscape.

As archaeological excavations only shed light on a small slice of any given historical period, remote sensing offers enormous help in interpretation and reconstruction of palaeo- and present landscapes from prehistoric times to the period of the World Wars. For heritage managers, archaeologists and remote sensing experts it is essential to understand how the excavated elements are related to the historic environment. The authors represent almost the full palette of archaeology and within this volume the reader will inevitably meet certain opinions that can diverge from the position of the official heritage management in these countries.

The thoughtful reader of this volume will learn about a wide range of good practice. Although the general expectation of such symposia is to encourage the adoption of the 'best practice', each country has to elaborate its own methodology and should establish its own optimum approaches from an awareness of the existing opportunities. The articles published in this volume on remote sensing set out a professional benchmark of different approaches to support this aim, and the EAC is, therefore, delighted to publish these important papers in its Occasional Papers series.

We offer our congratulations to the contributors of the papers and I'd like to express our special thanks to Kristín Huld Sigurðardóttir for hosting the conference, to the EAC/AARG Working Party and personally to Dave Cowley for inviting the contributors and overseeing the production of this volume within a year of the symposium to be launched at the 12th EAC Heritage Management Symposium in 2011. We believe this volume will find a place in the libraries of the state heritage agencies, in universities, museums, excavation services, research centres and also on the bookshelf of the interested professional.

Budapest, 11th February 2011

Acknowledgments

The symposium on which this publication is based was funded by EAC, English Heritage, the Royal Commission on the Ancient and Historical Monuments of Scotland and the Aerial Archaeology Research Group. Thanks to the officers and staff of these organisations for their support in general, but also specifically in securing funding and helping develop the programme of lectures and in progressing this publication. Special mention is due to Dirk Krausse, Martin Kuna, Chris Musson, Adrian Olivier, Rog Palmer, Bernard Randoin, Barney Sloane and Katalin Wollák. The symposium benefited from in-kind support of many organisations who provided their staff time free of charge or funded their attendance. The financial support of the Riksantikvarieämbetet (The Swedish National Heritage Board) for publication costs is gratefully acknowledged.

The symposium was held in Iceland as Eyjafjallajökull began to erupt, providing the delegates with an opportunity to view the volcano (from a safe distance) on a memorable fieldtrip. Kristín Huld Sigurðardóttir was a perfect host, ensuring a smooth-running enjoyable event, which was supported by the Icelandic Ministry Education, Science and Culture and the Archaeological Heritage Agency of Iceland.

For editorial assistance and other help thanks to Sharron Corder, Johanna Dreßler (German abstracts), Catherine Fruchart (French abstracts), Ralf Hesse, Rachel Opitz and Steve Trow; and to Erzsébet Jerem, Dorottya Domanovszky and Gergely Hős at Archaeolingua for maintaining a tight production schedule with good humour.

David C Cowley
Edinburgh, January 2011

Opening address

Katrín Jakobsdóttir, Minister of Education,
Science and Culture, Iceland

I would like to welcome all the participants in the EAC Symposium to Iceland. Many of you are here for the first time, while some have come to our country before. I hope we will be warm and generous hosts to you all, and that you will leave Iceland with good memories and a favourable impression of the land and its people. I know that you have a very busy schedule ahead of you for the next two days, but I hope that the field trip on Saturday will give you an opportunity to see some of the country and the work that is being done here in the field of archaeology.

Your organization, the Europae Archaeologiae Consilium, is devoted to the managing of Europe's archaeological heritage. After thousands of years of human habitation Europe is covered with the remains of former times, some of them more obvious than others, some of them more interesting than others, but all worth consideration and protection, at least until they have been discovered, recorded and a decision has been made on their preservation. This is a tremendous task that you have undertaken.

One of the most pressing tasks for government in our times is to balance the need for preserving our nature and heritage as well as trying to create the best possible conditions for ourselves and future generations. These objectives do not always go hand-in-hand, and a good deal of balancing is needed to meet the needs of preservation on the one hand and social evolution on the other. The development in recent decades of remote sensing for archaeological remains has made this task a great deal easier than before, as it can help us to do research without the damage and cost associated with more traditional excavation. We now appreciate more than ever how destructive excavation, however carefully it is conducted, can be for delicate archaeological sites; we may destroy more than we save – and we simply do not have the time or the financial resources to properly carry out all the excavations that we feel are needed.

The title of this symposium, Remote Sensing for Archaeological Heritage Management in the 21st century, points out that there are a number of methods already available and in development that are revolutionizing traditional archaeological research. The names of these methods would have sounded completely alien to archaeologists 50 years ago: Satellite imagery, aerial photography, 3D surveying, airborne laser scanning, multi/hyper-spectral data, high resolution LiDAR – to most people this still sounds more like something out of a George Lucas space-age film than the names of the tools modern archaeologists must become familiar with.

Dear guests,
As the topics of this symposium show, the future of archaeological research is already here. You, the professionals, are aware of this, but the general public is not. I hope that this symposium organized by EAC here in Iceland may serve among other things to bridge this gap, and that you may also all learn a little from each other, sharing information, visions and plans for the future benefit of all those interested in the managing of Europe's Archaeological Heritage in the future.

Reykjavík, Iceland, 25th March 2010

1 | Remote sensing for archaeological heritage management

David C Cowley and *Kristín Huld Sigurðardóttir*

Abstract: This introductory paper defines the scope of remote sensing in this volume and introduces the key themes of landscape, management, integration and communication. The broad structure of the volume is outlined.

Introduction

This collection of papers arises from a symposium held in Reykjavik, Iceland, between the 25th and 27th of March 2010, organised by the Europae Archaeologiae Consilium (EAC) and the Aerial Archaeology Research Group (AARG) to review remote sensing for archaeological heritage management at the start of the 21st century. Remote sensing techniques, here taken to include airborne, underwater and geophysical recording, underpin much of what is known about the past, in both terrestrial and maritime contexts. Traditional aerial reconnaissance in light aircraft and recording using hand-held oblique photographs, for example, has revolutionised our knowledge of the past in many lowland arable areas. These well-established techniques continue to demonstrate their value by increasing what we know, while new approaches, for example in multi- and hyper-spectral data collection, challenge the ways in which we see the past, opening up massive new potential. Airborne Laser Scanning has rapidly established its credentials in recording the earth's surface rapidly, and clearly enough to reveal fine archaeological detail, even under a woodland canopy. Equally, the vast – and untapped – collections of aerial images from the past 70 years hold a massive potential to document the historic environment, some of it now destroyed by development in the second half of the 20th century.

These recording and prospection techniques, new and long established, are part of the very foundations of our knowledge of the past and should inform its effective management. However, this statement cannot be taken for granted, as there are often major dislocations between different parts of the profession. For example, the outcomes of national survey and mapping programmes are not routinely regarded as of any direct relevance by academics and other parts of the profession, for whom excavated and published material is key. Equally the relationship between data collection and informed management may not be as strong as it could be, even though remote sensing (data collection) is central to informing cultural resource managers of what is there. There should be a widespread understanding of these connections amongst archaeologists and heritage managers, but this is clearly not always the case and informed dialogue is needed. This can promote mutual understanding, where on the one hand cultural resource managers appreciate the technical issues and potential of the various techniques, and on the other the application of remote sensing properly supports management in its wider sense. At present, this connection is most often made in a site/area specific context, and is not as common at a landscape scale, though this volume amply illustrates that it is happening.

Central to this relationship is the connection between *knowledge* (ever more data) and *understanding* (interpretation) – and how this informs management. Effective and sustainable management, the central remit of EAC, has to be the key driver in developing these remote sensing approaches. Without this focus the acquisition of aerial photographs or other data sources may become little more than 'stamp collecting', and while academic research pays dividends in other ways, without a clear connection into management of the archaeological resource in the present – for the future – it runs the risk of being an essentially self-indulgent and introspective pursuit.

From site to landscape

Remote sensing techniques support the basics of discovering sites and monuments and creating reliable records, but also sustain registration and large-scale mapping, without which the environment cannot be effectively understood, characterised and managed for the future. These developments in the remote sensing techniques used by archaeologists have been accompanied by changing perspectives, especially in a shift of focus from traditional archaeological *sites* to *landscapes*, and to more broadly-based inventories of monument types that include recent military and industrial heritage.

This is valuable as it feeds directly into wider perspectives grounded in landscape, principally the European Landscape Convention (ELC) which aims to promote integrated landscape protection, management and planning. Landscape is explicitly recognised as a basic component of cultural heritage, which should be integrated with government policy. It enshrines the importance of the common landscape, as well as those that might be designated as rare or special, and is applicable in both maritime and terrestrial environments.

Management

The ELC, as a pan-European directive that encourages active engagement in understanding our complex and diverse landscapes, has as much to recommend it to remote sensing archaeologists as a means of contextualizing and establishing the management relevance of their work, as it does to Cultural Resource Managers tasked with implementing strategies to manage this resource. It is in force and commits its signatories to integrated landscape protection and management built on understanding, especially, for example, the in-depth knowledge illustrated by Dominic Powlesland in the work of The Landscape Research Centre (this volume). This type of approach is driven by a desire to improve knowledge and to develop the tools to see and analyse the landscape. This is especially important as considerations of landscape are often dominated by, or at least biased towards, those with a 'natural landscape' agenda – a concept that has no legitimacy anywhere in Europe, where all landscapes are cultural, even if the degree to which this is the case varies.

This concept is at the heart of attempts to understand and manage the past and the historic environment in which we live. Multi-scale analyses, from site level considerations to landscape-scale characterisation, are required to grasp what may be there to manage, and to prioritise qualitatively and quantitatively what is protected and how mitigation strategies can be designed if preservation *in situ* is not an option. Broad-brush characterisation of landscapes to outline general attributes can be valuable to create an overview. The historic environment is a mosaic of spatial and chronological attributes producing often complex palimpsests, all of which pose challenges to understanding. Characterisation of the landscape may provide a useful way to help match survey techniques to contexts, identifying strengths and weaknesses and adapting accordingly. Equally, assessments of information content (i.e. the potential contribution to knowledge that sites may have now or in the future) based on an appreciation of monument condition, for example, may enhance the future value of preserved monuments, while also helping to determine responses to threat through protection or planning processes. Heritage professionals cannot save everything, but by drawing on data that has both time-depth and spatial breadth created in a real-world context, they can promote robust frameworks for managing the environment.

Integration, communication and dissemination

A key theme of this volume is the importance of integrating techniques and data in understandings of the archaeological resource base from which both narratives and management strategies can be built. Beyond this basic requirement to integrate and utilize complementary approaches and datasets, there is a clear need for effective communication and working between different heritage professionals including the academic world, and across national and regional boundaries. These issues were aired during round-table discussion at the symposium where the differing motivations for policy makers and academics, for example, were made clear. However, various projects reported in this volume are excellent examples of the benefits that arise from good connections between researchers, policy makers and heritage managers, and are cause to be optimistic. There remains a strong need to extend this good practice more widely, effectively integrating research priorities with the demands of heritage management.

It will be clear from the papers in this volume that the change in approaches has often been dramatic, but that progress is not always even across Europe, as should be expected with differing landscapes and traditions. The symposium aimed to promote dialogue and understanding between different areas of the heritage field and to communicate the possibilities of differing approaches to colleagues across Europe. This aim is fundamental to the published volume – to promote mutual understanding and successful dialogue, and to encourage the continued adoption and development of best practice.

'Best practice' does not refer to a 'one size fits all' approach – but rather the informed application of techniques that are appropriate to particular contexts and fit for the purpose in hand. Thus detailed mapping can sit alongside broad-brush characterisation in supporting differing management issues, from the level of individual monuments to entire landscape areas. A critical view of what is fit for purpose is vital – guarding both against the danger of embracing 'new' approaches simply because they are new, and the inertia of embedded traditions in heritage management. Ready accessibility to data is vital if it is to feed effectively into planning, and effective dissemination requires imaginative and integrated approaches that break down the 'silos' into which some information is consigned.

The importance of challenging established practice and in particular breaking down compartmentalised approaches that reflect vested interests, entrenched positions and inter-institutional competition is illustrated in papers from Italy (Ch 3 – *Campana*) and Poland (Ch 13 – *Rączkowski*). In the former case an innovative and cost-effective approach developed for a large-scale infrastructure project, and validated by a national committee, was rejected by the autonomous regional heritage administration. In Poland, the difficulties of integrating compartmentalised institutional datasets are well illustrated. In both cases the cost effective recording and management of the historic environment is not necessarily to the fore, as tradition, vested interest and resistance to change exert an undesirable influence. This is an important area where the EAC can promote the creation of guidance in heritage management policy that supports the implementation of best practice. Just demonstrating best practice and potential gains from its use, as in the Italian case, is not enough. The political dimension of decision-making, especially in compartmentalised management structures, also needs to be pushed forward.

Structure of the volume

The majority of the papers in this volume were presented in Reykjavik though they have been supplemented by several specially commissioned contributions. The contributors come from a variety of universities and national and regional heritage institutions, and include academics as well as data-collectors and those involved in everyday conservation and protection of the historic environment. Thus, it should be noted that the views of authors do not necessarily represent a 'national' or institutional view, and may also include critical remarks which reflect the authors' own opinion. These are, however, presented in a spirit to foster constructive dialogue.

The structure of the volume reflects the general approach taken at the symposium, beginning with general papers, and then exploring specific techniques, data sources and contexts before concluding with a series of case studies from the length and breadth of Europe.

In Part 1 the importance of making remote sensing work for the profession is explored in three papers that deal with both general issues and the place of integrated suites of remote sensed (and other) data in national and local contexts. The integration of techniques in identifying and mitigating development or otherwise managing often highly complex archaeology is discussed in studies from the UK (*Powlesland*) and Italy (*Campana*). The work of The Landscape Research Centre described by Powlesland may be a pinprick on the map of Europe, but is one of best-studied and understood archaeological landscapes in the world, and its results provide lessons from which we can all learn. The potential to develop best practice drawing on integrated suites of remote sensed and other data for major infrastructure projects is powerfully demonstrated by Campana, in a paper that also illustrates the problems of ensuring that best practice is adopted. This is not something that can be taken for granted, no matter how good a case is presented because such decisions are always at risk of being tainted by considerations of vested interests and traditions in power. Powlesland and Campana demonstrate the strength of integrated techniques in understanding the environment and providing a solid basis for management. In the third paper of this section our ability to explore, understand and interpret the landscape are further examined in a paper that stresses the importance of broad-brush characterisation to help match techniques to context (*Cowley*).

In Part 2 the challenges and potential of new technologies are discussed, which in exploring both terrestrial and maritime contexts show how new environments – under water and under woodland canopy in particular – have become ever more accessible. The engagement of archaeologists with new approaches has often been irregular, with costs and a lack of appropriate skills amongst the principal problems. This is gradually becoming less of an issue with the now well-established 'new' technology of Airborne Laser Scanning (*Doneus & Briese*, *Shaw & Corns*) but remains a significant challenge for the highly complex data presented by multi- and hyper-spectral data (*Beck*) that is on the cutting edge of archaeological prospection. The effective use of such data is a keynote of a case study of integrated approaches from the UK (*Bennett* et al.), while the importance of 3D recording is also clearly stated (*Remondino*). The final two papers in this section explore geophysical survey techniques, in both terrestrial (*Gaffney & Gaffney*) and maritime (*Firth*) contexts, both papers emphasizing the importance of visualisations in understanding and communicating the past.

Part 3 turns to the fundamental contribution that remote sensing approaches can make to the creation and exploration of the archaeological resource base. Reliable inventories are central to informed management of cultural heritage, and papers from England (*Horne*), Poland (*Rączkowski*) and Germany (*Bofinger & Hesse*) illustrate differing approaches drawing on both traditional and cutting edge data sources to create extensive and consistent baseline datasets. The tremendous challenge of exploring and managing submerged palaeo-landscapes is laid out by *Fitch* and his collaborators, illustrating the potential value to archaeologists of data recorded by offshore industries for non-archaeological purposes. The dividend of opening up archival data as sources of otherwise unrecoverable heritage information is also underlined with reference to historic aerial photographs (*Ferguson*).

In Part 4 the importance of interpretation and understanding is explored in case studies from across the varied landscapes of Europe. The creation of broadly-based datasets that can be relied on by heritage managers and archaeological researchers requires understanding of source data in regional and local contexts. Exemplars of integrated approaches are illustrated from the Czech Republic (*Gojda*), France (*Georges-Leroy*) and Hungary (*Czajlik*), while the role that aerial photographs are playing in defining landscapes of national importance is discussed in an Icelandic case study (*Einarsson & Aldred*). The need for effective integration of planning and management with field strategies is further developed in a Slovenian case study (*Rutar & Črešnar*), while the importance of 'new' types of heritage contexts is shown through a consideration of the remains of World War I in Belgium (*Stichelbaut* et al.). Preliminary results from a five-year aerial archaeology project in Denmark (*Helles Olesen*) show the benefits of integrating traditional aerial reconnaissance and inspection of historic aerial photographs to record the historic environment, and in developing the use of the aerial perspective for monitoring monument condition. The penultimate paper, by *Palmer*, examines how expertise and experience is vital to securing the best use of aerial data (or any data for that matter). Illustrating the way in which he teaches/trains presents a model of best practice, but also raises the central theme of the final paper (*Musson*) where the needs for training and development are examined, in part taking an historical perspective but also laying out the challenges addressed in the European Union funded *ArchaeoLandscapes* project.

Remote sensing for archaeological heritage management

This volume, and the symposium on which it is based, originates in the work of the AARG/EAC working party established in early 2007 to work on the main areas of common ground between the organisations and in particular on the development of broad-based strategies, especially with reference to heritage management, standards and statements of best practice.

In compiling up-to-date statements of current practice in archaeological remote sensing this volume should help to encourage best practice across Europe. The key themes that run through the contributions are the importance of a management focus in a landscape context, the integration of techniques and thinking and the promotion of approaches that are fit for defined purposes and specific contexts. We will conclude by stressing the need for statements of best practice to develop into published guidelines and standards that may be adopted even where the will is not yet present.

Acknowledgements

Our thanks to Noel Fojut, Chris Musson, Adrian Olivier, Rog Palmer and Katalin Wollák for their comments on this text and to Anthony Beck, both for his comments and for taking notes during the round table discussion at the end of the symposium on which parts of this introduction are based.

PART 1

Making remote sensing work for archaeological heritage management

Integrated suites of prospection techniques are required to create reliable information that can support heritage management and academic research – here illustrated in the 'total archaeology' approach pioneered by the University of Siena.
© *Stefano Campana.*

2 | Identifying the unimaginable – Managing the unmanageable

Dominic Powlesland

Abstract: This paper identifies the roles that remote sensing has in the discovery and recording of archaeological sites, contexts and – by linking these together – landscapes. Drawing on over 30 years of work by the Landscape Research Centre in northern England as a model, the paper examines how large-scale excavations and a variety of remote sensing and survey techniques should be integrated to build an understanding of the archaeological evidence, in particular creating awareness of zones differing in their degree of visibility. Different approaches and sensors record varying information, which both confirm evidence from other sources and compliment them by providing unique data, contributing to a vast multi-layered map of this landscape. This complex landscape challenges established views of the past and requires detailed interpretation before effective strategies of management can be planned. Establishing how much of the remotely-sensed landscape is undamaged by recent agriculture or other means is a vital issue for planning, preservation or other mitigation strategies.

The landscape resource identified through the multi-faceted and long-term remote sensing programme in the Vale of Pickering highlights the challenges to current statutory designation arrangements presented by extensive complex archaeological deposits where the relationships between visibility, information return and future value are not straightforward.

Introduction

'... there are known knowns; there are things we know we know. We also know there are known unknowns; that is to say we know there are some things we do not know. But there are also unknown unknowns — the ones we don't know we don't know.'
Donald Rumsfeld (USA Secretary for Defense, 12 February 2002, unfortunately not speaking about remote sensing).

Had Donald Rumsfeld been an archaeologist his words, used here with reference to the existence or otherwise of weapons of mass destruction in Iraq, might not have been viewed by most as lunatic ramblings but rather as an insight into the problems of identifying the past through remote sensing. The archaeological landscape is like a book – the cover is formed from geology and topography, the pages are composed of soils, and the words from potsherds, bones, environmental and other evidence. The grammar is defined by the ditches, pits and other features that can frequently be identified through remote sensing and examined through excavation. The challenge is to read the book in spite of thousands of years of erosion, deposition, reworking, damage and decay.

If we wish to read the book then we must ensure that we have all the pages, that they are in the right order, that the words are on the page and that the grammar allows us to correctly interpret the words. Even then we need to address the multi-dimensional nature of the human past located most simply within three dimensional space but also framed within the temporal, environmental and social dimensions. If the book is to be of any relevance then it must be readable by many, without an obscure vocabulary, and open to revision and reordering unconstrained by 'established' theoretical paradigms, known 'knowns' and social and temporal frameworks derived from minimal and untested data-sets.

This paper is substantially based upon the search for such a book, a book targeted not so much at changing the past, for the past is as it is, but on changing our understanding of the past, the holistic objective of the field archaeologist. It draws upon more than 30 years of fieldwork and remote sensing in a single landscape, the Vale of Pickering in Eastern Yorkshire. The impact of this work is in no small way a consequence of a singular focus on a single landscape and the long term support of English Heritage and its predecessor organisations, through a multitude of different interlinked research and rescue archaeology driven projects, undertaken by a small team over more than three decades. The twin themes of remote sensing and heritage management, the focus of the EAC symposium in Reykjavik, are not inherently close bedfellows although, as will be argued here, heritage management without the underpinning of remote sensing and the understanding that this brings is, for landscapes at least, an unachievable concept.

Our subject, like most academic pursuits, is partially driven by fashion, fashion that is most often driven by objectives handed down by government where the objectives are frequently poorly linked to the realities necessary to fulfil them. Clearly we are all dedicated to the sustainability of the archaeological resource and securing its survival through its management. Equally we are all aware that we cannot secure anything but a tiny fraction of global heritage and thus our responsibilities are amplified in that we must make every attempt to secure the future of the 'best bits'; past experience has however revealed that the degree to which the 'best bits' maintain this status over time is not unequivocal. Many of us primarily concerned with active discovery, whether through excavation or remote sensing, have looked on in despair whilst 'theoretical archaeology' as a sub-species of our subject has carved a large and disconnected niche within the University based community, not because we are not also driven by ideas and theories but because we have observed so many pages of obscure language and theory, often derived from other subjects, developed without the underpinning of secure data often undermining our role as communicators and protectors of the past for the people who live and work within the present day landscape.

Ways of seeing: the nature of the evidence

Archaeologists, like most in the environmental sciences, rely heavily on distribution maps, distribution maps which claim to portray knowledge; undoubtedly the distribution map is an important vehicle for presenting and articulating distributions of data on a map; rarely, however, do we deal with the empty space that exists between the dots; we allow people to assume that the absence of evidence does indicate the evidence of absence. In almost every case the dots on the map reflect not reality but only what we know (Figures 2.1 & 2.2). If we accept that our distribution maps reflect not fact but only the known facts we must also consider the source of those facts, not so much where the information has come from but how the information has been secured.

The extraordinarily limited budgets available to national archaeological bodies such as English Heritage, and the historical development of our subject, has led us to a situation where the majority of our data has been secured not through proactive research but through serendipity. Our Sites and Monuments Records are dominated by 'sites' discovered as a consequence of other activities whether that be past building projects or casual discovery of material brought to the surface by ploughing. The distributions are skewed, reflecting those aspects most noticeable such as cemeteries demonstrated through the observation of skeletal material, and these distributions often reflect the distribution of modern human activity. In many areas, for instance on the top of the Yorkshire Wolds, the survival of some upstanding elements of past landscapes such as prehistoric earthen barrows, the frequent target of early antiquarians, give the impression that what we see is what we have, so that the lack of contemporary settlement evidence is effectively overlooked or not even considered. Even well constructed proactive research programmes, such as the annual air photographic campaigns supported through the government agencies in the UK, rely heavily upon the chance that the plane and viewer be in the right place at the right time, over the right crops, at the right stage of growth, in the right long-term climatic conditions, with the right lighting to observe and record crop-mark formations. The multitude of factors that influence the formation of crop-marks, whether recorded by air photography or by sophisticated multi-spectral sensors, are such that we can only hope to discover a tiny percentage of those features that are susceptible to aerial observation at any one time. Nevertheless the importance of aerial reconnaissance in building an evidential framework for past landscapes cannot be overestimated.

During the last three decades attitudes and approaches to aerial reconnaissance have radically altered, a situation influenced firstly in Britain and more recently in Europe by the Aerial Archaeology Research Group

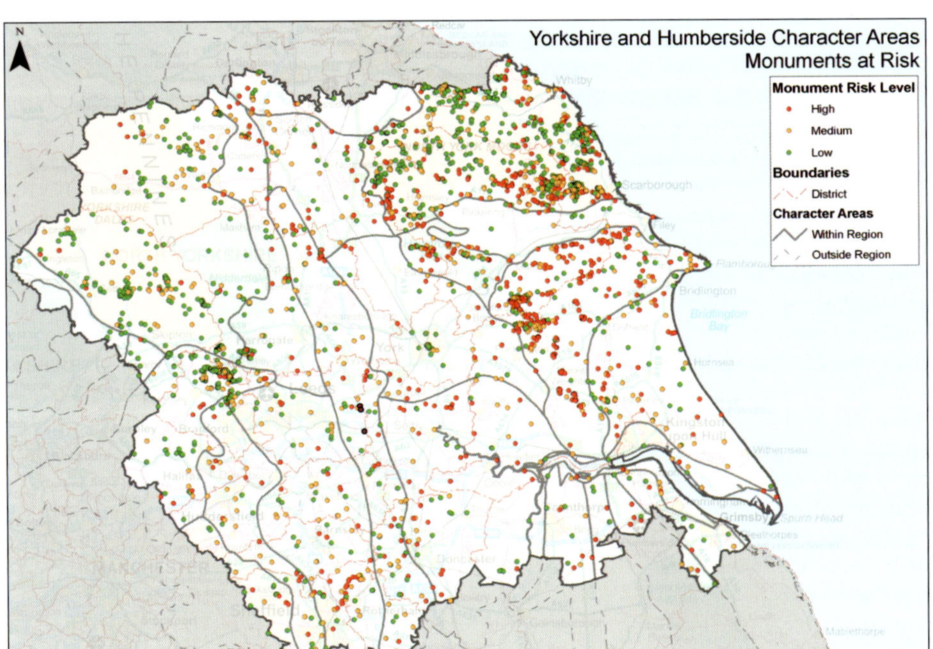

Figure 2.1: Distribution map showing the designated monuments at risk in Yorkshire & Humberside. *Reproduced with the permission of English Heritage.*

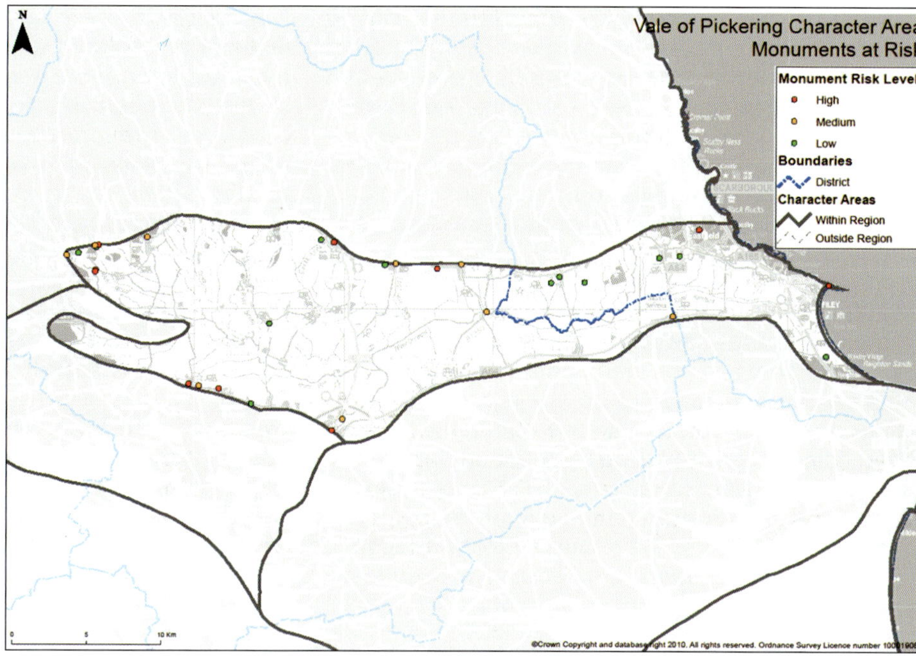

Figure 2.2: Distribution map showing the designated monuments at risk in the Vale of Pickering. *Reproduced with the permission of English Heritage.*

through its annual meetings and conferences like the EAC symposium in Reykjavik; presentations that used to be dominated by often stunning but random individual pictures of 'great sites' have been replaced by more considered regional surveys drawing upon a range of technologies rather than solely 35mm oblique photographs with the odd vertical large format image thrown in for effect. Changes in the nature and the role of airborne remote sensing increasingly reflect the importance of a landscape rather than a site based view. Similarly the practice of ground based remote sensing is increasingly being driven by the need to undertake large area multi-sensor surveys rather than targeted site specific surveys, a situation that has been influenced by changes in hardware that facilitate rapid surveying of large areas.

An integral component of Heritage Management is the need to monitor the status of what is being managed, but at a more fundamental level the resource cannot be managed or monitored without first identifying it; this may seem obvious but all too often we have seen Heritage Management schemes that are directed towards the most obvious heritage components without clear regard for context or the totality of the managed 'monument'. We are all aware of World Heritage Sites surrounded by fences that define the boundaries not of the 'site' in context but of those 'best bits' referenced above. The role of remote sensing in this environment must now be patently clear; in general terms remote sensing provides a variety of complementary methods that allow us to identify and characterise the archaeological resource and do this both relatively quickly and relatively cheaply. Regardless of improvements in technology and our ability to examine large areas rapidly, we need to remain fully aware that the complex combination of factors that contribute towards our ability to detect features is such that we are unlikely ever to develop a one-off approach that will return a comprehensive picture of past activity, and thus we must accept that landscapes will need to be constantly re-visited in different conditions and with different instruments to build up a comprehensive picture of the buried past. At the same time we must also remain aware that much of the evidence we seek may not be responsive to any current remote sensing technique and thus physical intervention through excavation may be the only way to assess or even identify the buried resource.

Research background: The Landscape Research Centre

The Landscape Research Centre (LRC) was established as a registered charity in 1980 to manage ongoing archaeological fieldwork projects and promote research into the evolution of the landscape, with particular reference to the Vale of Pickering in North Yorkshire. One of the benefits of the research undertaken by the LRC, with a team working and living in the same landscape for such a long time, is the ability to look reflexively at the results secured over a long period and the changes in landscape comprehension that have been required to accommodate the changing evidence base of the landscape that has emerged beneath our feet. With work spanning more than 30 years combining evidence from rescue excavations with multiple oblique air photographic sorties, two vertical air photographic and multi-spectral sorties and extensive airborne LiDAR, alongside more than 1,250 hectares of ground based geophysics, the combined dataset is amongst the most comprehensive gathered for a single landscape anywhere in Europe.

Origins

Our work was instigated following the chance discovery of an Early Anglo-Saxon or Anglian cemetery during sand and gravel extraction. In response to this discovery in the autumn of 1977 a salvage excavation was undertaken over the area threatened with immediate quarrying, by John Dent of the Humberside

Archaeology Unit, whilst arrangements were put in place for a program of rescue excavations designed to operate ahead of the quarry workings which were begun in 1978. Between 1978 and 1984 seasonal excavations in advance of sand and gravel extraction covered the larger part of an area measuring 600m × 100m, and revealed extensive multi-period activity from the Late Mesolithic to Early Medieval periods (Powlesland *et al.* 1986).

It was quickly realised that perhaps the most important characteristic of this site was the presence of deposits of blown sand that had preserved upstanding but buried Late Neolithic and Early Bronze Age round barrows as well as large areas of Late Bronze Age/Early Iron Age domestic activity in a way rarely found in Britain. The Early Anglo-Saxon cemetery was just one aspect of this site which indicated intensive land use spanning more than 6000 years. The rescue excavations at West Heslerton, funded from the public purse on the basis of hard fought for annual grant applications, reflected the rapid growth of rescue archaeology in Britain during the 1970s, and pre-dated the introduction of developer funded archaeology driven by Planning Policy Guidance Note 16 (PPG16) by more than a decade. Without the action of an observant quarry worker, walking his dog one lunchtime in 1977, it is pretty certain that none of what followed would have taken place and that the Vale of Pickering would have remained a distributional blank in archaeological maps. It should be noted here that Cook's Quarry site has over the last 30 years proved remarkably resistant to producing informative results from all forms of remote sensing; it is possible that even with the sort of evaluation techniques widely used to assess quarry sites today that the buried archaeology would most probably have been overlooked beneath the rather misleading sterile blown sands which seal the archaeology.

Within the regional setting in the late 1970s, the Yorkshire Wolds held a pre-eminent archaeological position particularly with reference to prehistory, largely on account of the many upstanding earthen monuments, particularly the barrows which had been the subject of investigations by Canon Greenwell and J R Mortimer (Greenwell 1877; Mortimer 1905) in the closing decades of the 19th century, a situation that still prevails today when viewed through a plot of monuments at risk (Figure 2.1). With the singular exception of the Late Palaeolithic and Early Mesolithic site of Star Carr, the Roman Fort and Vicus at Malton and a number of monastic institutions controlling river crossings, the Vale of Pickering appeared to have been little settled in the past. The fact that the picture as articulated through the monuments at risk map shows little change from the late 1970s to today reflects the peculiarities of past designation rather than any current reality of understanding, and yet this unfortunately tends to be reflected in government agency policy issues where non-archaeological agencies wish to deploy quick fix information in decision making rather than true knowledge. This situation is further exacerbated by the failure of so many to engage with the concept of archaeological landscapes rather than 'sites'.

Large scale rescue excavations

During the 1980s and 1990s our attention was primarily focussed upon large scale rescue excavations driven by research objectives and conducted within a context of annual large scale seasonal excavations undertaken with a volunteer workforce. These excavations fundamentally changed our comprehension both of the later Prehistoric and Anglo-Saxon activity in the Vale of Pickering, but also highlighted the need to develop the contextual evidence from the landscape beyond the trenches. The excavation of the Anglian cemetery, which had initially brought us to the area, was followed with the total excavation of the associated Early to Middle Saxon settlement at West Heslerton, extending over more than 20 hectares. Initially the excavation was approached within a research framework based upon the established view of small scale shifting settlement; it became rapidly clear that the established model of Early Anglo-Saxon settlement development that had arisen from the interpretations of the excavations at Mucking and West Stow was inappropriate for the evidence we were finding in Yorkshire (Hamerow 1993; West 1985). The evidence from all the excavations we had undertaken pointed towards much higher populations than we might have anticipated both in later Prehistory and in the Post-Roman period. Similarly, the accruing evidence from remote sensing indicated a densely utilised landscape reflecting deep chronological depth and continuity requiring regular re-assessment and re-interpretation based not so much upon established interpretations as upon a return to first principles.

Aerial survey

A programme of air photography was begun during the first season of excavation, both to secure aerial views of the large area excavations and to attempt to identify some context for the archaeological evidence within a landscape which was essentially devoid of documented evidence. With major multi-period and high density activity emerging at the Cook's Quarry site it was clear that the apparent emptiness in the landscape beyond was not a reflection of reality but of the parlous state of knowledge. By 1980, when the Landscape Research Centre was established as a charitable trust to manage and promote the work at West Heslerton, crop-mark evidence began to accrue indicating that far from being isolated the evidence being unearthed through excavation belonged within an extensively and intensively used landscape, although the prevailing conditions in the area immediately around the Quarry site produced only limited crop-mark returns. Initially flights were made through a local flying club, and with a tiny budget only limited sorties could be undertaken; however, thanks to the generosity of a local farmer who loved flying and who had his own Cessna 172 which he was happy to fly with the door removed, and who quickly developed an interest in aerial archaeology, we were alerted to the presence of crop-marks as they formed, and with the landing strip only a couple of kilometres away we could get airborne at very short notice, often multiple times in a week. Thanks to the generosity of our farmer flyer, Carl Wilkinson, we were able to document a large number of crop mark sites throughout our research area throughout the 1980s and early 1990s (Figure 2.3).

Figure 2.3: Combined and overlain air photographs showing a large multi-period cemetery with multiple types of barrows and flat graves in the left hand half and the slight crop-mark of a settlement enclosure in the upper right.

We adopted a policy of block flying, repeatedly covering our defined research area in all directions, circling as we moved along our flight-path to give the maximum chance of seeing features and minimise view-angle effects. The few known crop-mark sites recorded by St. Joseph and others over the Vale or other evidence recorded in the emerging Sites and Monuments Record gave us no reason to believe that the density of activity throughout much of the Vale of Pickering was anything other than very low. In many ways it could be argued that we were taken in by the lack of knowledge and we too imagined a landscape dominated by 'empty wetlands', 'sparsely utilised poor sandy soils on the valley margins' and 'extensively wooded areas' both on the heavier soils and on the scarp slopes of the Yorkshire Wolds and the North Yorkshire Moors, a landscape that was wholly in contrast with the established picture of the Yorkshire Wolds with its wealth of prehistoric evidence which had largely been accrued in the 19th century. In essence, at the start of our research we had no evidence of absence and a large absence of evidence, and thus shortcomings in our imagination of what may lie in the landscape reflected a poor understanding of the evidence base. Each year as the excavations preceded the operations of the quarry company it became clear that this picture was absurd and, by 1980, when the LRC was established as a permanent research base within the research area rather than running short seasons from the remoteness of my then home in Manchester, it became patently clear that the Vale had supported an equal or greater level of activity to the Yorkshire Wolds (Powlesland 2003). Whilst the aeolian sands identified during excavation at Cook's Quarry had led to extraordinary levels of physical preservation, they reduced the effectiveness of air-photography, in part because the sandy soils were at the time considered of limited agricultural potential and thus were rarely planted with appropriate crops to yield a crop mark record. Thirty years later, changes in crop strains and the very extensive use of irrigation, particularly for growing turf and root-crops, has changed the status of this land; large areas of archaeological deposits formerly protected by aeolian sands are now being badly damaged as a consequence of intensive root-crop production.

With regular flights over the same landscape over many years, necessary if we were to attempt to recover a full and incremental crop-mark record, it has been possible to identify, even if largely from anecdotal evidence, that changes in crop strains, increasing intensity of farming with resultant plough damage, and climate change, have combined to change the nature and timing of crop-mark formation. In 1984, for instance, one field (Heslerton Site 20), which must have been flown in ideal circumstances many times, produced an outstanding series of previously unknown crop-marks; an investigation on the ground confirmed that the field had been ploughed more deeply than in previous years, with the plough cutting through the protective layer of blown sand and increasing the contrast between the cut ditches and other features and the natural subsoil into which they had been cut. In this case crop-marks are now present in this field in most years in a variety of crops; more worryingly, some of the major features are now visible as soil-marks in the ploughed field (Figures 2.4, 2.5 & 2.6).

In the last decade we have become increasingly aware that crop-marks are appearing much earlier than we had noticed in the past with brilliant results accruing from air-photography conducted in late May and early June, about a month earlier than was the case 20 years ago. Not only has the time at which crop-marks are

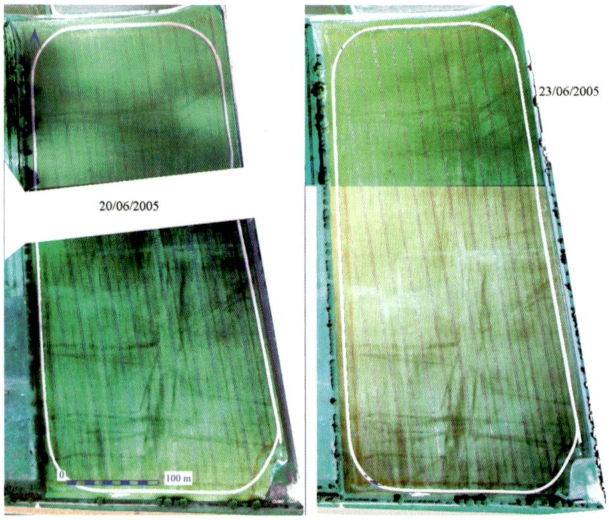

Figure 2.4: Crop-marks recorded on two dates, three days apart, showing trackways and settlement enclosures, first seen in 1984 and now showing almost every year (Heslerton Site 20).

visible apparently shifted, but also the nature of the crop-marks themselves has changed; this is particularly the case with cereal crops which now grow to a height that is considerably less than half of that found in the past. Following the observation of extraordinarily clear crop marks in barley during late May in 2006, it was realised that they showed with most clarity when viewed vertically and hardly showed at all from a low view angle; a visit to the field on the ground revealed that the colour differences observed from the air were concentrated in the leaf at the base of the crop stem and thus could only be clearly resolved when seen from directly above.

Although the returns, in terms of new 'sites' discovered each year, did diminish over time, by the early 1990s the flying programme was reduced with only one or two flights per season undertaken primarily to record the large excavations which were then in progress. Occasional flights in the last decade have added detail in some areas, revealing crop marks that had only been seen once or twice before with others that were entirely 'new'. By 1992 it was clear that whilst we had identified a remarkable density of activity in the landscape there were clearly areas where features should have existed but none had been seen, and that we should investigate alternate approaches to extending the cropmark record. By looking at a landscape of crop marks rather than at individual fields, gaps in the record became very obvious. One of the principal features in the crop mark record was a linear settlement running for more than 10km through the research area and bounded to the north by the former wetlands that had been present in the base of the Vale of Pickering; this series of features, identified as a 'ladder settlement' on account of the layout of the ditched enclosures and track-ways, appeared as a discontinuous ribbon of settlement activity along the valley side.

Multispectral imaging

Following discussions with Danny Donoghue of the Department of Geography, University of Durham it was felt that there might be some potential for the identification of new features and filling the gaps in the accumulated crop mark record by using multi-spectral imaging. It was hoped that by deploying a multi-spectral scanner, recording images from both visible through to thermal wavelengths, we might be able to identify features invisible to the eye in areas of pasture. A joint research proposal to the Natural Environment

Figure 2.5: Geomagnetic survey over Heslerton Site 20 and adjacent fields enhancing the data recovered from aerial photography.

Figure 2.6: Heslerton Site 20 seen from the edge of the Yorkshire Wolds with clear soil-marks of major track-way ditches.

Research Council (NERC) was accepted and we were given a data grant covering the collection and delivery both of high resolution large format photographs and 12 band multi-spectral imagery (Donoghue *et al.* 1992; Powlesland *et al.* 1997; Powlesland 2001).

Both multi-spectral data and vertical large format photographs were recorded on two flights as the first flight had to be abandoned due to cloud cover. The flight dates were largely determined by the availability of the NERC plane and undertaken at an earlier date than we might have selected on the basis of past evidence. The results were in short extraordinary, particularly from the air-photography; crop mark conditions were almost perfect and the combined results from the two flights recorded more than 75% of the known crop marks recorded in the previous 14 years. In addition to recording features from the visible wavelengths, the data gathered by the Deadelus 12 band scanner revealed a number of features in the infra-red and thermal images that were not visible in the conventional photography (Figure 2.7).

The splendid results during this first combined multi-spectral and vertical photography flight may have given us a slightly false impression of the potential of the multi-spectral imaging and indeed a second NERC data award, involving night and day flights as well as LiDAR and CASI scanning undertaken in 2002, proved far less valuable in terms of the identification of new features. Although the second NERC flight revealed very few crop and soil marks, the collection of LiDAR data made a significant contribution in its own right. It is quite clear from the two NERC surveys that the essential factors influencing the identification of archaeological features in multi-spectral and hyper-spectral data are broadly similar to those affecting conventional air photography and, given the scale of the features we are trying to identify, there are considerable limitations in the potential of the scanners we used dictated by the resolution of the instruments deployed. In both the NERC surveys the maximum resolution of the spectral scanners was between 1.75m and 2m, whereas the vertical photography had an effective resolution of between 8cm and 12.5cm. Both theoretically and through demonstration it is clear that multi-spectral and hyper-spectral imaging can allow us to image features from beyond the visible spectrum; however, the limitation on ground resolution and the cost and inflexibility of these instruments will constrain the returns from these systems for some time to come. Sub metre resolution satellite images are now available covering much of the planet and it is inevitable that the resolution will continue to improve and, whilst Google Earth™ and Microsoft's Virtual Earth web based world viewers are providing free on-line access to a vast array of imagery, we need to remain aware that the highest resolution imagery presented through both systems are gathered through conventional air photography rather than by satellite imaging. It will be a very long time before we are able to secure high resolution multi-spectral imaging with a flexibility and economy that remotely compares with what is possible from conventional air photography.

The writing of this paper has been delayed by the fact that in much of England this year we have witnessed an extraordinary crop-mark season, and the once-in-a-lifetime opportunity to view and record crop-marks running for many kilometres through multiple crops, including new features never observed before in a landscape very intensively flown over more than

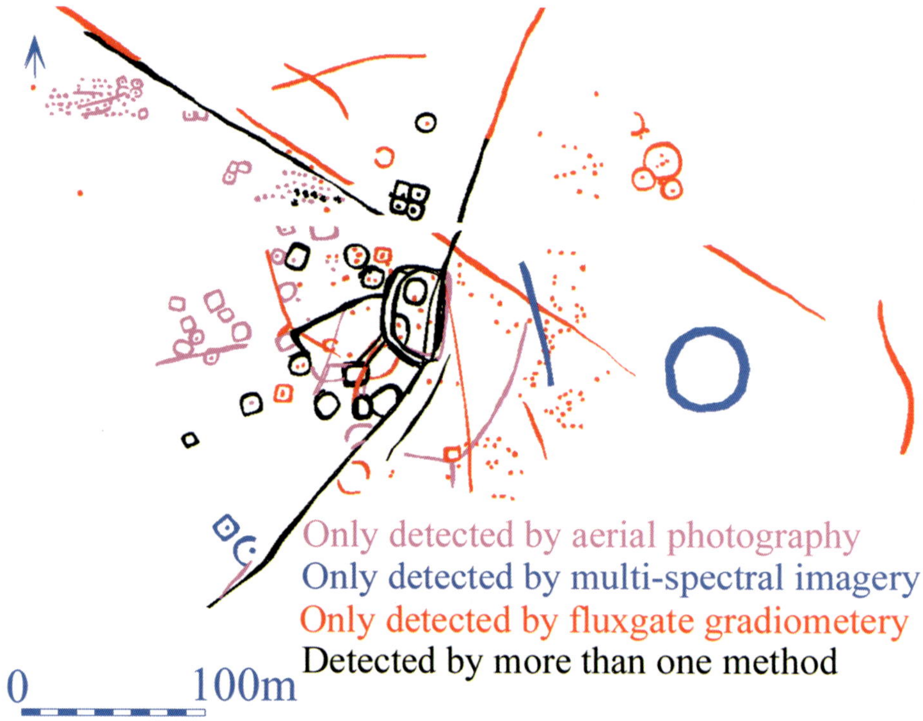

Figure 2.7: Comparing the returns from air photography, multi-spectral imaging and gradiometer survey in a single field.

30 years, has been an entirely appropriate distraction. Sadly, much recent literature concerned with remote sensing grossly overemphasises the importance of satellite imaging over air-photography (Parcak 2009; Wiseman & El Baz 2008). This is unfortunate as within Europe, particularly as restrictions upon domestic air photography in many countries have been lifted, the contribution of vertical and oblique air photography is unlikely ever to be matched by satellite imaging. Whilst one can appreciate that in countries where restrictions on flying or matters of accessibility make it impossible to undertake dedicated archaeological air-photography, the use of sub-metre satellite images can contribute to the identification of archaeological features, particularly in desert areas where features are identified through soil features rather than vegetation; however, limitations dictated by image resolution will restrict the type of features or level of detail that can be observed.

With reference to the examination of landscapes, the 2010 crop mark season in England has re-emphasised the phenomenal returns that can be gained from speculative oblique air photography in perfect conditions, covering the same ground over many flights over a number of weeks as crop-marks emerge, change their nature and finally disappear or are harvested (Figure 2.8). Had we had the resources to undertake large area vertical flights on the same occasions it is likely that we would have recorded even more; however, the returns from air photography whether vertical or oblique do not reveal everything.

Geophysical survey

The remarkable air-photographic record for Heslerton, even allowing for new additions and additional features identified in the multi-spectral images, gives only a partial view of the buried archaeology. It was already realised by the late 1990s that if we were to improve our understanding of the buried landscape, and particularly those areas where gaps resulted from the unsuitability of the crops being grown or the sub-soils, a singular reliance on ad-hoc air photography was unlikely to generate the results needed to generate a detailed picture of landscape over time. Alternative methods would be needed to enhance the picture that had emerged from airborne remote sensing techniques. Initial tests employing ground based geophysics had been conducted to assist the planning of excavations at Cook's Quarry, West Heslerton, in the early 1980s; the results were far from promising and neither resistance nor magnetic surveys produced a useful return. By the late 1980s improvements in the reliability and response of fluxgate gradiometers, and excavations on chalky sub-soils rather than the pure sands encountered at Cook's Quarry, encouraged us to try again. Gradiometer surveys undertaken ahead of excavation on our behalf by English Heritage, and our own very high resolution surveys integrated with the excavations, produced exceptional results indicating that the failure of the original surveys to reveal any features reflected local soil conditions rather than widespread limitations in the local environment (Lyall & Powlesland 1996).

The realisation that geophysical and particularly gradiometer survey could radically enhance the picture emerging from the various aerial surveys and fill in the apparent gaps in the evidence base led to the development of a series of geophysical survey projects funded by English Heritage from the Historic Environment Enabling Programme and the Aggregates Levy Sustainability Fund (ALSF). During the last decade these projects have enabled us to undertake more than 1200 hectares of contiguous geophysical survey; the results of this work have been astonishing (Powlesland 2003, 2009). The geophysical survey work has been driven by the desire to identify the scale of the archaeological resource, the potential risk from aggregates extraction, and to identify areas where aggregates could be sourced with a minimal archaeological risk. If we are to approach landscapes then simply identifying vast areas of archaeological evidence is not enough; we cannot secure the future survival of the whole resource even if we wished to. It would be easy to argue that the archaeology of the southern side of the Vale of Pickering, centred upon West Heslerton is unique; however, this uniqueness derives more from the fact that it has been identified through decades of research and repeat observation rather than any inherent exceptional uniqueness. There is one characteristic that is indeed exceptional, the fact that very large areas of the surveyed landscape are buried by varying thicknesses of aeolian sand which has facilitated exceptional preservation of the buried deposits; in the intensively farmed rural landscape of lowland Britain the survival of intact floor deposits within Prehistoric and Roman settlement sites is very rare.

Understanding the resource and the risk

Clearly, if we are to use the evidence identified through remote sensing, which at best has to be interpreted on the basis of analogy and experience without the support of conventional archaeological data to provide the chronological framework, we need to try and identify the potential of the resource to change our comprehension of the past. If we are to address the sustainability of the resource then we must avoid the mistakes made in the 1970s when so many crop-mark sites were afforded some protection through designation as scheduled monuments, not because they were necessarily the best preserved examples but because they produced clear crop-marks; the clarity of the crop-marks was almost certainly indicative of sites that had been truncated by ploughing. This comment is not raised as a criticism but a reflection on the crop-mark formation process arising from many years of observation over the same landscape; to have suggested at the time that we should preferably schedule the poor quality crop-mark sites would probably have received very limited support and would have been difficult to argue with the landowners. Without some degree of qualitative assessment the results of thousands of hours of geophysical surveying would be diminished.

The importance of the aeolian sands that are distributed along the southern side of the Vale of Pickering at the base of the scarp slope of the Yorkshire Wolds cannot be underestimated as they have sealed old ground surfaces and landscape features over very large areas (Figure 2.9). The formation processes which led to the accrual of a deposit which, in some areas where blown sands have built up against hedge boundaries can be several metres thick, is still poorly understood. Archaeological evidence and investigation through excavation indicates that some sort of shifting dune formation had begun to develop during the Mesolithic at the latest; most likely the origins of the blown sand are likely to be found in the late glacial context, with surfaces being sealed for long periods and later being re-exposed and then buried again. Excavation at a number of sites within the area of blown sand indicates that as early as the Bronze Age monuments such as a round barrow at West Heslerton were completely buried beneath blown sands, and in other areas the blown sands continued to accrue from the Roman period onwards burying intact floor deposits and protecting them from later plough damage.

Figure 2.8: Stunning crop-marks showing across multiple different crops in exceptional conditions in 2010.

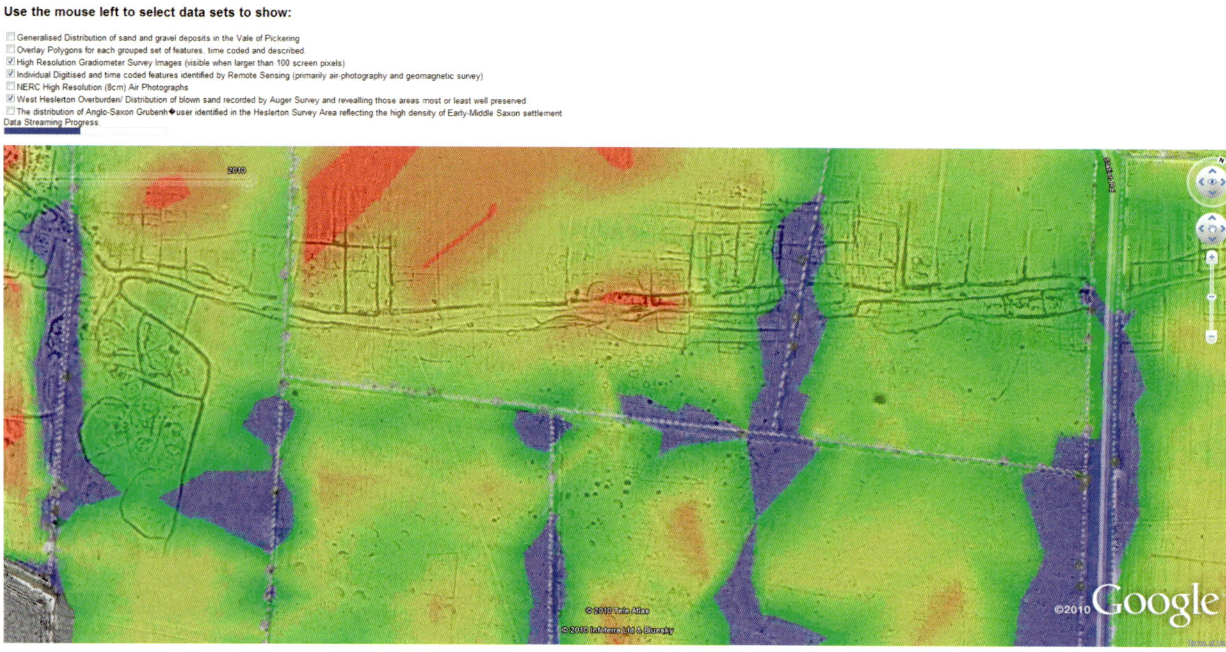

Figure 2.9: Geophysical survey results overlain by a deposit depth model showing the depth of topsoil and blown sand covering the buried features (scale red <0.25m to blue maximum >5m).

The relatively sterile agricultural environment that prevailed in the areas covered with iron-enriched blown sands meant that they were not subjected to intensive farming with consequent plough damage until the last 20 years, when intensive irrigated root crop production was introduced. The intensive root crop production is having a devastating effect on what was once an exceptionally well preserved archaeological landscape; deeper ploughing, and harvesting of potatoes in particular, is eroding the preserved sub-surface archaeology as the plough cuts through the hard blown sands into the much softer archaeological deposits beneath. Clearly it is important to identify those areas where the blown sand covering remains substantially intact; with this objective in mind an extensive auger survey has allowed us to map the depth of the overburden on top of the archaeological horizons and allowed us to identify those areas most at risk from deep ploughing and those with most potential for the recovery of exceptional archaeological deposits. If we are to work towards aspects of sustainability and long term management at a landscape scale it is essential that we are proactive in trying to identify those areas that hold the most potential for the return of good quality information, and accept that in some areas the archaeological deposits are already compromised and are thus worthy of preservation by record through the normal planning process and excavation prior to mineral extraction or other development.

Surface collection: to walk or to sieve

In order to gain insight into the buried remains surface collection is often advocated as this can secure some dating evidence; in the intensively ploughed rural landscape of England where the majority of the buried resource relates to cut features, earthen monuments and settlements in which the majority of the structural components were comprised of post-hole structures rather than stone or brick and tile structures, the returns from surface collection campaigns can frequently be very limited. Experience in Heslerton has shown that large amounts of cultural material are brought to the surface in years when plough damage occurs, but in the following year the plough turns the soil back leaving the high densities of material in the base of the plough-soil. In one example, observed in the 1980s, a field that one year was covered with vast amounts of pottery, animal bone and other material such as quern stones was found two years later to have almost no cultural material on the surface at all. It discouraged us from undertaking surface collection surveys which are time consuming, labour intensive and unreliable; perhaps if we had a policy of regular repeat surface collection over the same fields over a number of years the reliability of the results may be improved. An alternative approach has been tested which secures not only cultural material which can indicate the nature and chronological range of the buried features, but also allows us to assess the condition of the buried resource through the degree of fragmentation of the ceramics in particular. By gridded plough-soil sieving, based on metre square sample trenches cut to the base of the plough soil and sieved in their entirety, spaced at 20m intervals, the gathered material culture evidence can reveal the chronological depth, intensity of activity and quality of preservation of the buried resource without any bias related to recent agricultural activity or reflecting differences in collection by different individuals (Figure 2.10). Experience shows that fragmentation and sherd abrasion can give a clear indication of active plough damage. By targeting the absolute return from measured volumes of plough-soil statistically valid comparison can be made across sites and the landscape as a whole. Admittedly this approach can also be resource intensive and we have never been able to apply this approach over the large areas that deserve it.

Assembling evidence

The LRC have been actively involved in the application of GIS technologies in field archaeology since the mid 1980s, driven more by the need to quantify, manage and articulate the archaeological resource than by any need to undertake spatial analysis beyond that which can be rapidly undertaken by eye. With respect to our remote sensing activities it has always been important to be able to map and interpret the results of our work; vast plots of the geophysical surveys, georeferenced air-photography and multi-spectral images, whilst they do demonstrate the levels of activity in the landscape, are static and remain non-interactive. To make these data useful nearly 30,000 anomalies have been individually digitised and give a unique identifier linked to full supporting data on feature source, class, chronology, and interpretation which have then be grouped into larger classes representing individual cemeteries, settlements, field systems etc. As a digital resource these data need to serve a number of different functions related to fundamental research into the evolution of the landscape and its populations over time, as well as to underpin long term management of the resource through planning and development control.

Using the evidence: responding to the threats from aggregate extraction

We are as concerned about those areas examined which show intensive activity as those in which activity levels seem low particularly in the context of aggregate extraction. Planning-related archaeology tends to be entirely reactive and site based rather than landscape based so that the planning process is targeted at areas pre-selected for extraction by the developer rather than by assessing the potential aggregate resource as a whole with a view to identifying those areas where mineral extraction will cause least archaeological impact. The economic reasons behind current procedures are entirely understandable but, if we are to seriously take on board aspects such as sustainability of the resource, then perhaps we should look towards an alternative model where we employ landscape survey techniques to identify those areas of landscape that hold appropriate mineral resources with low levels of or poorly preserved archaeology and encourage mineral operators to target those areas. In the case of the Vale of Pickering, the research investment by English Heritage and the ALSF over more than a decade means that we are in a position to steer future mineral extraction towards areas where the archaeological impacts are likely to be less and thus more affordable to the extraction industry. During the last few years we have seen radical changes in geophysical survey equipment, particularly in continental Europe where instruments towed by vehicles can gather high density data over more than 10 hectares per day rather than the 1 hectare gathered by a team of two people using conventional gridded and walked surveys. If we are to take a strategic approach to the assessment of landscapes then it is imperative that we adopt these new technologies which, despite large initial capital costs, will allow vast areas to be covered at a very much reduced cost per hectare. Using towed multi-probe instruments with spatial control generated by high precision real-time GPS instruments it is entirely conceivable that we can look towards surveys covering more than 10,000 hectares.

Figure 2.10: Ceramics recovered through 1m square topsoil sieving experiment: The area on the left has suffered recent plough damage with more than 1.5kg of pottery in a single sample; the area on the right shows a less damaged and more normal profile. Underlying geophysical survey plot in grey.

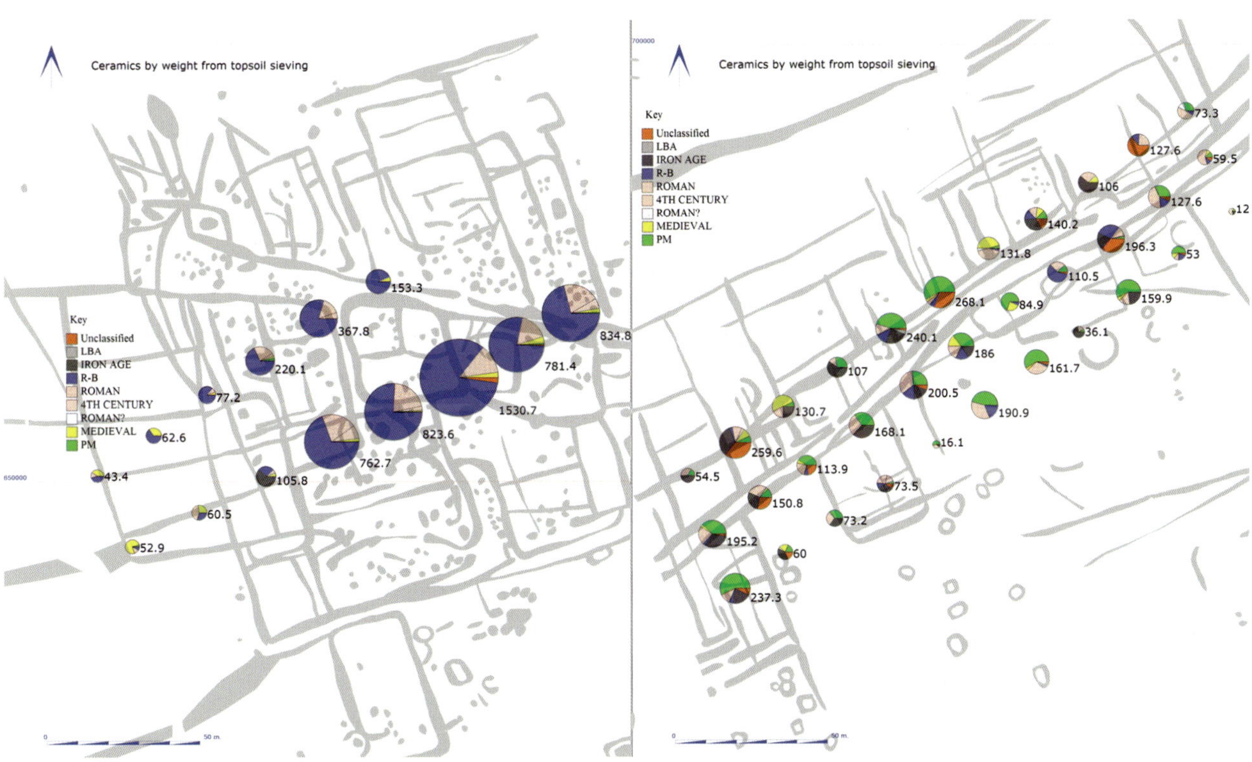

Figure 2.11: High resolution vertical photography overlain with time-coded feature plots in Landscape Research Centre Digital Atlas (Heslerton Site 5).

The landscape beyond: how much do we need to see?

When we excavate we deploy the concept of compare and contrast in relation to other similar sites within the analytical and interpretive process as standard archaeological practice; at present this is impossible at the landscape scale and we are often reduced to the facile comparison of the presence or absence of particular monument types. One of our objectives in trying to survey such a large contiguous landscape has been to try and identify a base level of understanding which will allow us to identify the scale of past landscape units which can be considered a single entity, and thus identify what scale of sample needs to be understood to allow us to compare archaeological activity at the landscape scale. The intensive surveys undertaken in the Vale of Pickering have not only filled gaps in the initial record based on oblique air photography, but also allowed us to identify what appear to be genuine gaps in the landscape which apparently are largely devoid of settlement or burial evidence and are interpreted as the boundaries between individual blocks of land which are likely to have been established before the end of the Neolithic. As a landscape archaeologist working without chronological constraints it is impossible not to function within an environmentally deterministic framework and question the deployment of random sampling schemes applied to the assessment of archaeological landscapes; human settlement in the past was far from random in its distribution and, if we are to be proactive in developing landscape management strategies, we need to get away from the position where the majority of archaeological knowledge is gained through serendipity or chance discovery. By identifying both the scale of the individual landscape blocks and their relationship to the geology, topography, the environment and the level of feature repetition from one block to another we are in a far stronger position to develop both long-term management and research strategies.

Figure 2.12: Geophysical survey results overlain on air photography (Heslerton Site 5).

Figure 2.13: Geophysical survey results with a transparent overlay showing the depth of topsoil and the plotted features in the Landscape Research Centre Digital Atlas (Heslerton Site 5). The topsoil depths indicate that the area is being actively damaged by ploughing.

Delivering, enhancing and disseminating the data

More than 30 years of research in the Vale of Pickering has totally transformed our perception of the landscape, the density of past occupation and population levels, making what was formerly almost an archaeological void into what now appears, incorrectly, to be one of the most densely utilised landscapes in Britain; it is wholly incomprehensible that the levels of activity in the Vale of Pickering are not matched in other similar environments both locally, as for instance in the Great Wold Valley, and in other river valleys elsewhere in lowland Britain. Whilst the digitised images with their digitised feature overlays and supporting databases will be transferred to the North Yorkshire Historic Environment Record which will serve the needs of the environmental planning community, there is clearly a need to seek dynamic ways to deliver these data to the wider community. Vast plots measuring more than 5 × 1.15m have proved very useful when presenting the results at seminars and conferences or for museum based exhibitions but, even printed at 1:2,000, much of the detail is difficult to see; conventional publication enabling the reader to appreciate the scale of the results is similarly out of the question. This is a case where the internet provides a perfect publication environment, by delivering the data through an internet based earth viewer. After considering many systems we selected Google Earth as the most appropriate delivery system that would permit the publication of a comprehensive data-set combining all the LRC geophysical survey data, the high resolution vertical air-photography with other selected air photographs, and the complete cover of vector polygons of each identified feature with basic supporting data. The decision to use Google Earth™ (GE) was largely driven by the fact that basic air-photographic mapping is available freely within GE; the data can be draped upon the topography and viewed in real time at any scale in plan and perspective views. More importantly, following the introduction of GE version 5.0, the program supports time based animation so that features can be tagged with a suggested start date, an end date and, more importantly, with a date to reflect continuity of the features' survival within the landscape. The flexibility of the system means that the user can switch on and off different data sources or look at particular time periods at any scale without the need to learn any complex software (Figures 2.11, 2.12 & 2.13).

In order to use the data within GE it had to be transformed from the Ordnance Survey Grid System into the Latitude Longitude projection used by GE, and a considerable amount of work was needed to make the very large images load at an acceptable speed. In order to expose the underlying intelligence of the features they have been grouped into higher level feature clusters such as settlements or cemeteries, whilst each individual feature has been assigned chronological attributes indicating the start, end and continuity dates, work that is analogous to the post excavation analysis of excavated data. It must be remembered that the data-set is derived from remote sensing and thus the chronological framework adopted lacks the precision gained from excavated data but is of course held within a comprehensive database which can change over time when new or improved dating evidence is available. This work is as yet incomplete but the pilot example, which includes more than half of the geophysical survey images and overlying data, can be viewed and interrogated successfully even on a wireless connection linked to a low speed broadband connection (Figure 2.14).

Whilst publication via GE can make the results of remote sensing projects available to all, and therefore make explaining management decisions regarding the archaeological resource far easier than simply referencing a Sites and Monuments Record and a static print out, it exposes a number of problems with regard to sustainability and management which are difficult to resolve. The combined evidence identified and presented through GE relates to several thousand of hectares, land owned by a multitude of farmers, land on which for the most part there is very little or no surface indication of what lies beneath. By exposing this evidence on the internet the potential threat from treasure hunters is easy to appreciate, as is the resulting impact on landowners, without whose co-operation much of the evidence could not have been collected, and whose crops are damaged by 'nighthawks', treasure hunters operating under the cover of darkness. The damage caused by metal-detectorists both to crops and, more significantly, to the archaeological resource both in the fields and during excavation has been devastating, removing metalwork and coinage, much of which has limited intrinsic value but provides the primary dating evidence from the Late Iron-Age onwards, from buried landscape features that we would hope to preserve for investigation by future generations. In an attempt to limit the risk to the resource access to the GE dataset, which is held on a secure server, requires user login and verification in order to view the data. This is not necessarily a perfect solution; however, it does provide access to educational and academic institutions, the planning community and *bona fide* users. At present the active Digital Atlas Project can be viewed by registering in the members' area of the LRC website; once registration has been validated access can be gained to the password protected area (http://www.landscaperesearchcentre.org/html/members_area.php).

Designation, protection, sustainability and management?

In Britain, whilst we have designation systems designed to afford structural remains, whether buried or upstanding, a degree of protection, there is no designation system that allows us to protect landscapes. Even where 'sites' have been designated as Scheduled Monuments and thus theoretically afforded some degree of protection by the state, the number that have been fundamentally damaged to the point where their long term value is so diminished as to make them irrelevant is quite frankly embarrassing.

The landscape resource identified through the multi-faceted and long-term remote sensing programme in the Vale of Pickering (Figure 2.15) is effectively un-manageable within current statuary designation arrangements. More significantly, this work has changed our comprehension of the landscape to such a degree that it reflects a need to adopt new paradigms for approaching landscape than the site based studies with which we are all familiar. The realisation that very substantial damage has affected the buried resource over the long period during which the evidence has been assembled both highlights the need for new approaches to designation and land management as well as the need for comparative surveys at a regional, national and European scale; projects which by deploying more efficient and faster data collection systems such as towed geophysical instruments could examine landscapes at a similar scale but in a tenth of the time. Only by developing such projects elsewhere can we set the long-term research in the Vale of Pickering within its true context and enhance our understanding of its significance; this is necessary if we are to develop novel and sustainable approaches to landscape management. In the case of the Vale of Pickering one approach may

Figure 2.14: The Landscape Research Centre Digital Atlas showing Iron Age to Anglo-Saxon activity overlain on the geophysical survey images.

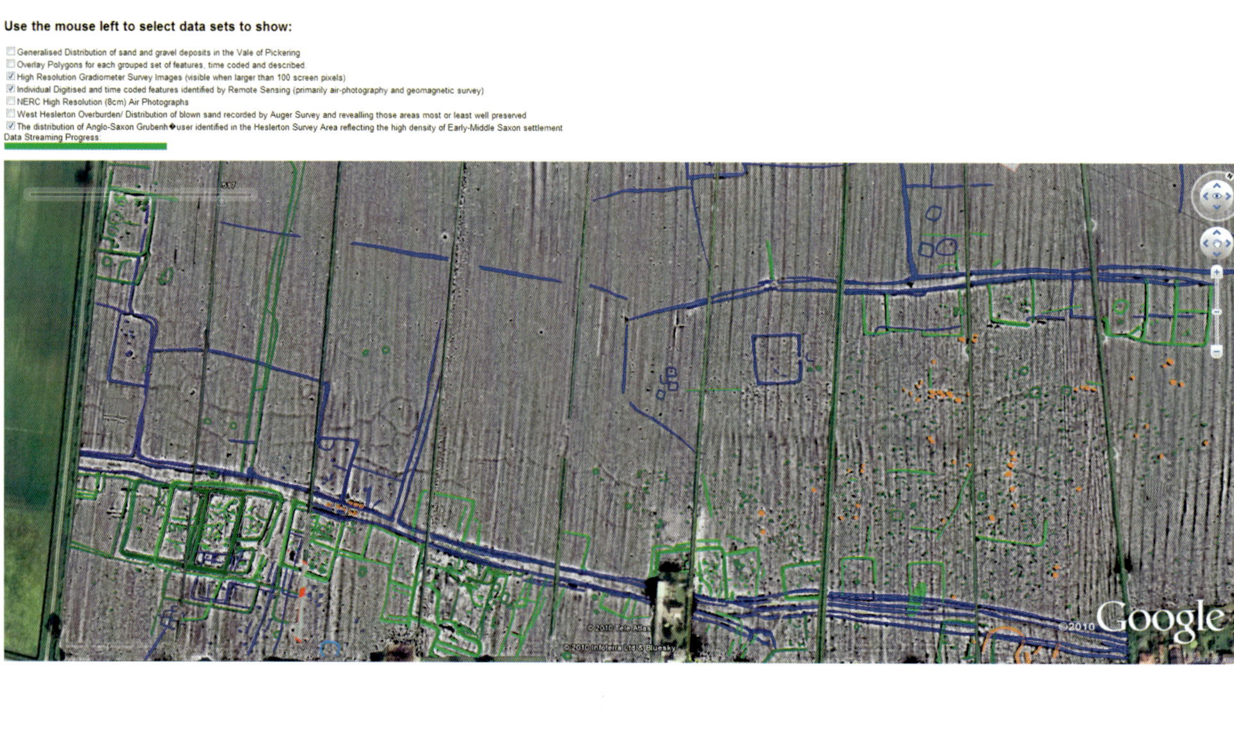

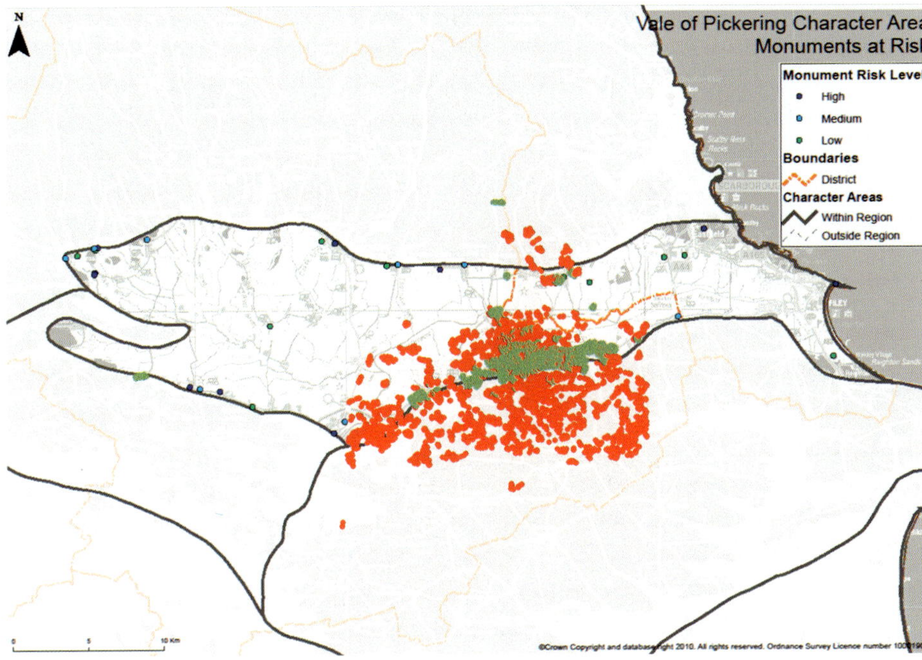

Figure 2.15: Vale of Pickering Monuments at Risk Map overlain with the distribution of features identified through remote sensing; features identified from the air in red and those from ground based geophysics in green (>27,000 anomalies altogether). This map should be compared with Figure 2.2 above.

be to develop an archaeological and environmental research park which could both protect a substantial physical area as well as provide a setting for long-term research and public engagement. Without such a development the gradual attrition of the identified, if not understood, resource is inevitable. Whilst we have been able to identify physical attrition of the buried resource through industrialised agriculture, more subtle decay is affecting the environmental evidence that has been preserved in the former wetlands that bound the most intensive settlement zone observed through remote sensing. The impacts of global warming, increased drainage and irrigation are leading to rapid loss of the environmental evidence that has rarely survived in the adjacent dry-land environments that have proved to be so advantageous for long term settlement. Only by pro-active intervention through excavation and targeted environmental sampling can we hope to recover an understanding of the environmental context in which settlement thrived from the prehistoric to medieval periods. Not wishing to present too bleak a picture of the future sustainability of the resource and thus the potential to properly understand and interpret the more than three decades of research in the Vale of Pickering, this work highlights a fundamental disconnection between the way that society and government approach the human as opposed to the natural environment, where designation and management at a landscape scale is becoming the norm. Clearly there is an urgent need to integrate the management of archaeological, cultural and natural landscapes and to remind ourselves that whilst we can increase biodiversity and encourage the survival, for instance, of rare species of birds, the evidence of the past cannot be replaced once it is lost. On the 10th anniversary of the opening of the European Landscape Convention for signatory states it is time the joined up talking of 'natural' and 'cultural' landscapes became the protection and planning reality – to return to my opening, at the moment how we best develop and plan this protection is a 'known unknown'.

References

Donoghue, D.N.M., Powlesland, D.J. & Pryor, C. 1992: Integration of Remotely Sensed and Ground Based Geophysical Data for Archaeological Prospecting using a Geographical Information System. In Cracknell, A.P. & Vaughan, R.A. (eds): *Proceedings of the 18th Annual Conference of the Remote Sensing Society, University of Dundee* 1992. 197–207.

Greenwell, W. & Rolleston, G. 1877: *British Barrows.* Oxford, University Press.

Hamerow, H. 1993: *Excavations at Mucking 2: The Anglo-Saxon Settlement.* English Heritage Research Report 21.

Lyall, J. & Powlesland, D.J. 1996: The application of high resolution fluxgate gradiometery as an aid to excavation planning and strategy formulation. *Internet Archaeology* 1 (http://intarch.ac.uk/journal/issue1/index.html).

Mortimer, J.R. 1905: *Forty Years Researches in British and Saxon Burial Mounds in East Yorkshire.* London, Brown, A. and Sons.

Powlesland, D.J., Haughton, C.A. & Hanson, J.H. 1986: Excavations at Heslerton, North Yorkshire 1978–82. *Archaeological Journal* 143, 53–173.

Powlesland, D., Lyall, J. & Donoghue, D. 1997: Enhancing the record through remote sensing: the application and integration of multi-sensor, non-invasive remote sensing techniques for the enhancement of the Sites and Monuments Record. Heslerton Parish Project, N. Yorkshire, England. *Internet Archaeology* 2 (http://intarch.ac.uk/journal/issue2/pld/index.html).

Powlesland, D. 2001: The Heslerton Parish Project: An integrated multi-sensor approach to the archaeological study of Eastern Yorkshire, England. In Forte, M. & Campana, S. (eds): *Remote Sensing in*

Archaeology, XI Ciclio di Lezioni sulla. Firenze 2001, 233–35.

Powlesland, D. 2003: The Heslerton Parish Project: 20 years of archaeological research in the Vale of Pickering. In Manby, T.G., Moorhouse, S. & Ottaway, P. (eds): *The Archaeology of Yorkshire An assessment at the beginning of the 21st century*. Yorkshire Archaeological Society, Leeds, 275–92.

Powlesland, D. 2009: Why Bother? Large scale geomagnetic survey and the quest for 'Real Archaeology'. In Campana, S. & Piro, S. (eds): *Seeing the unseen, geophysics and Landscape Archaeology*. Papers presented at the XVth International Summer School for Archaeology, University of Sienna. Taylor & Francis, 167–82.

Parcak, S. 2009: *Satellite Remote Sensing for Archaeology*. Routledge, London.

West, S.E. 1985: *The Anglo-Saxon Village at West Stow, vols 1 and 2*. East Anglian Archaeology 24.

Wiseman, J. & El-Baz, F. (eds) 2007: *Remote Sensing in Archaeology*. Springer.

3 | 'Total Archaeology' to reduce the need for Rescue Archaeology: The BREBEMI Project (Italy)

Stefano Campana

Abstract: In Italy there is growing debate about the methods that can or should be adopted within the fields of preventive and rescue archaeology in the face of major infrastructure projects. Greater immediacy has been added by the potential (but not yet fully realised) impact of new domestic legislation dealing with procedures to be adopted to assess the potential archaeological implications of development projects. This contribution describes the methods brought to bear in advance of a major motorway development in Northern Italy, along with some of the insights gained from new approaches, both in drawing information from existing sources and from the deployment of a wide range of survey and investigation methods in the field. It is clear, however, that conflicts remain between 'traditional' approaches and the new opportunities presented by looking afresh at maximising gains and minimising losses in the course of such developments.

Introduction

My mentor, the late professor Riccardo Francovich, started collecting data on the landscape of Tuscany in the early 1970s, mainly through a systematic programme of field-walking survey and archaeological excavation, along with the collection of information from written sources and the examination of historical aerial photographs (Francovich 2006). After about 30 years of work within the Tuscan landscape it became clear that, despite the huge database of information assembled by that time (about 25,000 individual items), a large amount of essential evidence remained effectively undetectable, with the result it was impossible to answer important archaeological questions within the confines of this methodological approach (Campana & Francovich 2009).

There were at least three contributory factors. Firstly, the nature of the landscape itself, with its clay soils, large areas of forestry, intensive agricultural exploitation and distinctive morphological patterns. Secondly, the peculiarity of the material culture, constantly changing over time and with related difficulties caused by post-depositional processes. And last but not least the definition and range of the evidence in terms of artefacts but also ecofacts and environmental factors. It is important to emphasise here that the main aim of the University's research is to write a historical narrative in which the emphasis is on the interpretation of archaeological sites in relation to their surrounding landscapes. In pursuit of this we have moved progressively from a 'thematic' approach to a more 'global' consideration of historical processes and interpretation. In essence this has meant switching from a site-based approach to a landscape perspective: in the study of castles, for instance, this has brought into play the consideration of related field systems, communication routes, production centres and cemeteries as well as geomorphology, palaeo-environmental factors and archaeo-zoology etc. A central tenet of our research has been the need to view the context as a *system* developing over time rather than as something that can be related, as in so many earlier approaches, to specific themes, time-periods or site types such as castles, Roman villas, oppida and so on. We feel that this broader perspective on the past will enable us to describe historical patterns in a more complex, balanced and representative way.

Over the past decade LAP&T, the Laboratory of Landscape Archaeology and Remote Sensing established by Riccardo Francovich at the University of Siena, has focused particular attention on issues of archaeological visibility, principally by developing a more integrated approach to instrumentation and operational strategies in archaeological prospection (Campana 2009). Today the University's archaeological maps contain features that would previously have been 'invisible' to field-walking survey and other traditional research methods (Figure 3.1). While the one-time gaps are thus being filled to one extent or another we have come to realise that our 'total archaeology' approach has not yet addressed a challenge that has for some time lain before our very eyes, that of preventive and rescue archaeology.

Rescue archaeology in Italy is synonymous with rescue *excavation*, giving rise to a vast number of small-scale 'test' excavations (Guzzo 2000; Guermandi 2001; Ricci 1996, 2006). It is only in the last five years that the scenario has begun to change to any significant extent, thanks mainly to the work of a few individual archaeologists and the establishment of two ministerial commissions, one of which has drafted a new domestic law on 'preventive archaeology', obliging the initiators of every public construction project, whether for buildings or for infrastructure developments, to commission and

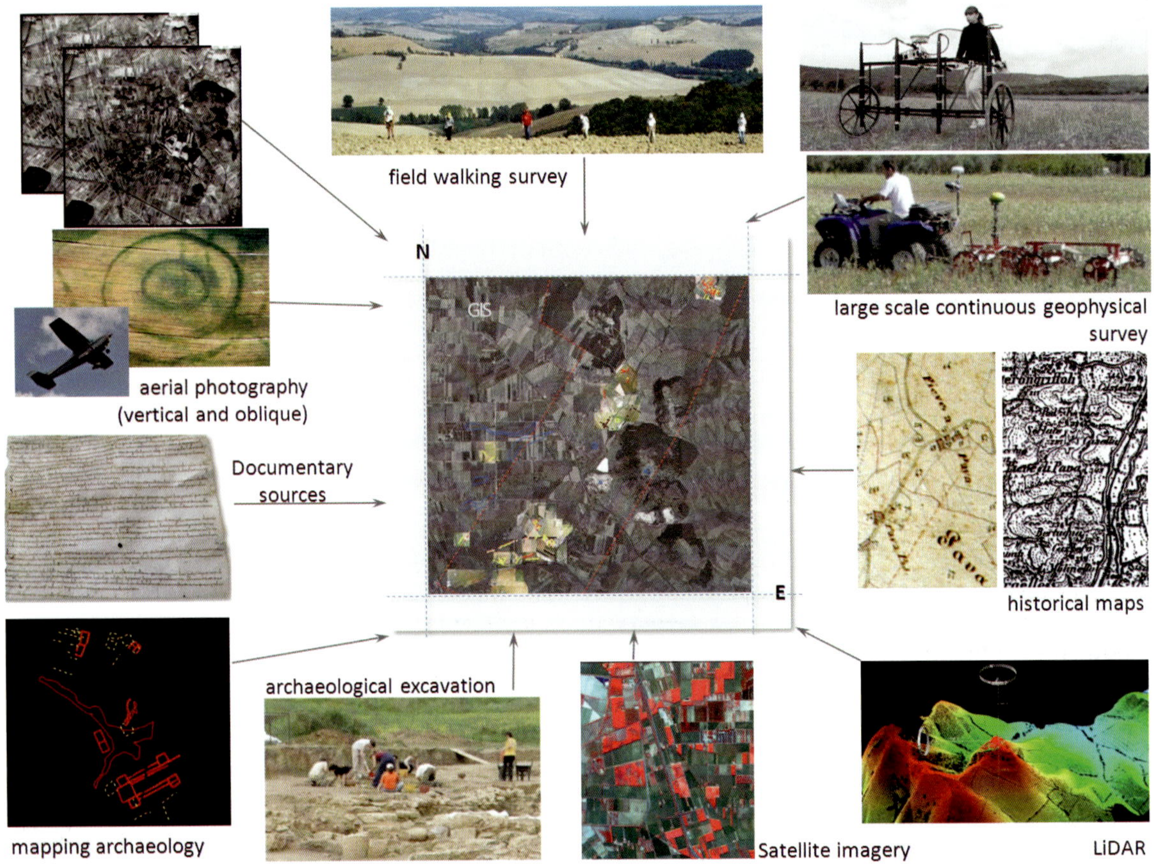

Figure 3.1: The 'total archaeology' approach, based on the integration of a variety of research methods matched against the physical and palaeo-ecological characteristics of the context and the nature of the material culture likely to be encountered.

present a report setting out an 'archaeological impact assessment' (Carandini 2008). Compiling this kind of report involves three main steps:

- The collection of all known data from the archaeological literature and from historical cartography, along with place-name and palaeo-morphological studies.
- The analysis of vertical aerial photo evidence (without, unfortunately, any reference to oblique photography from exploratory aerial survey) and, when possible or potentially useful, the collection and analysis of LiDAR data. In some cases there is a requirement for more intensive work on particular areas through such methods as geophysical prospection or small-scale test excavation.
- The mapping of 'archaeological risk', followed by targeted test excavation or in some cases larger scale investigation through mechanical stripping of the surface deposits.

The example presented in this paper formed part of the so-called BREBEMI project in northern Italy, BREBEMI being the acronym for a motorway construction project linking the cities of **BRE**scia, **BE**rgamo and **MI**lano over a total distance of approximately 100km. The project was initiated before the new law on rescue archaeology came into force and in this case the Archaeological Superintendency of Lombardy, armed with virtually unlimited power within its own region, required the motorway contractors to carry out 'excavation by surface stripping' over the *whole* of the area affected by the motorway construction. Naturally, this approach made nonsense of the contractor's financial and logistical planning, increasing the total cost of the project by a completely unrealistic amount. The construction company therefore called on the writer and his colleagues at the University of Siena to act as consultants in the design of an alternative approach that might be acceptable to the Superintendency.

Background: landscape, research design and project team

The motorway will be constructed through the typical landscape of the Po Valley, with its extremely flat morphology and sand-and-gravel soils, heavily affected by intensive arable cultivation involving the systematic use of heavy-grade tractors and deep ploughing over at least the last sixty years. The area also has substantial concentrations of industrial and related residential development (Figure 3.2).

For the first time in Italy the influence of the new law gave an opportunity to make systematic and innovative use of a range of non-invasive techniques to minimise the risk of archaeological damage in advance of large-scale motorway construction. The project design (Figure 3.3) thus envisaged the systematic collection of historical and geographical data and interpretations from documentary sources,

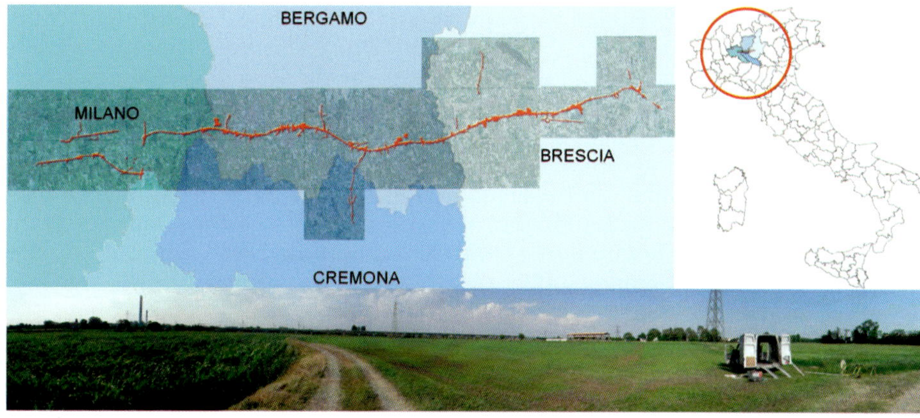

Figure 3.2: General view of the motorway path and overview of the landscape pattern.

along with geomorphological studies, the analysis of vertical historical aerial photographs and the initiation of oblique aerial survey and LiDAR acquisition along the whole of the motorway corridor, in some cases including a substantial buffer zone on either side. Also included was the systematic collection of geophysical data, both magnetic and geo-electrical, across large and contiguous areas of between 200 and 750 hectares respectively, building on an approach successfully tested in Italy, France and above all the UK (Campana & Piro 2009; Dabas 2009; Powlesland 2006, 2009). Systematic test excavations were also planned to verify anomalies identified by any or all of these techniques and independently, the regional Superintendency designed a pattern of random test trenches amounting to a 5% sample of the motorway corridor.

A GIS environment was designed to manage and integrate the collected data at all stages of the project, from data acquisition in the field to interpretation and field checking, so as to assess any significant trends in the collected data and to develop archaeological models. The aim of the project was to reduce the degree of uncertainty about the presence (or *potential* presence) of archaeological remains by identifying areas that ought *not* to be subjected to disturbance by the construction works in the light of the demonstrated presence of either surface or sub-surface archaeological remains.

The Laboratory of Landscape Archaeology and Remote Sensing already had experience in using each of these survey methods but saw the BREBEMI project as an extraordinary opportunity to add its weight to an important culture-change in the theory and practice of preventive and rescue archaeology in Italy. A decision was therefore taken to involve some of the most highly skilled and specialized companies, institutes and research workers from across Europe. The Laboratory used Archeolandscapes Tech and Survey Enterprise (ATS), a spin-off company of the University of Siena, to act as project coordinator and to manage the following activities:

- Aerial survey, in collaboration Klaus Leidorf, of Luftbilddocumentazion from Germany, and Chris Musson from the UK.
- Interpretation and mapping of information from vertical aerial photographs, by the Laboratory's own staff.
- LiDAR processing and interpretation in collaboration with Prof. Dominic Powlesland of the Landscape Research Centre and University of Leeds in the UK.

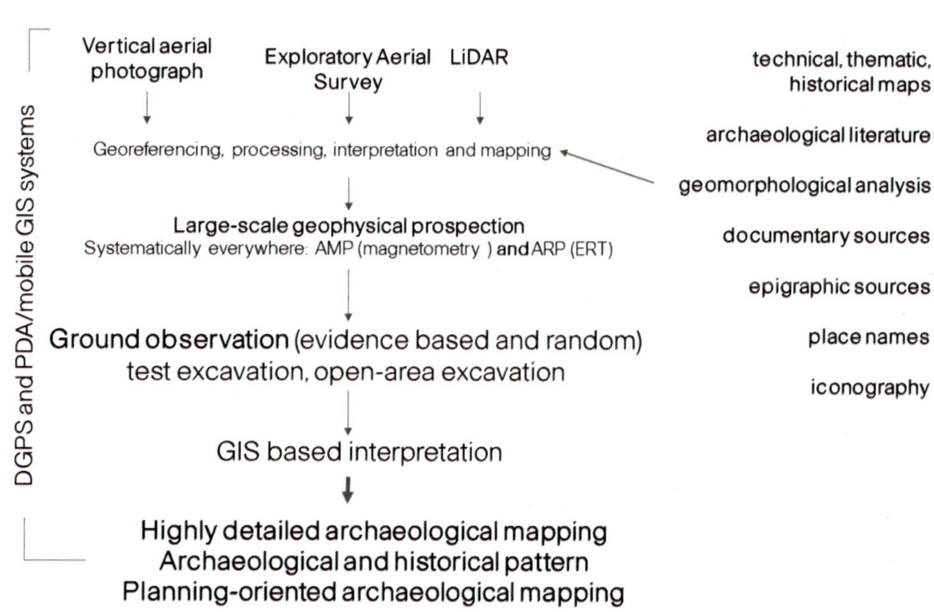

Figure 3.3: Flowchart of information and activities within the BREBEMI project.

- Processing and interpretation of magnetic data, again in collaboration with Prof. Powlesland.
- The collection and interpretation of geo-electrical and magnetic data, by SoIng (Italy)
- GIS and topographical survey, integrated archaeological data interpretation, selective ground observation and test excavation, by ATS.
- The collection of information from historical and geographical documentary sources was carried out by the University of Bergamo under the direction of Prof. J. Schiavini, as were place-name and geomorphological studies.

The geophysical prospection (Figure 3.4) involved the use of magnetic and geoelectrical instruments (respectively ARP and AMP, Automatic Resistivity Profiling™ and Automatic Magnetic Profiling™) developed by Geocarta, a French spin-off company of CNRS, the National Centre for Scientific Research. Geocarta, under the scientific direction of Michel Dabas, also exercised quality control over the collected data and remained on call to provide general assistance throughout the whole process from fieldwork to data processing and interpretation (Dabas 2009). The initial collection of the data was undertaken by SoIng of Livorno, an official partner of Geocarta with long-standing experience in geophysical survey for environmental projects.

Altogether, the project management involved the co-ordination of a team of about 25 research workers from Tuscany, Northern Italy, France, Germany and the UK, carrying out a wide variety of inter-linked work in a very short period – about 4 months or 80 working days.

Results

Bearing in mind the large size and peculiar shape of the survey area this paper will concentrate for the most part on a sample area which is representative of the landscape as a whole in terms of known archaeological data, geomorphological complexity, the availability of geophysical and other survey data and ground observation. This sample, measuring about 20km in linear extent, lies between Caravaggio and Urago d'Oglio, roughly bounded by the Rivers Oglio and Serio. The research work itself can be divided into two main steps: the collection of existing knowledge, and the survey work in the field.

The first step involved the collection and entry into a GIS environment of all the available information about a 2km-wide buffer zone centred on the motorway corridor, from archaeological sites and finds to geomorphology and the evidence of existing aerial photographs etc. This involved the collection of the following information and material (Figure 3.5):

- Place-name registers and historical maps, including historical cadastral maps and the national maps of the Istituto Geografico Militare (University of Bergamo – Centre for Territorial Studies (CST).
- The Archaeological Map of Lombardy, with related updates (University of Bergamo – CST).
- Maps of springs, palaeo-river channels, fluvial ridges and fluvial terraces (University of Bergamo – CST).
- The interpretation and mapping of information from historical and contemporary vertical aerial photographs, principally the GAI series of 1954 and the CGR series of 2007 (LAP&T and the University of Bergamo – CST).
- New aerial prospection and aerial photography along the motorway route in the spring and summer of 2009 (ATS in collaboration with Klaus Leidorf from Germany and Chris Musson from the UK).
- The capture, processing and interpretation of LiDAR data (collection and initial processing by CGR of Parma, further analysis and interpretation by ATS in collaboration with Prof. Dominic Powlesland in the UK).

The collection and mapping of the sites published in the Archaeological Map of Lombardy (Poggiani Keller 1992), with subsequent and updates, produced evidence of 118 already-known archaeological sites within the 2km-wide buffer zone, representing a density of about 2.38 sites per square kilometre, a relatively high figure in comparison with the national average. Even so, this obviously constituted only the tip of the

Figure 3.4: Geophysical instruments used during the survey. *Left* – the Automatic Magnetic Profiler *(AMP © Geocarta)*, capable of recording up to 20ha/day. *Right* – the Automatic Resistivity Profiler *(ARP© Geocarta)*, capable of recording up to 4ha/day. To increase productivity within the project two ARP instruments were often used simultaneously.

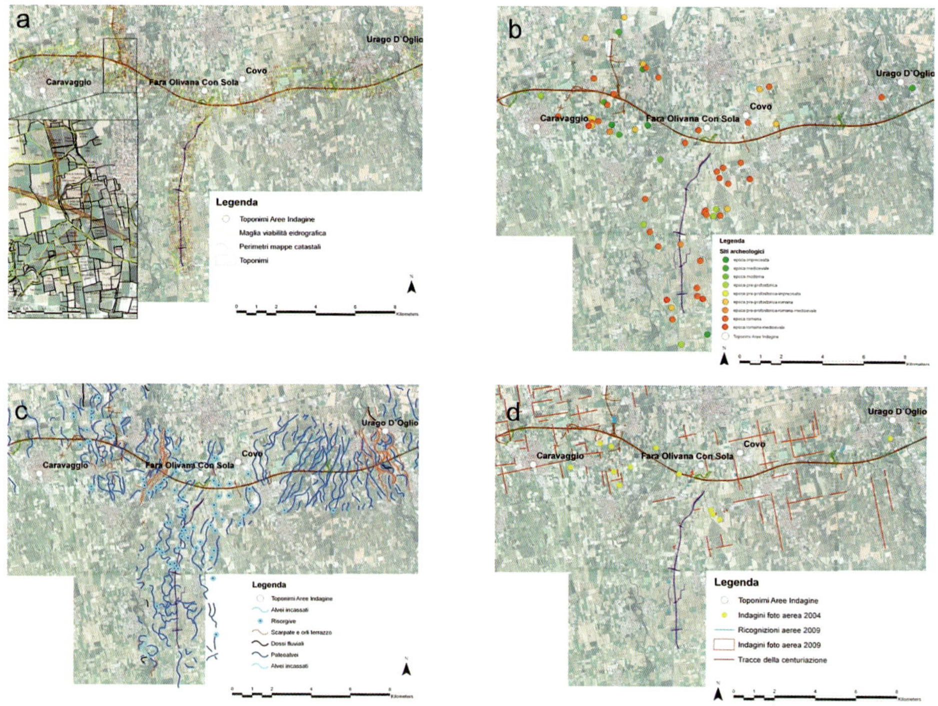

Figure 3.5: Mapped evidence for part of the survey area. *Top left* – historical cadastral map recording 3650 potentially relevant place-names and 154km of field boundaries within the 1km-wide motorway buffer zone. *Top right* – distribution of known sites and related archaeological evidence (118 in all, including 50 within the sample area). *Bottom left* – springs, palaeo-channels, fluvial ridges and fluvial terraces, clearly showing the hydro-geological volatility of the area. *Bottom right* – distribution map of features detected through exploratory aerial survey and oblique aerial photography.

iceberg in terms of the potential number of sites within the survey area. Recent studies in Tuscany, Lazio and Puglia (Campana 2009; Guaitoli 1997) have suggested that, in the absence of systematic survey projects, the 'published' archaeology as represented in the archives of the Archaeological Superintendency, represents no more than 1% to 5% of the 'real' archaeological potential. If applied to the BREBEMI motorway this would suggest the possibility of between 2,000 and 12,000 archaeological points of interest within the buffer zone!

The first stages of the analytical work went some way towards confirming this suspicion. For instance, the new aerial survey and the analysis of the historical aerial-photographs added another 76 'sites' of various kinds, substantially enriching the landscape picture and in some cases providing very detailed information about the sites concerned. An equally important contribution from the aerial-photo studies lay, as expected, in the reconstruction of the centuriation grid, knowledge of which is essential to the better understanding of the landscape and settlement patterns of the Roman and later periods.

In some cases, for example at a location close to Bariano (Figure 3.6), oblique aerial photography produced really striking results, bringing to light very detailed evidence of post holes, graves, round barrows and other previously unknown archaeological features but at the same time allowing the motorway construction company to take protective measures so as to avoid major logistical problems and significant waste of money during the eventual construction works.

The project also involved the capture 150 square km of LiDAR data at a resolution of 4 points per square metre, covering the full length of the motorway corridor along with the 1km buffer zone on either side. As noted earlier, the morphology of the area is to all intents and purposes completely flat and the land-use devoted for

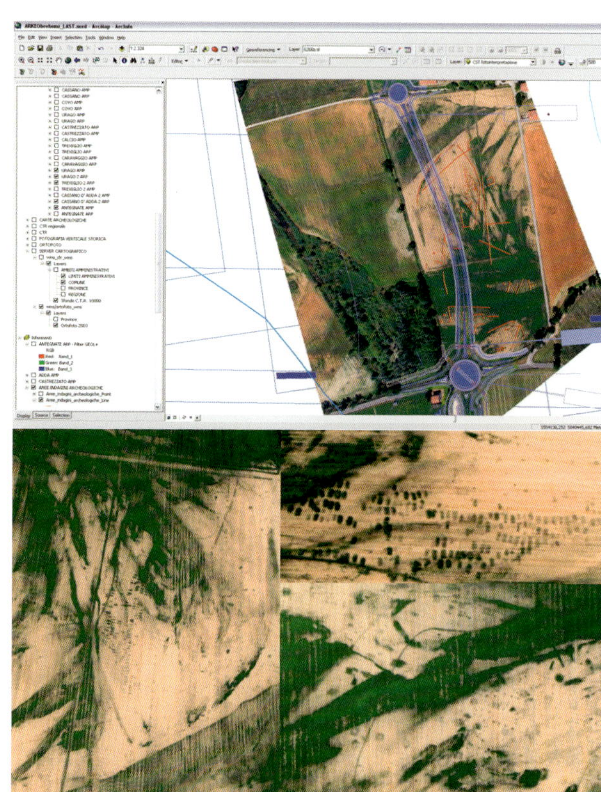

Figure 3.6: Newly discovered cropmark sites near Bariano
Top – the relationship between site and the motorway.
Bottom left – archaeological features associated with ancient road systems, the centuriation pattern, large round barrows and graves. *Right* – details of the cemeteries and settlement evidence including a ditch, post holes and probable sunken floored buildings.

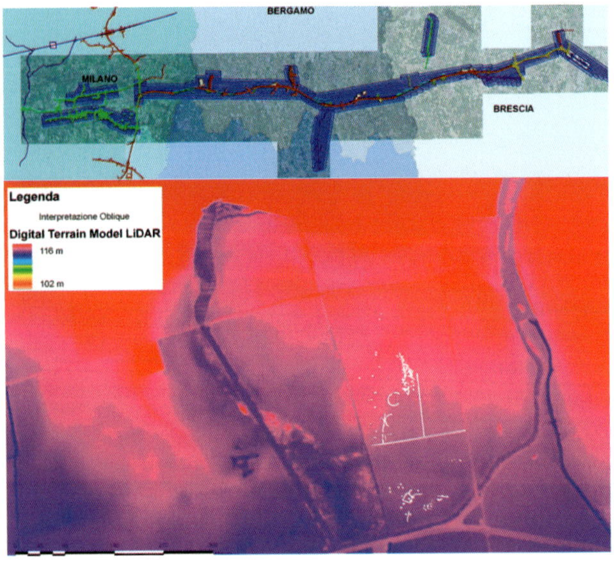

Figure 3.7: LiDAR survey. *Top* – data acquisition pattern for the project as a whole. *Bottom* – overlay on LiDAR data of features detected through aerial survey in the Bariano area (white). There is a clear correspondence between the aerial evidence and a terrace bordered on the east and west by two shallow valleys.

the most part to intensive cereal and maize production. The collection of LiDAR data was essentially aimed at identifying barely perceptible ridges, elevated areas and depressions, many of them perhaps related to former watercourses. The first stage of data processing, to create a basic digital terrain model, was carried out by CGR of Parma, the survey company that undertook the initial data capture. The second step involved collaboration between ATS and Prof. Dominic Powlesland in the UK, using his own visualization software, LidarViewer. This allowed the identification of 509 potentially significant features, consisting of 173 depressions, mainly interpretable as palaeo-river channels on the basis of their size, continuity and sinuous shape, along with 336 ridges or 'elevated' areas, at least some of them interpretable as fluvial ridges.

The information currently available shows a clear tendency for known archaeological 'sites' to occupy fluvial ridges and other 'elevated' areas within the plain. This is not to imply that these 366 raised areas correspond to a similar number of archaeological sites, only that these areas have a higher potential for the recovery of traces of past human activity. For instance, overlaying the LiDAR data on the aerial survey results for the area illustrated in Figure 3.6 shows that there is a clear correspondence between the features detected from the air and a terrace or ridge bordered on either side by two shallow depressions or 'valleys' (Figure 3.7). An alternative interpretation would see the aerial photographic features at Bariano as potentially continuing across the *whole* of the fields concerned but only being *visible* as cropmarks on the thinner and potentially drier soil of the ridges compared with the deeper and less responsive soil in the flanking depressions.

There can be no clear rule of interpretation about such situations but there are many other instances within the survey area where there is a clear relationship between topographical features in the LiDAR data and known or suspected archaeological sites established through documentary, place-name and cartographic research or through geophysical prospection or aerial-photo studies. With all due caution it is fair to stress the importance of carefully analysed LiDAR data, even in apparently 'unpromising' situations, in the process of archaeological prospection and indeed within the archaeological process as a whole.

Turning now to the second part of the process, and in particular the collection of geophysical measurements and related ground observation, both parties to the project, BREBEMI and the Superintendency, demanded a high level of reliability in the interpretation of the geophysical data. This is what prompted LAP&T and ATS to involve Geocarta in the systematic collection of ARP (magnetic) and AMP (geo-electrical) data on a field-by-field basis across the whole length of the project area. A total of 217 hectares of magnetic data and 215 hectares of geo-electrical data was collected, processed and interpreted (Figure 3.8). Ground observation of the first 150 hectares has been carried out through more than 200 test excavations, to a linear extent of about 5,220m (2.6 hectares) of 'targeted' interventions and a further 5,000m (2.2 hectares) of random excavations. Before looking at the results it is worth making some general comments on the kind of high-speed geophysical prospection involved in this case.

Figure 3.8: Geophysical prospection: quantification of detected features.

MAGNETIC ANOMALIES – AMP© METHOD ARCHAEOLOGICAL AND PALAEO-GEOMORPHOLOGICAL FEATURES	
Type of feature	Number
Road system	31
Field system	355
Centuriation	46
Ditch	33
Palaeochannel	204
Generic structure	90
Fence	1
Dipole Cluster	1389
Generic dipole	1770
Generic metal feature	151
Not id.	70
Generic alignment	449
Palaeo-morphology	444
Generic atrophic feature	25
Noise	142
Agricultural pattern	79
Grave	18
TOTAL	5297

ELECTRICAL ANOMALIES – ARP© METHOD ARCHAEOLOGICAL AND PALAEO-GEOMORPHOLOGICAL FEATURES	
Type of feature	Number
Road system	23
Field system	36
Palaeochannel	264
Generic alignment	152
Generic structure	15
Not id.	219
Centuriation	5
High conductivity	124
Low conductivity	13
Generic atrophic feature	35
Noise	2
Palaeo-morphology	454
Channel	8
Agricultural pattern	26
TOTAL	1376

Figure 3.9: Extracts from the magnetic map, with related interpretation and ground-observation by excavation. *Top left* – a structure related to water management. *Top right* – a feature that was expected to be (and was) a kiln. *Centre* – circular features with numerous parallels throughout Europe as (Bronze Age) round barrows, with ground-observation confirming this interpretation. *Bottom* – a feature of a size and shape that finds parallels within Italy and elsewhere as a medieval mound or motte; the site still awaits verification by excavation.

- High-speed prospection instruments demand high-speed processing and (more problematically) high-speed archaeological interpretation and mapping.
- The process of archaeological interpretation was more difficult in this case because of the peculiar shape of the survey area, a strip 100–150m wide along the full 100km length of the motorway.
- The prospection instruments for the most part performed extremely well but the use of a prototype instrument for collecting the magnetic data appears to have introduced a certain amount of noise into the dataset.
- This background noise, along with the physical and cultural peculiarity of the survey area, in particular the low magnetic contrast and perhaps other factors not yet identified, resulted in the identification of a large number of dipole clusters that were difficult to interpret, reducing the perceived reliability of the geophysical results.

Despite these problems we remain convinced that the systematic high-speed collection of geo-electrical and magnetic data is theoretically correct within such projects. In practice, however, there were too many occasions in this particular physical and cultural context where the magnetic data did not materially help archaeological interpretation.

Even allowing for these problems the geophysical prospection allowed the identification of a large range of both positive and negative evidence for the presence or likely absence of buried archaeological features, as shown in Figures 3.8 and 3.9. Despite the problems encountered it should be emphasised that the interpretation of the geophysical data in most cases achieved a higher level of interpretative reliability *when combined with* information from other datasets such as those derived from documentary sources, cartographical studies, aerial photography and LiDAR prospection. In the most favourable cases it is undoubtedly possible to achieve a full and detailed interpretation of the survey data. Despite degrees of uncertainty in other instances it is certainly possible to construct a reasonably reliable map of archaeological risk and potential which can then be subjected to ground-observation by test excavation or more substantial stratigraphical investigation in advance of the construction of the motorway.

Conclusions

Over a period of no more than four months of multi-faceted investigation it proved possible to collect and interpret a vast amount of data, greatly enriching our understanding of this particular stretch of landscape. The collected evidence and its interpretation also helped the motorway contractor to plan in advance for archaeological work which might otherwise have necessitated delays and extra expenditure during the construction work through the discovery of unforeseen archaeological sites and deposits.

The first 438 hectares of geophysical prospection and ground-observation have shown up some critical comparisons with the 'excavation by surface stripping' prospection system adopted by the regional Superintendency. In this context it is important to stress that while geophysical prospection and interpretation improve in reliability every year it is not possible to say the same for the method of rescue investigation adopted by the Superintendency, using mechanical stripping rather than prior survey and targeted stratigraphic excavation. Another key point is that it is not possible to verify the results of the excavation work initiated by the Superintendency – every archaeologist knows that excavation destroys the evidence upon which it relies, especially if it is not carried out within a suitable methodological framework. By contrast it is entirely possible – and desirable – to use stratigraphic excavation to verify and interpret potential archaeological features recorded initially through geophysical or other forms of non-invasive prospection.

There is a clear contrast here between the approach of LAP&T and ATS within the BREBEMI project compared with the traditional approach advocated by the regional Superintendency. Fortunately an 'outside' assessment of the relative merits of the two approaches, based on depositions in writing and in person by both parties, was made by the Technical and Scientific Committee for Italian Archaeology, consisting of leading academics along with the General Director of the Superintendency at national level. After a detailed analysis of the two approaches the Committee was unanimous in its conclusion that the strategy proposed by LAP&T and ATS, and the survey and ground-observation work subsequently undertaken, represented the most advanced approach to this kind of preventive archaeology so far attempted in Italy and that this case study should represent an example for future projects of infrastructure and building development.

One final observation is perhaps in order. The greatest improvement in rescue and preventive archaeology will surely come not from technological development alone but from a more consistent application of the kind of 'total archaeology' and 'global' historical approach advocated at the beginning of this paper. This change of approach is imperative because we need first to understand the local context by working closely with local archaeologists and historians in the attempt to improve our capacity to interpret and test the 'global' dataset assembled from multiple survey techniques. Only then will it be possible to reduce the archaeological risk and maximize the archaeological returns from preventive and rescue archaeology.

Postscript

Sadly, the regional Superintendent Dr Raffaella Poggiani Keller – as is its right within the present organizational structure in Italy – ignored the national Committee's opinion, suspending further work by the consultancy and applying its own 'method of surface stripping' to the rest of the motorway. On the basis of this example it will clearly take time for more advanced methods to attain a widespread application elsewhere. Nevertheless, through the impact of the new law and the example of this and other projects over the past few years the ground has surely been prepared for a culture-change in the official approach to preventive and rescue archaeology within Italy.

Acknowledgments

First of all the author owes a huge debt of gratitude to the late Professor Riccardo Francovich, of the University of Siena, who gave him the cultural background and the intellectual vigour to face a challenge like that of the BREBEMI project. Special thanks are also due to two good friends who have followed and inspired so much of the writer's research work since early in his career, Chris Musson and Dominic Powlesland from the UK; as ever, they helped with constructive criticism and comments throughout all stages of the project.

Sincere thanks are also due to the BREBEMI Company for the great opportunity and trust provided by the president Dr Francesco Bettoni, the general director Prof. Bruno Bottiglieri, the chief of the rescue archaeology bureau Dr Paola Rigobello and the company's chief engineer Dr Lorenzo Foddai. The author is further indebted to the Solng Company, in particular to Annalisa Morelli, Gianfranco Morelli and Giovanni Bitella, as well as to Iacopo Nicolosi at the Italian National Institute of Geophysics and Vulcanology. Grateful thanks also go to the Geocarta Company and Michel Dabas, as well as to Klaus Leidorf of Luftbilddocumentazion in Germany. All of these friends and colleagues contributed to the successful conduct and management of the field investigations and to the overall outcome of the project.

Special thanks are also due to the team of the Laboratory of Landscape Archaeology and Remote Sensing at the University of Siena and of the spin-off company ATS Enterprise: Cristina Felici, Matteo Sordini, Francesco Pericci, Lorenzo Marasco, Barbara Frezza, Anna Caprasecca and Francesco Brogi.

Finally, heartfelt thanks also go to the president of the Superior Consiluim of Cultural Heritage Prof. Andrea Carandini, to director Prof. Giuseppe Sassetelli and the members of the Scientific Committee for Italian Archaeology, and lastly to the General Director of the Superintendency, Dr Stefano De Caro, for providing the opportunity to discuss our project and for having the courage to present a report which will hopefully see the start of a cultural revolution in Italian archaeology moving from rescue to a real preventive approach.

References

Campana, S. 2009: Archaeological Site Detection and Mapping: some thoughts on differing scales of detail and archaeological 'non-visibility'. In Campana, S. & Piro, S. (eds): *Seeing the Unseen. Geophysics and Landscape Archaeology.* Proceedings of the XVth International Summer School, Taylor & Francis, London, 5–26.

Campana, S. & Francovich, R. 2009: Seeing the Unseen. Buried Archaeological Landscapes. In Campana, S. & Piro S. (eds): *Seeing the Unseen. Geophysics and Landscape Archaeology.* Proceedings of the XVth International Summer School, Taylor & Francis, London, 67–76.

Campana, S. & Piro S. (eds) 2009: *Seeing the Unseen. Geophysics and Landscape Archaeology.* Proceedings of the XVth International Summer School, Taylor & Francis, London.

Carandini, A. 2008: *Archeologia Classica.* Vedere il tempo antico con gli occhi del 2000. Torino.

Dabas, M. 2009: Theory and practice of the new fast electrical imaging system ARP©. In Campana, S. & Piro S. (eds): *Seeing the Unseen. Geophysics and Landscape Archaeology.* Proceedings of the XVth International Summer School, Taylor & Francis, London, 105–26.

Francovich, R. 2006: Archeologia e Territorio. In Detti, T. (ed.): *La terra dei Musei. Paesaggio, arte, storia del paesaggio senese.* Florence, 12–39.

Guzzo, P.G. 2000: Legislazione e tutela. In Francovich, R. & Manacorda, D. (eds): *Dizionario di Archeologia.* Bari, 177–83.

Guaitoli, M. 1997: *Metodologie di catalogazione dei beni archeologici.* Bari.

Guermandi, M.P. 2001: *Rischio Archeologico. Se lo conosci lo eviti, atti del convegno di studi su cartografia archeologica e tutela del territorio.* Ferrara, 24–25 marzo 2000, Firenze.

Poggiani Keller, R. (ed.) 1992: *Carta Archeologica della Lombardia, II. La Provincia di Bergamo*, Vol.1, 2, 3 (Saggi, Schede, Cartografia), Modena.

Powlesland, D. 2006. Redefining past landscapes: 30 years of remote sensing in the Vale of Pickering. In Campana, S. & Forte, M. (eds): *From Space to Place, Proceeding of the 2nd International Conference on Remote Sensing Archaeology.* Rome 4–7 December 2006, Archaeopress, BAR International Series, 197–201.

Powlesland, D. 2009: Why bother? Large scale geomagnetic survey and the quest for 'Real Archaeology'. In Campana, S. & Piro S. (eds): *Seeing the Unseen. Geophysics and Landscape Archaeology.* Proceedings of the XVth International Summer School, Taylor & Francis, London, 167–82.

Ricci, A. 1996: *I mali dell'abbondanza. Considerazioni impolitiche sui beni culturali.* Roma.

Ricci, A. 2006: *Attorno alla nuda pietra. Archeologia e città tra identità e progetto.* Roma.

4 | Remote sensing for archaeology and heritage management – site discovery, interpretation and registration

David C Cowley

Abstract: Understanding the impact of factors such as past and present land use and the interests of archaeologists on what we know is vital to creating reliable knowledge that supports effective management. This paper explores these issues, drawing on examples from Scotland to identify the challenges of understanding particular landscapes and highlighting the importance of broad-brush characterisation in providing a framework for matching survey methodology to local contexts.

Introduction

The contemporary landscape is formed by complex interactions of many factors and processes, the evidence of which is manifested in its physical character and in the variability of that character from place to place. Such variability in landscape character not only reflects its history (i.e. the palimpsest) however, but also its archaeological potential, which in its turn has a direct effect on the effectiveness of different survey techniques. Thus, broad-brush characterisation of the contemporary landscape and its historic components is the essential framework for matching survey methodology to landscape contexts, and for identifying the limitations of existing survey techniques. Only by taking these factors on board can we hope to understand general patterns of remains occurring in the landscape and ensure the application of cost-effective techniques that challenge the boundaries of our knowledge. This perspective comes from a view of the landscape as a mosaic of complex attributes, including the nature of the existing landscape, its archaeological potential and the appropriate techniques to understand it.

Central to this is understanding *how we know what we think we know* and identifying *what we think we don't know* (see Powlesland this volume). An important starting point for addressing these issues is the recognition that the factors which structure archaeological data are not limited to what happened in the distant past. They include land use, the interests of archaeologists and survey methodology, to name only a few. Without rigorous analysis of such factors our understanding of what happened and where it happened in the past will more likely reflect biases in data coverage than anything else. This is directly relevant to effective heritage management as it informs considerations of how representative or rare monuments and landscapes may be, how their condition and information value may be assessed, and how effective survey and recording techniques are likely to be in providing the basis for management.

These broad themes will be illustrated in a Scottish context, not necessarily as an example of best practice, and certainly not to advocate a 'one size fits all' approach, but to show how the structure and origins of data, and its strengths and weaknesses, can help in the design of effective survey and recording strategies.

Landscape characterisation

Broad-brush characterisation is not widely applied in Europe, but where it has been used it provides a mechanism to take an overview of the development and nature of the landscape. In Scotland, characterisation is being developed through a national programme of Historic Land-use Assessment (HLA: http://hla.rcahms.gov.uk/). Jointly funded by Historic Scotland and the Royal Commission on the Ancient and Historical Monuments of Scotland (RCAHMS), HLA evaluates and records land use from the prehistoric to the present. Using contemporary and historic maps and aerial photographs the landscape is examined and broken down into areas that have particular characteristics, dependent on present and past land use, with a minimum area that can be defined of one hectare. HLA traces the origins of the present landscape, recognising that it is always evolving as its use changes and develops. It identifies historical features (i.e. relict types) and assesses how these continue to influence the present use of land and its appearance (Figure 4.1).

In England, Historic Landscape Characterisation (HLC), run as a partnership between English Heritage and County Council Sites and Monuments Records, offers a similar broad-brush approach to recording and interpretation of the landscape (Clark *et al.* 2004; Fairclough & Macinnes 2003). This approach has been trialled in the Mediterranean (Crow & Turner 2009; Turner & Crow 2010) and in the inundated landscapes of European coastal shelves (Fitch *et al.* this volume). Its use has been advocated more widely (see Einarsson &

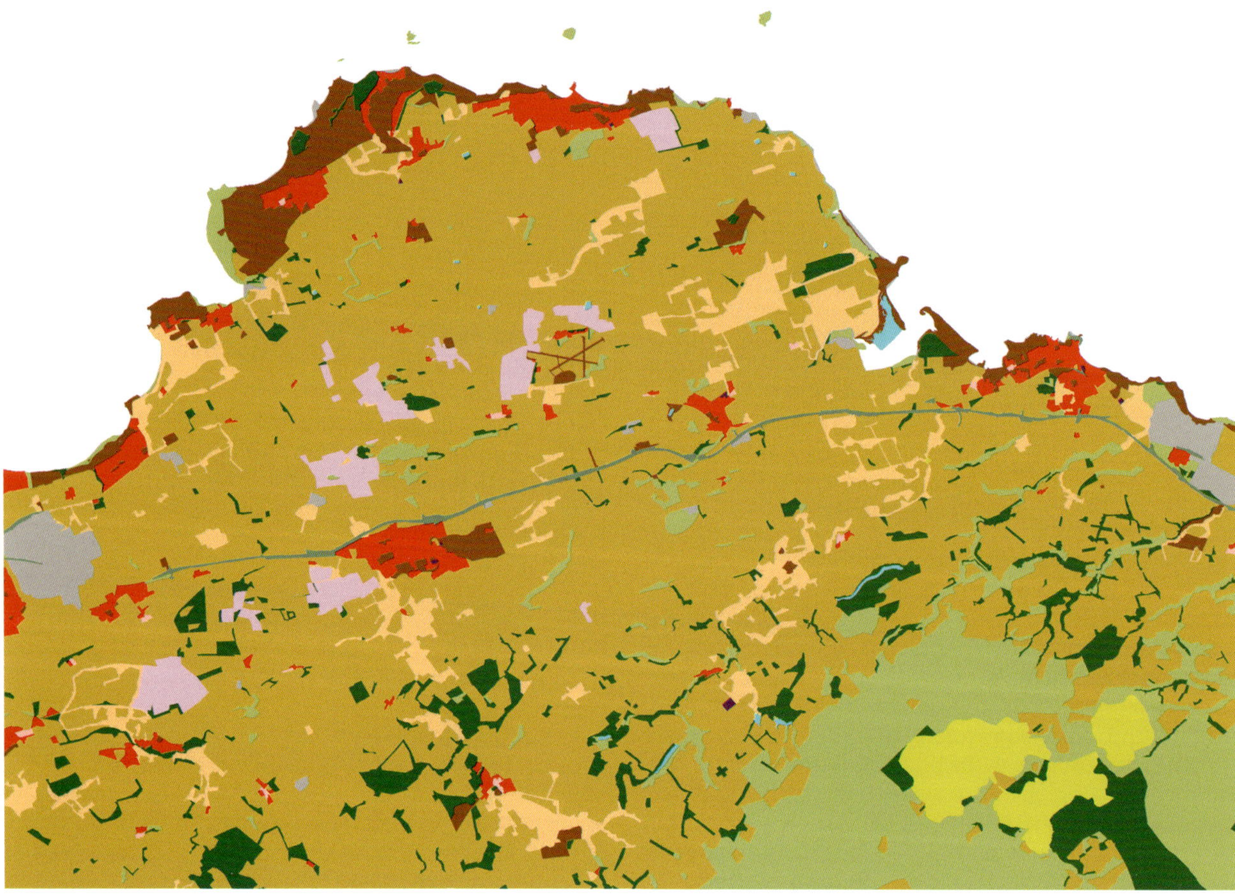

Figure 4.1: Historic Landuse Assessment data of East Lothian displayed for current landscape character (e.g. built up, recreation, woodland, pasture etc.). These broad characteristics have implications for archaeological potential and the effectiveness of different survey techniques. All areas also have their relict characteristics recorded as attribute data where possible, indicating, for example, where medieval or earlier landscape features may be visible. *GV004748, © Crown copyright: RCAHMS, reproduced under licence.*

Aldred this volume). Such broad-brush characterisation provides a means of achieving the aims of the European Landscape Convention (ELC): to promote integrated landscape protection, management and planning (Council of Europe 2000). The ELC explicitly recognises landscape, whether common or designated as rare or special, as a basic component of heritage, applicable equally to maritime and terrestrial environments. Characterisation should also provide a way to deal with the ongoing challenges of changing climate and land use (e.g. Wordsworth 2010) and the threats that these pose to historic landscapes.

Defining approaches to survey

Beyond its role in support of the aims of the ELC, characterisation can help define generalised archaeological potential and the applicability of existing survey methodologies, in particular identifying where data-collection strategies are not effective. These themes can be illustrated with reference to three different landscapes in Scotland. In lowland areas intensive cultivation, especially since the 18th century has swept the remains of earlier features away and little survives on the surface beyond a few large unitary monuments. Where these areas coincide with well-drained soils and arable land use buried sites can be recorded from the air as cropmarking. In Scotland this approach has placed about 7,500 otherwise unrecoverable site on record (Figure 4.2 A). By contrast in many upland areas there are extensive prehistoric, medieval and post-medieval remains surviving because of the contraction of intensive cultivation to lowland areas since the mid-18th century. In some areas, this has resulted in extensive areas of early landscape features surviving as surface remains (Figure 4.2 B). A third contrasting landscape is provided by Arran, an island off the west coast of Scotland, with a blend of dramatic highland scenery and lowland fields, often referred to as 'Scotland in miniature'. An aerial view (Figure 4.2C) of the northwest of Arran illustrates its topographic extremes, rising from sea level to 875m at the top of Goat Fell; it also shows major changes in vegetation. The pasture in a green band along the shore marks the extent of the predominantly 19th century pattern of settlement and land use. Relatively intensive land use in these areas has tended to remove the surface features of most earlier remains. This limits the likely archaeological return from aerial or field observation, either as earthworks visible in low relief or as cropmarks, which are unlikely to form due to the combination of pasture and a wet climate. The hillside above the pasture fields is a zone of potential for surviving archaeological remains, depending on the extent of earlier settlement and land use beyond the footprint of the 19th century fields. Further up the hill the relative extremes of topography militate against

Figure 4.2: Contrasting landscapes.
A – in lowland areas centuries of cultivation have levelled earlier landscape features, but they have been recorded in large numbers when revealed by cropmarking. For many periods, including the Neolithic, this data underpins our understanding of the period and its monuments, such as this cursus monument at Holywood in the southwest.
B – Neolithic and Bronze Age remains survive on the surface, though party enveloped in peat, on the rolling moorland of Shetland.
C – Winter conditions highlight the stark contrasts in the landscape on the northwest coast of Arran.
SC622986, DP082172, DP056902, reproduced under licence. RCAHMS: The National Collection of Aerial Photography.

extensive exploitation in the past, while areas of deep peat and rank heather vegetation make the detection of potential remains more difficult.

These landscapes have contrasting histories and archaeological potential, and represent very different survey and recording challenges. As noted above, many aspects of the varying archaeological potential and the choice of survey strategy are conditioned by contemporary and past land use, and therein lies the importance of broad-brush characterisation. In the well-drained lowland soils of eastern Scotland that are set to arable, aerial reconnaissance and recording of cropmarks are highly effective in revealing plough-levelled sites. Equally, in upland areas extensive surface archaeological remains are responsive to both field observation and to recording from the air under low vegetation and oblique raking light in winter or early and late in the day at other times of year.

However, while these are undoubtedly strengths, there are large parts of the country where such approaches are not effective. For example, about 28% of Scotland is now permanent pasture, most of which has been formerly cultivated. Here, historic cultivation has largely removed earlier surface remains and existing large scale prospection techniques are ineffective (i.e. no cropmarks, few surface remains). So while approximately 11% of the country has the potential to produce archaeological cropmarking in any given year and can be examined in a reasonably systematic way, nearly three times that area remains a largely closed book. This is due to a combination of factors including broad environmental patterns and contemporary and historic land use, aspects of which may be summarised in the broad-brush characterisation described above. These examples illustrate the connections between land use, archaeological potential, the applicability of survey techniques and broad-brush landscape characterisation, providing a framework for articulating these issues. In addition it may help to frame the research questions that, in time perhaps, can help to shed light on the black holes of knowledge, such as the archaeology of the country's extensive areas of pasture.

Understanding the archaeological record

Archaeological sites and historic landscapes require identification, documentation and understanding if they are to be effectively managed. The role of broad-brush characterisation has been discussed above and attention is now turned to another central theme – understanding the way in which the archaeological record is constituted. This is fundamental to effective use of the record of the historic environment, since biases inherent in data coverage, produced by varying land use patterns, along with the knowledge of archaeologists and their survey techniques, have significant impacts.

Data coverage

Maps of monument distributions are one of the commonest ways in which archaeological information is conveyed, and while these are rarely simple or necessarily well understood, they are often presented as true representations of past activity (see Halliday 2011 for discussion of these issues). That this is rarely the case can be illustrated by the distribution of (mostly) Bronze Age burnt mounds in Scotland. These heaps of fire-cracked stones from cooking sites and/or saunas (Buckley 1990) are well known and the RCAHMS database contains 1808 records of such features (at January 2011).

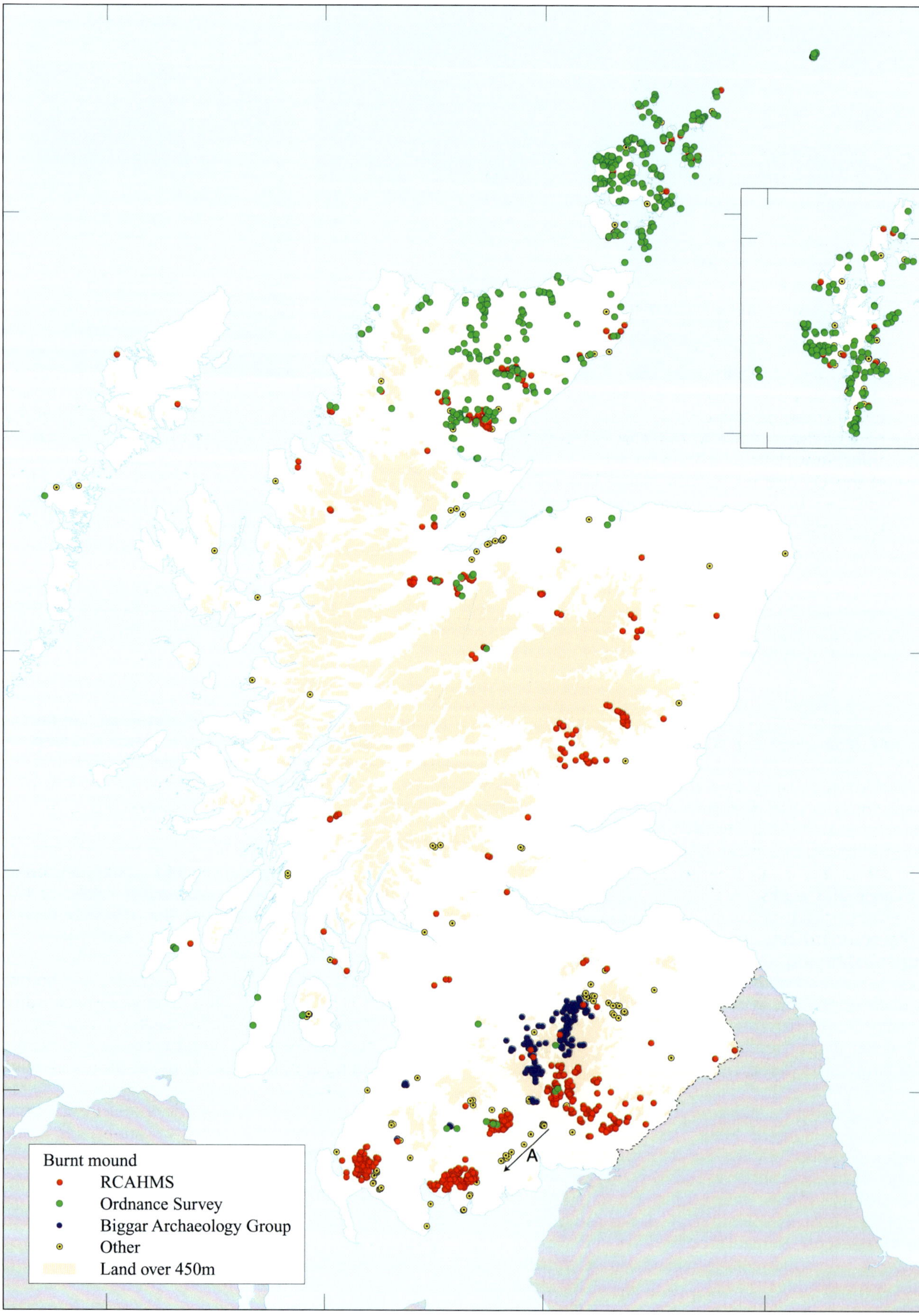

Figure 4.3: The distribution of burnt mounds known in Scotland at January 2011, classified according to the agency that identified them. This is a powerful example of the influence that experience and knowledge has on datasets. 'A' and the arrow indicates the line of a gas pipeline. *GV004749, © Crown copyright: RCAHMS, reproduced under licence.*

Figure 4.4: Most burnt mounds are recognised as earthworks in unimproved ground, such as this one defined by a C-shaped mound around a small hollow (with spade) at Auchenfedrick in southwest Scotland. © *David C Cowley.*

However, it is immediately evident that their distribution (Figure 4.3) is uneven nationally and marked by discrete dense clusters of sites. This pattern is a direct result of the distribution of archaeological fieldwork by individuals with particular experience, interests and knowledge, along with varying land use types. The site locations on the map are coded to indicate the agency which first discovered the monument, divided fourfold between the Archaeology Division of the Ordnance Survey (OS), RCAHMS, Biggar Archaeology Group, and 'others' (contracting units, local societies, antiquarians etc.). Building on antiquarian records the OS began to identify these monuments as its surveyors moved into new fieldwork areas in the north of Scotland. At this point Strat Halliday, a fieldworker from RCAHMS, visited the OS and was shown some burnt mounds. On returning to his fieldwork area in the southwest he began to recognise them (Halliday 1990). The subsequent development of the distribution of known burnt mounds reflects RCAHMS field survey projects and Halliday's role in educating his colleagues in their identification. This experience was also passed on to Tam Ward of the Biggar Archaeology Group, an active local voluntary group, who expanded the distribution across the counties of Lanarkshire and Peeblesshire. Both counties had seen some previous archaeological fieldwork in the 1960s and 1970s but this did not identify any of these monuments as it focussed on traditional site-types such as 'forts', 'settlements' and 'burial cairns' (RCAHMS 1967, 1978). In effect this work predated an awakening interest in the wider landscape. The other sites on the map are mainly chance discoveries, with the exception of a clear line of burnt mounds in the southwest (Figure 4.3 A) which relates to monitoring in advance of a works for a gas pipeline (Maynard 1993). The sites noted by Maynard are some of a very few recorded in lowland areas and were identified only after topsoil stripping. This reflects the other major bias in the burnt mound distribution – that they are generally visible in unimproved rough ground where ancient monuments have survived as earthworks (Figure 4.4) beyond the limits of medieval and modern agriculture which have otherwise destroyed or covered them up. Levelled burnt mounds should be recoverable by field walking of ploughed land, but there is not a widespread tradition of this method in Scotland. However, where it has been applied systematically, it has demonstrated its value, for example in the discovery of the first open-air Upper Palaeolithic site in Scotland (Ballin *et al.* 2010).

Though this example is based on ground-based survey and observation its lessons are equally relevant to other datasets or survey methods, where bias may be introduced by factors such as the type of sensor used or the personal interests of the observer. The known distribution of burnt mounds is heavily biased by fieldwork patterns and modern land use, and grasping their place in Bronze Age settlement systems or their rarity from area to area, for example, is dependent on an appreciation of these factors. While this may seem like an extreme example, these or similar biases are inherent in all archaeological information-gathering. Academic discourse and management strategies that fail to properly understand them are ultimately flawed.

Survey bias

The application of aerial reconnaissance to archaeology is another case where the relationship between data and how it may reflect the past is not direct. Over the last 50 years this method of survey has revolutionised our understanding of much of lowland Scotland, recording many thousands of otherwise unknown levelled monuments revealed through differential cropmarking and vegetation (Figure 4.5). In some areas of southern and eastern Scotland the *majority* of known sites and monuments have been recorded in this way (Figure 4.6). However, the effectiveness of aerial reconnaissance is conditioned by rainfall patterns, crop phenology, soil types and land use, producing best results over well-

Figure 4.5: This sub-rectangular Neolithic pit-defined enclosure was recorded for the first time during aerial survey in 2008. *DP044522, reproduced under licence. RCAHMS: The National Collection of Aerial Photography.*

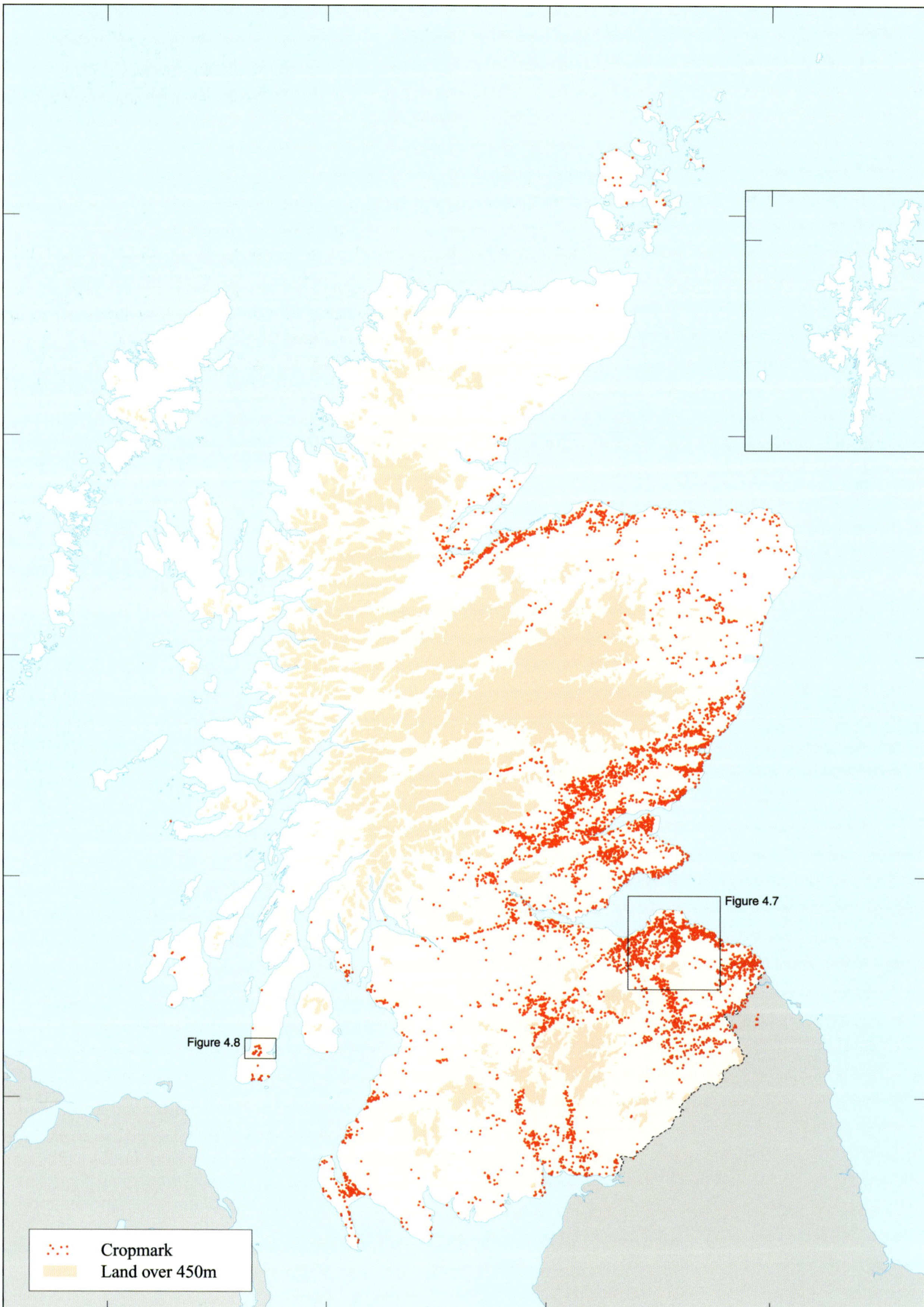

Figure 4.6: Map of plough-levelled archaeological sites known from cropmarking at 2010. The gross distribution reflects the locations of arable crops and well-drained soils. *GV004748, © Crown copyright: RCAHMS, reproduced under licence.*

drained soils set to cereal crops in drier parts of the country. Beyond these physical factors there are the biases introduced by the interests and methodology of the airborne observer that clearly pattern the data strongly (papers in Brophy & Cowley 2005; Cowley 2002; Cowley & Dickson 2007). Thus, while the general processes of cropmark formation are understood (e.g. Wilson 2000) any close examination of the cropmarked information invariably raises questions about how representative it is. The map of these sites recorded in East Lothian during 1977 and 1992, for example, shows how survey over the same area in two different years can produce markedly differing returns, conditioned by many factors including localised cropping and rainfall patterns (Figure 4.7). This example highlights not only the potentially unpredictable variability of the survey data, but also the importance of sustained programmes of aerial reconnaissance.

The Scottish experience shows the importance of maintaining a critical review of survey programmes to ensure they remain fit for purpose and able to react to changing circumstances. Key to this is the role of survey in pushing boundaries rather than simply reinforcing established perceptions. Intermittently, summer weather patterns in Scotland have reversed their usual pattern of dry in the east and damp in the west, and this occurred in 2008, 2009 and for a short period in 2010. In all these years the unusual conditions prompted reconnaissance in the west that produced major discoveries. While much of the survey work contributed to filling in site distributions in areas of known potential, it was reconnaissance during 2008 that really produced significant changes in our knowledge. A speculative sortie to the Mull of Kintyre prompted by unusually high soil moisture deficit figures (i.e. dryness) generated a remarkable return of

Figure 4.7: The comparison of sites recorded in 1977 and 1992 shows how variable the visibility of cropmarked sites may be from year to year, and highlights how important sustained programmes of survey are. See Figure 4.6 for location. *GV004751, © Crown copyright: RCAHMS, reproduced under licence.*

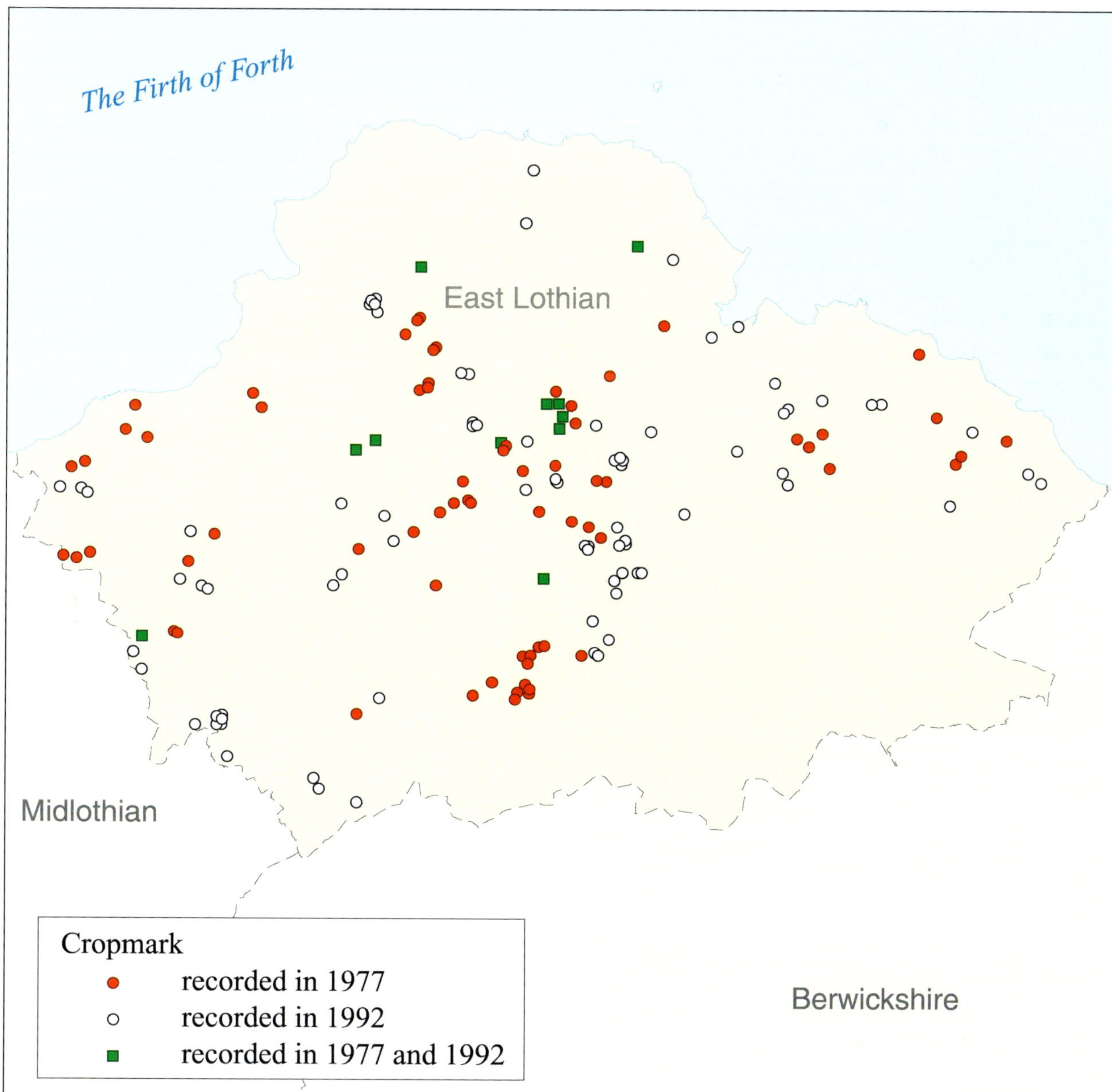

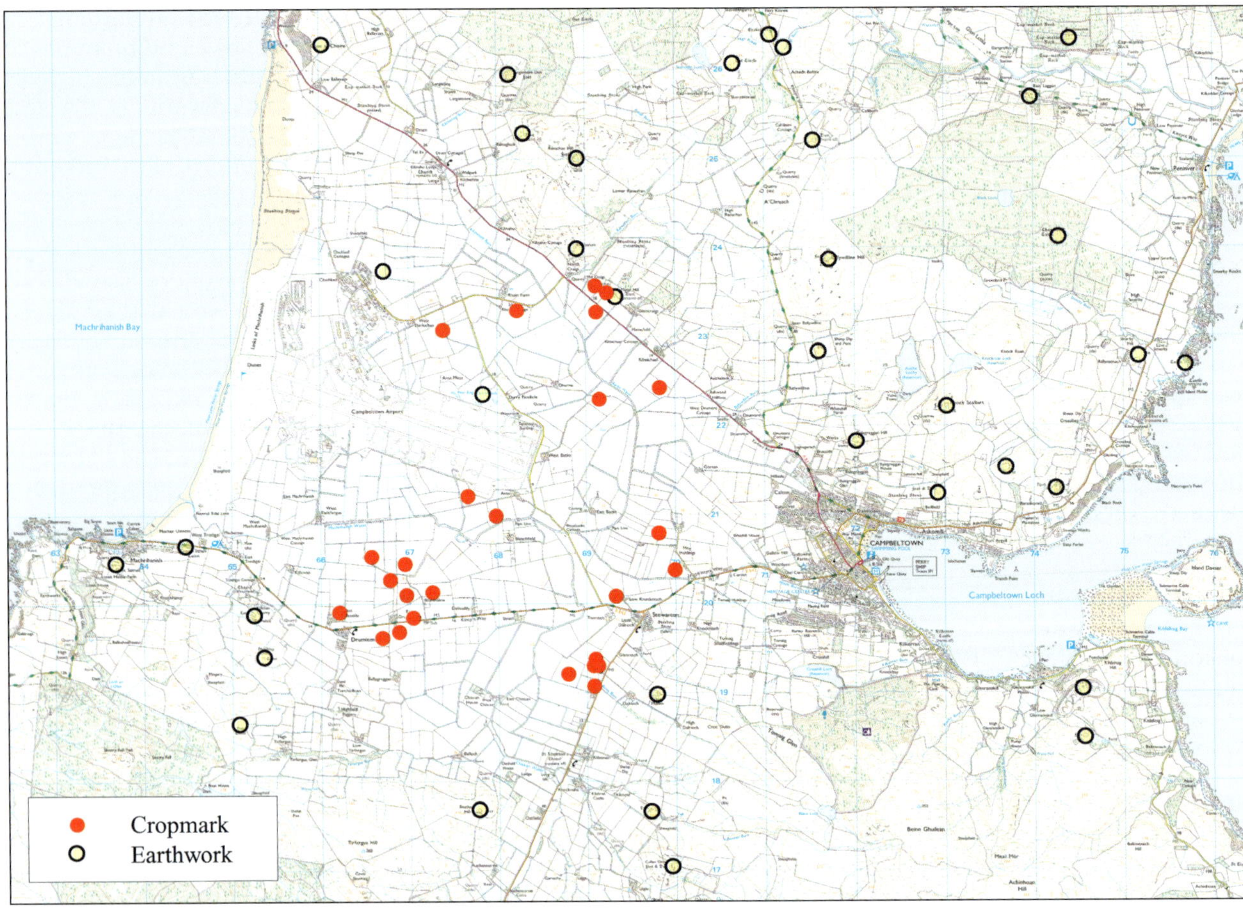

Figure 4.8: The distribution of plough-levelled sites recorded as cropmarking from the air in 2008 and on 1970s vertical aerial photographs (e.g. Figure 4.9) – shown against known earthwork monuments. See Figure 4.6 for location *GV 004750, © Crown copyright: RCAHMS, reproduced under licence. Background map © Crown Copyright and database right 2011. All rights reserved. Ordnance Survey Licence number 100020548.*

previously unrecorded sites scattered across the arable fields in an area that had never before been subjected to such survey (Figure 4.8).

These returns were well beyond our expectations and prompted a structured search of the RCAHMS holdings of historic aerial photographs, which identified images taken to aid map revision by the OS in 1976 and 1977 – years that were very dry with a high potential for cropmarking. On examination, these too resulted in the identification of a number of sites, duplicating some of those recorded in 2008, but also adding otherwise unrecorded sites (Figure 4.9). Together, the 2008 aerial reconnaissance and the structured inspection of the 1970s photographs have challenged the ways in which the archaeology of Argyll is regarded. The orthodox view of the archaeology of much of the west is dominated by stone built monuments (e.g. RCAHMS 1971), and there was little expectation that buried sites more familiar in the east of the country would be present, much less that they would be revealed by cropmarking. This example highlights the

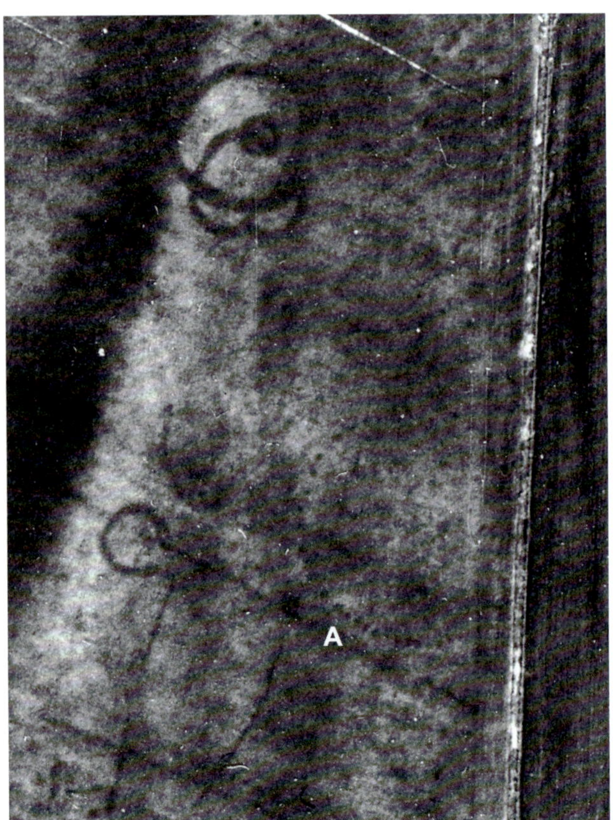

Figure 4.9: A Neolithic pit-defined circle (A) and other monuments recorded by chance on an aerial photograph taken to help map making. *OS 77 081, © Crown copyright Ordnance Survey. All rights reserved.*

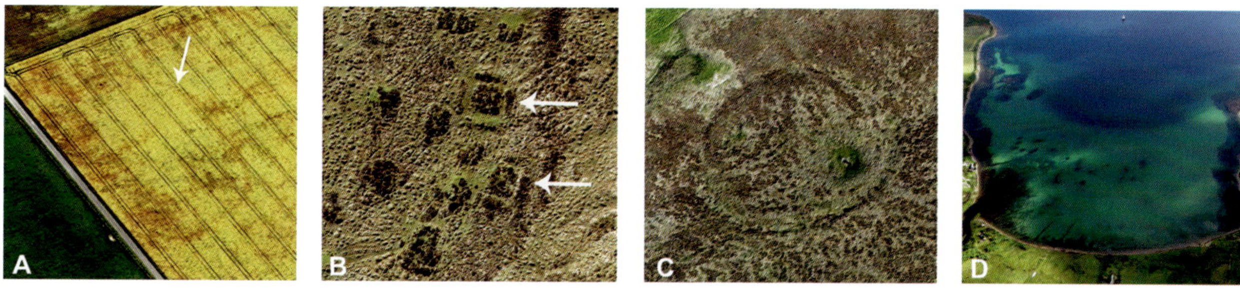

Figure 4.10: Contrasting sites on Orkney. A – an oval enclosure recorded as a cropmark represents a new type of site. B – a cemetery including square barrows of 1st millennium AD date on Hoy. C – the earthwork enclosure on Hundland Hill is also a previously unrecognised site type that may be prehistoric in date. D – potentially prehistoric features inundated by rising sea levels in Mill Bay, Hoy. *DP066519, DP060149, DP059774, DP066524, reproduced under licence. RCAHMS: The National Collection of Aerial Photography.*

value of targeted aerial reconnaissance in recording sites for the first time, but also shows the enormous potential locked up in archives of aerial photographs (see Ferguson this volume; Cowley & Ferguson 2010 and papers in Cowley *et al.* 2010). There are some 1.6 million mainly vertical aerial photographs of Scotland in the RCAHMS archive, ranging in date from the 1940s to the last decade, the majority of which have yet to be inspected systematically for archaeological purposes. The Mull of Kintyre case and others illustrate that survey undertaken in as yet unexamined archives has huge potential.

Condition and potential information return

Plough-levelled sites recorded through differential cropmarking can present something of a contradiction in terms of their potential information content. It has become increasingly evident that 'nice cropmarks' – i.e. those that look good on photographs – are often heavily truncated by ploughing and may therefore be poor reservoirs of information (e.g. Dunwell & Ralston 2008a). Many cropmark sites do in practice produce useful information on excavation but there is rarely any horizontal stratigraphy or significant deposits of sediments that might hold palaeo-environmental information (outside the truncated ditches etc.). This presents a basic paradox that those sites that we are best equipped to find through aerial reconnaissance are generally heavily damaged and, if excavated, have a potentially limited information return. On the other hand chance discoveries of deeply buried sites demonstrate that there are areas in the landscape with good preservation, but which we are ill-equipped to identify in advance through survey (e.g. Barclay 1985; Pollock 1985). On excavation such sites, with preserved stratigraphy and potentially informative sediments, will generally have a much higher information return than heavily truncated cropmarked sites. Appreciation of information return clearly has major implications for what is considered worthy of protection and on the strategies that may be employed to mitigate threats and retrieve information when sites are going to be destroyed. It is also a challenge in terms of how we approach archaeological prospection in those areas that we cannot see effectively – either from the ground or from the air – because of deep soils, recognising that these parts of the landscape may contain information rich-sites, with good potential preservation and informative sediments.

Extending what we know

The discussion above has identified some of the issues of using existing survey data, whether remote-sensed or ground-based. In this next section the challenge of extending what we know is discussed. The example of the Mull of Kintyre illustrates the value of survey – in that case a combination of aerial reconnaissance and targeted exploration of historic archive images – in putting sites on record. In this section the benefits of integrated data sources, the strengths and weaknesses of that record, and the translation of data to understanding and knowledge are discussed.

Integrated survey

It will be clear to anyone reading the papers in this volume that integrated approaches to archaeological prospection are necessary to generate reliable information for understanding and managment. At a basic level a landscape approach provides the framework for integrated thinking, allowing differing contexts and environments to be explored and issues of scale to be addressed. Recent work in the Orkney Islands, which lie just off the northern tip of the Scottish mainland, provides a case in point. This is not an area that had been considered worth systematically exploring from an aerial perspective, but as in the Mull of Kintyre an unusual weather pattern during 2009 provided the conditions for widespread formation of cropmarking that revealed numerous previously unknown plough-levelled monuments. These sites include some that have analogies amongst known sites but others are of completely new types (Figure 4.10 A). Likewise reconnaissance across upland areas on the islands what is presently the only recorded example of a type of cemetery of 1st millennium AD date on Orkney, characterised by square ditched barrows (Figure 4.10 B), as well as a scatter of previously unknown earthwork enclosures (Figure 4.10 C). Furthermore, guided by archaeologists exploring the potential for submerged landscapes around the shoreline, prospection across shallow water recorded targets that may include inundated prehistoric landscapes (Figure 4.10 D). These are notable successes and highlight the added-value that the critical application of remote sensing techniques in appropriate contexts provides.

The identification of these sites on Orkney has not been an end in itself, and follow-up work has developed what we know. Field observation by the author and Dan Lee of Orkney College has reinforced the preliminary interpretation of the square barrow on Hoy, while inspection of upcast from a rabbit hole in the boundary bank of an enigmatic earthwork on Hundland Hill (Figure 4.10 C) produced burnt human bone (Ingrid Mainland pers. com.), hinting the site there may be prehistoric in date. Coring in the shallow waters of Mill Bay (Figure 4.10 D) by Sue Dawson (University of Dundee) and Caroline Wickham-Jones (University of Aberdeen) has identified a peat under the sand – unequivocal evidence for an inundated land surface. Finally, survey by the Orkney College Geophysics Unit on a site discovered from the air has clarified the nature of the features recorded on the aerial photograph, but also provided additional, complementary information (Figure 4.11).

The examples from Orkney illustrate the role of scale or resolution in survey and observation. Aerial survey operates effectively at a landscape scale, providing a highly cost-effective broad perspective, but at a relatively coarse resolution. The enhancements gained by field observation, geophysical survey and palaeo-environmental work provide detail and potential dating and have a finer resolution, but come at a greater cost. This is the essence of a multi-scaled approach to the landscape, ranging from intra-site work to the overview of national datasets (e.g. Figure 4.3) and broad-brush characterisation.

Strengths and weaknesses – challenges and potential

The landscape of Scotland, in common with other parts of the world, is one of often dramatic contrasts and variety. This variety carries with it implications for the suitability of particular survey techniques, the character of surviving remains and the strategies employed to understand and manage them. In many areas of upland Scotland, for example, the relict earthwork remains of monuments from the Neolithic to the post-medieval period routinely survive above the maximum extent of 19th and 20th century cultivation, which in other areas has levelled earlier remains. In some lowland contexts plough-levelled sites have been recorded as cropmarking across large areas. At a national scale, then, these are two strengths of the archaeological record in Scotland – a *smorgasbord* of differing types and preservation of archaeological traces. However, while these data are undeniably valuable they are also heavily biased. Aerial reconnaissance allows us insights into lowland areas of well-drained soils and arable cropping, but is frustrated by heavy, poorly drained, and deep soils (Cowley & Dickson 2007). At a national scale, with its reliance on arable cropping (*c.* 11% of the country), this survey method only addresses a relatively small proportion of the land mass, though these areas appear to have been a major focus for prehistoric and historic settlement and land use. Earthwork remains survive only where complex patterns of later expanding and retreating land use have allowed, sandwiched between lowland truncated areas and the limitations imposed by extremes of altitude and topography.

While the wealth of cropmarked and earthwork records are undoubted strengths, they also highlight how little is known about equally large areas of other land cover and landscape histories. This is particularly true for the approximately 28% of the country that is permanently set to pasture, much of which is formerly cultivated (SNH 1998). Here, earthworks have been ploughed out and cropmarking or parchmarking is unlikely in all but the most extreme conditions. Thus, at present there is no cost-effective way of undertaking primary reconnaissance in these areas. The same is true of the extensive peat lands, which are virtually a closed book to archaeological prospection. This then is one of the challenges for remote sensing approaches: balancing a well-proven role exploring the known and revealing the unknown against awareness of the weaknesses of those approaches. Put crudely this concerns *where* resources are best expended, based on analysis of information return, *how* the challenges for developing survey methodology are defined and *what* is required for effective management. This requires a critical approach to data sources, identifying challenges, pinpointing strengths and weaknesses and exploring landscapes in a way that is appropriate to context. If archaeological prospection is to develop, then these processes must feed directly through to cutting edge research into prospection and analytical techniques, as is happening for example, at the Ludwig Boltzmann Institute for Archaeological Prospection and Virtual Archaeology in Vienna (http://archpro.lbg.ac.at), within the DART project (Beck this volume; http://dartproject.info) and at Bournemouth University (Bennett *et al.* this volume). These research projects share an explicitly stated motivation to improve the basis on which archaeological sites and the historic environment are managed.

From data to understanding and knowledge

The integration of aerial reconnaissance with other approaches discussed above ensures that new discoveries are further explored, contextualised and built into archaeological narratives and inform effective management. However, this is not a uniform state of affairs and there is a large amount of archaeological survey data that has not been contextualised in this way. For aerial survey in Scotland, this is a product of a sustained focus on primary data collection. This may have been understandable in the early years of reconnaissance when there were no frameworks within which to analyse material and relatively few excavated sites to provide context. However, over time this narrow focus has generated a large number of poorly understood site records, employing basic classifications such as 'enclosure' and 'cropmark', with no textual records or mapping. Of course, it is vital that these sites are on record, but this basic 'data' needs interrogation to generate understanding and knowledge. Scotland has the benefit of over five decades of aerial reconnaissance, but paradoxically interpretation of this information has been sporadic and has come relatively late in the process. Systematic mapping of cropmarked sites, for example, has only been in progress over the last five years, for the first time providing accurate locations and consistent depictions and interpretations across large areas (Cowley *et al.* 2009). The broader analysis of the results of aerial reconnaissance has also made a slow start. Sporadic work in the 1970s and 1980s (e.g. Macinnes

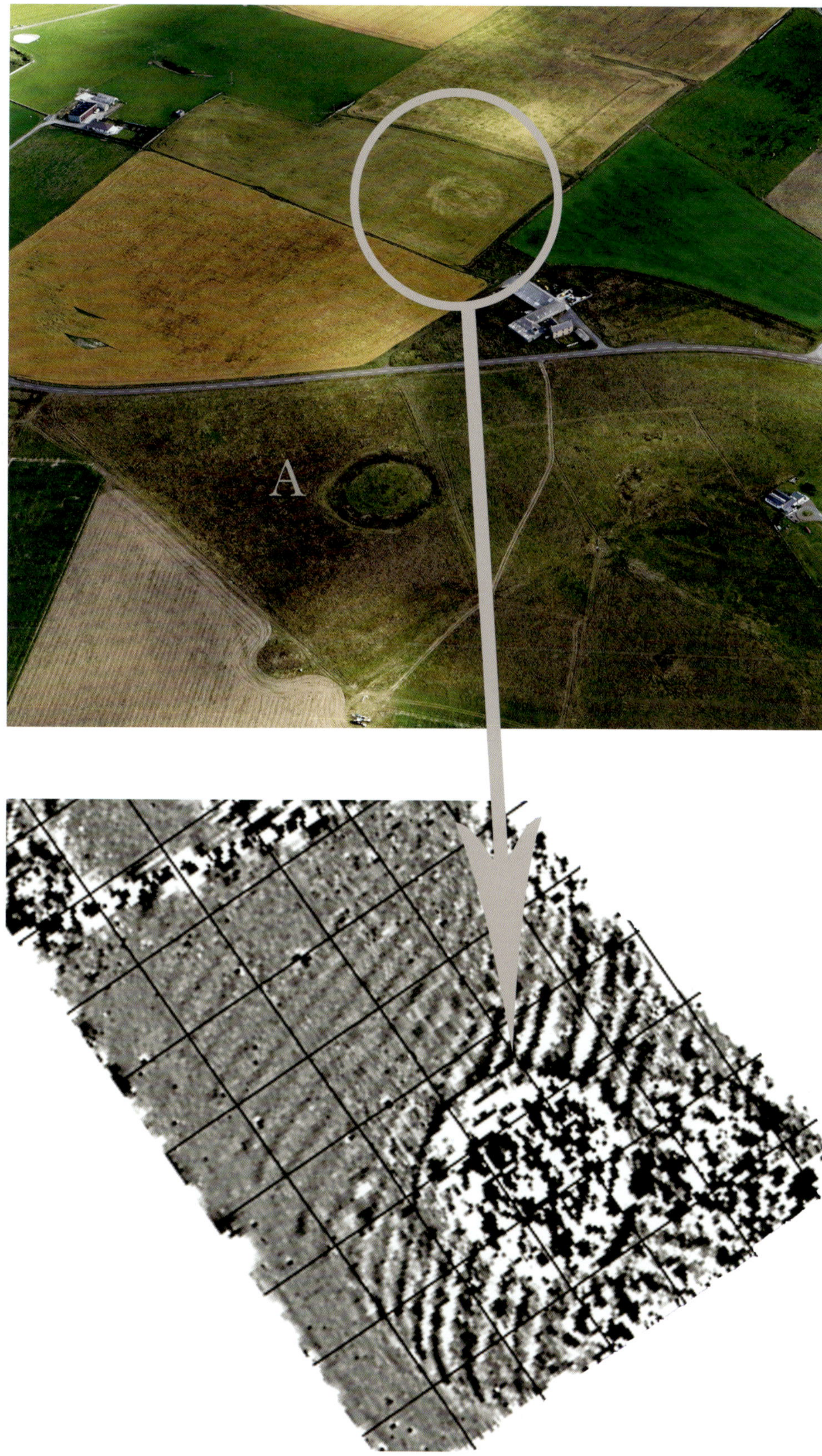

Figure 4.11: The ringed monument (top) was recorded for the first time in 2009, lying within the buffer zone for the Heart of Neolithic Orkney World Heritage Site and just across the road from the well-known Ring of Bookan (A). Geophysical survey by ORCA Geophysics has considerably enhanced our understanding of this monument, adding detail and recording additional features, such as the ridge and furrow cultivation. Top: *DP083340, reproduced under licence. RCAHMS: The National Collection of Aerial Photography*. Bottom: © *ORCA Geophysics*.

1982; Pollock 1985) demonstrated the potential, but it is only in the last decade that significant progress has really been made, with regional and thematic studies that set some of this material in context (Brophy 1999, 2011; Cowley 2009; Dunwell & Ralston 2008b; Jones 2011; Millican 2007; RCAHMS 1994). There remains much to be done, but the main lesson is that a dislocation between primary survey and broader-based research is undesirable as it does not help to bridge the gaps between data, understanding and knowledge (is it a late Iron Age settlement? it is rare? etc.). The early integration of data collection strategies with research and management issues should be standard practice, providing a solid foundation to future progress.

Conclusion

In all archaeological information there is a close interplay between data, experience, knowledge, theory, practice and policy. Recognising these complex relationships between what we know, how we know it, and the weaknesses and gaps in knowledge, requires archaeologists to keep an open mind to new approaches and information. This is especially important in dealing with the many types of site documented primarily, or solely, from survey, because the predisposition of many archaeologists is to attach a higher value to excavated or published material, which therefore generates an imbalanced view of the past.

This paper has illustrated the importance of understanding how archaeological data and our knowledge of the historic landscape are structured – to inform understanding and management, and to define challenges for the future. Broad-brush characterisation of the landscape can provide a framework to define approaches, to identify strengths and to highlight the challenges of exploring other parts of the country where there are currently no effective means of extensive reconnaissance and the primary record is poor as a result. A multi-scaled and integrated approach to landscape prospection and characterisation is advocated, but one where techniques and approaches must be appropriate to the local context and where integrated data create reliable information and support management effectively. Central to the effective use of data is an understanding of how useful archaeological information (over and above basic location and classification) is created and impacts on practice, policy and theory.

Acknowledgments

My thanks to Ant Beck, Noel Fojut, Strat Halliday, Chris Musson, Rog Palmer and Robin Turner for comments on the text, to Kevin MacLeod for producing the distribution maps and to Nick Card and ORCA Geophysics for permission to use the survey at Bookan.

References

Ballin, T.B., Saville, A., Tipping, T. & Ward, T. 2010: An Upper Palaeolithic flint and chert assemblage from Howburn Farm, South Lanarkshire, Scotland: first results. *Oxford Journal of Archaeology* 29(4), 323–60.

Barclay, G.J. 1985: Excavations at Upper Suisgill, Sutherland. *Proceedings of the Society of Antiquaries of Scotland* 115, 159–198.

Brophy, K. 1999: The cursus monuments of Scotland. In Barclay, A. & Harding, J. (eds): *Pathways and ceremonies: the cursus monument of Britain and Ireland*. Neolithic Studies Group Seminar Papers 4, Oxford. 119–29.

Brophy, K. 2011: *Reading between the lines: the Neolithic cursus monuments of Scotland*. Amberley.

Brophy, K. & Cowley, D. (eds) 2005: *From the Air: understanding aerial archaeology*. Tempus, Stroud.

Buckley, V. (complier) 1990: *Burnt Offerings – International Contributions to Burnt Mound Archaeology*. Wordwell, Dublin.

Clark, J., Darlington, J. & Fairclough, G. 2004: *Using Historic Landscape Characterisation*. English Heritage & Lancashire County Council.

Council of Europe, 2000: *European Landscape Convention*. Council of Europe (Treaty series No. 176), Florence.

Cowley, D. 2002: A case study in the analysis of patterns of aerial reconnaissance in a lowland area of southwest Scotland. *Archaeological Prospection* 9(4), 255–65.

Cowley, D.C. 2009: The Traprain environs in a regional perspective. In Haselgrove, C. (ed.): *The Traprain Law Environs Project – Fieldwork and Excavations 2000–2004*, Society of Antiquaries of Scotland Monograph, 205–23.

Cowley, D.C. & Dickson, A.L. 2007: Clays and 'difficult soils' in eastern and southern Scotland: dealing with the gaps. In Mills, J. & Palmer, R. (eds): *Populating Clay Landscapes: Recent advances in archaeology on difficult soils*. Stroud, Tempus. 43–54.

Cowley D. & Ferguson L. 2010: Historic aerial photographs for archaeology and heritage management. In Campana, S., Forte, M., & Liuzza, C. (eds): *Space Time and Place: Proceedings of the III International Conference on Remote Sensing in Archaeology*, 17–21 August 2009, Tiruciarapalli, India, BAR. 97–104.

Cowley, D.C., Hale, D.N., Hunter, F. & Macleod, K.H.J. 2009: Survey in the Traprain Law Environs Project area. In Haselgrove, C. (ed.): *The Traprain Law Environs Project – Fieldwork and Excavations 2000–2004*, Society of Antiquaries of Scotland Monograph, 11–21.

Cowley D.C., Standring R.A. & Abicht M.J. (eds) 2010: *Landscapes through the lens: aerial photographs and historic environment*. Oxbow.

Crow, J. & Turner, S. 2009: Silivri and the Thracian hinterland of Istanbul: an historic landscape. *Anatolian Studies* 59 (2009), 167–81.

Dunwell, A.J. & Ralston, I.B.M. 2008a: *The Management of Cropmark Archaeology in Lowland Scotland: a case study from the Lunan Valley, Angus*. Historic Scotland.

Dunwell, A.J. & Ralston, I.BM. 2008b: *Archaeology and Early History of Angus*. Stroud, Tempus/The History Press.

Fairclough, G. & Macinnes, L. 2003: *Understanding Historic Landscape Character, Topic paper 5, Landscape Character Assessment Guidance for England and Scotland*. Countryside Agency, Scottish Natural Heritage, Historic Scotland and English Heritage.

Halliday, S. 1990: Patterns of fieldwork and the distribution of burnt mounds in Scotland. In Buckley, V. (complier) 1990: *Burnt Offerings – International Contributions to Burnt Mound Archaeology*. Wordwell, Dublin. 60–1.

Halliday, S. 2011: Pattern, perception and preconception: recognising regionality in Scottish prehistory. In Jones, A.M. & Kirkham, G. (eds): *Beyond the Core: Reflections on Regionality in Prehistory*. Oxbow, Oxford.

Jones, R. H. 2011: *Roman Camps in Scotland*. Society of Antiquaries of Scotland, Edinburgh.

Macinnes, L. 1982: Pattern and purpose: the settlement evidence. In Harding, D.W. (ed.): *Later Prehistoric Settlement in South-East Scotland*. Edinburgh University Dept. of Archaeology. 57–73.

Maynard, D. 1993: Burnt Mounds around a Pipeline in Dumfries and Galloway. *Transactions of the Dumfriesshire and Galloway Natural History and Antiquarian Society* series 3 LXVII, 33–52.

Millican, K. 2007: Turning in circles: a new assessment of the Neolithic Timber circles of Scotland. *Proceedings of the Society of Antiquaries of Scotland* 137, 2–33.

Pollock, D. 1985: The Lunan Valley Project: medieval rural settlement in Angus. *Proceedings of the Society of Antiquaries of Scotland* 127, 339–58.

RCAHMS, 1967: *Peeblesshire: an inventory of the ancient monuments*. Edinburgh.

RCAHMS, 1971: *Argyll: an inventory of the ancient monuments, Volume I, Kintyre*. Edinburgh.

RCAHMS, 1978: *Lanarkshire: an inventory of the prehistoric and Roman monuments*. Edinburgh.

RCAHMS, 1994: *South-East Perth: an archaeological* landscape. Edinburgh.

SNH, 1998: *Land Cover Change in Scotland*. http://www.snh.org.uk/publications/on-line/education/advances5/5-land-cover-change.asp, accessed 20 Jan 2011.

Turner, S. & Crow, J. 2010: Unlocking historic landscapes in the Eastern Mediterranean: two pilot studies using Historic Landscape Characterisation. *Antiquity* 84, 216–29.

Wordsworth, J. 2010: Changing the land – the implications of climate-change policies, actions and adaptations for Scotland's rural historic environment. In Trow, S., Holyoak, V. & Byrnes, E. (eds): *Heritage Management of Farmed and Forested Landscapes in Europe*. EAC Occasional Paper No. 4. Archaeolingua, 163–8.

PART 2

New environments and technologies: challenges and potential

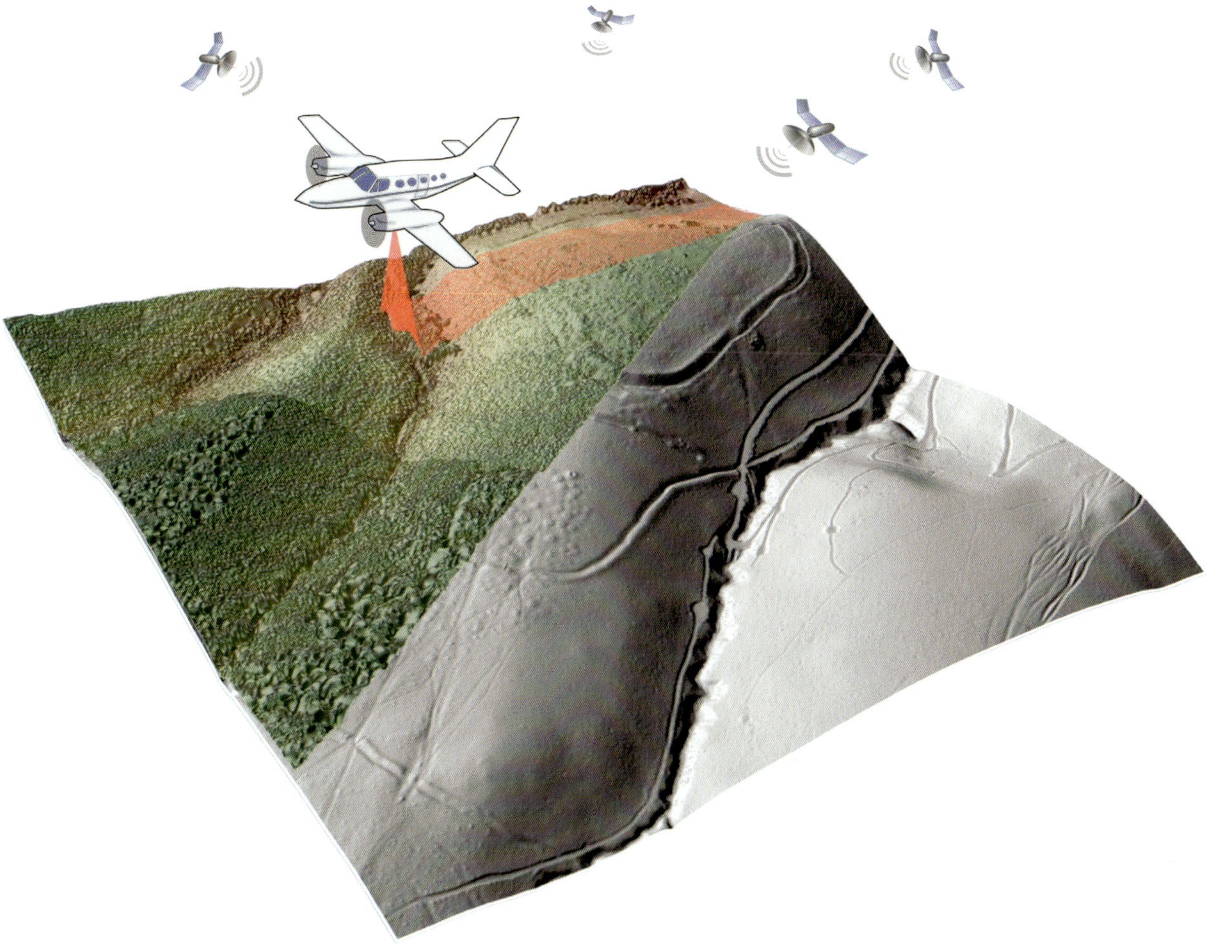

Technologies such as Airborne Laser Scanning have revolutionized archaeological survey in forested land. © *Martin Fera, Aerial Archive, Dept. for Prehistoric and Medieval Archaeology, Vienna.*

5 | Airborne Laser Scanning in forested areas – potential and limitations of an archaeological prospection technique

Michael Doneus and *Christian Briese*

Abstract: Archaeological applications using Airborne Laser Scanning (ALS) are increasing in number. Since the production of ALS-derived digital terrain models (DTM) involves a considerable amount of money, most applications use general purpose ALS data, which are usually cheaper and sometimes even provided for free for scientific applications. The main problem that comes with this kind of data is the frequent lack of meta-information. The archaeologist often does not get the information about original point density, time of flight, instrument used, type of flying platform, filter and DTM generation procedure. Therefore, ALS becomes a kind of 'black box', where the derived DTM is used without further knowledge about underlying technology, algorithms, and metadata. Consequently, there is a certain risk that the data used will not be suitable for the archaeological application.

Based on the experience of a two-year project 'LiDAR-Supported Archaeological Prospection in Woodland', the paper provides a review of archaeological ALS, explains its the basic process, demonstrates its potential for landscape archaeology especially in densely forested areas, and draws attention to some critical parameters of ALS, which should be known to the user. Finally, further issues, which need to be solved in the near future, are discussed.

Introduction

During the last decade, Airborne Laser Scanning (ALS) using the technology of Light Detection And Ranging (LiDAR) (Hug *et al.* 2004; Wagner *et al.* 2006) has revolutionized archaeological prospection of forested areas. In many countries, archaeologists get free access to ALS-derived digital terrain models (DTM), collected country-wide by order of governmental organisations. The number of archaeological publications using these detailed topographic models in one way or another is therefore constantly increasing. Country-wide interpretation of ALS-derived DTMs has become possible and is in some cases already being undertaken (see e.g. Bofinger & Hesse this volume).

These developments seem to be very promising and certainly will have a positive impact, especially on cultural heritage management and landscape archaeology. While this is regarded by the authors as a very positive development, there seems to be a certain risk attached, which has two aspects.

Firstly, one has to consider that most applications use general purpose ALS data. In most cases this is born out of necessity, since the production of ALS-derived terrain models involves a considerable amount of money. General purpose data are usually available for large areas, cheaper and sometimes even provided free for scientific applications. However, the production of a DTM using the technology of ALS is a complex process, which involves several assumptions and decisions throughout the workflow of project preparation, data acquisition and subsequent analysis. Therefore, the resulting DTM and its applicability for a certain archaeological purpose will depend largely on the original intention of the ALS data acquisition campaign. This is reflected in the meta-information, which becomes an important aid to understand the archaeological potential and limitation of the DTM. However, the archaeologist often does not get the meta-information about original point density, time of flight, instrument used, type of flying platform, filter and DTM generation procedure etc.

Furthermore, the importance of this kind of meta-information is often underestimated or even ignored. This is the second dangerous aspect: due to a lack of knowledge about underlying technology, algorithms, and metadata, ALS becomes a kind of 'black box', where the derived DTM is used without source criticism, and the archaeological potential and limitation of the ALS data again remain unconsidered.

Therefore, there is a certain potential that general purpose data will not be suitable for certain archaeological applications, and that this is not recognized. Based on the experience of a two-year project 'LiDAR-Supported Archaeological Prospection in Woodland', the paper will demonstrate several issues, which can account for the success or failure of an archaeological application, and should be considered when using ALS-derived DTMs. Moreover, it will demonstrate the potential of ALS for landscape archaeology especially in densely forested areas, before discussing future developments.

ALS and forests – the state of the art

Vegetation protects archaeological remains to a certain degree from erosion. Therefore, in forested areas, archaeological sites often survive in relief and can be detected and mapped on the ground. If the relief is distinct, sites are detected easily in-situ, but their layout can only be understood after a considerable amount of time spent walking across the features and mapping them. However, there are many instances where relict features had been subject to erosion during periods when the forest had been cleared. Figure 5.1 is a view towards round barrows of a probably Iron Age cemetery. The barrows have a diameter of 15m, but protrude only between 30cm and 1m from the general trend of the terrain surface. Therefore, from the ground, they are hard to find and even more difficult to map.

Aerial reconnaissance and photography is also of limited use in forested areas. The visibility of structures depends upon a variety of factors (see among others Riley 1987; Scollar *et al.* 1990; Ille 1993; Wilson 2000). Most of them cannot be influenced and therefore only the choices of when to fly and the viewing direction can make a difference. Experience has shown that usually only linear structures (e.g. hollow ways and earthwork structures) can be identified if they are still relatively well-preserved and if they run more or less perpendicular to the direction of the sunlight. Furthermore, shadows casted by trees mask these structures. Therefore, in forested areas only large, relatively well-preserved ditch systems can usually be recognized (see also Scollar *et al.* 1990).

To make the structures visible, one would typically need a very detailed DTM with at least one point per square meter. The resulting DTM can be digitally enhanced using virtual light sources, exaggeration of relief, colour coding and similar techniques. However, for a long time, this idea was impractical because of the enormous resources of time and money that would have to be spent for the terrestrial survey using, for example, a total station. As a consequence, only ten years ago, large-scale prospection of archaeological remains in forested areas would have seemed to be impossible. At that time, the extensive identification of sites within forested areas was one of the unresolved issues for archaeological prospection.

Over the past few years, ALS has, due to its active sensing principle (in contrast to aerial archaeology utilising passive imagery that relies on the actual sun light), turned out to be a potential tool for recognition and measurement of archaeological features that survive as topographic features in open and wooded areas (Ackermann 1999; Wehr & Lohr 1999; Turner 2009). The ALS sensor is usually mounted below an aeroplane or helicopter, where it emits short infrared pulses (in the presented case-study: 60,000 per second) towards the earth's surface fan-shaped across the flight path. Each pulse will result in one or more echoes, where the last echo is typically returned from the ground surface.

While the aircraft moves forward, a differential global navigation satellite system (GNSS, e.g. like GPS) and an inertial measurement unit (IMU) determine the position and attitude of the scanner in a global co-ordinate system. With this information, the position of each single echo can be calculated. The high density of the measured points (up to several points per square meter) and their even distribution are crucial for the determination of a highly precise and accurate surface model.

In contrast to passive imagery (aerial photography), where a point of the surface has to be visible from at least two images (viewpoints) for its 3D determination, for ALS only one line of sight is necessary (direct 3D measurement technique). Furthermore, the technique of ALS can differentiate between the first and the last light echo that can be detected from one single measurement pulse (within one line of sight). The first echo (often also called 'first pulse') is reflected by the earth's surface (also from treetops, high voltage

Figure 5.1: Mannersdorf am Leithagebirge. The image is centred on a round barrow with a diameter of 15m. It is almost entirely eroded and therefore hardly discernible from the ground.

transmission lines or roof edges), while the last echo (often also called 'last pulse') usually represents the ground beneath. For the time being, the application of ALS seems to be the only possibility to gain terrain surface information at an acceptable cost to systematically search for unknown archaeological traces in woodland.

In September 2000, archaeologists (at least aerial archaeologists) first time became aware of airborne LiDAR as a tool for detailed documentation, visualisation and even detection of archaeological earthworks (Holden 2001; Motkin 2001). Early applications used general purpose data provided e.g. by the British Environment Agency, which had been produced for flood risk assessment and was available as a georeferenced and filtered grid with a resolution of 2m (Holden *et al.* 2002; Challis 2006; Crutchley 2006).

In The Netherlands, the entire country had been scanned between 1996 and 2004 with a resolution of 1 point per 16 square meters (van Zijverden & Laan 2005). This rather low resolution was not sufficient to provide detail on archaeological structures, but archaeologists had soon realised its potential to predict sites based on geomorphological reconstructions (van Zijverden 2002). Similar applications can be also found in Flandres (De Man *et al.* 2005) and Great Britain (Brunning & Farr-Cox 2005; Carey *et al.* 2006).

All of the so far mentioned applications were using rather crude datasets from data produced for non-archaeological purposes. A few early projects, however, commissioned scans driven by archaeological research questions (e.g. Shell & Roughley 2004; Bewley *et al.* 2005; Bofinger *et al.* 2006; Harmon *et al.* 2006; Powlesland 2006). A very recent focus is the application of ALS producing datasets of extremely high-resolution with 30 to 50 points per square meter (Shaw & Corns this volume; Corns & Shaw 2009; Lasaponara *et al.* 2010) resulting in extremely clear representations of the archaeologically relevant topography.

In most cases, the areas investigated were mainly non-forested zones. Therefore, filtering the data (i.e. classification into terrain and off-terrain points) was not among the reported issues. As far as it can be gathered from the publications, the used datasets had already been filtered by the providers. The classification of ALS-derived points into surface and non-surface points does become an important issue once vegetated areas (especially forests) are subject of archaeological investigation. One of the first examples of an archaeological site revealed by ALS under a forest canopy was the visualisation of an approximately 500 hectare large medieval ridge and furrow system in Germany by the geographer Benoit Sittler (2004). He had been using the county-wide available general purpose data provided and pre-filtered by the county of Baden-Württemberg (see also Sittler *et al.* 2005; Sittler & Schellberg 2006). In a later stage of the project, further sites were investigated using the same dataset (Sittler *et al.* 2007).

Another early publication investigated the archaeological potential of ALS within a very dense tree cover (deciduous trees with understorey and a mature, thinned conifer plantation). A prehistoric earthwork was scanned in the forest of Dean using a conventional sensor with a density of 4 points per square meter. Filtering produced a good representation of the ramparts (the process of filtering is not described) and even subtle linear features from a field system could be identified. Significantly, forest clearance remains could not be filtered from the conventional scan data and are therefore evident as 'fishbone' patterns in the filtered DTM (Devereux *et al.* 2005, 658 & Fig. 4).

The same dataset was revisited to assess the impact of different kind of forest canopy on the ALS-based detection of archaeological remains (Crow *et al.* 2007). Paradoxically, the quality of surface representation was less in cleared areas, because of dense and low secondary vegetation (Crow *et al.* 2007), which could not be effectively removed. Similar problems were reported by Ole Risbøl *et al.* (2006), when trying to locate industrial features, (slag heaps, charcoal pits, and tar production sites) in a mixed forest in Norway (see also Risbøl *et al.* 2007; Risbøl 2009). Identification of cultural remains was difficult in dense vegetation (small and bushy birch), where even an increased point density in the overlapping area of two perpendicular scanning swaths did not yield better results (Risbøl *et al.* 2006). Low and dense vegetation was also identified as problematic by Julie M. Gallagher *et al.* (2008) while working in the Isle Royale National Park in Michigan, USA.

The reported problems regarding areas with low and dense vegetation seem to be the result of the scanner equipment used. All of these projects had been using the first generation of ALS instrumentation, so called discrete echo scanners (see also below). The first archaeological application using the second generation of so called full-waveform scanners (in this case RIEGL LMS-Q560) was by Michael Doneus and Christian Briese (2006a). In the paper they demonstrated the potential of these new scanners to distinguish between extremely dense low vegetation and solid ground (see also Briese *et al.* 2007). In another paper, the ability of full-waveform scanners to detect subtle topographic features in areas with dense and low vegetation was proven (Doneus *et al.* 2007).

The only other archaeological project mentioning the use of full-waveform ALS is based on Monte Irsi, a deserted medieval village in southern Italy (Lasaponara & Masini 2009; Lasaponara *et al.* 2010). While Arlen F. Chase *et al.* (2011) did use the latest Optech scanner with waveform option (Optech GEMINI ALTM), they do not mention its application. Anyway, using an extremely high data sampling rate (20 points per square meter), they managed to get a good representation of the ground in a jungle environment, revealing an ancient Maya settlement and its surrounding field terraces (see also Chase *et al.* 2010).

This summary of the state of the art in the archaeological application of ALS clearly demonstrates that the topic found a rapidly growing field of research. Recently, at least three books and booklets have been published summarizing technology and applications

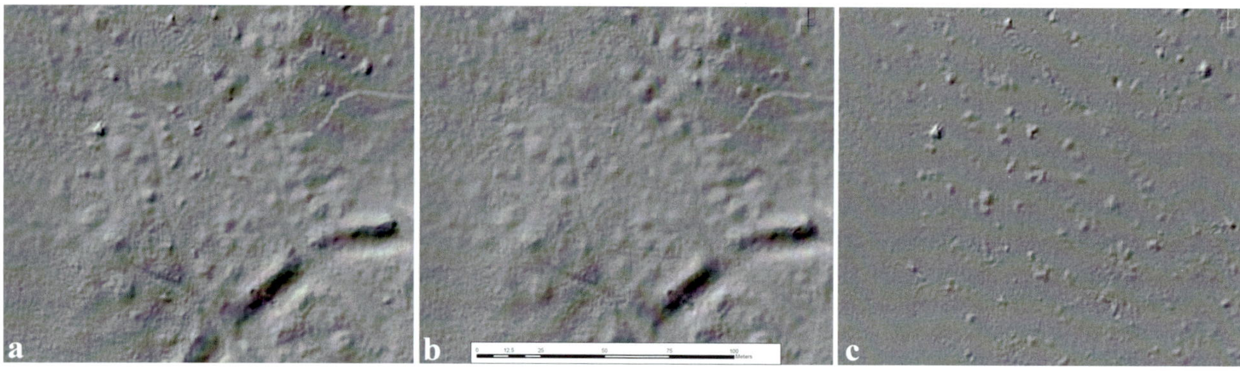

Figure 5.2: (a) Shading of a DTM derived from classified terrain echoes. Low vegetation and clearance piles are still represented in the DTM. (b) Shading of a DTM derived from the classified terrain echoes after removing points with large echo widths. (c) Shading of the difference model (a) – (b). The elevation differences indicate removed low vegetation and clearance piles.

of airborne and terrestrial LiDAR in archaeology and the environmental sciences (Jones 2007; Heritage & Large 2009; Crutchley 2010), which will help to spread knowledge and applications further throughout the archaeological community.

Although some of the papers are quite technical, the importance of the meta-information coming along with the ALS data so far remains unconsidered. A good deal of the papers reviewed in this chapter do not mention any metadata at all. These are important to evaluate the archaeological potential of the published examples. Otherwise, ALS becomes a kind of 'black box', where the derived DTM is used without source criticism, and the archaeological potential and limitation of the ALS data again remain unconsidered. Therefore, it seems to be important to have a closer look at some of the parameters and technical issues of ALS before demonstrating its archaeological potential.

Technical issues

During the workflow of an ALS project, from data acquisition to the visualisation for the final interpretation, several considerations and decisions have to be made, which do have a considerable impact on the quality and usability of the final result. They should be reflected in the metadata, which therefore need to be considered when trying to archaeologically evaluate the resulting DTM. In the following some of the parameters will be discussed.

Choice of sensor

At the time of writing (2011), there are two different types of sensor systems available: discrete echo scanners (so called conventional scanners) and full-waveform recording systems. While conventional systems use analogue detectors to record multiple echoes (most detectors deliver only the first and last echo) in real time, full-waveform (FWF) scanners digitise the entire analogue echo waveform for each emitted laser beam (typically with an interval of 1 ns). In previous publications, we assumed that FWF-ALS systems will show considerable advantages for the generation of DTMs in vegetated areas for various reasons:

(1) The digital data stream of a FWF system has to be post-processed (Wagner *et al.* 2006, 105), but this can be seen as an advantage as the user is not restricted to a group of discrete echoes controlled by a conventional detector but can choose their own algorithms that are best suited for the individual application.

(2) During post-processing the full-waveform can be modelled as a series of Gaussian functions (Hug *et al.* 2004; Wagner *et al.* 2006), each representing an individual object – laser interaction. This allows further physical observations of the reflecting surface elements to be gained, which can be useful for subsequent object classification. In that way, it was for example possible to distinguish piles of brushwood and branches from solid ground (Doneus & Briese 2006a). As a result, in comparison with the conventional ALS data, a more reliable classification of the laser points and a higher accuracy of the terrain points can be expected (Figure 5.2).

(3) The interval between two consecutive echoes, which the analogue detectors of conventional systems can discern, typically equals a height difference of 1.5m (Kraus 2004). This is good enough to clearly distinguish trees from the terrain surface, but will not allow discrimination of low level vegetation from the ground. So, in areas with low vegetation the resulting DTM will often not be a satisfactory representation of the terrain surface. This makes the interpretation of micro-topographical archaeological features difficult (Figure 5.3). With the analysis of the FWF signal a separation of these echoes (objects) can be feasible.

Archaeologically speaking, the DTM derived from a conventional scan may show considerable problems in areas with low and dense vegetation, as can be seen in Figure 5.3. While the left and middle parts (Figure 5.3a, b) show a scene scanned using a conventional system (Table 5.1, column 1), the right image (Figure 5.3c) depicts the same area as scanned using FWF-ALS (Table 5.1, column 2). All scans have a comparable original point density (4 points per m^2, see Table 5.1) and were filtered using the same software (SCOP++, see Kraus & Pfeifer 1998; Kraus & Otepka 2005). In Figure 5.3a and Figure 5.3c, the same filter parameters were applied. Due to the low and extremely dense vegetation, the

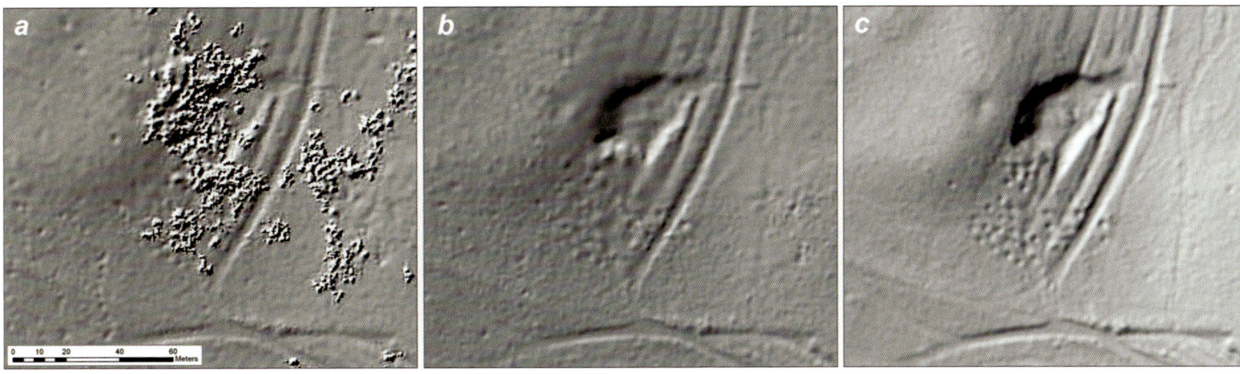

Figure 5.3: (a) Shading of a DTM derived from a conventional ALS using the filter (classification) settings of Figure 5.3c – areas with low and dense vegetation cause considerable problems (remaining vegetation echoes). (b) After tuning the filter settings, the dense vegetation could be removed and small pits become partly visible. (c) The same area scanned using a FWF-ALS with similar point density: the low and dense vegetation could be removed; the small pits are now clearly visible. They have been verified on the ground as badger's earths.

small pits in the centre only become visible in the FWF-ALS data. Tuning the filter-settings, the low and dense vegetation could also be removed in the conventional data (Figure 5.3b). However, the quality of the DTM is highly reduced – only a few of the pits can be discerned and the remaining terrain point density is quite low. They could be verified on the ground both as presently used and remains of past badger's earths.

Data acquisition

Important metadata concerning data acquisition are date, point density and field of view. There is still a lack of research on limitations due to the type of tree cover and the seemingly short time-frame available for scanning wooded areas. The best time for data collection is the dormant period, when deciduous trees and most of the understory have lost their leaves. Freshly fallen leaves at the beginning of the period (which in Central Europe can be between October and November) tend to fill up the shallow depressions archaeologists are interested in. Only when the leaves have subsided after some rainfall or snow will micro topographic features to become visible again. There should be no 'wet' surfaces, because they will reflect the laser pulses poorly toward the sensor. Because of the dependency on vegetation and climatic conditions, the optimal timeframe for data acquisition can vary from year to year and region to region. For example, because of the specific situation in Norway, with long snowy winters, flights have to be scheduled in late spring, when deciduous trees usually have become foliated again. Therefore, results tend to be better in areas covered with conifers (Risbøl *et al.* 2006).

The point density is a function that mainly relies on the pulse rate, field of view, flying height above the terrain, overlap between two neighbouring scan stripes and speed of the aircraft. In a forest, only part of the laser pulses will pass through to the ground surface. To get a high density of ground points, a slow moving platform with a high pulse rate has to be used. Using a wide scan angle and a large overlap (e.g. >= 50%) can increase the number of points per square meter.

Although it is often argued that scanning a forest demands a narrow opening angle of the scanner to

Table 5.1: Parameters of the ALS flights used in this paper. The parameters listed here are considered by the authors as a minimum standard.

	1: County of Lower Austria	2: FWF-Project Testflight	3: FWF-Project Total Scan
Purpose	Topography	Archaeology	Archaeology
Time of Data Acquisition	March & April 2008	08. April 2006	26. March – 12. April 2007
Point-Density (pt per sq m)	min. 4	min. 4	Ca. 7
Scanner Type	Conventional	Full-Waveform	Full-Waveform
Scan angle (whole FOV)	15°	45°	45°
Flying height above ground	1100m	600m	600m
Speed of aircraft (TAS)	65 m/s	36 m/s	36 m/s
Laser Pulse Rate	77.500 Hz	100.000 Hz	100.000 Hz
Scan Rate	60 Hz	66 Hz	66 Hz
Strip Adjustment	Yes	Yes	Yes
Filtering	Robust interpolation (SCOP ++)	Robust interpolation (SCOP ++)	Robust interpolation (SCOP ++)

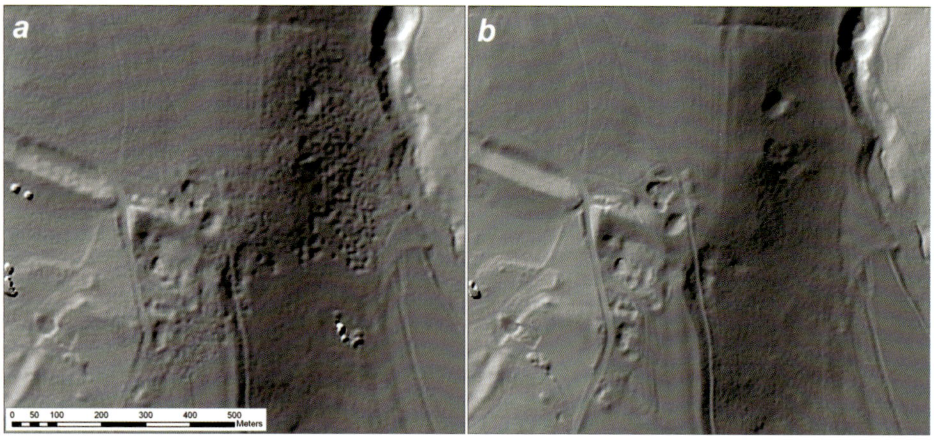

Figure 5.4: Schwarzenbach, Iron-age hillfort. (a) Roundish depressions are introduced to the DTM shading due to a missing strip-adjustment (errors in the geo-referencing). (b) the same area after a simultaneous 3D strip adjustment. The quality of the DTM could be improved significantly.

receive only nadir or near-nadir returns, using the maximum scan angle (up to 45–60 degrees) with a large overlap (e.g. 50%) seems to be advantageous. Furthermore, the large overlapping area of two neighbouring strips will be an advantage during the process of strip-adjustment for advanced geo-referencing. With a large overlap it will also be certain that every object on the ground would be hit at least twice from two view points, and that there is a good chance that some of the oblique laser pulses will hit the ground below conifers where the almost vertical laser pulses of a system operating with a narrow scan angle might not get through. Also in steep terrain, a wide scan angle will result in a higher number of points hitting the slopes almost perpendicular.

A possible drawback of a wide opening angle with a large overlapping area is an increasing number of last echoes, which are not returning from the terrain surface. Especially near the borders of the scanning strip, a high number of last echoes will be returned from tree trunks and consequently have to be filtered out later on.

Geo-referencing

Geo-referencing of the laser data is usually done by the data providers, where a calibration over asphalt or flat horizontal areas with sparse, very low vegetation is usually performed (Pfeifer *et al.* 2004). The geo-referencing process still causes a few problems (Kager 2004) introducing inaccuracies into the resulting point cloud. These inaccuracies will lead to problems within the overlapping area of two strips, where the co-ordinates of a single object point, which was scanned (approximately) twice, will deviate from each other horizontally and vertically. This results in doubled objects, noise and formation of non existing structures (sinusoidal curves, but also edges). During the process of archaeological interpretation these can be considerably irritating. This is demonstrated by an example from the Iron-age hillfort in Schwarzenbach (Figures 5.4 & 5.5). In the displayed area a large number of roundish depressions can be observed, which resemble archaeological features as for example lime kilns. These features are, however, non-existent. They are errors resulting from an inadequate geo-referencing.

These inaccuracies can be improved by doing a simultaneous 3D strip adjustment (Kager 2004). For this, tying features between the overlapping stripes are used analogously to tie points in aerotriangulation. As tying features, homologous planes and straight lines with low noise from covering vegetation are used (e.g. roofs, meadows, roads). In the example from Figure 5.4a, the errors visible in the area with strip overlap could be reduced by strip adjustment, resulting in a significantly improved DTM (Figure 5.4b).

The method of geo-referencing used in most cases is not communicated via metadata. Therefore it is important that an interpreting archaeologist at least knows about the problem and takes care in areas with strip overlap. In that way the roundish depressions depicted in Figure 5.4 and sinusoidal curves (Figure 5.5) can be identified as virtual features due to discrepancies between overlapping strips.

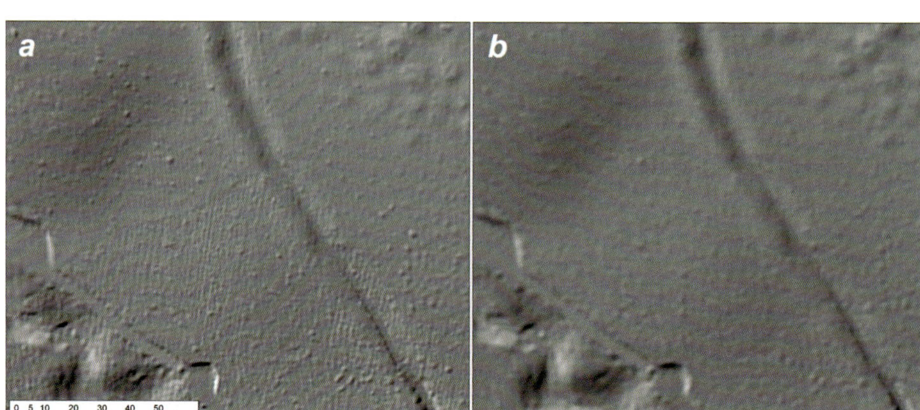

Figure 5.5: Purbach, hillfort. The sinusoidal curve crossing the area horizontally in the DTM shading on the left image (a) is removed after strip adjustment (b).

Figure 5.6: St. Anna in der Wüste. Comparison of (a) general purpose DTM and (b) DTM derived from ALS data scanned and filtered for archaeological purposes. For further information, refer to the text.

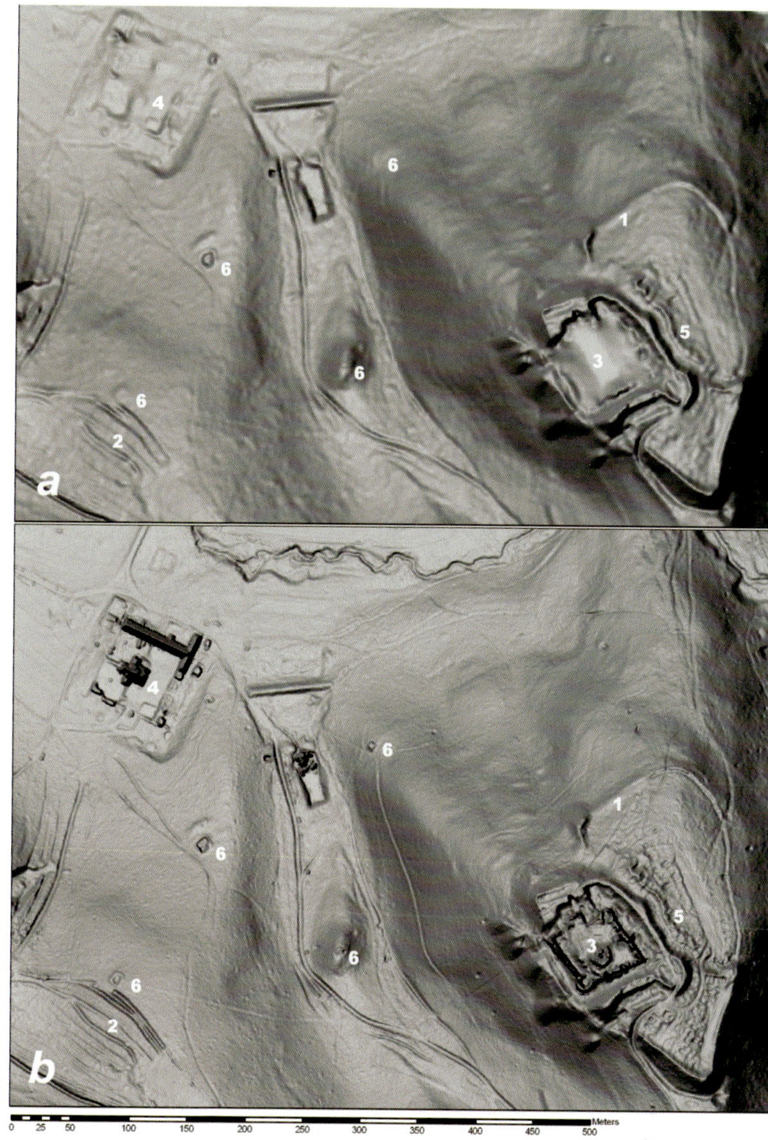

Choice of filtering software and parameters

Another important piece of information which usually does not find its way into metadata is the method used for filtering the ALS data (i.e. classification of the ALS points into terrain and off-terrain points). As stated above, a single laser pulse will result in several echoes. Only the last echo may typically originate from the terrain surface. However, very often, the laser pulse will not be able to reach the ground surface. In these cases, the last echo will be returning from cars, tree-trunks, dense vegetation or even animals, for example. Therefore, any last echo can either be a terrain or an off-terrain point.

Since the archaeological interpretation of an ALS campaign is typically based on the derived DTM, it is critical that all off-terrain points are removed from the data set used for DTM generation. Therefore, the last echo data has to be classified (typically the term filtering is used) into terrain and off-terrain points. Although filtering the vegetation is of prime importance in forested areas, only a few projects take care to get their hands on the filtering process (Cifani *et al.* 2007; Crow *et al.* 2007; Doneus *et al.* 2007; Lasaponara & Masini 2009; Millard *et al.* 2009; Heinzel & Sittler 2010; Lasaponara *et al.* 2010). Even if data acquisition was commissioned for the special purpose of the archaeological project, in several cases DTMs were derived from pre-filtered data. Also, the method of filtering used is mentioned only in a very few publications. Most of them use TerraScan (http://www.terrasolid.fi/). Gabriele Cifani *et al.* (2007 – see also Opitz 2009) were using their own implementation of a morphological and spatial autocorrelation filter to extract archaeological information from a rather crude, general-purpose dataset covering a densely vegetated region north of Rome.

There are a large number of filtering techniques available. Each of them was produced for a special purpose and each is best in certain environments. Depending on the methods and settings used, whole buildings can be deleted from the data, or small-scaled features, like round barrows and lime-kilns, flattened and removed (e.g. see Figures 5.6 & 5.7). The examples clearly show that the process of filtering is of crucial importance for the archaeological interpretability of a DTM. All images display DTMs filtered using the same software (SCOP++, see Kraus & Pfeifer 1998; Kraus & Otepka 2005). Figure 5.6a displays general-purpose data from the County of Lower Austria. These are high-quality data fit for the purpose of topography. Therefore their filtering technique within SCOP++ uses parameter settings which removes building structures and to a certain degree flattens micro-topographic structures. Above all, the DTM is delivered with a grid width of 1m.

While the general purpose data of Figure 5.6a do show important archaeological information, such as the rampart of a prehistoric hillfort (5.6a/1) or hollow ways (5.6a/2), the building structures of the ruined castle of Scharfeneck (5.6a/3) and a monastery (5.6a/4) have been completely removed due to the filtering parameters. Finer archaeological detail, such as pit-structures surrounding the castle (5.6a/5) and three out of four hermit's cells (5.6a/6) are hardly to be recognized. The same area can, however, reveal more information when the original ALS point cloud is filtered in an archaeologically adapted way or special purpose ALS data is used. Figure 5.6b displays the same area filtered from special purpose ALS data. Besides the clearer view due to the higher resolution (0.5m), small structures of archaeological interest, like for example the hermit's cells (5.6b/6) become visible.

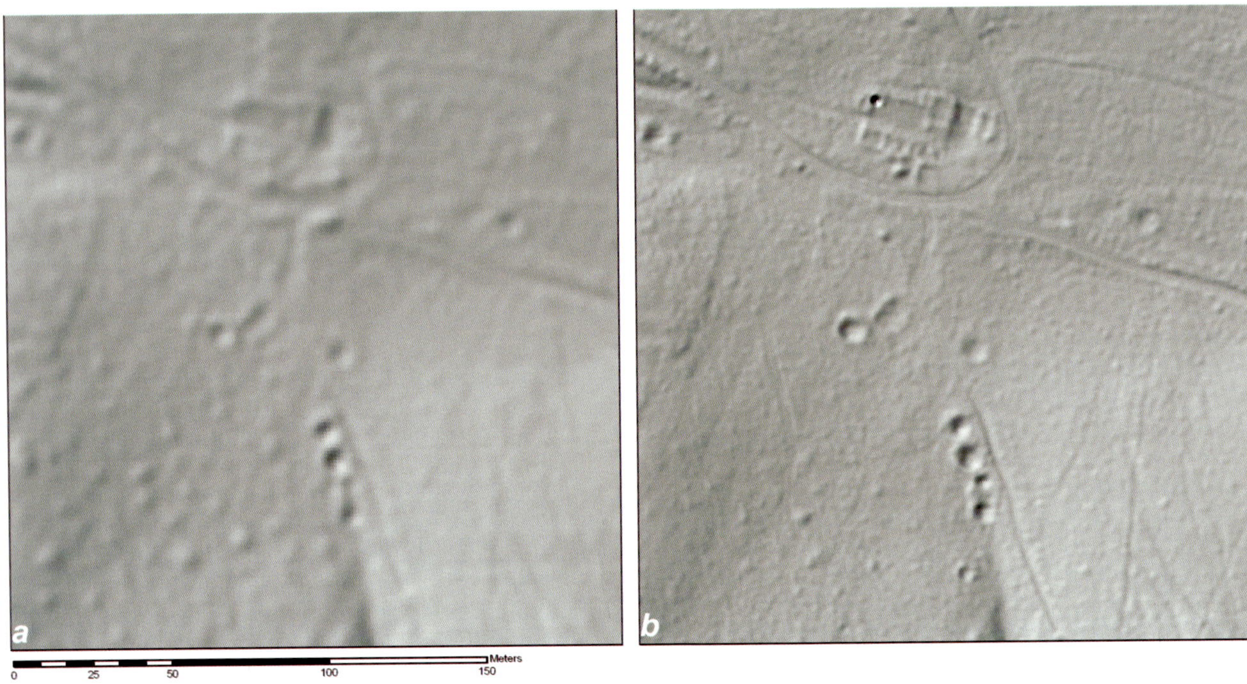

Figure 5.7: Mannersdorf. Remains of a post-medieval Knacker's yard and a row of lime kilns towards the south. (a) DTM shading of general purpose data with a resolution of 1m and (b) ALS data scanned and filtered for archaeological purposes with a resolution of 0.5m.

Similar results can be seen in the comparison of Figure 5.7. It shows the remains of a large building and a row of lime kilns south of it. While both building and lime kilns can be identified both in the general and special purpose data, details become only distinct in the latter one.

Visualisation and interpretation

One last point of caution concerns the representation of the data. If one tries to interpret a single shaded image, one will certainly miss linear features which run parallel to the rays of the light source. Therefore, at the very least, a second shading with perpendicular light-rays has to be provided. While Koreen Millard *et al.* (2009) have recently identified the problem of visualization of ALS-derived terrain models as a future challenge, only a small amount of publications have dealt with the problem of displaying and interpreting archaeological ALS data so far. Bewley *et al.* (2005) used a contour map with dense line interval to identify topographic relationships. Michael Doneus and Christian Briese (2006b) were discussing slope and the subtraction of a resampled DTM (5m grid width) from the high resolution DTM to get a model of the micro-relief. Astrid Humme *et al.* (2006) were applying geostatistical filtering to separate the archaeological structures in the micro-relief. Ralf Hesse (2010b) recently published a similar but advanced approach to derive local relief models from ALS-based DTMs, which seems to be particularly promising.

Other techniques comprise the use of a principal component analysis (PCA) creating an rgb-image from the three main components derived from altogether 16 shaded reliefs with varying illumination angles (Devereux *et al.* 2008). In a recent poster presented at the annual AARG meeting in Bucharest, Ziga Kokalji *et al.* (2010) presented various visualisation techniques including 'hill-shading, micro-relief topography, derivates of hill-shading from different directions (hill-shadings range, mean of hill-shadings, hill-shadings PCA), composite of hill-shaded relief and elevation differentiation, composite of hill-shaded relief and nDSM, composite of DOF and nDSM, and a sky view factor'. At the same meeting, Paolo Forlin (2010) gave a presentation on the use of a 'sky-view factor' to visualise LiDAR-derived DTMs.

Apart from these more sophisticated techniques, we have particularly good experiences using a very simple combination of hillshade and slope mapping. They are easy to create and especially easy to 'read' and interpret, while highlighting all features independent of their alignment. This is demonstrated in the depiction of a field of hollow ways crossing the Leithagebirge (Figure 5.8). The shaded DTM on the left part gives a good impression of the large number of paths running parallel to and partly intersecting each other. Since the light source is placed in the northwest (upper left corner), those tracks running exactly northwest-southeast become invisible. Combining this shading with a slope map (Figure 5.8b), all hollow ways are clearly visualized independent of their direction, while the image is still comprehensible. For extremely low relief features and for outlining the extent of the individual structures, local relief models as described by Hesse (2010b) seem to be a valuable addition.

While country-wide LiDAR derived DTMs become available, systematic archaeological interpretation of large datasets has started in various institutions (e.g. Schmidt *et al.*, 2005; Doneus *et al.* 2007; Boos *et al.* 2008; Challis *et al.* 2008; Opitz 2009; Ainsworth *et al.* 2010; Bofinger & Hesse 2010). Therefore, problems

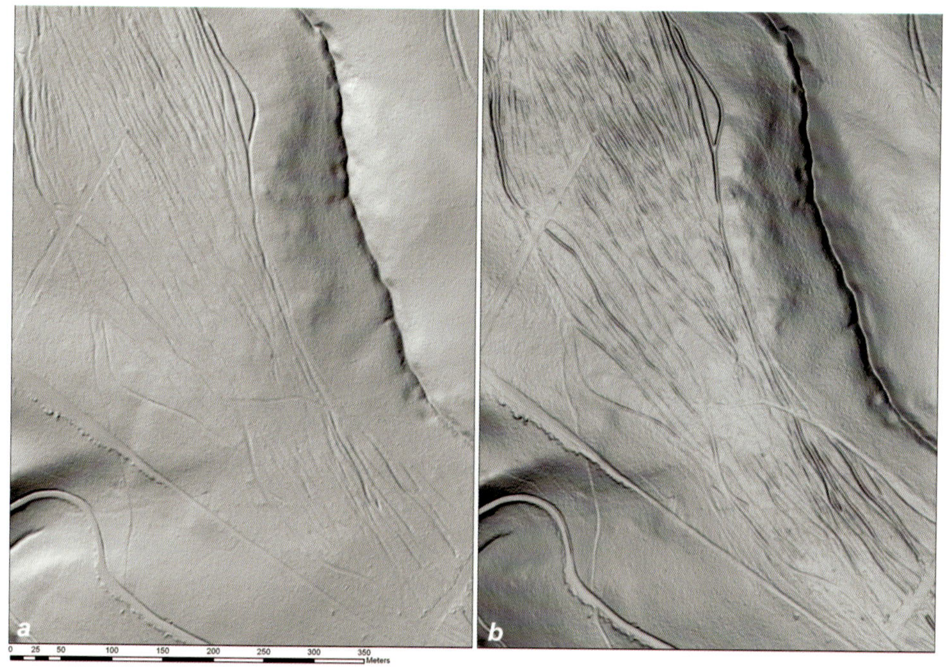

Figure 5.8: Mannersdorf - field of hollow ways. (a) Shaded DTM with light source placed in the northwest (upper left corner). Tracks running exactly northwest-southeast are invisible.
(b) Combination of shading with slope map. All hollow ways are clearly visualized independent of their direction.

of interpretation and of the archaeology beyond LiDAR are becoming more and more a focus of recent conference-sessions and publications (e.g. Doneus *et al.* 2008b; Ainsworth *et al.* 2010; Doneus 2010; Hesse 2010a; Poirier *et al.* 2010).

In Crete, various algorithms have been applied to a combined application of ALS, Airborne Thematic Mapper (ATM) and Compact Airborne Spectrographic Imager (CASI). Through various filters (anomaly and edge detection), archaeological remains have been classified (Rowlands *et al.* 2006). Rosa Coluzzi *et al.* (2010) were using edge detection algorithms to aid interpretation of linear palaeo-environmental structures.

Algorithms for automatic detection of archaeological features are still rare. De Boer (2005–2008) used a template-based algorithm in order to automatically detect burial mounds from ALS DTMs in the Netherlands. Johannes Heinzel and Benoit Sittler (2010) try to automatically detect ridge and furrow by applying self developed software. In Austria, a semi- and full automatic approach was designed to extract break-lines from LiDAR-based DTMs and in that way detect cultural features (Briese 2009).

Archaeological potential of ALS

By now it should be clear that appearance and quality of a DTM is influenced by quite a number of decisions and parameters. Therefore, it is not a digital copy of an existing relief but rather one of many possible representations. Whether it is fit for archaeological purposes has to be assessed from its meta-data.

The fact that a DTM derived from ALS data actually does hold a high archaeological potential has become clear already from the first published examples (Holden 2001; Motkin 2001; Sittler 2004). Various subsequent applications demonstrated that when using ALS it was possible to depict subtle features in relief, which are hardly detectable and comprehensible when surveying terrestrially. In several cases, previously unknown features were detected in ALS data. This expanded the potential of ALS from a plain measurement technique to a tool for archaeological reconnaissance (among others: Doneus *et al.* 2007, 2009; Hesse 2010b). Today we can state that regardless of whether ALS is used in open fields, meadows or forested and otherwise densely vegetated areas it has potential to serve three main aspects of cultural heritage protection: detection; documentation; and monitoring of sites, monuments and landscapes.

In the following, these three aspects will be presented using the experiences and results from a project which ran between 2006 and 2008. The case study was designed to investigate the potential of ALS for systematic large scale archaeological prospection, and to make a comparison between ALS and terrestrial surveying. The strategy of the project was to first perform a test scan over known archaeological sites within the project area. This allowed us to assess the suitability of FWF-ALS for archaeology, tune the scanner-settings and conceptualize a workflow including strip-adjustment, filtering, interpretation and verification for the total scan of the whole project area. Additionally, the data could be used to evaluate the problem of monitoring archaeological monuments and landscapes using repeated scans.

Detection of archaeological structures

To quantify the suitability of ALS to detect archaeological information under vegetation, a forested area of approximately $190 km^2$ was scanned with a high point density (a minimum of 3–4 terrain points per square meter after filtering) and archaeologically analysed. The area to be examined was the so-called 'Leithagebirge' south of Vienna, which rise some 200m to 300m above the valley of the river Leitha (Figure 5.9). It is covered by a forest of mixed deciduous trees, mainly oak and beech with varying degrees of understory.

In the project full-waveform ALS was applied. Both analysis and georeferencing of the data was done in cooperation with the Institute of Photogrammetry and Remote Sensing of the Vienna University of Technology. The ALS data were filtered applying robust interpolation (a technique based on Kriging) with an eccentric and asymmetric weight function (a brief description of the technique is given in Doneus *et al.* 2008a). This filter is implemented within the software package SCOP++ (www.inpho.de).

The resulting DTM was then systematically analysed in a GIS environment. The intention was to create an interpretation key including archaeological features as well as structures, which could be ascribed to geology, modern land-use or animal behaviour. Altogether, more than 10,000 features were mapped, but so far, only the most important ones could be inspected by in-situ validation. Often, archaeological features could be verified due to their morphology or due to artefacts (mainly ceramics) found on the ground, between the roots of fallen trees or in superficial soil-disturbances caused by boar or modern forest roads. However there were also many instances, where ground observation was inconclusive and further prospection techniques (foremost magnetics or ground penetrating radar) would be needed to verify a possible archaeological relevance.

During ground observation geo-tagged photographs were taken. The position of each photo was visualized in a GIS using a point layer with the corresponding images attached (Figure 5.10). Together with the information of the on-site observations, an interactive interpretation key was created.

At the end of the project, the mapped structures could be ascribed to roughly 400 sites (Figure 5.11). ALS can only document sites and features which are still surviving in (micro-) relief. The survival of features in relief is depending on a broad range of factors such as type and dimension of the original structure, attitude of later generations towards abandoned monuments and sites, environmental conditions (geology, vegetation, climate etc.) and time. Many factors will have erosive effects on various scales which add up through time. Therefore, the range of archaeology to be found in an ALS-based project is usually biased in terms of chronology and site-type. For example in this case study most of the mapped structures were medieval and post-medieval in date. Nevertheless, four late-Neolithic hillforts of the Baden culture (Boleraz group) are the oldest confirmed structures surviving in relief in this area (Figure 5.12).

All of the Iron Age hillforts which were evident in the ALS-derived DTM were previously known. However, more details could be drawn from ALS. In one case (Mannerdorf – Schloßberg), round barrows could be detected in the vicinity, which seem to be related with the hillfort (Doneus *et al.* 2008b). In Purbach, the

Figure 5.9: Map showing the eastern part of Austria with the project area of the 'Leithagebirge' identified by the white outline. *Graphics: Martin Fera.*

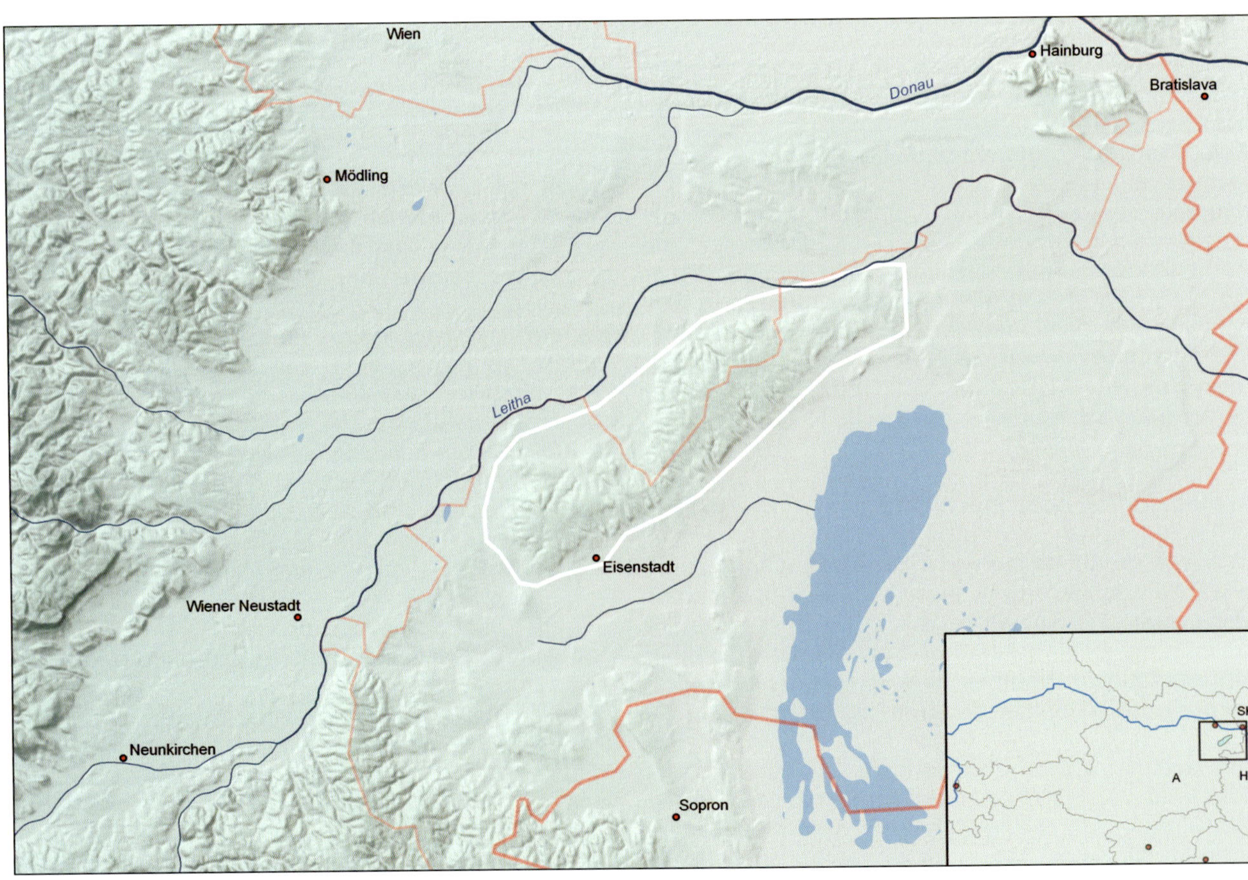

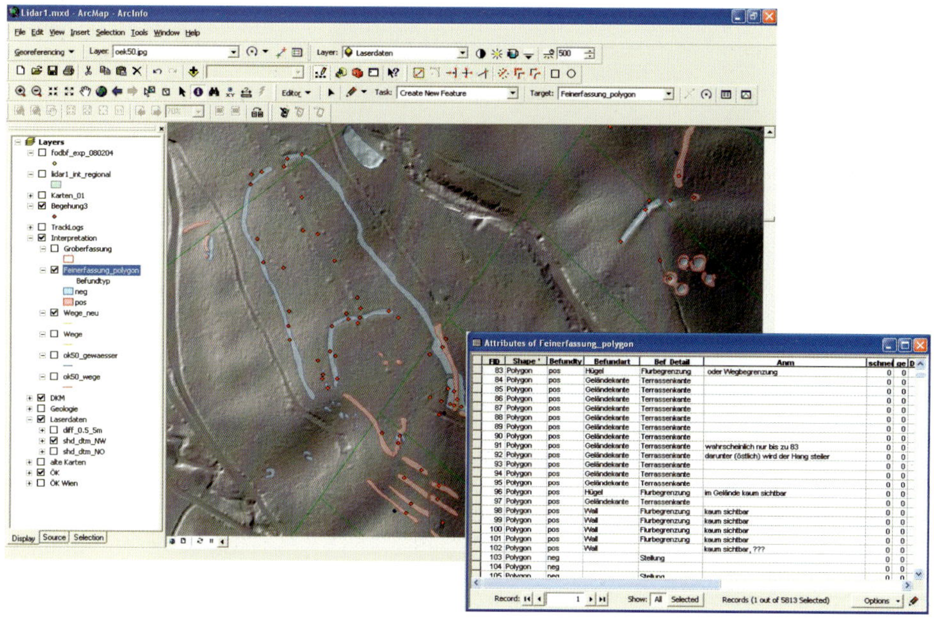

Figure 5.10: Screenshot of an interpretation session in Arc GIS. The red dots on the drawing area mark the locations of photographs taken on-site.

number of round barrows could be doubled (Doneus *et al.* 2008a) and in two other locations (Donnerskirchen and Jois) fortifications were discovered next to settlement areas which were previously thought to be undefended.

The list of site-types found during the project comprises also round barrows, building structures (see also Figure 5.7), stone quarries, fields of hollow-ways, medieval field systems, medieval border-markers (so called 'Hotter'), hundreds of lime-kilns, military trenches from the post-medieval period to World War II and a large number of bomb craters from World War II. Altogether, these prove the value of ALS for systematic, large-scaled archaeological prospection.

Other results clearly demonstrate the potential impact of ALS for landscape archaeology. Field systems and

Figure 5.11: Shading of a DTM of the total project area with the resulting sites. Note that the linear arrangements of sites in the centre of the Leithagebirge are rows of Medieval border-markers.

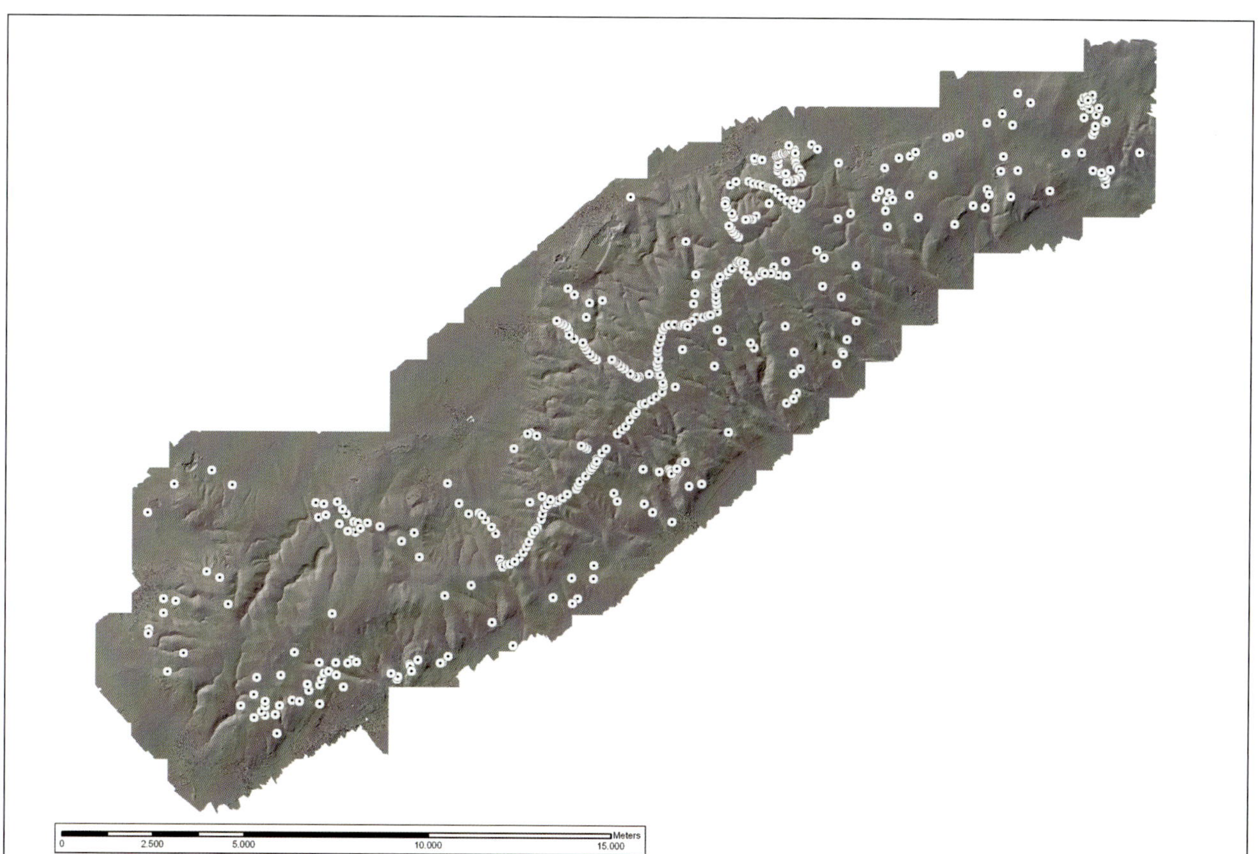

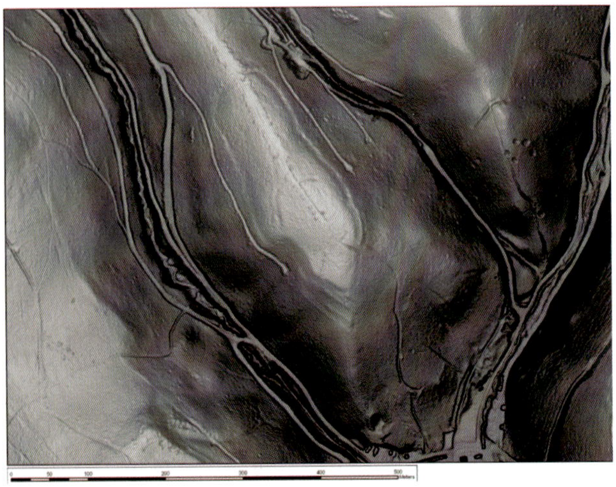

Figure 5.12: Donnerskirchen. Late Neolithic hillfort. 0.5m DTM displayed as a combined shading and slope map.

the large number of roads, tracks and hollow-ways are of special interest and some of them could be identified using old and modern maps. Other remains seem to be of older date, in some cases demonstrated by observations of where they are cut by other features. Of special importance seems to be a network of hollow-ways which interconnects three Iron Age hillforts (Figure 5.13). In some parts, the network consists of bundles and fields of intersecting and overlapping hollow-ways suggesting a long usage. In two instances, the hollow ways lead directly through entrance features into the interior of the hillforts. One is therefore tempted to interpret them as a path network which had its beginnings in prehistoric times.

In another area, the remains of a monastery and its surroundings were depicted on a contemporary engraving from the 17th century. The comparison illustrated ideas about former ideological concepts of space and landscape (Doneus *et al.* 2008b).

Documentation

The detection of a site is certainly the most important step in the archaeological process. To be able to take adequate measures for protection and to plan further archaeological investigations, detailed three-dimensional site and elevation plans must be produced. Only a few years ago this kind of documentation usually was difficult to obtain. Interpretation and measurements had to be taken terrestrially, which depending on the extent and complexity of the remaining earthwork structures could take even some weeks.

From the beginning it was apparent that in addition to the detection of sites in forested areas, ALS would also provide archaeologists with a detailed and accurate documentation of its features and its topography. But it was not clear whether the quality of the data would be sufficient to replace terrestrial surveying and a ground based interpretation. This was expressed by the authors in some general thoughts in 2006:

'The common point of ALS with surveying is that both methods take measurements and both methods should result in an interpretation map. The important difference lies in the fact that the archaeological surveyor is literally in touch with an already known site and for reasons of time usually only records what is interpreted on site as archaeologically relevant. During an ALS scan, the total area of interest will be documented with the same high point density, regardless if it contains archaeological features or not.

Throughout the process of terrestrially recording a site, an interpretation is already taking place. The interpretation of an ALS DTM is a desk-based analysis of the data later on in a remote place. Being in touch with the site has many advantages and it will be no difficulty for the surveyor to distinguish a pile of wood from a barrow. This can be a very difficult task when working with ALS data. On the other hand, what the surveyor doesn't see during the recording procedure at the site (e.g. because of the faint relief or the dense vegetation cover) will be lost in the final record. In an ALS point cloud, all features within the range

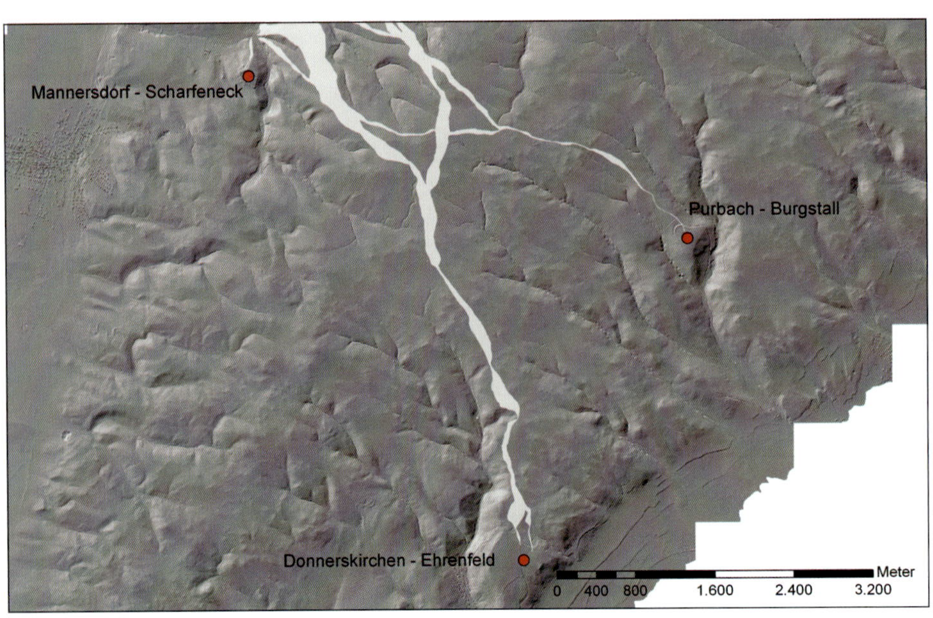

Figure 5.13: Mapping the outline of hollow ways (see also Figure 5.8) in the central part of the project area, a network of paths becomes evident, which interconnects three Iron Age hillforts.

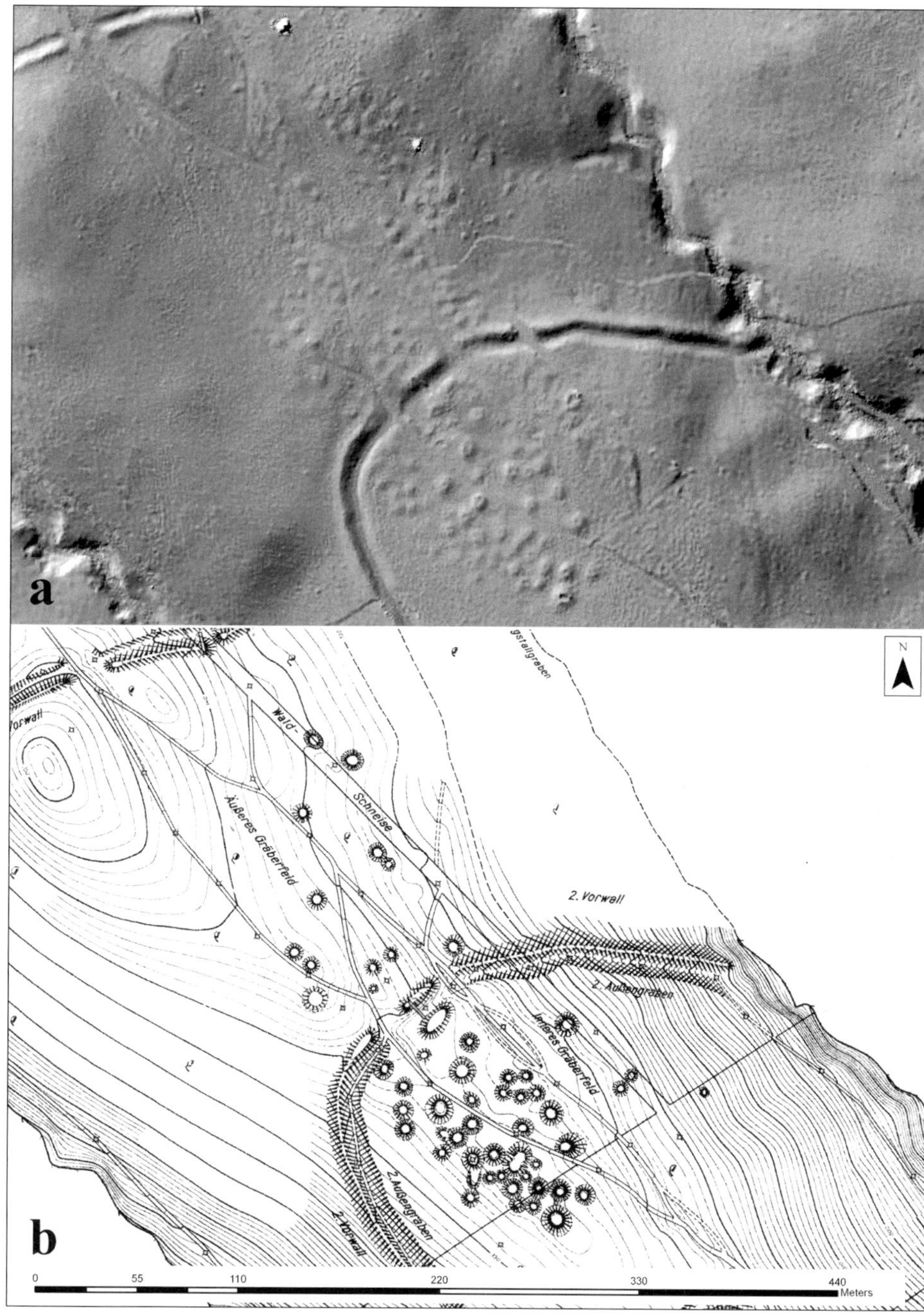

Figure 5.14: (a) Zoomed view to the area of the round barrows in a shading from the final DTM. The individual round barrows of the graveyard can now be seen in detail. Most of them show traces of looting in form of shallow depressions. (b) The same area as mapped in the 1960s. © *BEV 2007*.

discrimination of the instrument and which are sampled within the given point density are documented, regardless whether they are observed during the interpretation process or not. Anybody, who afterwards (re)interprets the data, will see the terrain in the same condition, regardless, if this happens the day after or years later. The data can be interpreted time and again and, as experience and knowledge increases and perception changes, more and more information will be probably found.

Re-interpretation of a site (i.e. re-measuring it) is theoretically also possible when surveying it terrestrially, but it is certainly not common practice. More often than wanted, sites cannot be re-measured because they were subject to damage or destruction. In these cases, previously not recorded features are lost forever.' (Doneus & Briese 2006b).

These arguments can be validated through a direct comparison of ALS and terrestrial surveying at the Iron Age hillfort in Purbach (see also Doneus *et al.* 2008a). Between May 1958 and May 1960 the site had been surveyed by the Austrian Federal Office of Surveying together with archaeologists, and a precise topographical map including the identified archaeological traces had been published (Ulbrich 1962).

The comparison clearly demonstrated the potential of ALS (Figure 5.14). Both ALS-derived DTM and topographical maps corresponded well in geometrical terms. The main features such as banks, ditches and round barrows which are depicted in Ulbrich's map can be seen in the laser-derived DTM.

The most marked differences occurred with the micro-topographical features as many of the very low earthworks were not identified by the trained surveyors and archaeologists in the field, but are clearly visible in the final DTM generated from the ALS-data. Therefore, the number of round barrows could almost be doubled and a field visit verified most of the newly identified barrows. In addition, it was seemingly not possible to identify the terminals of the two outer ramparts during the terrestrial survey. Even during a recent field visit it was impossible to see these areas because of the steep terrain and the dense vegetation cover. Nevertheless, they are quite evident in the shaded DTM (compare the end of the ditch in Figure 5.14a & b).

Monitoring

Heritage sites have to be monitored time and again to document their current condition and compare it with previous records. In that way, structural changes can be detected and potential threats identified. The process of monitoring sites in forested areas can be very time consuming and cost intensive. Therefore, quick, cheap, precise and accurate methods have to be developed that give the means to repeatedly compare the current state of a monument with previous documentations. Repeated terrestrial surveying is in most situations no option – especially with extensive and complex sites. Especially when it comes to monuments in forested areas and to questions of erosion, ALS, especially in combination with aerial photography seems to be a promising approach for this task.

The short time needed for data acquisition and the high detail and accuracy of the derived DTM also makes ALS a tool to monitor sites and map the occurrence and extent of damage or destruction (see also Bewley 2003). Ian Barnes (2003), archaeologist on the military Salisbury Plain Training Area, reported on a project designed to use ALS and Airborne Hyperspectral Scanning for archaeological management. An area of 40km² was repeatedly scanned at the end of winter 2001. However, due to the short time span between the flights, only the potential of change detection could be assessed. One central problem to be solved for this approach is the georeferencing of multiple datasets, which were produced at different times using different ALS sensor types. Furthermore, one has to deal with the problem of the time dependent influence of vegetation on ALS data (Hopkins *et al.* 2004; Pfeifer *et al.* 2004; RSPRS 2008).

Future aspects

ALS displays a high potential, but is archaeologically speaking still in its infancy. There are several key topics that need further detailed investigation regarding the archaeological requirements. Among the most important issues are the evaluation of the data acquisition time-frame, techniques for advanced classification of the acquired point cloud, methods of visualisation and archaeological interpretation of the resulting DTM and automatic extraction of archaeological features. Furthermore, the complex interaction of the laser beam with different types of vegetation cover (conifers, Mediterranean and Alpine vegetation) and the integrated geo-referencing of repetitive multi-seasonal ALS and terrestrial laser scanning data sets need further investigation to enhance the archaeological potential of the method.

These topics will be investigated within a special program line of the recently founded Ludwig Boltzmann Institute (LBI) for Archaeological Prospection and Virtual Archaeology (http://archpro.lbg.ac.at). The institute, which is situated in Vienna, is a novel and innovative research centre for the development and application of advanced non-destructive prospection methods. It employs scientists from archaeology and technology, and therefore will integrate two important aspects of archaeological prospection which are too often handled separately: (1) data acquisition, i.e. the development of new efficient techniques of non-destructive data capturing, processing, and visualization; (2) interpretation, i.e. new technological and methodological concepts to handle this data and to derive archaeologically relevant content by the means of integrated archaeological interpretation approaches. Further developments of archaeological ALS will combine these interrelated aspects.

Conclusion

Within a few years of development and application, ALS has revolutionized archaeological recording of forested areas. ALS has a huge potential for discovery, detailed documentation and monitoring of archaeological sites and structures under vegetation. As a result, applications using this technique are constantly increasing.

Meanwhile, general purpose data have become available for large areas in several parts of Europe. This gives archaeologists the opportunity to avoid the costly production of special purpose DTMs for their archaeological application. The authors regard this development as very positive and encourage the use of already available datasets. However, this implicates the risk of failure as each link in the process chain of ALS has an impact on the archaeological value of the resulting DTM. It is therefore crucial to understand that a DTM is not a given, 'objective' description of an existing relief. It is rather one possible representation, which more or less fits for the desired purpose. To evaluate its archaeological potential, the process chain has to be reviewed using the metadata that should come along with the product.

Therefore, it is of vital importance that the archaeological user understands the technology, issues and limitations coming with data collection, filtering and interpretation of ALS. This knowledge will help to assess the archaeological potential of a given general purpose dataset and therefore will assure a reasonable and successful application of ALS for archaeology.

Acknowledgements

This research has been supported by the Austrian Science Fund (FWF) under project No. P18674-G02. The authors thank the department 'Hydrologie und Geoinformation' of the County of Lower Austria for their support.

References

Ackermann, F. 1999: Airborne laser scanning: present status and future expectations. *ISPRS Journal of Photogrammetry and Remote Sensing* 54(2-3), 64–7.

Ainsworth, S., Oakey, M. & Kincey, M. 2010: Miner-Farmer landscapes of the North Pennines Area of Outstanding Natural Beauty (AONB): English Heritage research into extensive lead-mining landscapes. Paper presented at *AARG 2010: International Aerial Archaeology Conference*, 15–18 September 2010, Bucharest, Romania.

Barnes, I. 2003: Aerial remote-sensing techniques used in the management of archaeological monuments on the British Army's Salisbury Plain Training Area, Wiltshire, UK. *Archaeological Prospection* 10, 83–91.

Bewley, R.H. 2003: Aerial survey for archaeology. *The Photogrammetric Record* 18 (104), 273–92.

Bewley, R.H., Crutchley, S. & Shell, C. 2005: New light on an ancient landscape: LiDAR survey in the Stonehenge World Heritage Site. *Antiquity* 79, 636–47.

Bofinger, J. & Hesse, R. 2010: Mit dem Laser durch das Blätterdach – Der Einsatz von Airborne-Laserscanning zur Entdeckung von archäologischen Geländedenkmalen. In Bofinger, J. & Merkl, M. (eds): *Mit Hightech auf den Spuren der Kelten*. Landesamt für Denkmalpflege, Esslingen, 70–89.

Bofinger, J., Kurz, S. & Schmidt, S. 2006: Ancient Maps – modern data sets: different investigative techniques in the landscape of the Early Iron Age princely hill fort Heuneburg, Baden-Württemberg. In Campana, S. & Forte, M. (eds): *From Space to Place 2. International Conference on Remote Sensing in Archaeology*. CNR, Rome, Italy, December 2–4, 2006. Archaeopress, Oxford, 87–92.

Boos, S., Hornung S., Jung P. & Müller H. 2008: *Erfassung und Analyse von Lasercandaten zur Unterstützung der Landschaftsarchäologie*. Publikationen der DGFP 17, 219–27.

Briese, C. 2009: Automatic break line determination for the generation of a DTM along the river Main. In Bretar, F., Pierrot-Deseilligny, M. & Vosselman, G. (eds): *ISPRS Workshop Laserscanning `09. Paris, France, September 1–2, 2009*. Société Française de Photogrammétrie et de Télédétection, 236–41.

Briese, C., Doneus, M., Pfeifer, N. & Melzer, T. 2007: Verbesserte DGM-Erstellung mittels Full-Waveform Airborne Laserscanning. In Seyfert, E. (ed.): *Dreiländertagung; Von der Medizintechnik bis zur Planetenforschung – Photogrammetrie und Fernerkundung für das 21*. Jahrhundert, 215–22.

Brunning, R. & Farr-Cox, F. 2005: The River Siger rediscovered: LiDAR survey and relict landscape on the Somerset Claylands. *Archaeology and the Severn Estuary* 16, 7–15.

Carey, C., Brown, T., Challis, K., Howard, A.J. & Cooper, L. 2006: Predictive modelling of multi-period geoarchaeological resources at a river confluence: a case study from the Trent-Soar, UK. *Archaeological Prospection* 13(4), 241–50.

Challis, K. 2006: Airborne Laser Altimetry in Alluviated Landscapes. *Archaeological Prospection* 13(2), 103–127.

Challis, K., Kokalj, Z., Kincey, M., Moscrop, D. & Howard, A.J. 2008: Airborne LiDAR and Historic Environment Records. *Antiquity* 82, 1055–64.

Chase, A.F., Chase, D.Z. & Weishampel, J.F. 2010: Lasers in the Jungle. Airborne sensors reveal a vast Maya landscape. *Archaeology* 63 (4), 27–29.

Chase, A.F., Chase, D.Z., Weishampel, J.F. & Drake, J.B. 2011: Airborne LiDAR, Archaeology, and the Ancient Maya Landscape at Caracol, Belize. *Journal of Archaeological Science* 38(2), 387–98.

Cifani, G., Opitz, R. & Stoddard, S. 2007: Mapping the Ager Faliscus road-system: the contribution of LiDAR (light detection and ranging) survey. *Journal of Roman Archaeology* 20 (1), 165–76.

Coluzzi, R., Lanorte, A. & Lasaponara, R. 2010: On the LiDAR contribution for landscape archaeology and palaeoenvironmental studies: the case study of Bosco dell'Incoronata (Southern Italy). *Advances in Geosciences* 24, 125–32.

Corns, A. & Shaw, R., 2009: High resolution 3-dimensional documentation of archaeological monuments & landscapes using airborne LiDAR. *Journal of Cultural Heritage* 10S, e72–e77.

Crow, P., Benham, S., Devereux, B.J., Amable, G.S., 2007: Woodland vegetation and its implications for archaeological survey using LiDAR. *Forestry* 80(3), 241–52.

Crutchley, S. 2006: Light Detection and Ranging (LiDAR) in the Witham Valley, Lincolnshire: an Assessment of New Remote Sensing Techniques. *Archaeological Prospection* 13(4), 251–57.

Crutchley, S. 2010: *The Light Fantastic. Using airborne LiDAR in archaeological survey*. English Heritage, Swindon.

De Boer, A.G., 2005: Using pattern recognition to search LiDAR data for archeological sites. In Figueiredo, A. & Leite Velho, G. (eds): *The world is in your eyes, Proceedings of the XXXIII CAA2005 Conference*, Tomar, 245–54.

De Boer, A.G., Laan, W.N.H., Waldus, W. & van Zijverden, W.K. 2008: LiDAR-based surface height measurements: applications in archeology. In Frischer, B. & Dakouri-Hild, A. (eds): *Beyond illustration: 2D and 3D digital technologies as tools for discovery in archaeology*. Archaeopress, Oxford.

De Man, J., Cordemans, K., Verkeyn, J. & Mestdagh, H. 2005: A Laser Based Digital Elevation Model. New Possibilities for Flemish Archaeologists. In Bourgeois, J. & Meganck, M. (eds): *Aerial Photography and Archaeology 2003. A Century of Information*. Conference held at the Ghent University, December 10th–12th, 2003. Academia Press, Ghent, 151–61.

Devereux, B.J., Amable, G.S. & Crow, P. 2008: Visualisation of LiDAR terrain models for archaeological feature detection. *Antiquity* 82, 470–79.

Doneus, M., 2010. The Impact of Airborne Laser Scanning for Landscape Archaeology. Paper presented at *AARG 2010: International Aerial Archaeology Conference*, 15–18 September 2010, Bucharest, Romania.

Doneus, M. & Briese, C., 2006a: Digital terrain modelling for archaeological interpretation within forested areas using full-waveform laserscanning. In Ioannides, M., Arnold, D., Niccolucci, F. & Mania, K. (eds): *The 7th International Symposium on Virtual Reality, Archaeology and Cultural Heritage VAST (2006)*, 155–62.

Doneus, M. & Briese, C. 2006b: Full-waveform airborne laser scanning as a tool for archaeological reconnaissance. In Campana, S. & Forte, M. (eds): *From Space to Place 2. International Conference on Remote Sensing in Archaeology*. CNR, Rome, Italy, December 2–4, 2006. Archaeopress, Oxford, 99–106.

Doneus, M., Briese, C., Fera, M., Fornwagner, U., Griebl, M., Janner, M. & Zingerle, M.-C. 2007: Documentation and analysis of archaeological sites using aerial reconnaissance and airborne laser scanning. In Georgopoulos, A. (ed.): *AntiCIPAting the Future of the Cultural Past. Proceedings of the XXIst International Symposium CIPA*. Athens 2007, 275–80.

Doneus, M., Briese, C., Fera, M. & Janner, M. 2008a: Archaeological prospection of forested areas using full-waveform airborne laser scanning. *Journal of Archaeological Science* 35, 882–93.

Doneus, M., Briese, C. & Kühtreiber, T. 2008b: Flugzeuggetragenes Laserscanning als Werkzeug der archäologischen Kulturlandschaftsforschung. Das Fallbeispiel 'Wüste' bei Mannersdorf am Leithagebirge, Niederösterreich. *Archäologisches Korrespondenzblatt* 38(1), 137–56.

Forlin, P. 2010: Sky-view factor and automatic feature recognition applied on LiDAR data of a medieval silver mine area, Monte Calisio, Trento, northern Italy. Paper presented at *AARG 2010: International Aerial Archaeology Conference*, 15–18 September 2010, Bucharest, Romania.

Gallagher, J.M. & Josephs, R.L. 2008: Using LiDAR to Detect Cultural Resources in a Forest Environment: an Example from Isle Royale National Park, Michigan, USA. *Archaeological Prospection* 15, 187–206.

Harmon, J.M., Leone, M.P., Prince, S.D. & Snyder, M. 2006: LiDAR for Archaeological Landscape Analysis: A Case Study of Two Eighteenth-Century Maryland Plantation Sites. *American Antiquity* 71(4), 649–70.

Heinzel, J. & Sittler, B. 2010. LiDAR surveys of ancient landscapes in SW Germany: assessment of archaeological features under forests and attempts for automatic pattern recognition. In Forte, M., Campana, S. & Liuzza, C. (eds): *Space, Time, Place. 3rd International Conference on Remote Sensing in Archaeology, 17th–21st August 2009*, Tiruchirappalli, Tamil Nadu, India. Archaeopress, Oxford. 113–121.

Heritage, G.L. & Large, A.R.G. (eds) 2009: *Laser scanning for the environmental sciences*. Wiley-Blackwell, Chichester.

Hesse, R., 2010a: LiDAR prospection results: spatial patterns and cross-evaluation with known sites. Paper presented at *AARG 2010: International Aerial Archaeology Conference*, 15–18 September 2010, Bucharest, Romania.

Hesse, R. 2010b: LiDAR-derived Local Relief Models - a new tool for archaeological prospection. *Archaeological Prospection* 17(2), 67–72.

Holden, N. 2001: Digital airborne remote sensing: the principles of LiDAR and CASI. *AARGNews* 22, 23.

Holden, N., Horne, P. & Bewley, R.H. 2002: High-resolution digital airborne mapping and archaeology. In Bewley, R.H. & Rączkowski, W. (eds): *Aerial Archaeology: Developing Future Practice*. Proceedings of the NATO Advanced Research Workshop on Aerial Archaeology – Developing Future Practice, 15–17 November 2000, Leszno, Poland. IOS Press, Amsterdam. 173–80.

Hopkins, C., Lim, K., Chasmer, L.E., Treitz, P., Creed, I.F. & Gynan, C. 2004: Wet-land grass to plantation forest – estimating vegetation height from the standard deviation of LiDAR frequency distributions. In Thies, M., Koch, B., Spiecker, H. & Weinacker, H. (eds): *Laser-Scanners for Forest and Landscape Assessment – Instruments, Processing Methods and Applications*.

Hug, C., Ullrich, A. & Grimm, A. 2004: Litemapper-5600 – A Waveform-Digitizing LiDAR Terrain and Vegetation Mapping System. In Thies, M., Koch, B., Spiecker, H. & Weinacker, H. (eds): *Proceedings of Natscan, Laser-Scanners for Forest and Landscape Assessment – Instruments, Processing Methods and Applications*. 24–9.

Humme, A., Lindenbergh, R. & Sueur, C. 2006: Revealing Celtic fields from LiDAR data using Kriging based filtering. In Maas, H.G. & Schneider, D. (eds): *Proceedings of the ISPRS Commission V Symposium 'Image Engineering and Vision Metrology'*, Dresden.

Ille, P. 1993: Methoden der Luftbildarchäologie. In Herrmann, F.R. (ed.): *Zeitspuren-Luftbildarchäologie in Hessen*, Wiesbaden. 18–25.

Jones, D.M. (ed.) 2007: *3D Laser Scanning for Heritage. Advice and Guidance to users on laser scanning in archaeology and architecture*. English Heritage, Swindon.

Kager, H., 2004: Discrepancies Between Overlapping Laser Scanning Strips- Simultaneous Fitting of Aerial Laser Scanner Strips. In Altan, O. (ed.): Geo-Imagery Bridging Continents. XXth ISPRS Congress, 12–23 July 2004 Istanbul, Turkey, Istanbul, 555–60.

Kokalj, Z., Ostir, K., Veljanovski, T. & Kobler, A. 2010: Support of LiDAR advanced data visualisations in past cultural features recovery in vegetated areas: past water streams, bronze age hill forts, past settlements, WW1 trenches and shell craters.

Paper presented at *AARG 2010: International Aerial Archaeology Conference*, 15–18 September 2010, Bucharest, Romania.

Kraus, K. 2004: *Photogrammetrie. Band1. Geometrische Informationen aus Photographien und Laserscanneraufnahmen.* 7, vollständig bearbeitete und erweiterte Auflage. Walter de Gruyter, Berlin, New York.

Kraus, K. & Otepka, J. 2005: DTM Modelling and Visualization – The SCOP Approach. In: Fritsch, D. (ed.): *Photogrammetric Week*, Heidelberg. 241–52.

Kraus, K. & Pfeifer, N. 1998: Determination of terrain models in wooded areas with airborne laser scanner data. *ISPRS Journal of Photogrammetry and Remote Sensing* 53(4), 193–203.

Lasaponara, R., Coluzzi, R., Gizzi, F.T. & Masini, N. 2010: On the LiDAR contribution for the archaeological and geomorphological study of a deserted medieval village in Southern Italy. *Journal of Geophysics and Engineering* 7, 155–63.

Lasaponara, R. & Masini, N. 2009: Full-waveform Airborne Laser Scanning for the detection of medieval archaeological microtopographic relief. *Journal of Cultural Heritage* 10 (Supplement 1), e78–e82.

Millard, K., Burke, C., Stiff, D. & Redden, A. 2009: Detection of a low-relief 18th-century British siege trench using LiDAR vegetation penetration capabilities at Fort Beauséjour–Fort Cumberland National Historic Site, Canada. *Geoarchaeology* 24(5), 576–88.

Motkin, D. 2001: The assessment of LiDAR for archaeological use. *AARGNews* 22, 24–5.

Opitz, R. 2009: Integrating LiDAR and geophysical surveys at Falerii Novi and Falerii Veteres (Viterbo). *Papers of the British School at Rome* 77, 1–27, 335–43.

Pfeifer, N., Gorte, B. & Elberink, S.O. 2004: Influences of Vegetation on Laser Altimetry – Analysis and Correction Approaches. In Thies, M., Koch, B., Spiecker, H. & Weinacker, H. (eds): *Laser-Scanners for Forest and Landscape Assessment – Instruments, Processing Methods and Applications*. 283–87.

Poirier, N., Opitz, R., Nuninger, L. & Ostir, K. 2010: LiDAR integration in a well known archaeological context: what is the added value? Paper presented at *AARG 2010: International Aerial Archaeology Conference*, 15–18 September 2010, Bucharest, Romania.

Powlesland, D. 2006. Redefining past landscapes: 30 years of remote sensing in the Vale of Pickering. In Campana, S. & Forte, M. (eds): *From Space to Place 2. International Conference on Remote Sensing in Archaeology*. CNR, Rome, Italy, December 2–4, 2006. Archaeopress, Oxford. 197–201.

Riley, D.N. 1987: Air Photography and Archaeology. Duckworth, London.

Risbøl, O. 2009: Flugleperspektiv på kulturminner. Bruk av flybåren laserskanning i arkeologien. *Viking*, 211–26.

Risbøl, O., Gjertsen, K.A. & Skare, K. 2006: Airborne laser scanning of cultural remains in forests: some preliminary results from a Norwegian project. In Campana, S. & Forte, M. (eds): *From Space to Place 2. International Conference on Remote Sensing in Archaeology*. Proceedings of the 2nd international workshop, CNR, Rome, Italy, December 2–4, 2006. Archaeopress, Oxford. 107–12.

Risbøl, O., Gjertsen, K.A. & Skare, K. 2007: Flybåren laserskanning og kulturminner i skog – ny teknologi i arkeologiens tjeneste. *Kart og Plan* 2, 108–20.

Rowlands, A., Sarris, A. & Bell, J. 2006: Airborne multisensor remote sensing of exposed and subsurface archaeological remains at Itanos and Roussalakkos, Crete. In Campana, S. & Forte, M. (eds): *From Space to Place 2. International Conference on Remote Sensing in Archaeology*. Proceedings of the 2nd international workshop, CNR, Rome, Italy, December 2–4, 2006. Archaeopress, Oxford. 113–16.

RSPRS 2008: *Laser Scanning and Digital Aerial Photography. Today and Tomorrow. Proceedings of the 8th Annual International Conference and Exposition of the RSPRS, 10–11th December, Moscow.* http://www.rsprs.org/archive/conference_2008/materials_archive2008/.

Schmidt, S., Bofinger, J., Keller, R. & Kurz, S. 2005: LiDAR – High resolution raster data as a survey tool. In Figueiredo A. & Leite Velho, G. (eds): *The world is in your eyes, Proceedings of the XXXIII CAA2005 Conference*, Tomar. 255–60.

Scollar, I., Tabbagh, A., Hesse, A. & Herzog, I. 1990: *Archaeological Prospecting and Remote Sensing*. Cambridge University Press, Cambridge.

Shell, C. & Roughley, C. 2004: Exploring the Loughcrew landscape: a new airborne approach. *Archaeology Ireland* 18/2 (68), 22–5.

Sittler, B. 2004: Revealing Historical Landscapes by Using Airborne Laser Scanning. A 3–D Model of Ridge and Furrow in Forests near Rastatt (Germany). In Thies, M., Koch, B., Spiecker, H. & Weinacker, H. (eds): *Proceedings of Natscan, Laser-Scanners for Forest and Landscape Assessment – Instruments, Processing Methods and Applications*. 258–61.

Sittler, B. & Schellberg, S. 2006: The potential of LiDAR in assessing elements of cultural heritage hidden under forest canopies or overgrown by vegetation: Possibilities and limits in detecting microrelief structures for archaeological surveys. In Campana, S. & Forte, M. (eds): *From Space to Place. 2nd International Conference on Remote Sensing in* Archaeology. Proceedings of the 2nd international workshop, CNR, Rome, Italy, December 2–4, 2006. Archaeopress, Oxford, 117–22.

Sittler, B., Siwe, R. & Gültlinger, M. 2005: Airborne Laser Scanning as a New Remote Sensing Tool for Assessing Historic Landscapes. Generation of a Digital Elevation Model of Ridge and Furrow Fossilized under a Forest near Rastatt (Germany). In Bourgeois, J. & Meganck, M. (eds): *Aerial Photography and Archaeology 2003. A Century of Information*. Ghent University, December 10th–12th, 2003. Academia Press, Ghent, 213–19.

Sittler, B., Weinacker, H., Gültlinger, M. & Koupaliantz, L. 2007: The potential of LiDAR assessing elements of cultural hidden under forests. In Bochenek, Z. (ed.): *New developments and challenges in remote sensing. Proceedings of the 26th EARSeL*, Warsaw, Poland, 2006. Millpress, Rotterdam. 539–48.

Turner, M.D. (ed.), 2009: *Laser radar technology and applications XIV*. 15–16 April 2009, Orlando, Florida, United States. SPIE, Bellingham, Wash.

Ulbrich, K. 1962: Der Burgstall von Purbach im Nordburgenland. Burgenländische *Heimatblätter* 24, 138–48.

van Zijverden, W.K. 2002: Landschappelijke context. In Bulten, E.E.B., van der Heijden, F.J.G. & Hamburg, T. (eds): *Emmeloord: prehistorische visweren en visfuiken*, 17–30.

van Zijverden, W.K. & Laan, W.N.H. 2005: Landscape reconstructions and predictive modeling in archaeological research, using a LiDAR based DEM and digital boring databases. In Magistrat der Stadt Wien (ed.): *Archaeologie und Computer*, Workshop 9. Phoibos, Wien.

Wagner, W., Ullrich, A., Ducic, V., Melzer, T. & Studnicka, N. 2006: Gaussian decomposition and calibration of a novel small-footprint full-waveform digitising airborne laser scanner. *ISPRS Journal of Photogrammetry and Remote Sensing* 60(2), 100–12.

Wehr, A. & Lohr, U. 1999: Airborne laser scanning – an introduction and overview. *ISPRS Journal of Photogrammetry and Remote Sensing* 54(2–3), 68–82.

Wilson, D.R. 2000: *Air photo interpretation for archaeologists*. Tempus, Stroud.

6 | High resolution LiDAR specifically for archaeology: are we fully exploiting this valuable resource?

Robert Shaw **and** ***Anthony Corns***

Abstract: In Ireland the Discovery Programme has been pioneering the use of high resolution LiDAR data as a mapping and modelling resource for archaeological landscapes, working on a number of sites as part of its 3D modelling research interests. In general, the results have been well received, with the derived DTMs, DSMs and associated hillshade models having an instant 'wow' factor. As a result of this a number of government agencies have followed our lead, commissioning further archaeology specific LiDAR projects, and in some cases 're-discovering' data sets sidelined due to a lack of processing expertise.

The paper will use examples from Ireland to show the exceptional quality of models available from high resolution LiDAR data sets, how they compare with 'normal' LiDAR models, and how they are enhancing our understanding of landscapes. However the paper will also consider the problems of data management and access to LiDAR data which are seriously inhibiting the ability of agencies to fully exploit their investment. This could result in reluctance to commission future projects. The paper will consider whether the development of spatial data infrastructures (SDI) for cultural heritage data could play an important part in resolving this problem, and ensure that the opportunity exists for such valuable LiDAR data to be consumed by as wide a user community as possible.

Introduction

Over the past three decades the widespread availability of total stations has enabled archaeological surveyors to record the cultural components of the landscape in three dimensions (3D; see Corns & Shaw 2010; Remondino this volume). This has provided the opportunity to record the subtle morphological evidence of human activity on the landscape, a powerful tool to help archaeological investigators unravel the evolution and functions of historic sites using non destructive methods (Bowden 1999). Examples such as the topographic survey of the Hill of Tara completed in 1996 highlight the effectiveness of the techniques in identifying new monument features (Newman 1997). In this survey over 60,000 height points covering 60 hectares pushed the extents of archaeological ground surveying, providing a more naturalistic visualization of the archaeological complex in comparison to traditional hachure approach (Hale & Hepher 2008). The Tara model (Figure 6.1) was considered ground breaking at the time, but from a present perspective its limited extent and the artificial nature of the defining boundary give the impression of a surface model floating in space. Extending survey areas by further ground survey, whether by total station or dGPS, was not seen as an economically viable solution due to the expense in mobilising survey teams for weeks on end. For this reason the Discovery Programme looked to other technological approaches to enable more extensive landscape recording and modelling.

Aerial photogrammetry

For decades aerial photogrammetric surveys have been used by national mapping agencies in the production of cartographic products, orthoimages and associated height models, and was the first technique considered as a solution to modelling the wider archaeological landscape. The principle of this method is that through complex pixel matching processes within overlapping stereo digital images, a Digital Elevation Model (DEM) is extracted which, in turn, is used to generate orthophotographs for the survey extent. A number of exploratory aerial photogrammetric projects were undertaken by the Discovery Programme in a variety

Figure 6.1: Perspective view of the 3D Model generated from the total station survey of the Hill of Tara, completed in 1996. *© The Discovery Programme.*

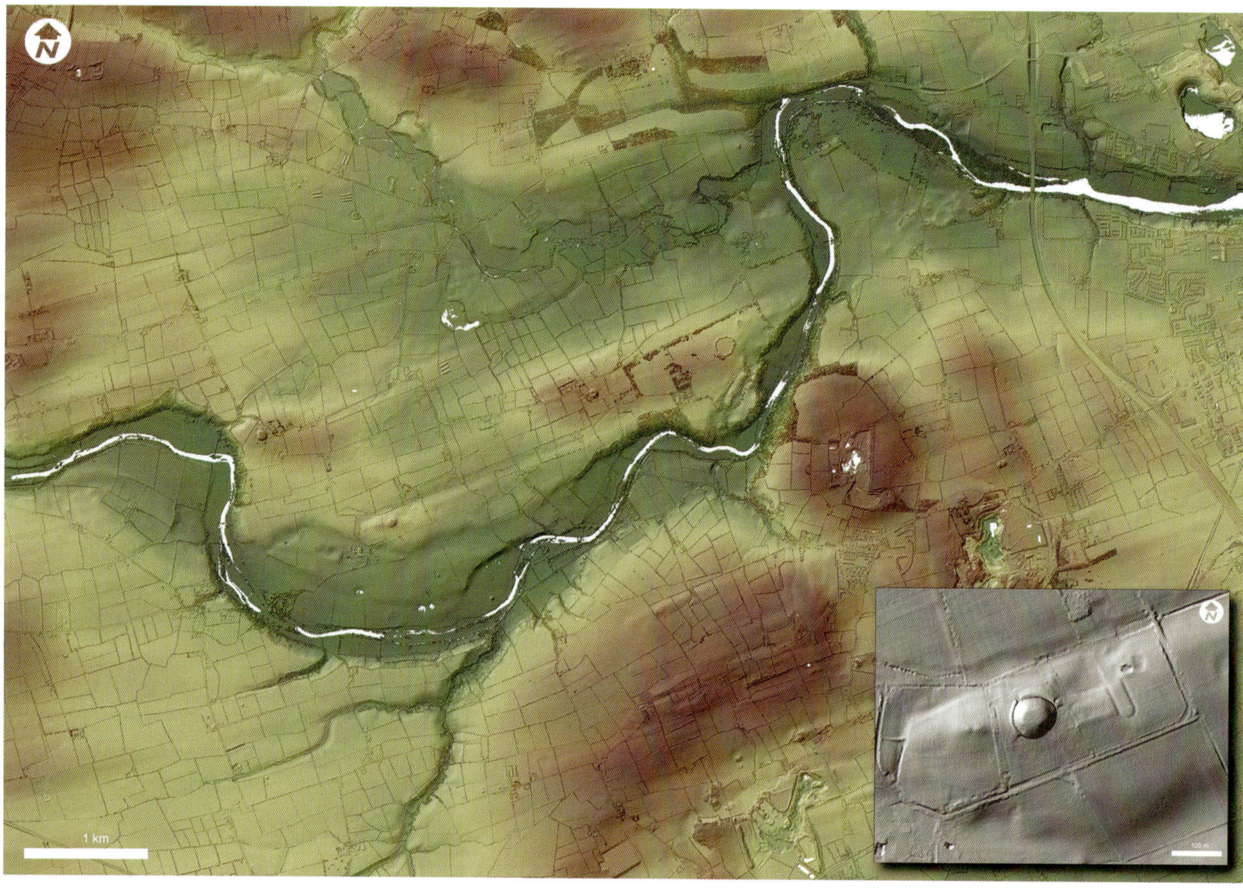

Figure 6.2: DSM generated from the Brú na Bóinne first return LiDAR data. Inset shows the DTM from last return data for an enlarged area around the Newgrange complex. © *The Discovery Programme / The Heritage Council / Meath County Council.*

of landscapes and employing a range of flying heights and photo-scales in order to assess the quality of the derived DEM. Whilst the orthophotographs proved to have enormous potential for landscape analysis, it was concluded that the derived DEMs were generally too coarse to reflect the subtleties required for modelling small-scale or complex archaeological sites. This was a result of difficulties with the automatic pixel matching routine, a core component of photogrammetric processing software. These algorithms use matrices to analyse the adjacent pixels on overlapping images in order to 'match' the pixels, from which parallax measurement enable the height component and resulting DSM to be extracted. However, there were problems with this technique in most of our project areas due to a predominance of grass or pasture land coverage. The automated pixel matching routines failed because of the large tonally indistinct areas of green. To resolve this large pixel correlation matrices had to be employed, significantly lowering the resolution of the final 3D models. However, a recent software development known as multi-ray matching which enables automatic terrain extraction from stereo imagery may effectively eliminate this problem (www.erdas.com, http://labs.erdas.com/blog_view.aspx?q=6074).

Although photogrammetry reduced the field survey time, considerable computer-based effort was required to create and edit the DSMs which somewhat negated this gain. Another limiting factor to using conventional image-based aerial survey was the ability of thick vegetation to obscure the presence of underlying topography and potential archaeological features. Where archaeological features did exist beneath vegetation cover, a process of field completion using total stations was required.

LiDAR

The next obvious step was to examine the potential of LiDAR (Light Detection And Ranging); for this we were fortunate that Meath County Council gave access to a recently gathered block of data for the Brú na Bóinne World Heritage Site. Many research projects (Doneus & Briese 2006a: Bewley *et al.* 2005) have already powerfully demonstrated the potential for LiDAR to record the archaeological landscape 3-dimensionally, including those areas of terrain beneath tree cover (Doneus & Briese 2006b, this volume; Bofinger & Hesse and Georges-Leroy this volume). Another factor that has made the application of this technique advantageous is the relatively short time period between commissioning of a survey and creation of the final functioning DSM / DTMs. Typical horizontal and vertical accuracies of the final data models are 0.6m and 0.15m respectively (Boyda & Hillb 2007). Values of this order are clearly suitable for topographic modelling of archaeological landscapes. Both DSM and DTM and associated hillshade models were generated for the Brú na Bóinne LiDAR survey (Figure 6.2), and graphically illustrate the powerful landscape models which can be generated. However, the ability of this technology to

Figure 6.3: Helicopter with the FLI-MAP 400 frame mounted on the fuselage. © *Fugro-BKS.*

successfully depict the subtle micro-topography of an individual monument is questionable, and this will be returned to in more detail later in this paper.

FLI-MAP 400 – helicopter based LiDAR

At the Discovery Programme we were alerted to a new method of survey being carried out by Fugro-BKS that appeared to advance LiDAR technology to a new level of precision and accuracy. Although the FLI-MAP 400 LiDAR system had been designed and developed primarily to meet the survey requirements of infrastructure projects including highways, railways and electricity distribution networks, its potential for 3D modelling of small-scale archaeological features presented some exciting possibilities.

System configuration and capabilities

The FLI-MAP 400 system can be mounted on a range of helicopter types without the need to modify the airframe and can therefore be mobilised quickly and cost-effectively to operate on a suitably specified aircraft near to the project area, without having to transit long distances (Figure 6.3).

The system consists of the following components which are all contained within a modular rigid frame that is attached to the cargo mounting point positioned on the fuselage beneath the helicopter and between the landing skids:

- A single Class I laser scanner operating at a frequency of 150 Khz that scans through a cycle inclined 7° forward to nadir and 7°aft along the line of flight;
- Two GPS receivers to supply accurate positional information when used in conjunction with ground-based GPS base stations;
- An Inertial Navigation System (INS/IMU) to continuously track the orientation and rotational elements of the sensor;
- An RGB digital line scanner to supply virtual colour attribution to the acquired LiDAR data and;
- Forward oblique and nadir facing medium format digital cameras and videos.

Unlike fixed-wing aircraft that are constrained by a minimum airspeed before which stalling occurs, the slower speed and lower flying heights at which a helicopter can operate facilitates the collection of data at a much higher resolution.

Filtering and processing the data

Most LiDAR systems are equipped to receive multiple returns from a single laser pulse that increases in diameter as it travels towards the ground. As the pulse strikes a branch or leaf in the canopy of vegetation, some of the energy of that pulse is reflected back to the sensor, whilst the remainder continues on its path to the ground. This process is repeated up to a further three times, or until a time when the ground surface is reached. This multiple return feature, combined with the ultra-high frequency of the system, enables effective penetration of even the most densely vegetated areas. However careful scheduling of data acquisition programmes to take account of seasonal variations in vegetation density help to maximise ground coverage. Similarly, the first return data can be used as a DEM.

Specialist processing software is used to remove the point cloud data that are classified as non-ground (vegetation, buildings and other elevation features) to leave a bare-earth terrain model. Complex algorithms have been developed to semi-automate this procedure requiring only limited manual editing to remove non-ground artefacts from the laser point cloud. The Discovery Programme liaised closely with the processing specialists to ensure that the special requirements of heritage and archaeology, specifically our particular interest in surface anomalies, were fully understood and incorporated into the generation of the bare-earth data set.
Simultaneously acquired aerial imagery is then mosaiced together and used in combination with the post-processed digital terrain LiDAR data, to produce digital orthophotographs that can be output at a suitable tile size and in a variety of formats to suit the specific client's application software.

Generating the DSM and DTM

The 2008 LiDAR project for the Hill of Tara provides a case study for the data handling and processing stages required to generate the DSM and DTM's. The data was acquired on a single day and involved acquisition using a Eurocopter AS355 (Twin Squirrel) helicopter operated at an altitude of 190m and a speed of 40knots. This generated in excess of 150 million individual LiDAR points to cover the survey area of 2.38km^2 equating to a point density in excess of 50pts per m^2. Digital orthophotographs were supplied at a ground resolution of 10cm.

The output ASCII files, simple X,Y,Z Irish National Grid coordinate files, were supplied as tiled data in order to facilitate data management and GIS processing

Figure 6.4: Plan view of the DSM generated for the Hill of Tara from FLI-MAP data at about 60pts/m^2. © *The Discovery Programme / The Heritage Council.*

procedures. The data, covering an area measuring 1,700m N-S by 1,400m E-W (approx. 240 hectares) were split into 12 tiles, each containing approximately 12 million Cartesian coordinates (3D data points). This ASCII data was firstly imported into a Microsoft Access database, from which it was displayed spatially within ArcMap 9.2 GIS system. The triangulated irregular network (TIN) surface models were created using the 3D Analyst application of ArcGIS and subsequently converted into raster grids to enable faster display times and processing. The grid tiles were then merged to form single composite DSM and DTM grids.

To effectively visualise the resulting DSM and DTMs hill-shade processing, based upon multiple light sources correlated to the frequency of relief features, was implemented (Loisios *et al.* 2007). The resulting model has optimal lighting conditions to enable

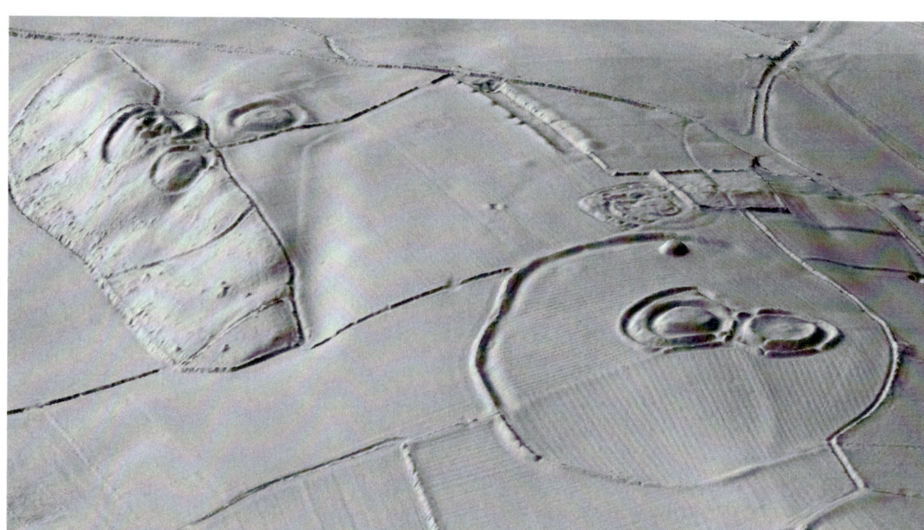

Figure 6.5: Perspective view of part of the Hill of Tara DTM. © *The Discovery Programme / The Heritage Council.*

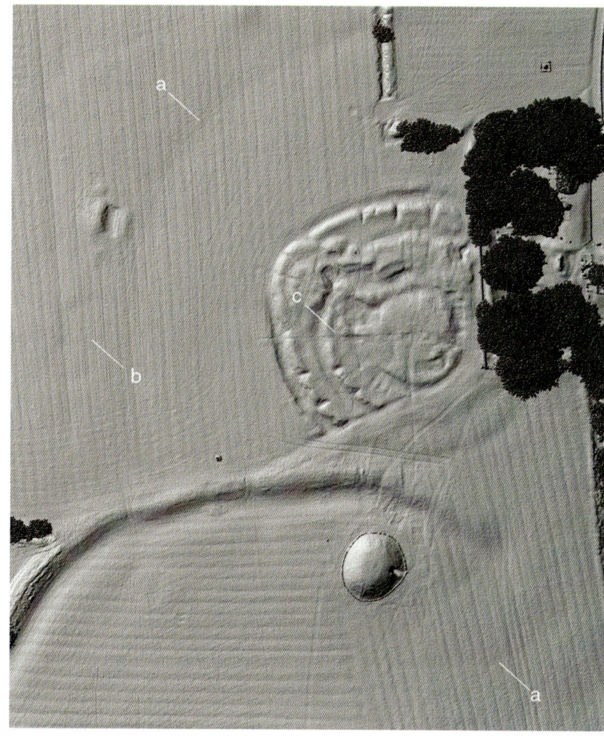

Figure 6.6: Hill-shaded DEM of Raith na Senad displaying the presence of the large ditched pit enclosure (a), conjoined barrows (b), and archaeological excavation trenches (c). © *The Discovery Programme / The Heritage Council.*

the identification of archaeological features. The exceptional detail of these hill-shaded models is readily apparent even at a cursory glance. Figure 6.4 displays the DSM (first return) for the whole survey area, with an enlargement of the conjoined earthworks of the Forrad and Tech Cormaic to illustrate the extraordinary detail and high level of resolution that exists throughout the whole model. This surface model includes all the vegetation and man-made features in the landscape, such as houses and barns and even captures overhead power-lines. By contrast, the DTM provides a 'bare earth' representation of the terrain and enables us to view the topographic detail of the ground surface beneath the obscuring vegetation (Figure 6.5).

Further insight, new discoveries

The assessment of the extent to which these models have enhanced our understanding of Tara and its surrounding landscape has only partially been completed but, from initial findings, new discoveries highlight the potential for high resolution LiDAR to unravel the archaeological landscape. Traces of the ditched pit enclosure and conjoined barrows, previously only visible on geophysical images (Fenwick & Newman 2002), are readily apparent as low relief topographical features to the northwest and west of Ráith na Senad respectively. The footprint of excavation cuttings from the 1950s can be observed running across the centre of Ráith na Senad. The series of irregular humps and hollows around the central area of the earthwork may be related to the less scientific activities of a group of British Israelites, who dug extensive holes in this site at the turn of the 19th and 20th centuries in an unsuccessful quest for the Arc of the Covenant (Figure 6.6).

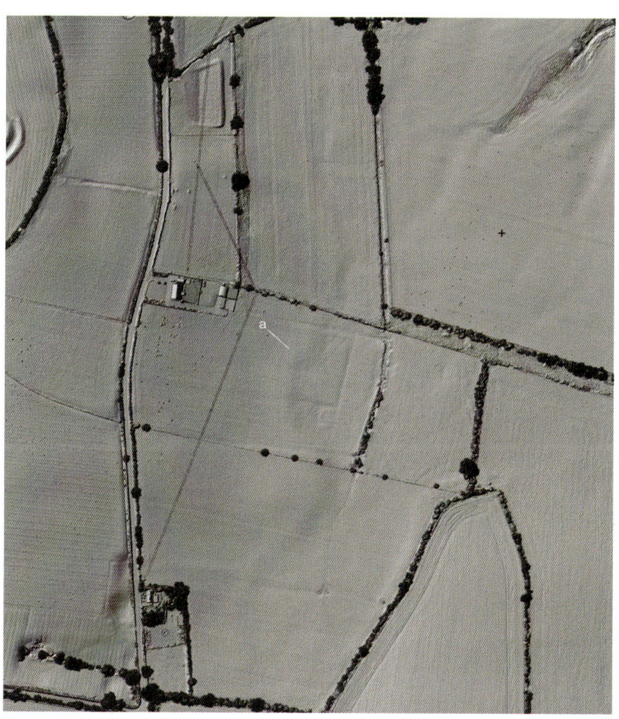

Figure 6.7: Hill-shaded DEM highlighting the presence of a palaeo-channel (a) of the River Gabhra. © *The Discovery Programme / The Heritage Council.*

Another feature identified from the surface models was the original palaeochannel of the River Gabhra. This can be traced meandering through the fields on the eastern flank of the Hill of Tara before descending through a narrow gorge, which was once landscaped to form a series of cascading waterfalls as part of the formal gardens of Tara Hall, before continuing on its course through the Gabhra Valley. The stream is now diverted along the drainage channels of modern field boundaries, but it is apparent from the surface model that the original source of the river was the spring identified in early documentary sources as Nemnach; according to legend, the site of the first water-powered mill in Ireland (Figure 6.7). To the east of the main archaeological complex, the surface model has enabled the identification of large enclosures and associated internal features and cultivation remains.

No evidence for these features is present on historic Ordnance Survey Ireland mapping, but their detection adds to our understanding of the activity over time in the area surrounding the main features of the Hill of Tara. Identifying such features will enable the Discovery Programme to better select areas for potential future research and geophysical survey.

Comparison with fixed-wing LiDAR

In order to draw definitive conclusions about the benefits of FLI-MAP over standard LiDAR, the two methods would have to be applied to a single chosen control site. Unfortunately the Discovery Programme does not have the financial resources to commission such a study. Without this our assertion that the models are of a higher resolution and accuracy are based on the

Survey Method	Project	Cos (Approx)	Resolution	Survey area
FliMAP (helicopter)	Hill of Tara	€ 30,000	60 pt/m^2	2.04km^2
LiDAR (fixed wing)	Brú na Bóinne	€ 30,000	1 pt/m^2	90km^2
Photogrammetry	Tara / Skreen	€ 10,000	2 pt/m^2	70km^2

Table 6.1: Approximate costs, resolution specified and area covered for three different techniques used by the Discovery Programme to generate DEMs.

specification of the systems and a visual comparisons of models at the same scale, albeit of different sites, such as in Figure 6.8. The potential to identify phasing and detailed morphology, and even to assess potential erosion tracks caused by visitors to the site, exists in the Tara example. This level of detail is not available in the Brú na Bóinne model. However, it should be borne in mind that these are rapidly emerging technologies; the sensor frequency of helicopter and fixed-wing systems have increased since the data was gathered for both of these surveys. The exciting consequences of such developments are the potential to cover larger areas for the same cost or at even higher resolutions.

Whilst it is tempting to present definitive project costs these can be misleading given so many unknown variables are involved. In Ireland most archaeological FLI-MAP projects have been carefully scheduled to coincide with other commercial projects to share and thus reduce the considerable mobilisation costs involved (the helicopter from Scotland, the sensor from the Netherlands). The resulting 'per km^2' cost is so dependent on this factor that we cannot make meaningful comparisons. However Table 6.1 summarises the approximate costs, coverage and resolutions of three surveys undertaken by the Discovery Programme using three different airborne techniques.

Survey Method	Estimated Cost
12.5cm FLI-MAP LiDAR	€ 40,000
1m ground based survey	> € 300,000
12.5cm ground based survey	> € 20,000,000

Table 6.2: Estimate of the costs, in 2008, to complete the 2.04km^2 survey area of the Hill of Tara by FLI-MAP and ground survey methods (based on an average 2,000 survey points per day).

Broad conclusions include the obvious compromise of coverage and resolution between the LiDAR systems. In simple terms fixed wing LiDAR will survey a considerably larger area for your money. The question which has to be addressed is does the survey objective for a project merit the higher resolution available with the FLI-MAP system. Such considerations may lead to the use of photogrammetry when terrain model resolution is not a high priority.

Comparisons with ground survey techniques have been made based on our experiences surveying Hill of Tara in the 1990s using total stations, and the more recent application of real-time differential GPS (dGPS) on sites in Ireland. Again we are without a definitive control site but, based on our dGPS estimated recording rate of 2,000 points/day on sites with average vegetation problems it is clearly not a cost effective method to cover areas as large as those we have undertaken with LiDAR (see Table 6.2).

However, it is difficult to place a value on the familiarization and understanding the field surveyor gains by spending time on the ground examining the landscape and features. This was a vital part in the older studies by the Discovery Programme at the Hill of Tara, and including a substantial component of field assessment is imperative in any remote sensing approach to archaeological landscape mapping.

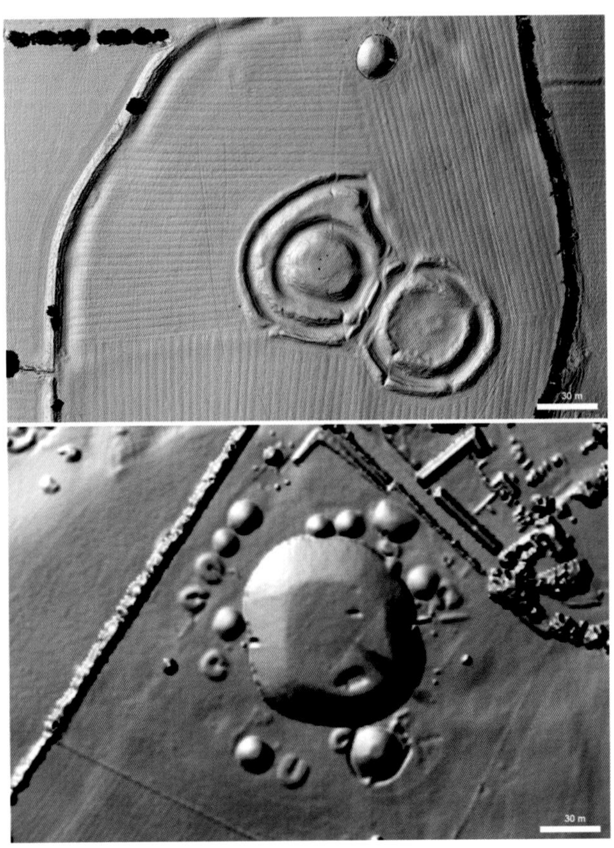

Figure 6.8: Top an extract from the Hill of Tara illustrating the level of detail reflected in a DSM generated from 12.5cm resolution data. Bottom shows an extract from the Brú na Bóinne DSM generated from 1m resolution data. © *The Discovery Programme / The Heritage Council.*

Figure 6.9: Map showing the location of recent FLI-MAP and LiDAR projects in Ireland. © *The Discovery Programme.*

Without this there is a serious risk of generating high resolution models of landscapes about which we understand little.

LiDAR proliferation in Ireland

The Hill of Tara LiDAR project generated significant publicity, both amongst academics and the general public (Corns *et al.* 2008). As a result a number of Irish heritage and archaeology agencies commissioned their own projects, often consulting with the Discovery Programme about defining specifications, or for advice with processing and modelling data sets.

The map (Figure 6.9) shows the distribution of some of the recent projects. It shows that government agencies are the major commissioning bodies for LiDAR, but includes some notable private sector projects with components of archaeological assessment, including road and port developments. The most extensive use of FLI-MAP has been on sites in the North of Ireland commissioned by the Department of Environment Northern Ireland.

In 2008 six sites were flown in Counties Fermanagh and Armagh, including Devenish Island, Kilterney Deer Park, Clogher, The Dorsey, Cornashee and Sheebeg. Although these sites are still being analysed a number of interesting discoveries have been made at Kilterney, a site that was in continuous settlement from the Neolithic to the late medieval period. From this analysis the archaeologists have found new medieval house structures and a previously unknown type of cultivation remains typically English medieval in character. Subsequently data for a further two sites have been acquired, Dunluce Castle on the North Antrim Coast and the promontory fort sites in the Townland of Linford, outside Larne. Two further surveys are planned for 2010, subject to funding.

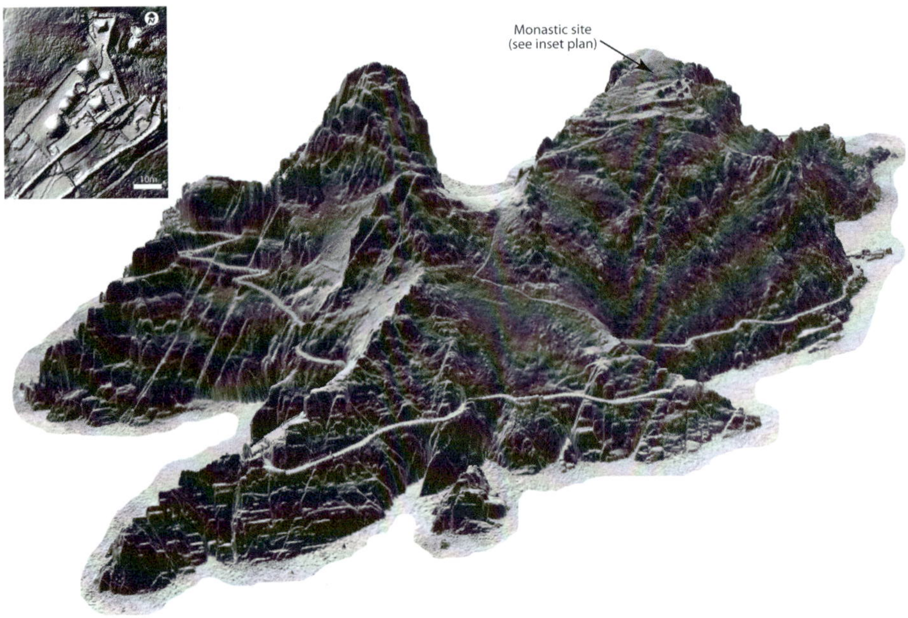

Figure 6.10: DSM generated from the first-return FLI-MAP data for Skellig Michael, 60pts per m² density. At its greatest extent this steep rocky island is approximately 900m long by 450m across, rising dramatically to 230m above the Atlantic Ocean. Inset shows plan view of monastic structures. © *The Discovery Programme / The Heritage Council.*

World Heritage Sites, conservation plans and research frameworks

A FLI-MAP survey of Skellig Michael, one of the three sites in Ireland currently on UNESCO's list of World Heritage Sites, was commissioned by the National Monuments Service, with the Discovery Programme giving advice on the specification and data modelling for the project. The DSM generated (Figure 6.10) was regarded as a primary resource for a project considering the maintenance and stabilisation of the islands heritage sites. It was used as the baseline survey by contractors working on the restoration of the monastic structures to tie in conventional surveys and drawn plans.

The Discovery Programme's FLI-MAP survey of the Hill of Tara has been identified as a core resource for the Tara/Skryne Landscape Conservation project (www.meath.ie/LocalAuthorities/Planning/TaraSkryneLandscapeProject/) initiated by Meath County Council. Even in the early consultation phase FLI-MAP outputs are being used in public meetings and presentations to engage with local residents and stakeholders (Figure 6.11). The increased awareness of the value of LiDAR data has resulted in renewed interest in existing data sets. The LiDAR data for the Brú na Bóinne – although a lower resolution from a fixed wing system – was modelled by the Discovery Programme and subsequently became an important resource during the development of a Research Framework for the World Heritage Site (Smyth 2009). It was identified as a vital data source particularly given the shift of research emphasis from sites to landscapes, and featured extensively in the final publication.

World Heritage Sites, conservation plans and research frameworks

The proliferation of projects utilising LiDAR in Ireland over the past three years has been extremely encouraging but during the research for this paper, in discussion with project managers and data users, a number of common issues were raised which were deemed to be hindering the full exploitation of the data. A fundamental problem in structuring time and resources to enable specialists to fully examine and study the data was identified as a major issue. The perception that the data capture and modelling phase was an end in itself was a result of not establishing a clearly defined research agenda in advance.

Issues of skills, training and technology were also raised as major concerns. The lack of dedicated GIS specialists in the various organisations, or staff with the appropriate level of GIS training or experience was identified as a primary problem. Data was often being delivered to staff members who did not know how to proceed to the modelling phase, and they did not have the GIS support services to call for help. For this reason some GIS data sets remained unprocessed

Figure 6.11: Members of the public examining the FLI-MAP DTM of the Hill of Tara during the consultation phase of the Tara/Skryne Landscape conservation project. © *Meath County Council.*

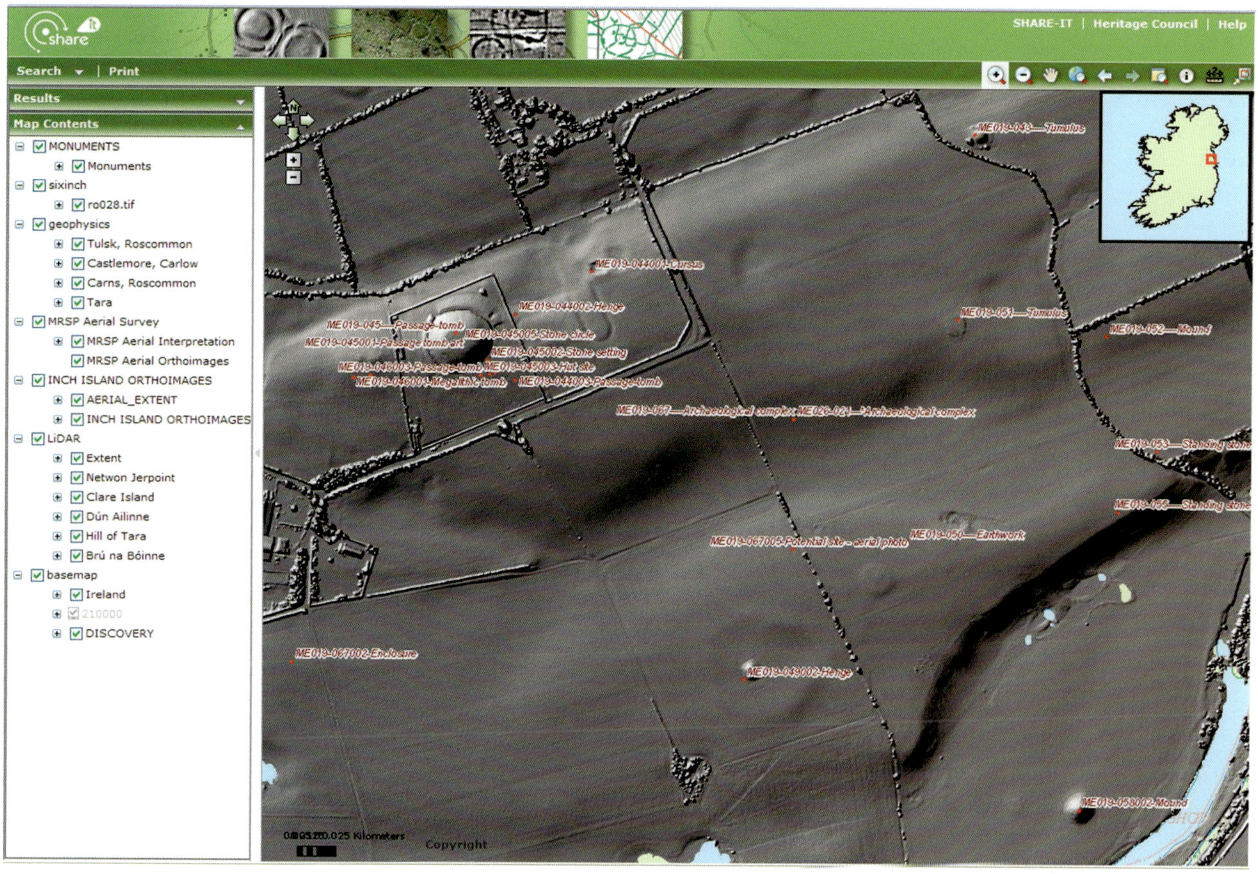

Figure 6.12: Screen image of the prototype web mapping application developed by the Discovery Programme. © *The Discovery Programme / The Heritage Council.*

for considerable time. Computing hardware issues, including the processing power of workstations and the lack of storage capacity, were also identified as problems in some organizations. The Discovery Programme sees a potential solution to these issues through the development of Spatial Data Infrastructure (SDI) for cultural heritage information.

Spatal Data Infrastructure (SDI)

Traditionally, the GIS technology used to view and interrogate spatial data (such as LiDAR DSMs and DTMs) has been expensive desktop-based software solutions but, in recent years, technological developments and the adoption of open standards has enabled the delivery and exploration of spatial data within a Spatial Data Infrastructure (SDI). SDI is the collective name for a group of technologies and supporting measures that enables access to spatial data. It is more than a single data set or database, and incorporates geographic data and attributes, documentation (metadata), a means to discover, visualise, and evaluate the data and provides access to data through web services.

Share-IT web mapping application

In 2008 the Discovery Programme undertook a detailed feasibility study to examine the potential development of an SDI for cultural heritage spatial data in Ireland, including DSM/DTMs derived from LiDAR data. Entitled Spatial Heritage and Archaeological Research Environment using IT (SHARE-IT) this study examined the long term access to digital spatial data within the cultural heritage domain. Potential solutions including open archives, metadata, data standards and the construction of an SDI were outlined. One of the components of an SDI – a web mapping application – was piloted during the project (Figure 6.12). Created using ESRI ArcGIS Server 9.3, this type of application with a selection of GIS tools built-in, enables the user to locate, view and interrogate spatial data of reliable quality from a remote securely archived location via the web. This may be the type of solution that will help organisations maximise the potential of their LiDAR investment.

Conclusions

This paper shows the effectiveness of FLI-MAP in generating DSM/DTMs of such spectacular resolution as to enable micro-topographic elements to be recorded and studied. However, it makes it clear that the selection of an appropriate LiDAR system requires careful consideration, balancing the proposed research agenda, size of area and available budget. It has been shown that even in the short time since the data sets discussed in this paper have been gathered, the resolution and coverage made available by each technique has improved significantly. For this reason, it should be emphasised that an assessment of the current 'state-of-play' of the technology should be made before any data is commissioned.

The paper has shown the enthusiastic take up of LiDAR – FLI-MAP in particular – by Irish heritage agencies, and that such data is contributing significantly to the research and management of heritage sites and landscapes. However, we have also identified some weaknesses in the skills, training and technology available, but suggested that solutions may lie in SDI developments and web mapping services.

Acknowledgements

A number of people need to be thanked for their support and help during our FLI-MAP research. Much of the data has been funded through grants from the Heritage Council, where the support of Ian Doyle is particularly appreciated. Assistance from the staff at BKS-Fugro, especially John McNally and Chris Boreland has been invaluable and helped us to realise the full potential of the data. Other important contributions came from: Claire Foley and James Patience, Department of Environment (Northern Ireland) with information on the Northern Ireland experience of FLI-MAP; Loretto Guinan, Meath County Council with the Brú na Bóinne and Tara/Skyrne Landscape applications; and Abigail Walsh, National Monuments Service for the Skellig Michael data. Finally we wish to acknowledge the support and help of colleagues at the Discovery Programme.

References

Bewley, R.H., Crutchley, S. & Shell, C.A. 2005: New light on an ancient landscape: LiDAR survey in the Stonehenge World Heritage Site. *Antiquity* 79(305), 636–47.

Bowden, M. 1999: *Unravelling the Landscape: An inquisitive approach to archaeology*. NPI Media Group, London.

Boyda, D.S. & Hillb, R.A. 2007: Validation of airborne LiDAR intensity values from a forested landscape using hymap data: preliminary analyses. *IAPRS* Volume XXXVI, Part 3 / W52.

Corns A., Fenwick J. & Shaw, R. 2008: More than meets the eye. *Archaeology Ireland* 22(3/85), 34–8.

Corns, A. & Shaw, R. 2010: Archaeological Survey: A Review of 3D Data Capture Techniques. In Niccolucci, F. & Hermon, S. (eds): *Proceedings of the International Conference on Computer Applications and Quantitative Methods in Archaeology, CAA 2004*, 13–17 April, 2004, Prato, Italy.

Doneus, M. & Briese, C. 2006a: Digital terrain modeling for archaeological interpretation within forested areas using full-waveform laser scanning. In Ioannides, M., Arnold, D., Niccolucci, F. & Mania, K. (eds): *The 7th International Symposium on Virtual Reality, Archaeology and Cultural Heritage VAST (2006)*, 155–62.

Doneus, M. & Briese, C. 2006b: Full-waveform airborne laser scanning as a tool for archaeological reconnaissance. In Campana, S. & Forte, M. (eds): *From Space to Place. Proceedings of the 2nd International Conference on Remote Sensing in Archaeology*. BAR International Series, 1568. Archaeopress, 99–106.

Fenwick, J. & Newman, C. 2002: Geomagnetic survey on the Hill of Tara, Co. Meath. 1998/99. *Discovery Programme Reports 6*, Royal Irish Academy, Dublin, 1–18.

Hale, A. & Hepher, J. 2008: 3D Data Fusion for the Presentation of Archaeological Landscapes: A Scottish Perspective. In Posluschny, A., Lambers, K. & Herzog, I. (eds), *Layers of Perception. Proceedings of the 35th International Conference on Computer Applications and Quantitative Methods in Archaeology (CAA), Berlin*. April 2–6, 2007. Koll. Vor- u. Frühgesch. 10 (Bonn 2008), CD.

Loisios, D., Tzelepis, N. & Nakos, B. 2007: A methodology for creating Analytical hill-shading by combining different lighting directions. *Proceedings of 23rd International Cartographic Conference, Moscow*, August 2007.

Newman, C. 1997: *Tara: an archaeological survey*. Discovery Programme Monographs 2. Royal Irish Academy, Dublin.

Smyth, J. (ed.) 2009: *Brú na Bóinne World Heritage Site: Research Framework*. The Heritage Council, Kilkenny Tara Skryne Landscape Project, http://www.meath.ie/LocalAuthorities/Planning/TaraSkryneLandscapeProject/, accessed 19th April 2010.

7 | Archaeological applications of multi/ hyper-spectral data – challenges and potential

Anthony Beck

Abstract: Multi/hyper-spectral sensors offer immense potential as archaeological prospection tools. The sensors are sensitive to emitted or reflected radiation over different areas (wavelengths) of the electromagnetic spectrum. Their two major advantages are that they have the potential to detect archaeological sites and monuments (henceforth archaeological residues) that are undetectable in the visible wavelengths and that they may extend the window of opportunity for their detection. For example, localised crop stress and vigour variations, which underpin crop mark formation, are sometimes better expressed in the near-infrared than in the visible. In addition, multi/hyper-spectral data collected from different platforms (aerial and satellite) under different conditions can be used to generate ancillary themes that aid interpretation (e.g. soil, geology and land-use layers) and are important for 'Total Archaeology'. However, multi/hyper-spectral sensors are relatively expensive and require systematic surveys under 'appropriate conditions' in order to be successful. It is this latter point which is critical: there is a poor understanding of the spatial, environmental and seasonal contrast dynamics that determine an 'appropriate condition' and therefore whether features of archaeological interest can be detected.

Introduction

In the UK, the practice of using remote sensing techniques for detecting archaeological sites and visualizing archaeological landscapes has traditionally been based on low altitude aerial photography using film emulsions sensitive at optical and sometimes near-infrared wavelengths. In the 1920s O. G. S. Crawford, the archaeological officer of the British Ordnance Survey, demonstrated that archaeological structures could be identified from shadow, soil and crop markings on panchromatic aerial photography (see Figure 7.1). Since that time, both oblique and vertical aerial photographs have been used extensively for archaeological reconnaissance and mapping all over the world (Bewley 2000; Donoghue 2001). Early aerial photographers helped to refine the instruments and establish methods that are still in use today. Crawford in particular established methods of site classification and wrote about the effects of weather, season, soil moisture and crop type on photographic return (Crawford 1923, 1928, 1929). Today, these aerial approaches are accepted as a cost-effective, non-invasive technique for the reconnaissance and survey of monuments.

However, recording using traditional observer directed reconnaissance and aerial photography is not without its problems. The reliance on a small component of

Figure 7.1: Factors leading to the detection of crop, shadow and soil marks. *Re-used under a Creative Commons Attribution-NonCommercial-ShareAlike 3.0 Unported License credited to Anthony Beck.*

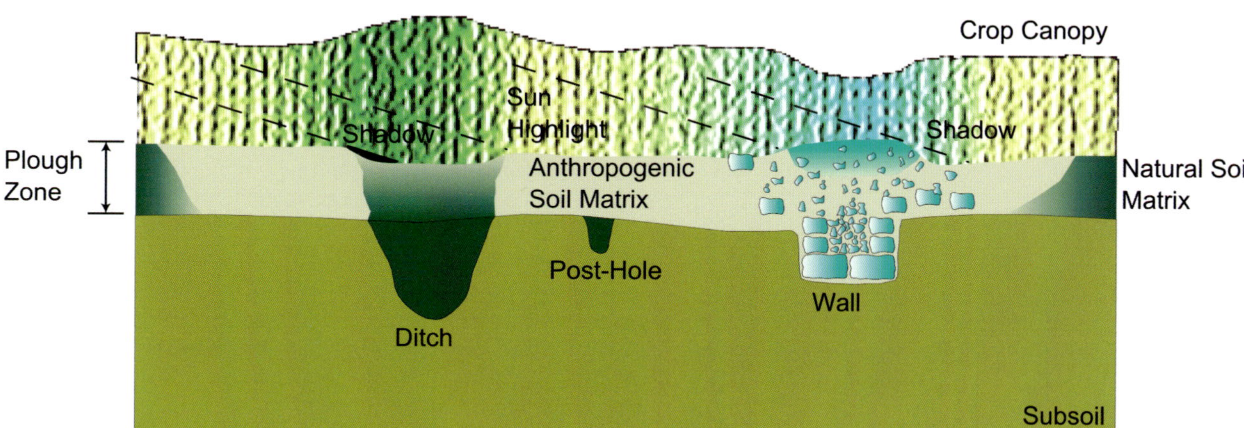

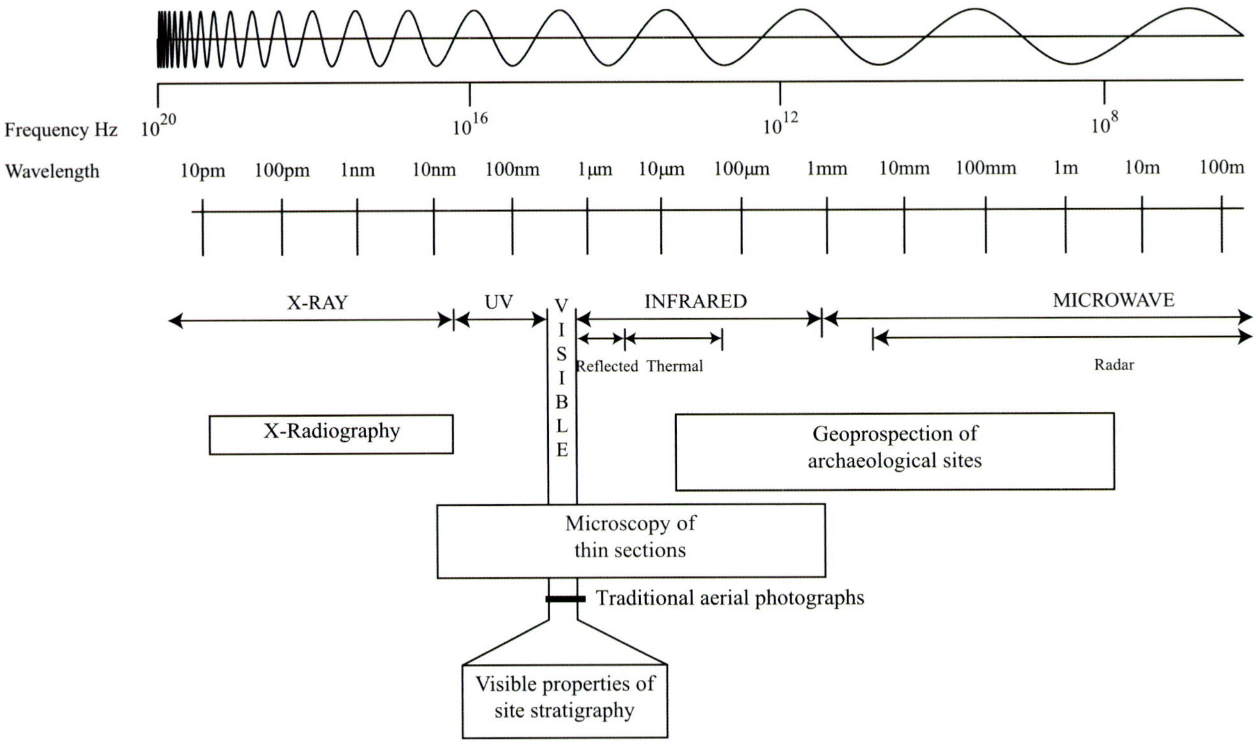

Figure 7.2: The archaeological exploitation of the electromagnetic spectrum. *Re-used under a Creative Commons Attribution-NonCommercial-ShareAlike 3.0 Unported License credited to Anthony Beck.*

the electromagnetic spectrum (see Figure 7.2) raises a number of issues. The small spectral window can induce a significant bias as only certain residues under specific conditions express contrasts in these wavelengths. The over-reliance on the visual component of the electromagnetic spectrum has had a significant impact on data capture. The collection technique and technology mitigate against using any other sensor (peripatetic surveys are directed by visual observation from a plane and collected using an optical system, a camera out of a window: this technique will never allow the detection of the multitude of archaeological residues whose contrast expression cannot be seen by the human eye – i.e. is outside the optical). In areas that have been intensively studied, such as the UK, a point of saturation can be reached. It is also possible that for some areas a century of flying, in different environmental conditions, has resulted in saturation: no previously unobserved features are being detected – this does not mean that there are no new archaeological residues to discover, rather that no more can be detected with that particular sensor configuration. In addition perceived wisdom can dominate policy which can impact on collection regimes. For example, until recently (Mills & Palmer 2007), UK aerial archaeological prospection in clay environments was thought not to work.

This paper will introduce multi and hyper-spectral remote sensing (including the important resolving characteristics of the sensors) and the nature of the archaeological problems to which they can be applied. A description of the challenges and potential of utilizing such data in the heritage sector then follows. Finally, a UK research project designed to improve the understanding of the application and the factors underpinning archaeological detection using multi/hyperspectral scanners is discussed.

Multi/Hyper spectral remote sensing

The underlying premise of remote sensing is that interpreters can extract information about objects and features by studying the measurements from a sensor system. Sensors fall into two main categories: passive and active. Passive sensors are the most common and record naturally occurring radiation that is reflected, primarily from the sun, or emitted, as thermal energy. Active sensors bathe the terrain in artificial energy and then record the amount scattered back to the sensor. The sensors themselves can be used on different platforms, commonly ground, aerial and satellite.

Every sensor has limitations most of which are based on the resolving characteristics, or resolution, of the sensor. Each remote sensing system has four major axes of resolution: spectral; spatial; radiometric; and temporal.

Spectral resolution (see Figure 7.3) refers to the dimensions and number of specific dimension units for which a sensor is sensitive. The limited spectral range of photographic film (350 – 950nm) is overcome by the use of photoelectric sensing devices where image data is recorded in a digital form. These devices are able to separate electromagnetic radiation into a number of discrete narrow wavebands, hence the term multispectral. Panchromatic (single band) imagery is generally sensitive to a broad spectral range normally in the visible and NIR wavelengths. By contrast, multiple band spectral scanners (or MultiSpectral Scannners) co-collect imagery from different regions of the electromagnetic spectrum. The multispectral scanner of Landsat Thematic Mapper contains 7 bands covering discrete wavelength ranges over different parts of the spectrum. Hyperspectral scanners co-collect data in many very narrow band passes. For example, the AVIRIS sensor can collect approximately 224 bands between 0.4 and 2.5μm at 10nm intervals. As shown in Figure 7.3, spectral resolution can increase from the 'broad band' panchromatic to the very narrow bands of hyperspectral. Increasing spectral resolution, at the appropriate areas of the electromagnetic spectrum, may help to improve image interpretation and classification by providing the spectral contrast necessary to distinguish between different objects.

Spatial resolution is dependent upon the resolving power of the sensor, the wavelength under consideration, the distance from the object and the size of the object to be identified. Many systems now represent this relationship as simply the ground dimension in metres for each pixel. In general terms spatial resolution decreases as sensor platform/altitude

Figure 7.3: Contrasting spectral resolutions of AVIRIS, Landsat and Panchromatic imagery. Note how closely the spectral curves in Landsat follow those in AVIRIS (based on Christophe & Pearlman (2008) and Beck (2007)). *Re-used under a Creative Commons Attribution-NonCommercial-ShareAlike 3.0 Unported License credited to Anthony Beck.*

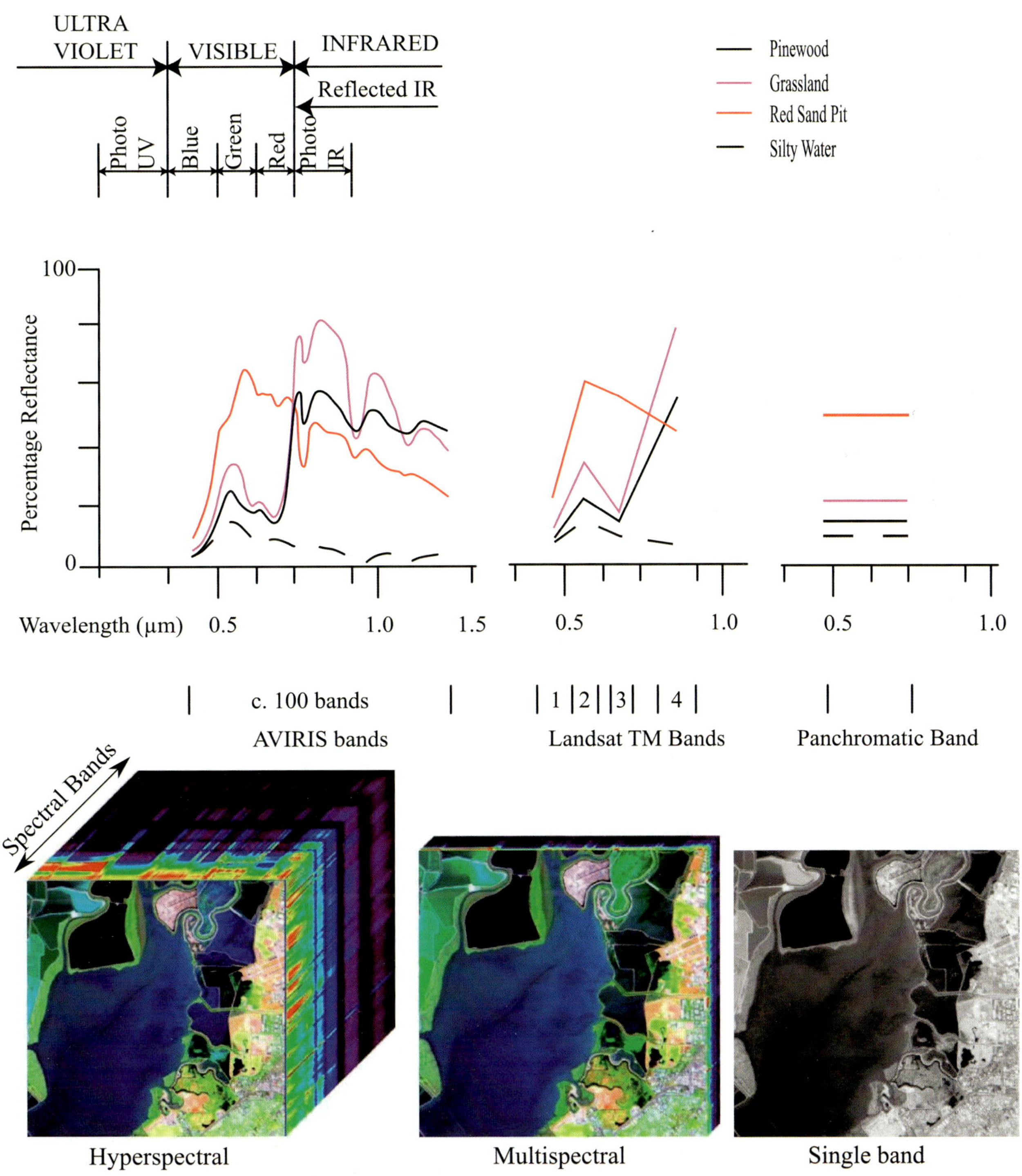

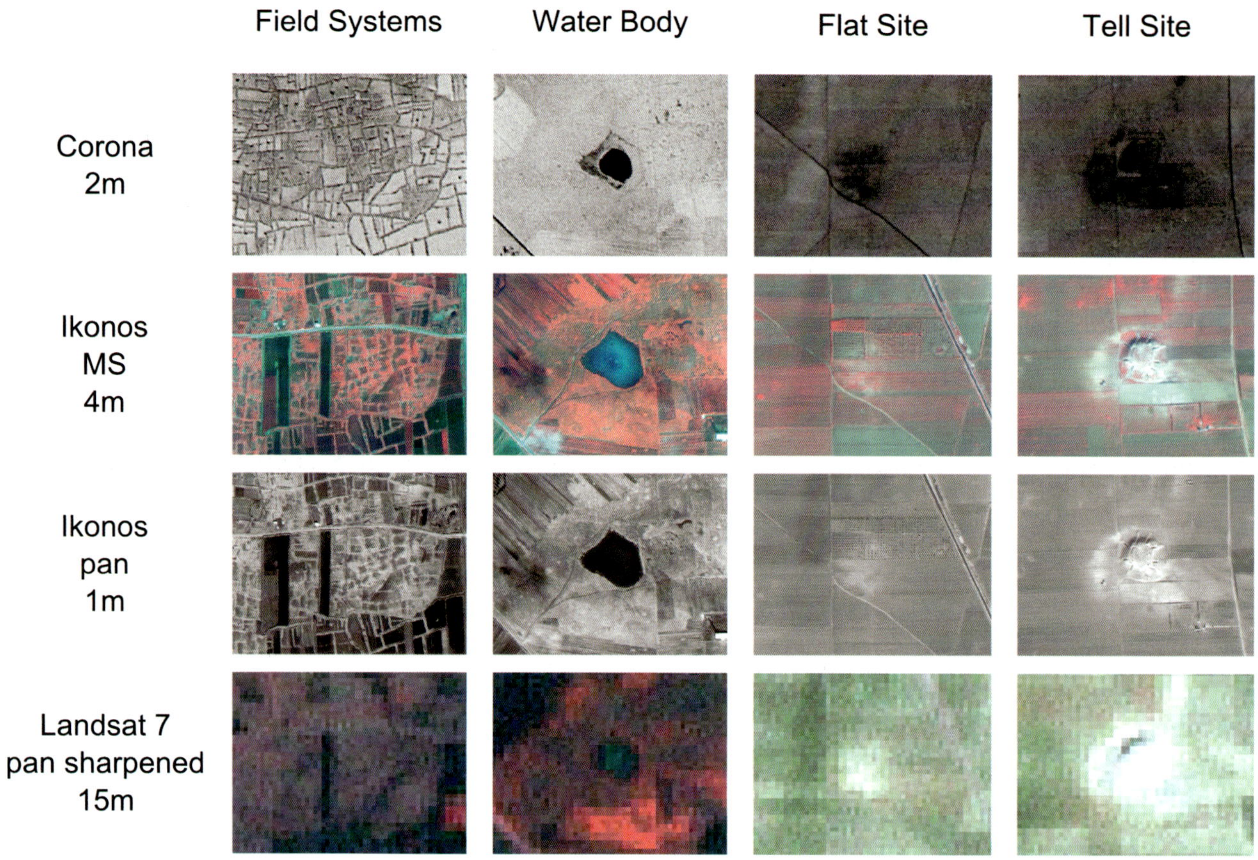

Figure 7.4: Comparison of sensors with different spatial and temporal resolutions (Corona, 1969; Ikonos, 2002: Landsat, 1999 after Beck (2004)). *Re-used under a Creative Commons Attribution-NonCommercial-ShareAlike 3.0 Unported License credited to Anthony Beck.*

increases. Increasing spatial resolution may help to improve image interpretation and classification by providing the spatial definition necessary to distinguish between different objects or to identify the form of specific objects (see Figure 7.4).

Radiometric resolution refers to the bit depth used during storage of the signal (e.g. 14-bit versus 8-bit) and is particularly important for archaeological prospection as it describes the subtlety of the sensor measurements which, in part, determines whether an object can be detected (see Figure 7.5). For example, if two panchromatic sensors with exactly the same spatial resolution, spectral resolution and dynamic range, but different radiometric resolutions, take a digital image of the same object from the same location (within, for example, the shadow of a building) at the same time, the sensor with the finer radiometric resolution is better for differentiating the object from the shadow. However, because under normal viewing conditions the human eye can discriminate only between 20 to 30 shades of grey[1], it is unlikely that the brain would be able to detect the object even though it exists numerically within the structure of the data. It is only with appropriate contrast manipulation of the finer radiometric resolution image that the object becomes apparent. This is analogous to trying to detect centimetre variations with one ruler that rounds measurements to the nearest millimetre and a different ruler that rounds measurements to the nearest decimetre. As archaeological residues commonly represent subtle shifts in radiance, much important archaeological information contained within an image can often go undetected. Increasing radiometric resolution may help to improve image interpretation and classification by providing increased recording sensitivity, and so allowing objects with subtly different reflectance characteristics to be identified.

Temporal resolution refers to how often a sensor system records a particular area. For all platforms except satellite, this value is likely to be infrequent. However, satellite imagery tends to cover the same area at the same time of day whereas all other sensor platforms can cover an area at different times of day. This is particularly significant for some residues where the nature of their contrast changes throughout the day (such as shadow marks or diurnal temperature variations (see Figure 7.6)) or under specific, temporally constrained, conditions. An increase in temporal resolution may help to provide a greater understanding on the nature of diurnal, seasonal and other temporal cycles and how these cycles impact on the expression of archaeological contrast. In addition, multi-temporal imagery is required for time-change analysis (see, for example, the change in the land-use between the Corona and Ikonos imagery in Figure 7.4).

[1] This value can vary significantly and is dependent upon the conditions of observation.

Archaeological remote sensing applications

Although there are many examples of upstanding architecture, the vast majority of archaeological residues are expressed on the ground surface or buried and essentially invisible to the human eye. However, traces can be identified via changes in chemical, physical and biological attributes (either directly or by proxy) through, for example, changes in phosphorous content, clusters of artefacts and cropmarks. Archaeologists employ two main approaches to identify and understand archaeological residues: survey (with no or limited destruction) and excavation (destructive). Each approach uses a variety of different techniques that sample different attributes of a landscape in progressively increasing detail.

Survey itself comes in many different forms, each providing different results. Survey is often the first stage of a long-term archaeological programme or management strategy and provides an overview of the range and nature of the extant archaeological record. However, survey is not just the preliminary stage for future intrusive excavations; a well-designed survey strategy will address questions that excavation can never answer (Banning 2002; Beck *et al.* 2007)

All surveys do not aim to achieve the same ends. There are many different survey objectives, but the majority can be generalised into the following groups:

- Reconnaissance survey: (Detection) primarily designed to detect all the positive and negative archaeological evidence within a study area.
- Evaluation survey: (Recognition) to assess the archaeological content of a landscape using survey techniques that facilitate subsequent field-prospection, statistical hypothesis building or the identification of spatial structure.
- Landscape research: (Identification) to form theoretical understanding of the relationships between settlement dynamics, hinterlands and the landscape itself.
- Cultural Resource Management (CRM): management and protection primarily designed for management of the available resources. CRM applications are not necessarily distinct from other survey objectives although they may be conducted as part of a more general information capture system.

Archaeological prospection using aerial photographs and other remote sensing techniques are generally part of reconnaissance survey but can be applied to all other forms of survey.

Thematic variables such as soil type, hydrology and topography are used to contextualise archaeological data. Thematic data is used during exploratory data analysis, as data layers in, for example, predictive modelling exercises or as a backdrop for CRM applications. From a landscape survey perspective these themes can be extremely useful, providing information on such diverse themes as land management (will archaeological residues be masked by vegetation?) and geomorphology (was this terrace formed before or after a certain date?). The thematic approach analyses different landscape components in an integrated manner. These landscape components commonly include the following themes:

- Digital terrain models
- Land use and cover (topography)
- Communication networks
- Hydrology networks
- Settlements
- Field Systems
- Soil/geology maps

The resultant thematic maps can be analysed independently or in conjunction with other data sets. Synthesis is achieved by composite mapping or multivariate statistical techniques between layers or individual layer analysis. Overlay techniques are

Figure 7.5: Differences in radiometric resolution: the 8-bit imagery does not record the objects in the oversaturated area. *Re-used under a Creative Commons Attribution-NonCommercial-ShareAlike 3.0 Unported License credited to Anthony Beck.*

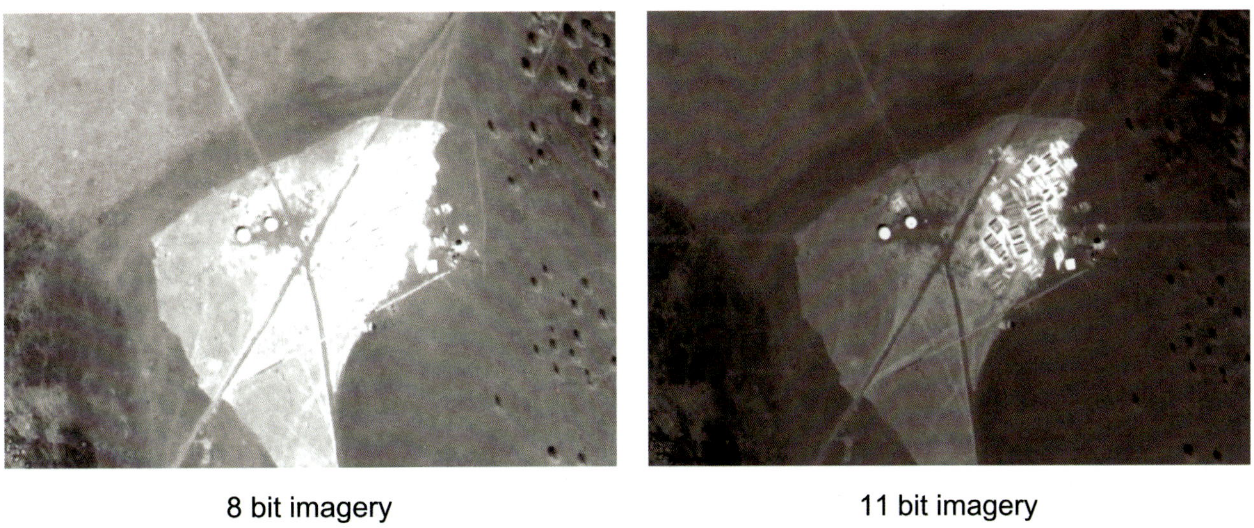

frequently used to look for spatial associations and relationships between the different themes (e.g. between settlement distribution and soil types using, for example, chi^2 analysis).

Archaeological challenges of utilising multi/ hyperspectral data

The main advantage to multi and hyperspectral imaging is that more of the electromagnetic spectrum is sampled at potentially finer spectral granularity; hence, more information about the objects in the image can be ascertained. The main disadvantages are cost and complexity. Fast computers, significant data storage capacities, specialist software and trained personnel are needed for analysing the data. Significant data storage is required as large multi-dimensional datasets generally exceed hundreds of megabytes. All of these factors greatly increase the cost of acquiring and processing hyperspectral data. However, as a relatively new suite of technologies the archaeological requirements are, as yet, uncertain.

Challenges for archaeological prospection

In recent decades, advances in sensor technology have led to a range of ground, airborne and spaceborne imaging instruments that can be applied to archaeological and heritage management problems. However, the development of the techniques associated with these technologies have evolved independently with little understanding of the physical, chemical, biological and environmental processes that determine whether archaeological residues will be identified in one or any sensor. Unfortunately, knowledge of which techniques will detect which components of the archaeological domain and under what conditions is poorly understood.

The majority of the techniques used in archaeological detection rely on visual interpretation, although a number of algorithms can be used to enhance contrast. Spectral signatures have been used to accurately identify different vegetation and geology types with multispectral scanners. Unfortunately archaeological sites do not exhibit spectral signatures that can be used for generic detection purposes (however, see Altaweel (2005) for use of spectral signatures in a constrained environment). Archaeological sites and features are created by localised formation and deformation processes (Schiffer 1996). For example, as a mud-brick built farmstead erodes, the silt, sand, clay, large clasts and organics in the mud-brick along with other anthropogenic debris are incorporated into the soil. In some environments this has been demonstrated to produce localised variations in soil particle size and structure (Beck 2007; Wilkinson *et al.* 2006). This impacts on drainage and localised crop stress and vigour, which in turn changes reflectance characteristics. Thus, archaeological contrasts can be expressed through, for example, variations in chemistry, magnetic field, resistance, topography, thermal variations or spectral reflectance (Bewley *et al.* 2005; Gaffney & Gater 2003; Scollar *et al.* 1990; Scudder *et al.* 1996).

Hence, it is hypothesised that archaeological residues produce localised contrasts in the landscape matrix which can be detected using an appropriate sensor under appropriate conditions. However, little is known about how different archaeological residues contrast with their local environment, how these contrasts are expressed in the electromagnetic spectrum, or how environmental, and other localised factors such as soil or vegetation, impact on contrast magnitude (over space and time). This requires an understanding of both the nature of the residues and the landscape matrix within which they exist.

Environmental and ambient conditions

Local conditions are a significant factor for site and feature detection. There is no point in trying to detect cropmarks (proxy residue type) using a hyper-spectral sensor dedicated to identifying crop stress (appropriate sensor) when there is no crop on the ground (unfavourable environmental conditions). A whole range of natural and anthropogenic factors play a part in how much contrast an archaeological feature may exhibit (for example soil type, crop type, diurnal temperature variations and soil moisture content).

Natural factors can be exacerbated by anthropogenic factors, particularly local land management practices. For example, irrigation may make crop features appear earlier; conversely any increased soil water content may reduce the visibility of soil marks. Agricultural intervention, in the form of harrowing, has been seen to cause a significant temporary disruption to magnetometry surveys (Maria Beck pers. comm.).

The challenge in this context is to have an appropriate understanding of the localised natural and anthropogenic factors that may impact upon contrast detection for varying residue types with different sensors (having different resolving characteristics). This information is crucial for programming flights or purchasing bespoke satellite imagery. Since multi and hyperspectral data is relatively expensive to acquire and processes it is critical that if bespoke data is acquired it is collected under the appropriate conditions to maximise the archaeological return. With a very few exceptions, in the UK all aerial archaeological surveys over the past century have been observer-directed – essentially relying on an airborne observer to see things. This form of prospection will not work for residues that express contrast outside the visual wavelengths. The decision to fly and systematically survey must instead be based upon quantitative models about archaeological contrast formation. The issue here is to devise a framework within which the relationship between an archaeological residue and the spatial, environmental and seasonal contrast dynamics of its immediate matrix can be modelled. This framework will allow the identification of appropriate sensors to detect contrast and timeframes when this potential contrast is more likely to be detected. In addition, this knowledge will also allow the user to determine which aspects of the archaeological domain are unlikely to be detected during a survey.

Archaeological potentials of utilising multi/hyperspectral data

Archaeological structures such as buildings, walls and ditches can be seen on conventional aerial photographs at an appropriate scale and viewing angle. Crop and soil marks are more difficult to detect with certainty. The visibility of crop marks often depends on vegetation type, soil conditions, sun-sensor geometry and the sensor characteristics and so it is extremely difficult to obtain photographs under optimal conditions. Soil marks develop due to subtle differences in localised soil properties and their expression is dictated by environmental conditions, particularly soil moisture. Multispectral sensors address some of these problems because they are able to 'look' simultaneously at a wide range of different wavelengths. Wavelengths in the near and short-wave infrared add important collateral information to the visual wavelengths and improve the ability to discriminate vegetation stress, soil, moisture and temperature variations than either the human eye or photographic film. Narrow band spectral imaging can often help to enhance or distinguish different features on the ground or provide information on their state of health or ambient conditions according to their particular absorption and reflectance properties or their spectral signature.

This increased sensitivity is crucial for contrast detection. For example, cropmarks are an instance of localised variations in vegetation stress or vigour correlated with subsurface archaeological features. Wavelengths outside the visible are also sensitive to changes in vegetation health. Theoretically, exploiting relevant areas of the electromagnetic spectrum at the appropriate degree of granularity will mean that crop stress or vigour relating to subsurface archaeological residues can be expressed more clearly and also that it can be detected both earlier and later in the growing cycle. Therefore, the window of opportunity for detecting archaeological features can be dramatically extended by using wavelengths outside the visible. It should also be noted that 'new' potential archaeological residues will be expressed in non-visible wavelengths (i.e. residue contrasts apparent outside the visible may never become apparent in the visible). This increased sensitivity means that archaeological contrasts can potentially be detected in soils and crops that have been traditionally categorised as marginal or unresponsive to aerial archaeological prospection. Hence, contrast signatures that are undetectable in the visible wavelengths can be detected.

Of particular importance is the thermal infrared which has traditionally been exploited through proxy indicators such as thaw marks. Thermal applications exploit the different emissivity characteristics of materials. The sun has a specific diurnal impact in the thermal wavelengths as described in Figure 7.6. Thermal prospection techniques have many important applications in geology, archaeology and environmental monitoring. Tabbagh and Hesse pioneered the evaluation of airborne thermography for archaeology and conducted a number of trial flights over different soil types in France (Scollar *et al.* 1990). Both pre-dawn and daytime thermal imagery proved valuable in detecting archaeology from surface soil patterns or from the thermal effect of a buried structure. Although this instrument provided promising results for several soil types, the relatively uniform temperature of the ground surface hides much of the information content in thermal imagery. This is a particular problem for vegetated surfaces where the plants regulate their own temperature through evapotranspiration which acts to create uniform canopy temperatures. However, archaeology is sometimes visible through vegetation at thermal wavelengths

Figure 7.6: Diurnal temperature variation of soil and water. *Re-used under a Creative Commons Attribution-NonCommercial-ShareAlike 3.0 Unported License credited to Anthony Beck.*

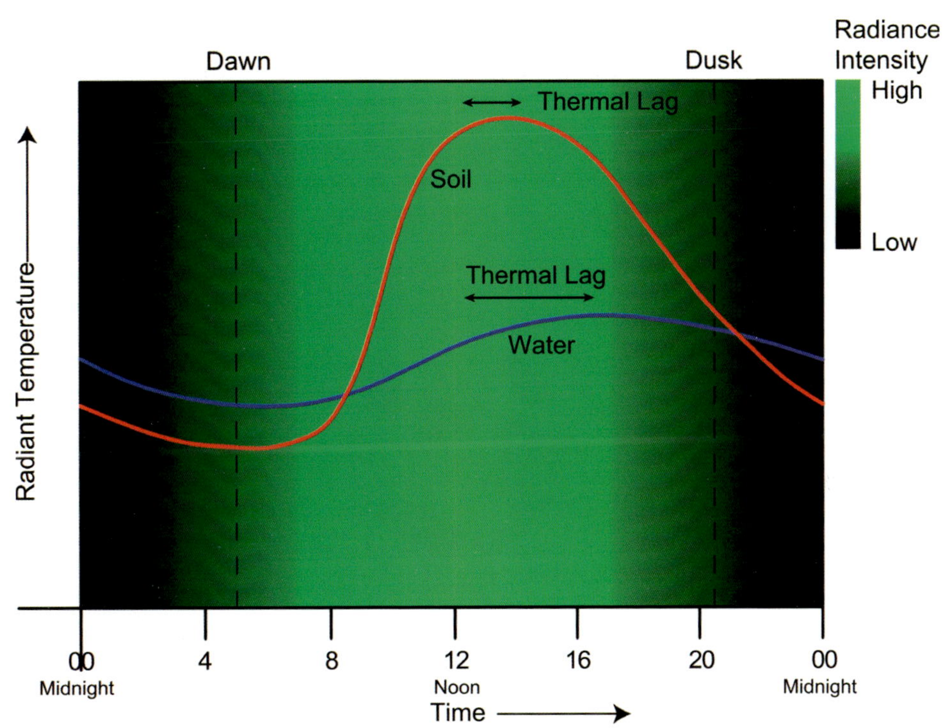

and invisible at other wavelengths (Bellerby *et al.* 1990). Apparent thermal inertia, differences between day and night-time radiant temperature, offers considerable potential for the detection of buried archaeology (Donoghue *et al.* 2006).

Detecting archaeological contrast

Another major advantage of multispectral imaging is that the data is produced in a digital form that can be modified using computer-based image processing techniques. The effect of image enhancement is to allow the user to experiment with different ways of adjusting the contrast, in an interactive way and in different parts of the image, to assist interpretation. The data in the archaeological domain can be visually identified directly within the raw data or by employing relatively simple histogram manipulations. However, although the feature has been detected it may be masked within the structure of the image. In such instances digital processing techniques are required in order to express the sometimes-subtle differences or contrast. This is particularly true for data with a medium to high radiometric resolution.

In most cases only a small component of the image domain from a sensor has archaeological significance (see Figure 7.7). Different sensing devices with varying resolving characteristics (spatial, spectral etc.) capture different elements of the archaeological domain. Therefore, to gain a greater understanding different detection techniques can be used. This is a multi-sensor approach where the different axes of resolutions from different sensing devices can be efficiently exploited. The utility of multi-sensor approaches cannot be stressed enough because combined sensor responses offer much finer granularity than the sum of their independent responses.

For their work in Syria, Beck *et al.* (2007) recommended using a range of different sensors aimed at identifying different landscape components each of which had a heritage dimension:

- Coarse spatial (>100m) and high spectral resolution imagery for broad-brush landscape theme identification (particularly soils and geology).
- Medium spatial (10–60m) and medium spectral (>6 bands) resolution imagery for refined landscape identification, e.g. Landsat ETM+ or equivalent.
- Fine to medium spatial (2–15m) and low to medium spectral (>3 bands) resolution to detect larger features (ploughed out sites, tells etc.), e.g. Quickbird MS, Ikonos MS or SPOT 5.
- Fine spatial (<1m) and low spectral (pan) resolution imagery to detect very small features (walls, cairns, linear soil marks etc.), e.g. Quickbird pan, Ikonos pan and/or Corona.

Multi-sensor approaches are particularly pertinent to those countries that have poorly developed national archaeological inventories and intend to use remote sensing techniques for rapid survey. In such situations a thorough understanding of the natural and anthropogenic factors that impact upon feature contrast is required for the most comprehensive survey.

The techniques presented in this paper will not allow archaeologists to automatically identify all the features of archaeological significance located within the structure of a digital image; rather it is argued that appropriate processing methodologies can only be applied when one has a thorough understanding of the nature of the archaeological residues, their relationships with the immediate matrix, the characteristics of the 'observing' sensor and the environmental conditions at the time of image capture. When the nature of the archaeological residue, the impact of natural

Figure 7.7: Conceptualised archaeological domain. *Re-used under a Creative Commons Attribution-NonCommercial-ShareAlike 3.0 Unported License credited to Anthony Beck.*

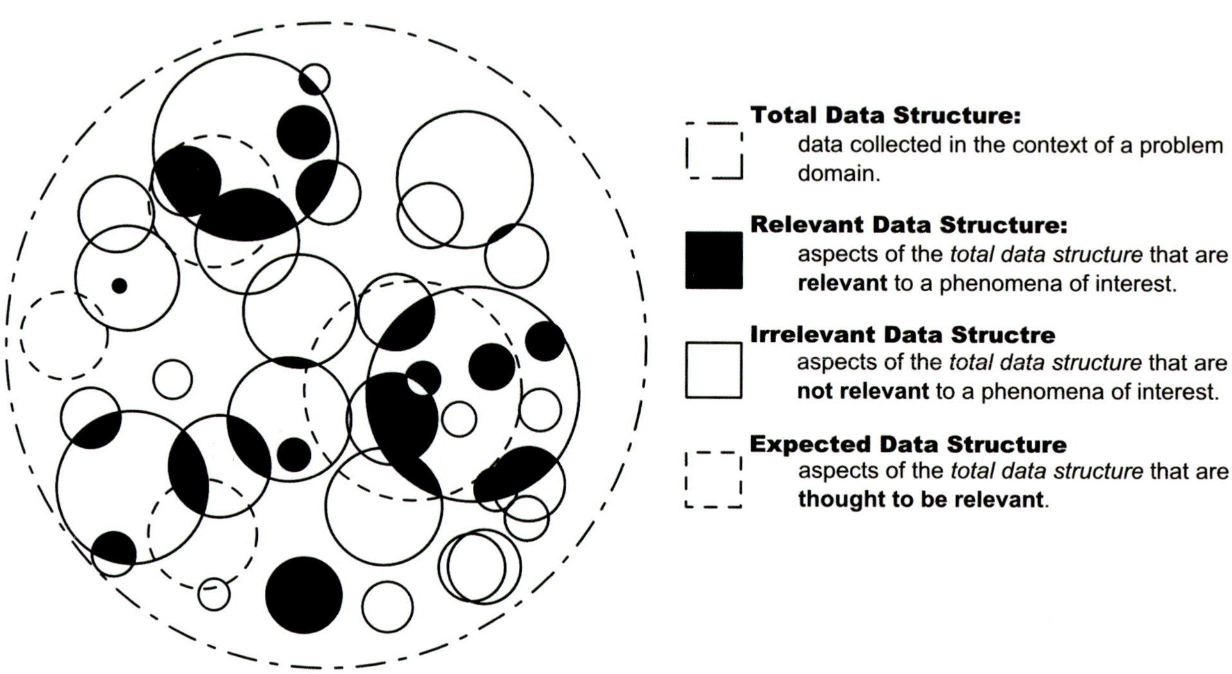

and anthropogenic factors over time and the sensor characteristics are understood, one can model how the archaeological feature will express contrast against any background value. This information can be exploited to develop contrast enhancement algorithms to improve recognition and identification.

Thematic modelling

The use of remote sensing for thematic modelling is dependent upon the type of 'theme' desired and the scale of interpretation. For example, in the absence of geological maps multispectral imagery can be used to identify different surficial soils and geologies. Landsat TM imagery is regularly used for this type of identification. Geological themes can be extracted from Landsat data which has a large ground footprint, a relatively high spatial resolution for the application (a 30m cell size is much smaller than any geological unit) and an appropriate spectral resolution. Ikonos and Quickbird imagery can supply information related to modern topography and can be used akin to aerial photography to update digital mapping. Corona historic photographs can supply information on broadly the same scale as modern Quickbird and Ikonos images but relate to a relict landscape and hence information can be elucidated regarding landscape change.

The DART Project

As discussed previously, there is variable understanding of the physical, chemical, biological and environmental factors which produce the archaeological contrasts that can be detected by sensing technologies. These factors vary geographically, seasonally and throughout the day, meaning that the ability to detect an archaeological residue with a specific sensor changes over time and space. Thus, current detection strategies are not fulfilling their potential, leading to sub-optimal heritage management. The Detection of Archaeological Residues using remote sensing Techniques (DART) project (www.comp.leeds.ac.uk/dart) will focus on analysing factors that influence archaeological residue contrast dynamics. DART aims to determine how different remote sensing technologies detect contrast caused by different underlying factors under dynamic environmental conditions. This understanding will allow the optimal deployment of the different sensors. By combining the results from a battery of sensors, each optimally deployed when the archaeological residues have the greatest likelihood of being detected, the maximal knowledge of archaeological residues can be achieved.

DART will address the following research issues:

- What are the factors that produce archaeological contrasts?
- How do these contrast processes vary over space and time?
- What causes these variations?
- How can we best detect these contrasts (sensors and conditions)?

The key will be to understand how archaeological residues differ from, and dynamically interact with, the localised soils/sediments and vegetation/crop and how these differences can be detected. Archaeological residue interaction models will be developed and tested under a range of different environmental, seasonal and crop conditions using the following procedure:

- Development of archaeological residue interaction models of soils/sediments and archaeological residues from multi-temporal sensor measurements and samples.
- Using models to understand the physical causation of archaeological residue contrast and its dynamics.
- Using fusion (data integration) techniques and models to map the different physical variations into measurable spectral, magnetic and electrical variations.
- Developing an interpretative and knowledge-based management framework including decision tools to:
- Assist curators in determining the condition of buried archaeological residues and sensor configurations appropriate for their detection.
- Enhance the discovery of archaeological residues from appropriate archival imagery.

In-situ measurements will be taken using probes and sensors, and samples will be taken for laboratory analysis. Standard geotechnical tests will be conducted such as density, grain size distribution, organic content, magnetic susceptibility, dielectric permittivity, geochemistry, pH and conductivity. Permanent in-situ probes will measure temperature gradient, density and soil moisture variations through a soil profile. In addition, each site will be visited on a monthly basis for measuring earth resistance, soil colour, conductivity, dielectric permittivity, hand-held spectro-radiometry, GPR transects and ambient climatic data. Regular aerial hyperspectral surveys will be commissioned; the aim is for a monthly flight.

The laboratory-based experimentation will establish links between routinely collected geotechnical data (from construction projects), for example, Atterberg limits, moisture content and particle size distribution, and geophysical properties of the soil. It is hypothesised that knowledge of the geotechnical soil characteristics will allow the prediction of the geophysical properties of the soil and hence guide the selection of the most appropriate survey technique. The laboratory tests will focus on determining electromagnetic signal attenuation/penetration for soils in different geotechnical states. Sub-samples with different compaction and moisture states will be measured with multi-frequency Time Domain Reflectometry (TDR) probes and vector analysers to gain an understanding of signal attenuation through the soil. In addition, spectroradiometry readings will be taken to understand the changes in surface reflectance due to changes in compaction (density) and soil moisture. If the link between geotechnical and geophysical properties can be made, then the British Geological Surveys' nationwide geotechnical database will provide information that will be used to calibrate geophysical sensors to local conditions.

Multi-temporal models will be developed that translate the geotechnical parameters into spectral, magnetic and electrical measures in order to determine contrast parameters. Environmental dynamics will be identified and their impact characterised. Data fusion techniques will be utilised to determine the factors that lead to contrast detection, the impact these factors will have on the sensor spectrum and the nature of any contrast dynamics. This knowledge base will underpin the decision support tools. This will provide a better understanding of soil, vegetation and environmental dynamics and their impact on detection within the sensor spectrum. These models will be benchmarked against the measurements from the in-situ probes and the laboratory data (empirical and simulated). By generating mappings between changing conditions and archaeological contrast this will provide a useful way of determining the suitability of different geophysical and remote sensing techniques.

Environmental and soil conditions affect the various archaeological prospection techniques differently. The decision support tools will formally integrate the knowledge acquired during the research into a reasoning system. Two approaches are envisaged for the knowledge management of the large quantity of information available. The first tool ('static') will utilise the knowledge base: general soil information for the sites; historical environmental and vegetation records; and metadata from aerial image archives in order to reduce the search space within those archives for the identification of archaeological residues. It will be tested against archive material with known archaeological residues. The second tool ('live') will utilize the knowledge base: live soil measurements; geophysical and remote sensing surveys; and satellite-derived environmental and vegetation data for the planning of prospection strategies. This latter tool will predict, based on environmental estimates, what archaeological residue types can be detected, with which techniques or sensors and at what times. It will be tested by deriving a programme of bespoke hyperspectral flights and geophysical surveys in an unstudied area and comparing results with known archaeological residues.

DART is not in isolation: a multi-disciplinary team of European scientists have recently founded the Ludwig Boltzmann Institute (LBI) for Archaeological Prospection and Virtual Archaeology (http://archpro.lbg.ac.at), which is based in Vienna. Researchers will develop advanced non-destructive prospection methods and apply them within a theoretical and methodological landscape archaeology framework. Techniques will include remote sensing, high resolution near surface geophysics, multi-sensor approaches, sophisticated computer science, and geomatics within a non-destructive conceptual framework for spatial archaeology. The LBI is based in Vienna, but integrates a Europe-wide partner consortium, representing academic and research institutions, archaeological service providers, and governmental authorities from Austria, Germany, Great Britain, Norway, and Sweden.

DART complements the objectives of the LBI, particularly on the points of methodological development. DART is an open-science project, which means that project data will be placed in the public domain as soon as is practicable. This means that colleagues at LBI will be able to benefit from the developments generated by DART in a timely manner, resulting in greater research impact and synergy.

Conclusion

Remote sensing can provide an impressive picture of the archaeological landscape without the need for invasive or expensive survey methods. The true potential of multispectral remote sensing, including thermal imaging, is still not clear and needs to be evaluated to test responsiveness under a broad range of climatic and ground conditions. Further research is likely to produce sensors capable of resolving relatively small features such as post-holes and shallow pits. When used appropriately, remote sensing provides a basis for testing hypotheses of landscape evolution that may be further explored by ground survey, geophysical survey or excavation. Large-scale airborne and satellite surveys can provide the framework on which planning policy and excavation strategies can be established. In addition, computer enhancement and the increased spectral resolution of the digital data places less dependency on the time of year for revealing archaeological features.

This paper has argued that archaeological remote sensing requires re-invigorating. However, there is an inherent paradox, particularly for aerial survey: technical developments have led to a range of new sensors that offer immense archaeological potential, but the scientific understanding of the factors that allow archaeological residues to be detected by these new sensors is poor. Archaeological residues represent modifications of a pre-existing landscape and are therefore strongly influenced by the local matrix. If the idea of a spectral signature can be applied at all, it will only work within a consistent background environment and for a specific form of archaeological residue (see for example Altaweel 2005). Hence, as archaeological contrast alters as environmental conditions change, there is inadequate understanding about the appropriate conditions under which the new sensors should be used, or what calibration techniques could significantly enhance the likelihood of detection. Increased understanding of the physical, chemical and biological factors that lead to archaeological contrast, and importantly how this contrast changes over time will lead to knowledge-led survey, using the most sensitive and appropriate sensors. This has the potential to dramatically improve the response in both marginal and well-understood landscapes, transforming both our understanding and our approaches to management and curation.

Remote sensing is increasingly important to many areas of archaeological enquiry from prospection through to management. It is therefore essential that it is not applied inappropriately. The inappropriate application of a single sensor could produce minimal results or the dogmatic application of that sensor will have diminishing archaeological returns. The combination of different sensors with different characteristics can produce profound interpretative synergies. Multiple sensors should be evaluated on the basis of 'fitness for purpose'. Fitness for purpose in this context refers to the cost/benefit returns of each sensor and should be based upon an understanding of the nature of the archaeological residues, the sensor characteristics and the environmental characteristics of the landscape.

Acknowledgements

The DART project (www.comp.leeds.ac.uk/dart) is supported by a Science and Heritage large research grant. The Science and Heritage programme (www.heritagescience.ac.uk) programme is funded by the Arts and Humanities Research Council (AHRC: www.ahrc.ac.uk) and the Engineering and Physical Sciences Research Council (EPSRC: www.epsrc.ac.uk). Many thanks to Professor Michael Doneus and Dr Geert Verhoeven for commenting on this text and providing information about the Ludwig Boltzmann Institute for Archaeological Prospection and Virtual Archaeology. Unless stated otherwise all images are re-used under a Creative Commons Attribution-NonCommercial-ShareAlike 3.0 Unported License credited to Anthony Beck.

References

Altaweel, M. 2005: The Use of ASTER Satellite Imagery in Archaeological Contexts. *Archaeological Prospection* 12(3), 151–66. DOI: 10.1002/arp.254.

Banning, E. 2002: *Archaeological Survey*. New York, Kluwer Academic.

Beck, A.R. in press: The role of satellite imagery in Desk Based Assessments: a case study from Homs, Syria. In Lasaponara, R. & Masini, N. (eds): *Satellite Remote Sensing: A New Tool for Archaeology*. Springer.

Beck, A.R. 2004: *The evaluation of Corona and Ikonos satellite imagery for archaeological applications in a semi-arid environment*. Unpublished PhD thesis, Durham.

Beck, A.R. 2007: Archaeological site detection: the importance of contrast. Newcastle. In *Proceedings of the 2007Annual Conference of the Remote Sensing and Photogrammetry Society*, Newcastle University, Sept. 11–14, 2007.

Beck, A.R., Philip, G., Abdulkarim, M. & Donoghue, D. 2007: Evaluation of Corona and Ikonos high resolution satellite imagery for archaeological prospection in western Syria. *Antiquity* 81, 161–75.

Beck, A.R., Shaw, J. & Stott, D. 2007: Best practice approaches for applying satellite imagery for landscape archaeological applications: a case study from the world heritage site of Sanchi, India. In Manfred, E. & Ulrich, M. (eds): *Remote Sensing for Environmental Monitoring, GIS Applications, and Geology VII*. SPIE Proceedings Vol. 6749. DOI: 10.1117/12.737553.

Beck, A.R., Wilkinson, K. & Philip, G. 2007: Some techniques for improving the detection of archaeological features from satellite imagery. In Manfred, E. & Ulrich, M. (eds): *Remote Sensing for Environmental Monitoring, GIS Applications, and Geology VII*. SPIE Proceedings Vol. 6749. DOI: 10.1117/12.736704.

Bellerby, T.J., Noel, M. & Branigan, K. 1990: A thermal method for archaeological prospection: preliminary investigations. *Archaeometry* 32, 191–203.

Bewley, R. 2000: Aerial Survey: Learning from a Hundred Years of Experience? In Bewley, R.H. & Rączkowski, W. (eds): *Aerial Archaeology. Developing Future Practice. NATO Science Series vol. 337*. Amsterdam, IOS Press.

Bewley, R.H., Crutchley, S.P. & Shell, C. 2005: New light on an ancient landscape: LiDAR survey in the Stonehenge World Heritage Site. *Antiquity* 79, 636–47.

Christophe, E. & Pearlman, W. 2008: Three-Dimensional SPIHT Coding of Volume Images with Random Access and Resolution Scalability. *EURASIP Journal on Image and Video Processing*, 2008, 1–14. DOI: 10.1155/2008/248905.

Crawford, O.G.S 1923: Air survey and archaeology. *Geographical Journal*, 61(5), 342–66.

Crawford, O.G.S 1928: *Air survey and archaeology*. Professional Papers, New Series 7. Southampton, Ordnance Survey.

Crawford, O.G.S. 1929: *Air photography for archaeologists*. Professional Papers, New Series 12. Southampton, Ordnance Survey.

Donoghue, D.N. 2001: Multispectral Remote Sensing for archaeology. *Consiglia nazionale Delle Ricerche Universita Degli Studi Di Siena*. 2001,181–92.

Donoghue, D.N., Beck, A. R., Galiatsatos, N., McManus, K. & Philip, G. 2006: The use of remote sensing data for visualising and interpreting archaeological landscapes. In Baltsavias, E., Gruen, A., Van Gool, L. & Pateraki, M. (eds): *Recording, modeling and visualization of cultural landscape*. London, Taylor and Francis.

Gaffney, C. & Gater, J. 2003: *Revealing the Buried Past: Geophysics for Archaeologists*. Stroud, Tempus.

Mills, J. & Palmer, R. 2007: *Populating Clay Landscapes*. Stroud, Tempus.

Schiffer, M.B. 1996: *Formation processes of the archaeological record*. Salt Lake City: University of New Mexico Press.

Scollar, I., Tabbagh, A., Hesse, A. & Herzog, I. 1990: *Archaeological prospecting and remote sensing. Topics in remote sensing*, 2. Cambridge: Cambridge University Press.

Scudder, S.J., Foss, J.E. & Collins, M.E. 1996: *Soil Science and Archaeology*. Advances in agronomy, 57.

Wilkinson, K.N., Beck, A.R. & Philip, G. 2006: Satellite imagery as a resource in the prospection for archaeological sites in central Syria. *Geoarchaeology*, 21(7).

8 | Making the most of airborne remote sensing techniques for archaeological survey and interpretation

Rebecca Bennett, Kate Welham, Ross A Hill **and** ***Andrew Ford***

Abstract: The use of airborne remote sensing has found increasing popularity in the historic environment sector over the past decade. Many landscape projects across Europe are incorporating the application of archive airborne survey and increasing numbers are commissioning bespoke survey. This is particularly true for Airborne Laser Scanning (ALS), but despite a number of promising applications, digital spectral surveys have been less frequently utilised. Our understanding of the full potential of these rich data sources is still in its infancy. This paper provides a summary of current applications and highlights the need for research in three key areas to improve our applications of airborne remote sensing for research into the historic environment. It concludes by introducing a project established at Bournemouth University to develop a multisensor approach to airborne survey of liminal environments.

Introduction

In the past decade the heritage sector across Europe has seen growth in the application of airborne remote sensing techniques, such as Airborne Laser Scanning (ALS) and digital spectral sensors. High profile projects such as the landscape survey of Loughcrew, Ireland (Shell & Roughley 2004) and the National Mapping Programme of the Stonehenge World Heritage Site (Bewley *et al.* 2005) have brought the application of ALS, based on the principle of Light Detection And Ranging (LiDAR), to national attention, while work in the Vale of Pickering (Powlesland *et al.* 2006) and at Aqueila, Italy (Traviglia 2006; Sterazi *et al.* 2008) have shown the potential of digital spectral imagery (referred to as either multi or hyperspectral imagery) for identifying archaeological features. It is clear that these sensors have great potential for improving our understanding of the quantity of features of archaeological interest in a landscape but their application poses a number of practical and theoretical challenges. Although often compared favourably with traditional aerial photography survey (see Beck this volume), each sensor by itself captures only a portion of what can be recognised in the shadow, soil, crop marks and earthworks as being of archaeological significance. As no single airborne sensor records all of the indicators we understand to represent archaeological remains, the power of their application has to be in their complementarity.

While the applications of multiple airborne sensors to record data for the same study area are becoming increasingly common, our knowledge of how to combine the vast information content efficiently to form an understanding of the features represented in it can still be regarded as in its infancy. This short paper will briefly review work undertaken to date and look at some of the themes to be addressed in the application of multiple airborne sensors for heritage management, concluding with details of a project at Bournemouth University that is pioneering multisensor methods for the survey of liminal landscapes in the UK.

Laying the foundations

In a landscape research context, the appeal of remote sensing techniques such as ALS and digital spectral imaging, lies in great part with their ability to complement the traditional ground-based and airborne techniques of walkover survey and low altitude aerial photography for large survey areas. Walkover survey is defined as the technique of surveying in transects to record archaeological features and can be undertaken with or without concurrent artefact collection (as per Fulford *et al.* 2006; RCHAMW 2009). This type of survey is time consuming, may be restricted by vegetation or landuse and is limited to identifying features with noticeable upstanding remains or artefact scatters caused by plough damage (Fulford *et al.* 2006). In contrast, aerial photography can identify features that have no upstanding traces but are typified by near-surface changes in soil moisture content and vegetation composition. However the identification of these features relies heavily on the differences in contrast between the matrix of an archaeological feature and that of its surroundings and/or the impact of this contrast on the structure and growth of vegetation. Additionally, variance of this kind is generally only visible under certain conditions, i.e. in recently ploughed soil or during long dry spells, limiting the time of acquisition and making aerial photographs far less useful for survey of certain environments including pasture and uncultivated land. It is well documented that aerial photographs only capture records of such features in specific circumstances and the visibility of crop and soil marks from the air is heavily affected by underlying soil type and geology, vegetation, agriculture and seasonal variance (Wilson 2000) and the bias that the interests of the airborne observer brings to the survey process (e.g. Palmer 2005). In addition, it has long been recognised that the infrared region of the spectrum is particularly sensitive to plant health, more so than the red, green and blue reflectance of the visible spectrum (Lillesand *et al.* 2008). The majority of aerial photographs capture only visible wavelengths and the implications of this small spectral window on both the techniques and the results of aerial survey are discussed in detail by Beck (this volume). While digital spectral sensors do not yet boast the high spatial resolution of aerial photography, they capture a wide range of wavelengths and are more sensitive to subtle changes in the spectral response of soil and vegetation.

In addition to variations in vegetation and soil, oblique aerial photography, especially in raking winter light (defined by Wilson (2000) as no more than 20° above the horizon) can be used to identify features with upstanding topography. Often this technique is more effective than observations from ground level, especially if the features are slight or the site covers an extensive area. However due to the bias in visibility introduced by differing illumination angles, mapping complex earthworks requires repeat acquisition in different light conditions and more often than not such repetition is impractical. The high accuracy, high resolution digital terrain and surface models (DTM and DSM respectively) that can be rendered from ALS surveys enable the identification of topographic features and can be artificially shaded from any angle to replicate optimum conditions (Devereux *et al.* 2008).

An increasing body of research has shown that ALS and digital spectral imaging have the potential to redress some of the weaknesses of the traditional techniques detailed above. Their complementarity to existing techniques has been amply illustrated both in the UK and elsewhere in Europe (Bewley *et al.* 2005; Powlesland *et al.* 2006; Campana *et al.* 2009; Gallo *et al.* 2009). Yet the potential of these sensors for improving our understanding of archaeological features extends beyond quantitative identification. In the detailed data that they collect about soil, vegetation and topography is a wealth of information that has the potential to enhance our management of archaeological landscapes, to monitor levels of degradation, to better understand the impact of landuse and vegetation cover on both the visibility and preservation of features and to aid the appropriate targeting of future research resources.

Airborne sensors in archaeological survey

Although the advantages of surveying vegetation, soil and topography from the air for archaeological prospection are well established, very few projects have had the opportunity to acquire contemporary ALS and spectral data for research in this field. This is in part due to the high cost of both deploying these sensors and processing the data collected, but is also a consequence of the restrictions of flying the sensors in tandem. Until recent improvements in the sensors themselves, an uncomfortable compromise had to be reached between the optimal operating parameters for the spectral and ALS systems, meaning that the best resolutions for either sensor were not attainable (Shell 2005). In addition the processing power required to handle both spectral and topographic data together exceeded the capacity available to many historic environment professionals. Where data from either of these sensors has been used for landscape research it has been predominantly subject to individual visual assessment rather than combined digital processing. Consequently it is important to briefly consider the independent issues surrounding the use and interpretation of spectral and ALS data for archaeological purposes.

Despite first being used to detect archaeological features over 20 years ago (Donoghue & Shennan 1988) and showing some promise for the UK and on the continent (Donoghue & Shennan 1988; Winterbottom & Dawson 2005; Powlesland *et al.* 2006; Traviglia 2006) uptake of digital spectral data for historic environment research has been limited. There are several reasons for this including the cost, spatial resolution and relative scarcity of equipment. In this volume, Beck identifies the issues associated with the application of spectral sensors in more detail, but to date there has been little understanding of the physical, biological and environmental variables that influence the visibility of archaeological features in airborne digital spectral data. Unlike many vegetation mapping and geological applications, archaeological features do not exhibit a spectral signature which can be consistently identified and much work is required to understand fully how and why features are represented in spectral imagery, and equally importantly, why known features may not be represented.

Many of the problems in resolving the physical details of features seen in spectral imagery have been magnified by the use of archive spectral data collected primarily for environmental and hydrological purposes, rather than archaeological survey (Challis *et al.* 2008). As a result, often no simultaneous ground observations or survey exists for the flights. As the spectral response of vegetation cover and soil can change on a daily basis, this lack of control data makes it extremely difficult to link the airborne data closely to either the observed ground conditions or the results of topographical and geophysical survey undertaken at other times. This lack of observational data can and should be partially mitigated by post-survey field visits and comparison to the known archaeological record. This is essential to verify the archaeological interpretations and feedback iteratively into the process of identifying features from the airborne imagery (see Winterbottom & Dawson 2005). However this can only correct interpretation at a broad visual level, rather than by looking at the physical, biological and chemical properties of the features at the time of survey. While this level of correction might be sufficient depending on the aims of the research being undertaken, there is little doubt that our understanding of the representation of archaeological features in digital spectral imagery could be improved immensely by qualitative links to ground observations, particularly geophysical survey. Fortunately there is increasing recognition of the need to compare airborne data to ground based spectral, soil and geophysical measurements, culminating in the first Natural Environment Research Council (NERC) supported simultaneous ground-based and airborne data acquisition for archaeological research in the UK in March 2010 in support of the current research project at Bournemouth University and the AHRC / ESPRC funded DART project (see Beck this volume.)

In addition to the complexities of understanding airborne spectral data, there is also the issue of the shear quantity of data derived from a single flight. Hyperspectral sensors such as the AISA Eagle are typically capable of recording spectral data in 244 bands across the 400–970nm range, but due to the narrow (2.3nm) bandwidth, much of the data in each of the bands is identical. The challenge is to find the bands within the data cube that best represent the archaeological features within the landscape. While of course the data themselves will vary between (and potentially within) landscapes, much progress has been made recently in trialling statistical autocorrelation methods such as Principal Components Analysis (PCA), Moran's I and Gi (Ciminale *et al.* 2009) and separability indices (Cavalli *et al.* 2009) to reduce data redundancy. In addition a number of indices, such as the Normalised Differential Vegetation Index (NDVI) and Tassled Cap Transformation that were developed specifically for vegetation mapping, and which work by targeting the spectral regions that reflect key biophysical properties, have been applied to archaeological research with varying degrees of success (Traviglia 2006; Challis *et al.* 2009).

Despite the challenges, it has been repeatedly shown that digital spectral imagery is of use for detecting sites that are not visible in conventional aerial photographs, particularly in areas dominated by arable farming (Powlesland *et al.* 1997; Rowlands & Sarris 2007; Challis *et al.* 2009). It is less clear how the increased spectral sensitivity can aid our interpretation of sites in liminal areas where the vegetation sensitivity to stress is reduced and pasture or scrub vegetation dominates, although Winterbottom & Dawson (2005) reported good results in machair dominated environments.

In contrast to digital spectral sensors, the application of ALS has enjoyed increasing popularity with historic environment professionals as it enables the creation of high resolution, high accuracy spatial models and the removal of vegetation from these models, allowing below-canopy modelling (Crow *et al.* 2007; Doneus *et al.* 2008). This, coupled with the increased availability of archive data from providers such as the Environment Agency in the UK has seen a surge in the number of research and commercial applications of the data. The explanation of the technique of ALS and examples of its use in a heritage management context are discussed elsewhere in this volume (see Doneus & Briese and Shaw & Corns this volume).

To date, studies using ALS for examining archaeological features in the UK have focussed on two main research areas; the potential for last-pulse return to record features beneath forest canopy (Crow *et al.* 2007) and the modelling of alluvial valleys (Challis 2006; Challis *et al.* 2009). While the use of archive data is still prevalent and provides a relatively simple route to the acquisition of terrain models avoiding the computing requirements of in-house data processing, commissioned surveys are becoming more common. Although these are expensive, it has been calculated that the cost of collecting airborne data is less than the equivalent cost of deploying a ground based survey team and is far more effective for some land cover types (Crow *et al.* 2007). Beyond the recently published English Heritage ALS guidelines (Crutchley 2010), there is still little in the way of guidance for those wishing to commission airborne remote sensing survey however and this is due in part to the fact that the use of these data is still an emerging discipline, changed continually by advances in sensor technology and processing techniques. Although the basic principles remain the same, far more technical research is required before we can state confidently what survey parameters are best suited for prospection, detailed feature mapping, monitoring and conservation purposes and how these may vary between and within landscapes. In this volume, Doneus & Briese discuss many of the issues surrounding survey and processing parameters and the availability of metadata with specific reference to forested landscapes. Even the simplest assumptions regarding ALS survey may not hold true in every circumstance. For example, it is thought that the highest possible spatial resolution will achieve the best results in terms of identifying archaeological features. However in recent work over heavily and uniformly cropped fields in Italy the noise produced by higher spatial resolution survey meant that the archaeological features were only visible when the resolution was reduced (S. Campana pers. comm.).

Like digital spectral imagery (e.g. Figure 8.1), there is little doubt as to the efficacy of ALS as a prospection tool, providing impressive increases in the number of known features even in previously well-researched landscapes (Shell & Roughley 2004; Bewley *et al.* 2005; Doneus *et al.* 2008; Gallagher *et al.* 2008; Weishampel *et al.* 2010). Despite this there is still a sense that the full potential of these data has yet to be achieved for historic environment applications. Although much work has been undertaken on the technical aspects of data processing (see Challis 2004; Challis 2005; Doneus this volume), assessments of digital terrain model accuracy rarely feature in published accounts. If the goal of the research is simply to indicate whether a feature is present or not then it might be considered that an assessment of the accuracy of the model is not a vital piece of metadata. However, given the capacity of modern sensors to produce surveys of high spatial resolution, it is feasible that models created from ALS data could be used to derive basic metrics of archaeological features. If you wish to estimate the diameter of an enclosure, the height of a bank or depth of a ditch from the data (which are feasible aims within the tolerances of the sensor), the accuracy of your model, whether it significantly smoothes or overshoots the original data points and how representative your surface model following vegetation removal is of the actual ground surface becomes more critical. To date projects have mostly used protocol designed for mapping features from aerial photographs, and while this provides an excellent and comparable basis for recording, given the increasing use of ALS data and the additional topographic data they can provide, we should perhaps look to review this.

Recently there has also been a move away from the traditional hillshaded views (Figure 8.2) of ALS data that primarily replicate the 'ideal' image that aerial photography might provide in low light levels. These views, though aesthetically pleasing are difficult to interpret and illumination from different angles (which is necessary if the viewer is to see features aligned with a single direction of illumination) provides a multitude of images with a great level of data redundancy. Even if this is reduced through the application of PCA (Devereux *et al.* 2008), visual analysis can still be hampered by the topography of the larger landscapes out-shadowing the archaeological features (Figure 8.3). The processing techniques recently developed in Germany (Hesse 2010; Bofinger & Hesse this volume) mark a move away from the creation of pseudo-aerial photograph views to a method of processing that improves the way we derive information from the data.

Two technical areas where research to date has indicated promising results involve the use of the intensity measurements of the returned laser beam. Challis *et al.* (2008) had some success in using the intensity values to enhance archaeological feature detection and also suggested that intensity data may be a potential proxy for, or comparable to, conventional geophysical techniques, namely earth resistance survey and potentially ground penetrating radar in it's reflectance of soil moisture values. This tentative link to geophysical survey techniques requires further research but is potentially an aspect of the ALS data that could be used to assess and monitor archaeological features and would provide improved comparability to traditional survey methods.

As the body of research into the technical and archaeological applications of airborne remotely sensed data grows, so too should our understanding of how to apply the data to specific research questions. The potential for the application of these data to heritage management issues, such as the identification and conservation of ephemeral remains, or research questions, such as how the change in upland management is impacting archaeological features, is limited by the lack of research in these areas to date.

Future directions – multiple remote sensor survey for the historic environment

From the work undertaken in the UK and elsewhere in Europe it is possible to outline a number of potential research directions that would improve the usability of airborne imagery for historic environment purposes. The first of these is improving our understanding of the physical, biological and chemical properties that both represent archaeological features in spectral imagery and affect the ALS intensity returns. The second lies in building on the existing body of research which aims to make the vast quantities of data derived from these types of survey more manageable. This should be in part achieved by a better understanding of the representation of archaeological features within these data but also by the continued improvement in processing techniques.

Figure 8.1: A typical 1.5m resolution digital spectral (CASI) image of Snail Down Barrow Cemetery, Everleigh, Wiltshire, UK.

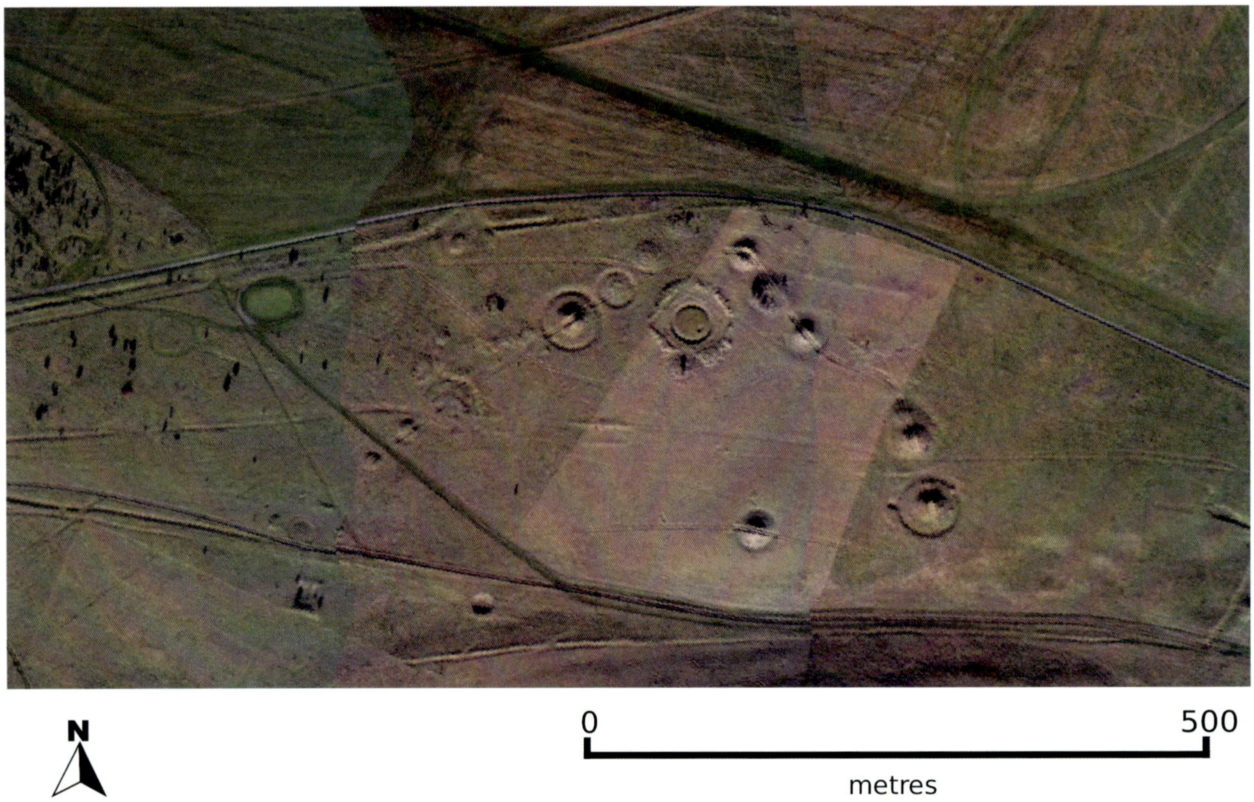

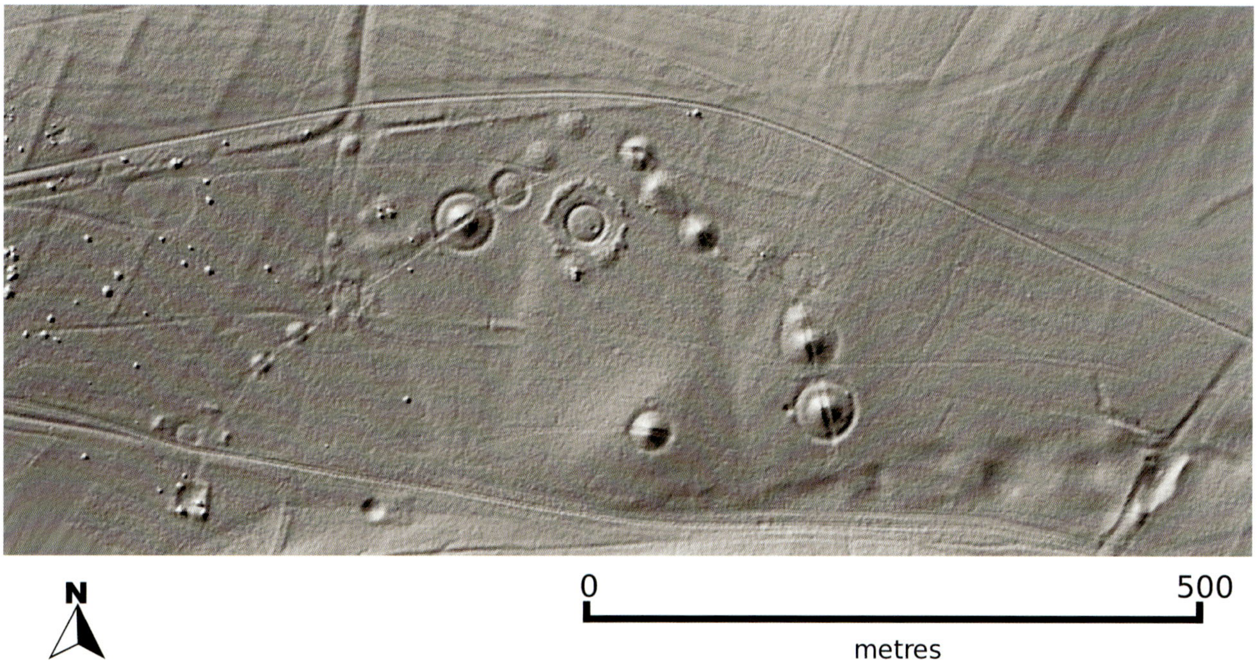

Figure 8.2: A typical 1m resolution hillshaded ALS terrain model of Snail Down Barrow Cemetery, Everleigh, Wiltshire, UK.

From the expanding body of research into airborne sensing techniques, the conclusion is that a combination of spectral and ALS survey has value that surpasses that of each survey alone. Shell notes that 'all imagery can be better employed for understanding landscape context by the use of a digital terrain mode' (2005, 282). While seemingly obvious, it is worth emphasising that using all of the data available about a feature improves our understanding of its form, and therefore our interpretation of it. It would seem that there is a need to tackle the amount of data available in a combined spectral and ALS survey in a manageable and semi-automated process while retaining confidence in the results of the classification and limiting the number of false positives. To date no research of this kind has taken place in the UK, although a similar method combining data from different sensors and exploring the use of filters to extract archaeological features was used in a study on Crete (Rowlands *et al.* 2007).

Finally, there is also a clear requirement to improve our understanding of how both digital spectral and ALS surveys can be applied to liminal areas where vegetation is less prone to stress and the terrain may pose greater challenges to the creation of accurate surface models. These areas are often much more difficult to access on the ground, so any improvement on traditional survey brought about by using airborne

Figure 8.3: Brent Hill Dartmoor, UK showing how macro topography can impede the identification of archaeological features in hillshaded (*left*), visualisations of ALS terrain models when compared with other techniques such as local relief modelling (*right*).

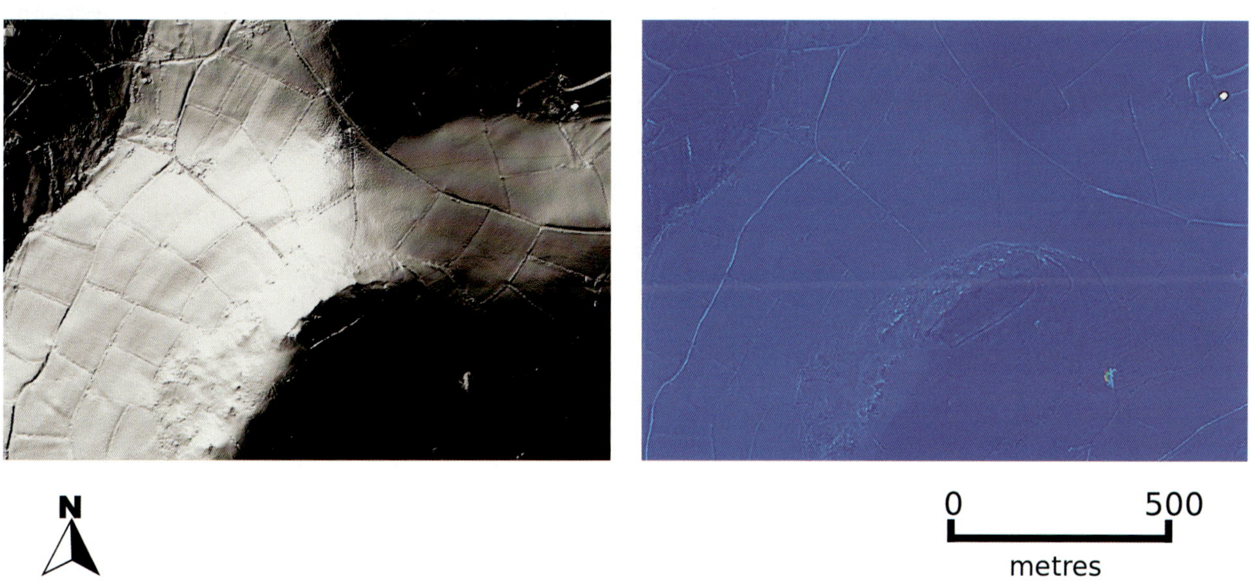

sensors will have a great impact on their management. The project established at Bournemouth University in 2009 seeks to contribute directly to this area of research by closely examining how multisensor techniques can be applied to an area not dominated by arable production.

Beyond the picturesque – analysing the full information content of remotely sensed images

The overall aim of the Bournemouth project is to assess the full information content of airborne remotely sensed data with respect to the detail that can be derived for archaeological features in a number of liminal areas. The initial study is being conducted on the chalk grassland of the Salisbury Plain, Wiltshire, England. Almost 39,000 hectares of the rolling chalk outcrops of the Plain is owned and managed by the Ministry of Defence as the largest military training area in the UK. The archaeological landscapes of this area are remarkable both for their location between the World Heritage status prehistoric landscapes of Stonehenge and Avebury and for the outstanding preservation of their archaeological features. Purchased by the War Office following the agricultural depression of the late 19th century (McOmish *et al.* 2002, 6), the Plain is the last area of chalk grassland in the UK that remains predominantly unaffected by agricultural intensification.

In addition to the preservation of the upstanding archaeological features, the Plain has been selected as a pilot study for the Bournemouth University project due to the quality of previous and on going investigations, which have characterised the nature of the archaeology through aerial, ground based metric and geophysical survey, providing an excellent baseline record. The area is generally well understood with a number of previous archaeological investigations, including full mapping of the aerial photograph archive by the National Mapping Programme (NMP) and the trial application of airborne remote sensing for environmental conservation (Crutchley 2002; Barnes 2003). Barnes' initial work looking at the application of digital spectral and ALS data established that many known archaeological features could be seen in the airborne data that was collected primarily for vegetation condition mapping (2003, 86). The study area also allowed for the use of a number of archive datasets of different sensor types (Table 1).

The project comprises two areas in the north east of the Plain, south of the villages of Everleigh and Upavon. At 4km^2 the Everleigh area forms the pilot study, using archive digital spectral and ALS data supplied by the Environment Agency and military 4-band vertical photography. For the larger 12km^2 Upavon area a successful application to the NERC Airborne Research and Survey Facility allowed the acquisition of hyperspectral data and high resolution LiDAR with parameters tailored for archaeological survey and with contemporary field spectroscopy and geophysical survey.

The principle aims of the pilot study are to investigate the combined use of archive spectral and ALS data to enhance the understanding of archaeological features within an area of mixed landuse including scrub, undisturbed chalk grassland, pasture and a single arable field. Spectral data collected with the CASI (Compact Airborne Spectrographic Imager) sensor were available for both January and May 2001. As so little was known about the relevant information content of the spectral data for this type of environment prior to the start of the project, the spectral data were analysed visually by band and in true and false colour combinations and using a series of common vegetation indices (selected for their biophysical rather than numerical basis) for each flight. The ALS was also visually analysed using a range of techniques including hillshading, PCA and Local Relief Modelling (after Hesse 2010). The project will focus on trialling a number of digital combination techniques for the remotely sensed data. Combination techniques range from simple addition of derived rasters to trialling of techniques such as pan-sharpening. These will be targeted at the portions of the spectral data identified as having the most information content for the archaeological features in the area.

Features were mapped using a GIS and a number of parameters were recorded for each feature identified including morphology, length, description, interpretation and land use. In this way the data could be analysed not just by the presence or absence of features in a given image but by percentage. Ancillary data enabled the results of the mapping exercise to be broken down by land use and feature type to give greater insight into the patterns of visibility. By comparing the information derived in this manner to the baseline recorded by the traditional mapping methods of the NMP, we can have some confidence in

Data Type	Resolution	Date Flown	Source
CASI	1.5m	January 2001	Environment Agency
	1.5m	May 2001	Environment Agency
LiDAR	2m	January 2001	Environment Agency
	1m	2005	Environment Agency
Aerial Photography (Oblique)	0.15m	Archive (c.1950–2002)	Wiltshire Sites and Monuments Record
Aerial Photography (Vertical)	0.15m	Modern, yearly summer coverage (2002–6)	Defence Estates
4-Band Aerial Photography (Vertical)	0.15m	Modern, yearly summer coverage (2006–7)	Defence Estates

Table 8.1: Airborne Digital Data Sources for the Everleigh Pilot Area.

whether they provide additional information to current common techniques.

The bespoke data acquired for the Upavon area by the ARSF in early 2010 will enable assessment of a number of key issues surrounding the potential for direct links between airborne remotely sensed data and ground observations. These will focus on the parameter of soil moisture differences measured both directly by the collection of soil samples and indirectly through earth resistance survey. It will also enable the comparison of the full data cube of hyperspectral data with the banded CASI data available for Everleigh and an assessment of whether improvement in spatial resolution has an impact on the useful information that can be derived from the ALS topography and intensity data.

Conclusions

While the individual value of digital spectral and ALS data for the interpretation of archaeological landscapes has been proven beyond doubt, significant work is required to exploit the full information content of these datasets. Only by developing both a robust understanding of what these surveys can tell us about archaeological features in a given landscape and techniques to streamline the combined data processing, will we actualise the full potential of airborne remote sensing for archaeology.

While no single sensor will ever be a panacea for archaeological feature identification it would seem the value of combining both ALS and digital spectral survey is greater than the value of each separate survey. In terms of identifying both the direct and proxy measures of archaeological features in a landscape the combined surveys are powerful tools, and given the ongoing improvement in sensor systems it would seem likely that this type of dual survey will become increasingly common. The combination of vegetation / soil and topographic measurements may prove particularly valid for areas that are hard to survey by traditional means and there is a real need for specific research into the application of multisensor survey in non-agricultural landscapes.

Acknowledgements

Archive CASI and ALS data were supplied by the Environment Agency and new data acquisition for the project was supported by the NERC Airborne Research and Survey Facility (GB10-07), Field Spectroscopy Facility and Geophysical Survey Facility. The Bournemouth University team would like to thank the Ministry of Defence and Defence Estates for facilitating access and especially Richard Osgood, Senior Historic Environment Advisor. Access to the archaeological archive was facilitated by the staff of the Wiltshire Sites and Monuments Record and we are especially grateful for the expertise and encouragement of Roy Canham. The research is supported by a Bournemouth University Doctoral Research Bursary.

References

Barnes, I. 2003: Aerial remote-sensing techniques used in the management of archaeological monuments on the British Army's Salisbury Plain Training Area, Wiltshire, UK. *Archaeological Prospection* 10(2), 83–90.

Bewley, R.H., Crutchley, S.P., & Shell, C.A. 2005: New light on an ancient landscape: LiDAR survey in the Stonehenge World Heritage Site. *Antiquity* 79(305), 636–47.

Campana, S., Dabas, M., Marasco, L., Piro, S., & Zamuner, D. 2009: Integration of remote sensing, geophysical surveys and archaeological excavation for the study of a medieval mound (Tuscany, Italy). *Archaeological Prospection* 16(3), 167–76.

Cavalli, R.M., Pascucci, S., & Pignatti, S. 2009: Optimal Spectral Domain Selection for Maximizing Archaeological Signatures: Italy Case Studies. *Sensors* 9(3), 1754–67.

Challis, K. 2004: *Trent Valley GeoArchaeology 2002 Component 2b: LiDAR Terrain Modelling*. York Archaeological Trust (unpublished report).

Challis, K. 2005: *Analysis of the Effectiveness of Aerial photography, LiDAR and IFSAR*. HP Vista Centre, Birmingham Archaeology (Unpublished Report).

Challis, K. 2006: Airborne laser altimetry in alluviated landscapes. *Archaeological Prospection* 13(2), 103–27.

Challis, K., Howard, A.J., & Kincey, M. 2008: *Airborne Remote Sensing of Aggregate Landscapes: Final Report*. Birmingham Archaeology.

Challis, K., Kincey, M., & Howard, A.J. 2009: Airborne remote sensing of valley floor geoarchaeology using Daedalus ATM and CASI. *Archaeological Prospection* 16(1), 17–33.

Ciminale, M., Gallo, D., Lasaponara, R., & Masini, N. 2009: A multiscale approach for reconstructing archaeological landscapes: Applications in Northern Apulia (Italy). *Archaeological Prospection* 16(3), 143–53.

Crow, P., Benham, S., Devereux, B.J., & Amable, G.S. 2007: Woodland vegetation and its implications for archaeological survey using LiDAR. *Forestry* 80(3), 241–52.

Crutchley, S. 2002: Understanding Salisbury Plain, England: analysis of the aerial evidence. In Bewley, R. & Rączkowski, W. (eds): *Aerial Archaeology: Developing Future Practice*. NATO Life Sciences, IOS Press. 256–61.

Crutchley, S. 2010: *The Light Fantastic – using airborne LiDAR in archaeological survey*. Swindon: English Heritage.

Devereux, B.J., Amable, G.S., & Crow, P. 2008: Visualisation of LiDAR terrain models for archaeological feature detection. *Antiquity* 82(316), 470–79.

Doneus, M., Briese, C., & Kuehtreiber, T. 2008: Airborne Laserscanning as a tool for landscape archaeology – the example of Wuste close to Mannersdorf

(Leithagebirge, Lower Austria). *Archaologisches Korrespondenzblatt* 38(1), 137–56.

Donoghue, D., & Shennan, I. 1988: The application of remote sensing to environmental archaeology. *Geoarchaeology* 3(4), 275–85.

Fulford, M.G., Powell, A.B., Entwistle, R., & Raymond, F. 2006: *Iron Age and Romano-British Settlements and Landscapes of Salisbury Plain*. Trowbridge: Wessex Archaeology Ltd.

Gallagher, J.M., & Josephs, R.L. 2008: Using LiDAR to Detect Cultural Resources in a Forested Environment: an Example from Isle Royale National Park, Michigan, USA. *Archaeological Prospection* 15(3), 187–206.

Gallo, D., Ciminale, M., Becker, H., & Masini, N. 2009: Remote sensing techniques for reconstructing a vast Neolithic settlement in Southern Italy. *Journal of Archaeological Science* 36(1), 43–50.

Hesse, R. 2010: LiDAR-derived Local Relief Models – a new tool for archaeological prospection. *Archaeological Prospection* Vol 17(2), 67–72. DOI: 10.1002/arp.374.

Lillesand, T.M., Kiefer, R.W., & Chipman, J.W. 2008: *Remote sensing and image interpretation.* 6th ed. Chichester: Wiley.

McOmish, D., Field, D., & Brown, G. 2002: *The Field Archaeology of the Salisbury Plain Training Area*. Swindon: English Heritage.

Palmer, R. 2005: 'If they used their own photographs they wouldn't take them like that'. In Brophy, K. & Cowley, D. (eds): *From the Air. Understanding Aerial Archaeology*. Stroud: Tempus, 94–116.

Powlesland, D., Lyall, J. & Donoghue, D. 1997: Enhancing the record through remote sensing: The application and integration of multi-sensor, non-invasive remote sensing techniques for the enhancement of the Sites and Monuments Record. Heslerton Parish Project, N. Yorkshire, England. *Internet Archaeology* 2. Available at: http://intarch.ac.uk/journal/issue2/pld_index.html [Accessed March 2, 2010].

Powlesland, D., Lyall, J., Hopkinson, G., Donoghue, D., Beck, M., Harte, A. & Stott, D. 2006: Beneath the sand: remote sensing, archaeology, aggregates and sustainability: a case study from Heslerton, the Vale of Pickering, North Yorkshire, UK. *Archaeological Prospection* 13(4), 291–9.

RCHAMW, 2009. *Upland Survey: The first 10 Years.* Available at: http://www.rcahmw.gov.uk/HI/ENG/Our+Services/Uplands+Archaeology+Initiative/.

Rowlands, A. & Sarris, A. 2007: Detection of exposed and subsurface archaeological remains using multi-sensor remote sensing. *Journal of Archaeological Science* 34(5): 795–803.

Shell, C.A. 2005. Digital airborne remote sensing. High-resolution digital airborne survey for archaeological research and cultural landscape management. In *Volo nel passato. Aerofotografia e cartografia archeologica*, Florence: Insegna del Giglio. 271–83.

Shell, C.A. & Roughley, C. 2004: Exploring the Loughcrew landscape: a new approach with airborne LiDAR. *Archaeology Ireland* 18: 20–3.

Sterazi, P., Coren, F., Creati, N., Vellico, M., & Pietrapertosa, C. 2008: Hyperspectral and LiDAR data fusion applied to archaeological studies: the Aquileia site. In Lasaponara, R. & Masini, N. (eds): *Advances in Remote Sensing for Archaeology and Cultural Heritage Management. 1st International Workshop on Advances in Remote Sensing for Archaeology and Cultural Heritage Management*, Rome, 30 September – 4 October 2008.

Traviglia, A. 2006: Archaeological usability of Hyperspectral images: successes and failures of image processing techniques. In Campana, S. & Forte, M. (eds): *From Space to Place, Proceeding of the 2nd International Conference on Remote Sensing Archaeology*. Rome 4–7 December 2006, Archaeopress, BAR International Series, 123–30.

Weishampel, A. F. Chase, D. Chase, J.B. Drake, R.L Shrestha, K.C., Slatton, Awe, J.J., Hightower, J. & Angelo, J. 2010: Remote Sensing of Ancient Maya Land Use Features at Caracol, Belize Related to Rainforest Canopy Structure. In Campana, S., Forte, M. & Liuzza, C. (eds): *Space, Time, Place. Third International Conference on Remote Sensing in Archaeology, 17th–21st August 2009, Tiruchirappalli, Tamil Nadu, India*. BAR -S2118.

Wilson, D.R. 2000: *Air photo interpretation for archaeologists*. Stroud, Tempus.

Winterbottom, S.J. & Dawson, T. 2005: Airborne multi-spectral prospection for buried archaeology in mobile sand dominated systems. *Archaeological Prospection* 12(4), 205–19.

9 | 3D Recording for cultural heritage

Fabio Remondino

Abstract: The importance of cultural and natural heritage documentation is well understood at an international level and the value of recording digitally in three dimensions (3D) is increasingly recognised. The development of new sensors, data capture methodologies and multi-resolution 3D representations, and the improvement of existing ones, is continuous, and research in these areas is growing, and can contribute significantly to the management, conservation and digital presentation of heritage. The article reviews some important documentation requirements and specifications, and 3D surveying and modelling techniques and methodologies, discussing their limitations and potential. Particular attention will be given to reality-based surveying techniques for structures and objects. Examples of heritage sites and objects digitally documented in 3D are presented throughout.

Introduction

The world's heritage (natural, cultural or mixed) suffers from on-going attrition – through war, natural disaster, climate change and human negligence. As defined by UNESCO, our heritage can be seen as an arch between what we inherit and what we leave behind, and in the last few years great efforts have been focused on what we inherit as cultural heritage and on its digital documentation. In particular the built environment and natural heritage has received a lot of attention and benefits from recording sensor and imaging advances. Indeed 3D data are today a critical component to permanently record the form of important objects so that, in digital form at least, they might be passed down to future generations. This has generated a large number of projects, mainly led by research groups, which have realized very good quality and complete digital models (Levoy *et al.* 2000; Beraldin *et al.* 2002; Stumpfel *et al.* 2003; Gruen *et al.* 2004; Guidi *et al.* 2006; Sonnemann *et al.* 2006; Ikeuchi & Miyazaki 2008; El-Hakim *et al.* 2008; Remondino *et al.* 2009). The technologies and methodologies for cultural heritage 3D documentation allow the generation of very realistic 3D results (in terms of geometric and radiometric accuracy) that can be used for many purposes, such as archaeological documentation, digital preservation and conservation, computer-aided restoration, virtual reality/computer graphics applications, 3D repositories and catalogues, web geographic systems, multimedia museum exhibitions, visualization and so on. But despite all these potential applications and the constant pressure of international heritage organizations, systematic and targeted use of 3D surveying and modelling in the cultural heritage field is still not yet employed as a default approach for different reasons:

- the high cost of 3D;
- the difficulties in achieving good 3D models by everyone;
- the view that 3D is an optional process of interpretation (an additional 'aesthetic' factor) and documentation (i.e. 2D is enough);
- the difficulty of integrating 3D worlds with other more standard 2D material.

However, the availability and use of 3D data opens a wide spectrum of further applications and permits new analysis, studies, interpretations, conservation policies or digital restoration. Thus 3D virtual heritages (Figure 9.1) should be more frequently used due to the great advantages that digital technologies and the third dimension offer to the heritage world and to recognize the documentation needs stated in numerous charters and resolutions.

3D Surveying and modelling

'*It is essential that the principles guiding the preservation and restoration of ancient buildings should be agreed and be laid down on an international basis, with each country being responsible for applying the plan within the framework of its own culture and traditions*' (The Venice Charter, i.e. The International Charter for the Conservation and Restoration of Monuments and Sites 1964). Even though this was stated more than 40 years ago, the need for a clear, rational, standardized terminology and methodology, as well as an accepted professional principle and technique for interpretation, presentation, digital documentation and presentation is still not established. Furthermore, '*...Preservation of the digital heritage requires sustained efforts on the part of governments, creators, publishers, relevant industries and heritage institutions. In the face of the current digital divide, it is necessary to reinforce international cooperation and solidarity to enable all countries to ensure creation, dissemination, preservation and continued accessibility of their digital heritage*' (UNESCO Charter on the Preservation of the Digital Heritage 2003). Therefore, although we may digitally record and produce models, we also require more international collaborations and information sharing to digitally preserve and make them accessible in all the possible forms and to all the possible users and clients. 3D surveying should be meant as the acquisition of unstructured 3D data (e.g.

Figure 9.1: Examples of possible products of reality-based 3D surveying and modelling of heritage sites and objects. These digital 3D models can be used in archaeological analyses and documentation, digital conservation or restoration, VR/CG applications, GIS, 3D repositories and catalogues, web geographic systems, visualization and animation, for example. These illustrations show the Angkor Wat Temple complex in Cambodia (*left*), a bass-relief in Bern, Switzerland (*centre*) and an underground Etruscan tomb in Cerveteri, Italy (*right*).

point clouds) using recording sensors. 3D modelling is instead the procedure to convert unstructured data into structured data (e.g. polygonal meshes).

The requirements for digital documentation and 3D modelling of cultural heritage have been elaborated as (Patias 2007):

- recording and processing of three (or possibly four) dimensional multi-source, multi-resolution and multi-content information;
- management and conservation of the 3D (4D) models for further applications;
- visualization and presentation of results for distribution of information to other users, allowing data retrieval through the internet or advanced online databases;
- creation of digital inventories and sharing for education, research, conservation, entertainment, walkthrough or tourism purposes.

Sensors, techniques and methodologies

Today a large number of geomatics data acquisition tools are available for mapping purposes and for visual cultural heritage digital recording. These include mid and high resolution satellite imagery (e.g. WorldView, Geo-Eye, Quickbird, IKONOS, SPOT, IRS, ASTER, Landsat), large and medium format or linear array digital aerial cameras (e.g. DMC, ULTRACAM, ADS40), space and aerial radar platforms (Radarsat, ERS, AirSAR, COSMO), airborne and terrestrial Laser Scanners (ALS/LiDAR), UAVs (e.g. model helicopters) with on-board digital cameras, panoramic linear sensors, still video cameras and even mobile phones. Moreover Global Navigation Satellite Systems (GNSS) and Inertial Navigation Systems (INS) allow precise positioning and navigation. Alongside great variety of digital sensors and data now available, new software has been developed in the last decade and many automated data processing procedures are now commercially available, in particular for image triangulation, surface model generation (Digital Terrain Model (DTM) or Digital Surface Model (DSM)), range data registration and feature extraction. Furthermore recent advances in Geographic Information Systems (GIS) and 3D repositories (e.g. BIM) allow new functionality for 3D data management, analysis and web-visualization (Manferdini *et al.* 2008) while rendering and animation software packages are now more affordable, with better functionalities and lower costs, for interactive or offline renderings. This continuous development of new sensors, data capture methodologies, multi-resolution 3D representations and the improvement of existing ones are contributing significantly to the documentation, conservation and presentation of heritage information and to the growth of research in the cultural heritage field. This is also driven by an increasing demand for digital documentation of archaeological sites at different scales and resolutions with new digital non-invasive techniques and methodologies (note that in this context a technique is regarded as a scientific procedure (e.g. image processing) to accomplish a specific task, while a methodology is a group or combination of techniques and activities combined to achieve a particular task). Reality-based surveying techniques (e.g. photogrammetry, laser scanning) employ hardware (passive or active sensors) and software to survey the reality as it is, documenting the site and reconstructing it from real measurements. 'Non-real' approaches are instead based on computer graphics software (e.g. 3D Studio, Maya, Sketchup) or procedural modelling approaches (Mueller *et al.* 2006; Haegler *et al.* 2009) allowing the generation of 3D data

Figure 9.2: Buildings like the Erechteion of Athens (Greece) can be surveyed and modelled from terrestrial images to retrieve the main architectural features and create a photo-realistic 3D model.

Figure 9.3: 3D survey and model of a Maya stele (c. 2.5m high) in Copan (Honduras) performed with a Time of Flight (ToF) laser scanner. The single scans (*left*) are aligned and registered together to produce a unique point cloud of the object. A geometric model is then created as polygonal mesh (*centre*) and rendered in a photo-realistic mode with a texture mapping procedure (*right*).

without any particular survey or knowledge of a site, and generally no metrical results, but really impressive 3D models.

Reality-based 3D surveying of heritage sites and objects is generally performed using methodologies based on passive sensors and image data (Remondino & El-Hakim 2006) (Figure 9.2), active optical sensors and range data (Blais 2004; Vosselman & Maas 2010) (Figure 9.3), classical surveying (e.g. total stations or GPS), 2D maps (Yin *et al.* 2009) or an integration of the aforementioned techniques (El-Hakim *et al.* 2004; De Luca *et al.* 2006; Stamos *et al.* 2008; Remondino *et al.* 2009). The choice or integration depends on the required accuracy, object dimensions, location constraints, the instrument's portability and usability, surface characteristics, the working team experience, the project budget, the final goal of the survey and so on. Although there is awareness of the potential of an image-based approach and its recent developments in automated and dense image matching (Hirschmueller 2008; Remondino *et al.* 2008; Hiep *et al.* 2009), its application by non-experts is not often possible and the reliability of the optical active sensor workflow (with related range-based modeling software) in certain projects is still much higher, although time consuming and expensive (Cignoni & Scopigno 2008).

Nevertheless there is still much discussion on which approach or technique is better in which surveying situation. So far the best answer to this question is given by the market which is more in favour of range sensors, but in many research projects the combination and integration of different sensors and techniques, in particular when surveying large and complex sites, is the ideal solution (El-Hakim *et al.* 2007; Guidi *et al.* 2009). Indeed the generation of digital 3D models of large sites for documentation and conservation purposes requires a technique with the following properties:

- *accuracy*: precision and reliability are two important factors of the surveying work, unless the work is done for simple and quick visualization;
- *portability*: a technique, in particular for terrestrial acquisitions, should be portable due to issues of accessibility for many heritage sites, absence of electricity, location constraints etc.;
- *low cost*: most archaeological and documentation missions have limited budgets and cannot effort expensive surveying instruments;
- *fast acquisition*: most sites and excavation areas have limited time for documentation so as not to disturb works or visitors;
- *flexibility*: due to the great variety and dimensions of sites and objects, the surveying technique should allow for different scales and it should be applicable in any possible condition.

As all these properties are not often found in a single technique, most of the surveying projects related to large and complex sites integrate and combine multiple sensors and techniques in order to achieve more accurate and complete surveying, modelling, interpretation and digital conservation results.

Archaeological surveying and documentation using non-invasive geomatics techniques and methodologies can be divided into three broad scales (Table 9.1; Lambers & Remondino 2007):

(1) *regional scale*: with the main goals of (i) recording and modelling the topography of archaeological landscapes and (ii) detecting and mapping archaeological artefacts and features. Spaceborne and airborne sensors provide suitable data for these tasks.
(2) *local scale*: where smaller sites with their architectures and excavated materials are recorded. The material remains of an activity at a defined location hold important clues about the social, cultural, technological and ideological background of the ancient inhabitants. Airborne as well as terrestrial active and passive sensors are very useful at this scale of recording and work.
(3) *object scale*: A thorough recording of artefacts, be they stone or metal tools, worked bones,

Archaeological survey	Object of interest	Available geomatics sensors and data
Regional scale	• Landscape • Topography • Sites	• Middle and high-resolution satellite imagery • Small scale aerial images • Radar and LiDAR • GNSS
Local scale	• Sites • Architectures • Excavation layers	• Large scale aerial images • Radar • ToF range sensors • Terrestrial images • Total stations • GNSS
Object scale	• Excavated artefacts • Museum objects	• Terrestrial images • Triangulation-based range sensors • GNSS

Table 9.1: Scales of archaeological survey with the related non-invasive geomatics sensors and data available for 3D digital documentation. From regional to object scale, the acquired geometric resolution and achievable level of detail gradually increase.

ceramic vessels, sculptures, and other pieces of ancient craftsmanship is not only a prerequisite for typological and chronological studies, but also for investigation of the exchange of goods and ideas, iconography, technology and a variety of other topics. While artefacts are usually recorded in 2D through drawings and photographs, 3D recording literally adds a new important dimension to archaeological studies at the object scale as it provides additional information that enables new kinds of investigation such as morphological comparisons and 3D fragment fitting. Furthermore, digital 3D copies of artefacts enable web-based exhibition and facilitate the production of physical replicas, thus helping to preserve the original artefacts or valorise them.

At each working scale, the GIS contribution is always fundamental, in the form of a local database or as webGIS accessible online to multiple users.

Range-based surveying and 3D modelling

Optical range sensors (Blais 2004; Vosselman & Maas 2010) like pulsed, phase-shift, triangulation-based laser scanners or structured light systems have recently received a great deal of attention for 3D documentation and modelling purposes, by experts and non-specialists alike. Range sensors directly deliver 3D distances (and thus 3D information in the form of unstructured point clouds) and are becoming quite common in the heritage field, despite their high costs, weight and the usual lack of good radiometric information for realistic renderings. The collected range data can be used for simple visualization, rectification, restoration, analyses, metric and mapping purposes (Figure 9.4), valorisation and conservation policies (Figure 9.5) for example. During survey, the range instrument should be placed in different locations, or the object moved in a way that the instrument can see it from different viewpoints. Successively, the acquired 3D raw data needs errors and outliers removed, noise reduction and sometimes holes filling before the alignment or registration of the data into a unique reference system is performed to produce a single point cloud of the surveyed scene or object. The registration is generally done in two steps: (i) manual or automatic raw alignment using targets or the data itself, and (ii) final global alignment based on iterative closest point methods (Salvi *et al.* 2007) or least squares method procedures (Gruen & Akca 2005). After the global alignment, redundant points should be removed before a surface model is produced and textured.

Terrestrial range sensors work from very short ranges (few centimetres) up to few kilometres, in accordance with surface proprieties and environment characteristics, delivering 3D data with positional accuracy ranging from a few hundreds of microns up to some millimetres. Range sensors, coupled with GNSS/INS sensors, can also be used on airborne platforms (e.g. ALS/LiDAR – Shan & Toth 2008), mainly for DTM/DSM generation, mapping and 3D city modelling. LiDAR data generally produce a DSM, and therefore a filtering and reduction is required to obtain a DTM.

Figure 9.4: 3D survey of a medieval castle in Ossana (NE Italy) and realization of floor plans and maps.

Figure 9.5: 3D surveying and modelling of a frescoed underground Etruscan tomb in Tarquinia (Italy) for conservation and valorisation purposes. A cross-section of the entire 3D model and two renderings of the interior chambers are shown.

Depending on the flying height and used sensor, an ALS survey can provide for point clouds with densities from 1 point/sqm up to 15–20 points/sqm. The new range sensors allow the storage of multiple echoes of the laser signal ('discrete return' or 'full-waveform') and open up a whole new range of possibilities in particular for archaeological identification and mapping of structures hidden below vegetation (Devereux *et al.* 2005; Doneus & Briese 2006; Doneus & Briese this volume).

The results of 3D scanning, in particular in terrestrial applications, are a function of a number of factors (Beraldin *et al.* 2007):

- intrinsic characteristics of the instrument (calibration, measurement principle, etc.);
- characteristics of the surveyed material in terms of reflection, light diffusion and absorption (amplitude response);
- characteristics of the working environment;
- coherence of the backscattered light (phase randomization);
- dependence from the chromatic content of the scanned material (frequency response).

Nowadays, the main research issues involved in range-based data processing and modelling are the automated extraction of features (like man-made objects) and the automated generation of structured 3D data (polygonal meshes) from the recorded unstructured point clouds.

Image-based surveying and 3D modelling

In contrast to range sensors, which are able to directly provide 3D data, passive sensors like CCD/CMOS frame or linear array digital cameras deliver image data which requires a mathematical formulation to transform the 2D image measurements into 3D information. Generally at least two images are required and 3D data can be derived using perspective or projective geometry formulations. Image-based modelling techniques (mainly photogrammetry and computer vision – see Remondino & El-Hakim 2006) are generally preferred in cases of lost objects, monuments or architecture with regular geometric shapes, small objects with free-form shape, cartographic applications, deformation analyses, low budgets, time or location constraints for the data acquisition and processing. The image-based workflow generally consists of (i) camera calibration, (ii) image orientation and (iii) 3D scene reconstruction and rendering. These steps can be performed in an automated or interactive way, according to the type of data, user experience and project specifications. Accurate 3D feature extraction from satellite and aerial images is still a manually driven procedure while in terrestrial applications more automation is possible for scene reconstruction. Fully automated methods based on a 'structure from motion' approach (Vergauwen & Van Gool 2006; Goesele *et al.* 2007; Agarwal *et al.* 2009) are becoming quite common in the 3D heritage community, although they are mainly useful for visualization, object-based navigation, annotation transfer or image browsing purposes and not for metric and accurate 3D reconstructions and documentations. Complete automation in image-based modelling is an active research topic, in particular for the 3D surveying and modelling of complex architectural scenes and man-made objects (Patias *et al.* 2008).

Photogrammetry (Mikhail *et al.* 2001; Luhmann *et al.* 2006) is considered the primary technique for processing image data, being able to deliver at any scale of application accurate and detailed 3D information with estimates of precision and reliability of the unknown parameters from the measured image correspondences (tie points). The image correspondences can be extracted automatically or semi-automatically according to the imaged scene and project requirements. Photogrammetry is employed in different applications like mapping, 3D documentation, conservation, digital restoration, reverse engineering, monitoring, visualization, animation, urban planning and deformation analysis for example. Photogrammetric 3D reconstructions of man-made objects or architectural structures are generally performed with interactive procedures (Figure 9.6a, b), where sparse point clouds and only a few geometric primitives are sufficient to describe the 3D geometry. Automated image matching procedures are instead employed for free-form objects (Figure 9.6c, d, e) where dense point clouds are required to correctly describe all the object discontinuities and features (Remondino *et al.* 2008). Digital surface models (DSM) can be used for detailed documentation, conservation, restoration, replica purposes and volume computation (Figure 9.7) and valorisation, for example.

Many researchers (Pomaska 2001; D'Ayala & Smars 2003; English Heritage 2005) have reported how the photogrammetric image-based approach allows surveys at different levels and in all possible combinations of object complexities, with high quality outputs, easy usage and manipulation of the final products, few time restrictions, good flexibility and low costs. Comparisons between photogrammetry and range sensors have been made by some workers (e.g. Boehler 2005; Remondino *et al.* 2005; Grussenmeyer *et al.* 2008).

Figure 9.6: Examples of structures digitally reconstructed with interactive (*top*) or automated (*bottom*) image measurements.

Sensor and data integration

As discussed above, the state-of-the-art for 3D documentation and modelling of large and complex sites uses integrated multiple sensors and technologies (photogrammetry, laser scanning, topographic surveying etc.) to (i) exploit the intrinsic strengths of each technique, (ii) compensate for weaknesses of individual methods, (iii) derive different geometric Levels of Detail (LoD) of the scene under investigation and (iv) achieve more accurate and complete geometric surveying for modelling, interpretation, representation and digital conservation issues. 3D modelling based on multi-scale data and multi-sensors integration provides the best 3D results in terms of appearance and geometric details (Figure 9.8). Each LOD shows only the necessary information while each technique is used where best suited. Since the 1990s multiple data sources have been integrated for industrial, military and mobile mapping applications. Sensor and data fusion were then applied also in the cultural heritage domain, mainly at terrestrial level (Stumpfel *et al.* 2003; El-Hakim *et al.* 2004) although some projects integrated satellite, aerial and ground information for a more complete and multi-resolution 3D survey (Gruen *et al.* 2005; Ronnholm *et al.* 2007; Guidi *et al.* 2009; Remondino *et al.* 2009).

The multi-sensor and multi-resolution concept should be distinguished in:

(i) geometric surveying and modelling (3D shape acquisition, data registration and further processing) where multiple resolutions and sensors are seamlessly combined and integrated to digitally reconstruct features with the most adequate geometric sampling step and derive different geometric LoD of the scene under investigation;
(ii) appearance modelling (texturing, blending, simplification and rendering) where photo-realistic representations are sought taking into consideration variations in lighting, surface specularity, seamless blending of the textures, user's viewpoint, simplification and LoD.

Beside images acquired in the visible part of the light spectrum, it is often necessary to acquire and integrate extra information from other sensors working in different spectral bands (e.g. IR, UV, X-rays) in order to study the object deeper. Thermal infrared (IR) information, for example, is useful for analysis

Figure 9.7: Excavation area at Pava (Siena, Italy) surveyed with oblique aerial images to derive a Digital Surface Model (DSM) for archaeological documentation and volume computation.

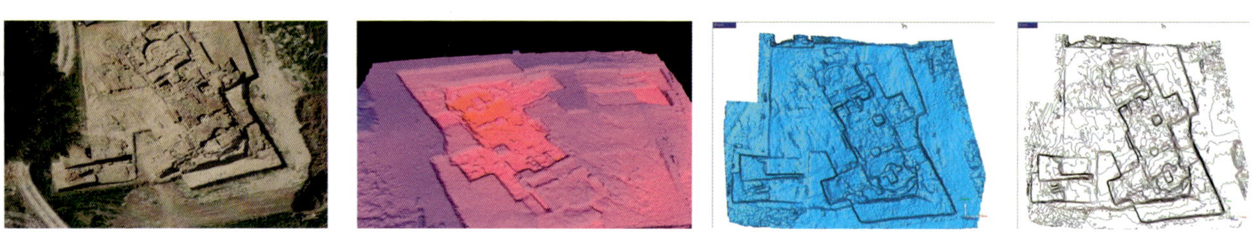

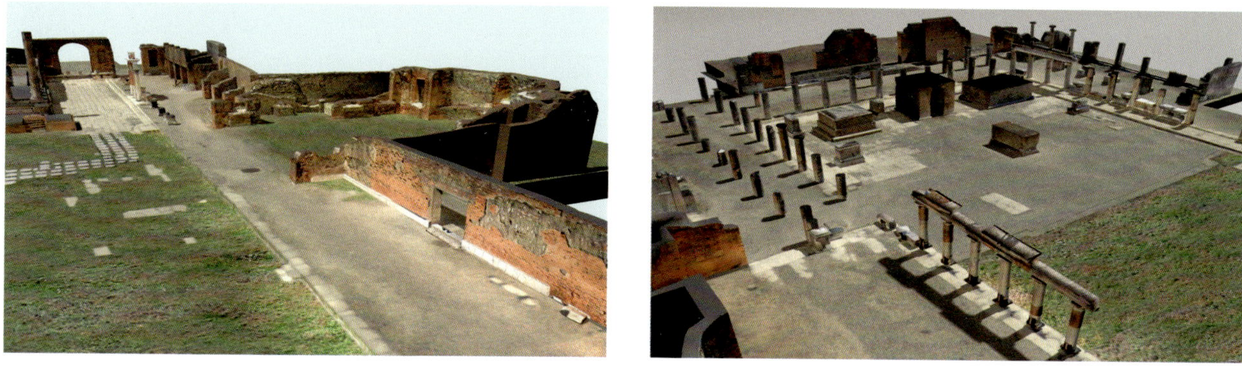

Figure 9.8: 3D surveying and modelling of the Roman forum in Pompeii, generated by integrating aerial images, terrestrial laser scanning and terrestrial images.

of historical buildings, their state of preservation and complexity of construction and phasing. Near IR is used to study paintings, revealing pentimenti and preparatory drawings. Ultraviolet (UV) radiation is useful for identifying different varnishes and over-paintings, in particular with induced visible fluorescence imaging systems (Figure 9.9; Pelagotti *et al.* 2006). All this multi-modal information needs to be aligned and often overlapped to the geometric data for information fusion, multi-spectral analysis or other diagnostic applications.

Conclusion

This article has reviewed 3D surveying and modelling methodologies for reality-based 3D documentation of heritage sites and objects. Examples and the potentials of the techniques have been presented and discussed. In the case of heritage sites and objects, photogrammetry provides accurate 3D reconstructions at different scales and for hybrid 3D models (e.g. digital terrain model plus archaeological structures as shown in Figure 9.5). Today 3D scanners are becoming a standard source for 3D data in many application areas, although image-based modelling still remains the cheapest, most portable and flexible approach. For large sites, the integration of images and range data is generally the best solution. Despite the fact that 3D documentation is not yet a standard in the heritage field, the reported examples show the potential of modern surveying technologies to digitally record and preserve our heritages as well as share and manage them. But it is clear that the image-based 3D documentation approach, together with active optical sensors, Spatial Information Systems, 3D modelling procedures, visualization and animation software are still all in a dynamic state of development, with even more powerful application prospects in the near future. Archaeologists and others in the heritage community should exploit much more the 3D domain as it is added-value able to provide unique and highly flexible information that can inform knowledge and effective management.

Acknowledgements

The author would like to thank Prof. A. Gruen and Dr H. Eisenbeiss (ETH Zurich, Switzerland), Prof. G. Guidi and Dr M. Russo (Politecnico of Milan, Italy), Dr S. El-Hakim and Dr A.J. Beraldin (NRC Canada), S. Girardi and A. Rizzi (FBK Trento, Italy), M. Rapana' (Univ. of Trento, Italy), S. Benedetti (Benedetti Graphics), A. Pelagotti (INO-CNR, Italy), Dr L. Marras (Art-Test, Italy), Dr S. Campana and M. Sordini (University of Siena, Italy), Dr B. Benedetti (SNS Pisa, Italy), Prof. M. Gaiani and Dr A.M. Manferdini (Univ. of Bologna, Italy), Dr J. von Schwerin (University of New Mexico, USA) co-authors in some publications, researches and field works for the reported projects and examples.

Figure 9.9: Metric digitization and vectorialization of frescoes in an Etruscan tomb in Tarquinia (Italy) for archaeological analyses and restoration or conservation policies.

References

Agarwal, S., Snavely, N., Simon, I., Seitz, S. & Szelinski, R. 2009: Building Rome in a day. In *Proceedings of the Internationa Conference of Computer Vision (ICCV) 2009*, Kyoto, Japan.

Beraldin, J.-A., Picard, M., El-Hakim, S.F., Godin, G., Valzano, V., Bandiera, A. & Latouche, D. 2002: Virtualizing a byzantine crypt by combining high resolution textures with laser scanner 3d data. *Proceedings of the International Conference on Virtual Systems and MultiMedia (VSMM) 2002*, 3–14.

Beraldin, A., Rioux, M., Cournoyer, L., Blais, F., Picard, M. & Pekelsky, J. 2007: Traceable 3d imaging metrology. In *Proceedings of Videometrics IX – SPIE Electronic Imaging*, 6491, B.1-B.11.

Blais, F. 2004: A review of 20 years of Range Sensors Development. *Journal of Electronic Imaging* 13(1), 231–40.

Bhoeler, W. 2005: Comparison of 3D scanning and other 3D measurement techniques. In Baltsavias, E., Gruen, A., Van Gool, L. & Pateraki, M. (eds): *Recording, Modeling and Visualization of Cultural Heritage*. Taylor & Francis / Balkema.

Cignoni, P. & Scopigno, R. 2008: Sampled 3d models for Cultural Heritage applications: a viable and enabling new medium or just a technological exercise. *ACM Journal on Computing and Cultural Heritage*, 1(1).

De Luca, L., Veron, P. & Florenzano, M. 2006: Reverse engineering of architectural buildings based on a hybrid modeling approach. *Computer & Graphics* 30, 160–76.

D'Ayala, D. & Smars, P. 2003: *Minimum requirement for metric use of non-metric photographic documentation*. University of Bath Report.

Devereux, B., Amable, G.S., Grow, P. & Cliff, A.D. 2005: The potential of airborne LiDAR for detection of archaeological features under woodland canopies. *Antiquity* 79, 648–660.

Doneus, M. & Briese, C. 2006: Full-waveform airborne laser scanning as a tool for archaeological reconnaissance. In Campana, S. & Forte, M. (eds), *Proceedings of the 2nd International Conference on Remote Sensing in Archaeology 'From Space to Place'*, BAR International Series 1568, 99–106.

El-Hakim, S.F., Beraldin, J.A., Picard, M. & Godin, G. 2004: Detailed 3D reconstruction of large-scale heritage sites with integrated techniques. *IEEE Computer Graphics and Application*, 24(3), 21–29.

El-Hakim, S., Gonzo, L., Voltolini, F., Girardi, S., Rizzi, A., Remondino, F. & Whiting, E. 2007: Detailed 3D modeling of castles. *International Journal of Architectural Computing* 5(2), 199–220.

El-Hakim, S., Beraldin, J.-A., Remondino, F., Picard, M., Cournoyer, L. & Baltsavias, M., 2008: Using terrestrial laser scanning and digital images for the 3D modelling of the Erechteion, Acropolis of Athens. *Proceedings of 'Digital Media and its Applications in Cultural Heritage' (DMACH) Conference*, 3–16, Amman, Jordan.

English Heritage, 2005: *Metric Survey Specifications for English Heritage*. English Heritage Report. Available at www.english-heritage.org.uk.

Gruen, A., Remondino, F. & Zhang, L. 2004: Photogrammetric Reconstruction of the Great Buddha of Bamiyan, Afghanistan. *The Photogrammetric Record*, 19(107), 177–99.

Guidi, G., Frischer, B., Russo, M., Spinetti, A., Carosso, L. & Micoli, L.L. 2006: Three-dimensional acquisition of large and detailed cultural heritage objects. *Machine Vision Applications* 17, 349–60.

Guidi, G., Remondino, F., Russo, M., Menna, F., Rizzi, A. & Ercoli, S. 2009: A multi-resolution methodology for the 3D modeling of large and complex archaeological areas. *International Journal of Architectural Computing* 7(1), 40–55.

Goesele, M., Snavely, N., Curless, B., Hoppe, H. & Seitz, S.M. 2007: Multi-view stereo for community photo collections. In *Proc. Int. Conference of Computer Vision (ICCV) 2007*, Rio de Janeiro, Brasil.

Gruen, A. & Akca, D. 2005: Least squares 3d surface and curve matching. *ISPRS Journal of Photogrammetry and Remote Sensing*, 59(3), 151–74.

Gruen, A., Remondino, F. & Zhang, L. 2005: The Bamiyan project: multi-resolution image-based modeling. In Baltsavias, E., Gruen, A., Van Gool, L. & Pateraki, M. (eds): *Recording, Modeling and Visualization of Cultural Heritage*. Taylor & Francis/ Balkema.

Grussenmeyer, P., Landes, T., Voegtle, T. & Ringle, K. 2008: Comparison methods of terrestrial laser scanning, photogrammetry and tacheometry data for recording of cultural heritage buildings. *International Archives of the Photogrammetry, Remote Sensing and Spatial Information Sciences*, 37(5), 213–8.

Haegler, S., Mueller, P. & Van Gool, L. 2009: Procedural modeling for digital Cultural Heritage. *EURASIP Journal on Image and Video Processing*, Volume 2009, Article ID 852392.

Hiep, V., Keriven, R., Labatut, P. & Pons, J. 2009: Towards high-resolution large-scale multi-view stereo. In *Proceedings of Computer Vision and Pattern Recognition (CVPR) 2009*, Kyoto, Japan.

Hirschmueller, H. 2008: Stereo processing by semi-global matching and mutual information. *IEEE Transactions on Pattern Analysis and Machine Intelligence*, 30(2) 328–41.

Ikeuchi, K. & Miyazaki, D. 2008: *Digitally Archiving Cultural Objects*. Springer.

Lambers, K. & Remondino, F. 2007: Optical 3D measurement techniques in archaeology: recent developments and applications. *Proceedings of the 35th Computer Applications and Quantitative Methods in Archaeology (CAA) Conference*, 27–35, Berlin, Germany.

Levoy, M., Pulli, K., Curless, B., Rusinkiewicz, S., Koller, D., Pereira, L., Ginzton, M., Anderson, S., Davis, J., Ginsberg, J., Shade, J. & Fulk, D. 2000: The digital

Michelangelo project: 3d scanning of large statues. *SIGGRAPH Computer Graphics Proceedings*, 131–44.

Luhmann, T., Robson S., Kyle, S. & Harley, I. 2006: *Close range photogrammetry: Principles, methods and applications*. Caithness, Whittles.

Manferdini, A.M., Remondino, F., Baldissini, S., Gaiani, M. & Benedetti, B. 2008: 3D modeling and semantic classification of archaeological finds for management and visualization in 3D archaeological databases. *Proceedings of 14th International Conference on Virtual Systems and MultiMedia (VSMM)*. Limassol, Cyprus. 221–28.

Mikhail, E.M., Bethel, J.S. & McGlone, J.C. 2001: *Introduction to modern photogrammetry*. John Wiles and Sons Inc.

Mueller, P., Wonka, P., Haegler, S., Ulmer, A. & Van Gool, L. 2006: Procedural modeling of buildings. *Proceedings of ACM SIGGRAPH 2006 / ACM Transactions on Graphics*, 25(3), 614–23.

Patias, P. 2007: Cultural heritage documentation. In Fryer, J., Mitchell, H. & Chandler, J. (eds): *Applications of 3D measurement from images*. Whittles Publishing. 225–50.

Patias, P., Grussenmeyer, P. & Hanke K. 2008: Applications in cultural heritage documentation. *Advances in Photogrammetry, Remote Sensing and Spatial Information Sciences*, 2008 ISPRS Congress Book 7, 363–84.

Pelagotti, A., Pezzati, L., Piva, A. & Del Mastio, A. 2006: Multispectral UV fluorescence analysis of painted surfaces. *Proceedings of 14th European Signal Processing Conference (EUSIPCO)*, Firenze, Italy.

Pomaska, G. 2001: Image acquisition for digital photogrammetry using off the shelf and metric cameras. *Proceedings of CIPA International Symposium*, Potsdam, Germany.

Remondino, F., Guarnieri, A. & Vettore, A. 2005: 3D modeling of close-range objects: photogrammetry or laser scanning? In *Procceedings of Videometrics VIII, SPIE-IS&T Electronic Imaging* 5665, 216–25.

Remondino, F. & El-Hakim, S. 2006: Image-based 3D modelling: a review. *The Photogrammetric Record*, 21(115), 269–91.

Remondino, F., El-Hakim, S., Gruen, A. & Zhang, L. 2008: Development and performance analysis of image matching for detailed surface reconstruction of heritage objects. *IEEE Signal Processing Magazine* 25(4), 55–65.

Remondino, F., El-Hakim, S., Girardi, S., Rizzi, A., Benedetti, S. & Gonzo, L. 2009: 3D Virtual reconstruction and visualization of complex architectures – The 3D-ARCH project. *International Archives of the Photogrammetry, Remote Sensing and Spatial Information Sciences*, Vol. 38(5/W1), Trento, Italy (on CD-Rom).

Ronnholm, P., Honkavaara, E., Litkey, P., Hyypp, H. & Hyypp, J., 2007: Integration of laser scanning and photogrammetry. *International Archives of Photogrammetry, Remote Sensing and Spatial Information Sciences* 36(2/W52), 355–62.

Salvi, J., Matabosch, C., Fofi, D. & Forest, J. 2007: A review of recent range image registration methods with accuracy evaluation. *Image and Vision Computing* 25(5), 578–96.

Shan, J. & Toth, C. 2008: *Topographic Laser Ranging and Scanning: Principles and Processing*. CRC Press.

Sonnemann, T., Saurbier, M., Remondino, F. & Schoretter, G. 2006: Reality-based 3D modeling of the Angkorian temples using aerial images. In Campana, S. & Forte, M. (eds): *Proceedings of 2nd International Conference on Remote Sensing in Archaeology*. Rome, Italy, 573–79.

Stamos, I., Liu, L., Chen, C., Wolberg, G., Yu, G. & Zokai, S. 2008: Integrating automated range registration with multiview geometry for the photorealistic modelling of large-scale scenes. *International Journal of Computer Vision* 78(2–3), 237–60.

Stumpfel, J., Tchou, C., Yun, N., Martinez, P., Hawkins, T., Jones, A., Emerson, B., Debevec, P. 2003: Digital reunification of the Parthenon and its sculptures. In *Proceedings of Virtual Reality, Archaeology and Cultural Heritage (VAST) 2003*, 41–50.

Vosselman, G. & Maas H. 2010: *Airborne and Terrestrial Laser Scanning*. CRC Press, Whittles Publishing.

Vergauwen, M. & Van Gool, L. 2006: Web-based 3D reconstruction service. *Machine vision and applications* 17(6), 411–26.

Yin, X., Wonka, P. & Razdan, A. 2009: Generating 3D building models from architectural drawings. *IEEE Computer Graphics and Applications* 29(1), 20–30.

10 | Through an imperfect filter: geophysical techniques and the management of archaeological heritage

Chris Gaffney and *Vincent Gaffney*

Abstract: Geophysical techniques have been part of the archaeologist's tool kit for over half a century. During that period methods have developed from small-scale survey using a single technique to a battery of methods for investigating the near surface. This change in emphasis has allowed the archaeological geophysicist to move from producing a 'context' for a site or feature towards prospecting, analysing and interpreting sites in a meaningful archaeological manner. The potential for managing, and therefore protecting, the buried archaeological heritage is evident especially in plough levelled agricultural areas or 'sensitive' zones where excavation is precluded.

As a consequence of recent technical developments it is now possible to collect multiple data sets in one sweep using vehicle or human powered carts or sledges. Usually the data sets are geo-referenced with on-board GPS and, additionally, some collecting procedures allow for grid-less survey, which has increased survey speed whilst reducing costs. A result of the implementation of new geophysical strategies is that the measurements can often be both large scale and data dense, which is a step-change from even five years ago when they were either large scale or data dense.

This paper reviews some recent technological developments and considers how geophysical results can be used to assess archaeological potential and inform on management issues relating to the archaeological resource. The issue of 'speed' versus 'minimal impact' will be discussed, as will the importance of the digital data environment within which geophysical measurements are evaluated. Sites with relevant management issues will illustrate the direction of current research.

Introduction

Geophysical survey has been an enduring aspect of 'archaeological science' for over 50 years and featured in the first issue of the journal *Archaeometry* (Aitken 1958). The use of ground based remote sensing for archaeological purposes dates back much further, for example during the 1930s in the USA (Bevan 2001) and in immediately post World War II England (Aitkinson 1953). The purpose of those early surveys was to locate significant archaeological deposits for excavation. Concepts such as 'context' came much later, primarily as the techniques and equipment started to advance in both complexity and reliability. This realignment of research interests clearly followed paradigm changes in archaeological thinking; primarily this reflects the change of interest from the traditional site, into the hinterland and beyond. Increasing coherence in data capture, display and interpretation created a significant foundation for geophysical work during this formative period although the results were inherently site or feature based (Gaffney 2008). Even 25 years ago it was clear that the plaintive cries for non-invasive techniques to step beyond 'wall-following' would be at best described as radical and the concept of managing the heritage using such technology was some time in the distant future (Heron & Gaffney 1987). Despite this conservative approach the technical advances of the 1980s and 1990s resulted in the worldwide application of archaeological geophysical techniques; although the uptake of techniques was actually limited to earth resistance and magnetometry rather than the whole range of available options. At a superficial level the images of, for example, Roman forts or prehistoric enclosed settlements that are prevalent in the published literature are a fundamental justification for the use of ground-based geophysical techniques for management purposes (e.g. Hopewell 2005, Fassbinder 2010). In such cases the ability to identify and delimit on a map where areas of significant archaeology exist may be adequate to inform the decisions of curators. However, there are many different types of applications within the published literature that can be linked to specific heritage management issues other than simply establishing the presence or absence of archaeology. For example, there are instances where geophysical techniques have been used within upstanding monuments and specifically for informing on management problems. The non-destructive nature of geophysical techniques has meant that confined spaces within churches and similar structures can be investigated without recourse to costly or inappropriate excavation (e.g. Nuzzo *et al.* 2009, Tsokas *et al.* 2008). Additionally, these techniques can be fine-tuned to surgically analyse individual features. A good example of this is the work to establish the structural nature and strength of the Acropolis wall in Athens (Tsokas *et al.* 2006).

Despite the ability to work in confined or relatively inaccessible areas, it is common to undertake geophysical survey in open ground. In suitable open conditions and depending on the contrast (in some remotely measurable property, see below) between the buried archaeology and the surrounding soil, an image can be produced (Scollar *et al.* 1990, Clark 1990, Aubrey *et al.* 2001, Gaffney & Gater 2003, Linford 2006). At first glance an image generated from the geophysical data can be regarded as a proxy for a traditional excavation plan. There is, therefore, an assumption, perhaps even a belief, that if the data are reasonably clear and then shapes that are imaged form archaeologically understandable patterns. Under such circumstances it is sometimes suggested that the image can be regarded as the basis for the justification of the survey. However, such a stance is insufficiently critical of the validity of the data and can promote the use of unsuitable methodologies, poor or insubstantial interpretations and, ultimately, dissatisfaction with the technique used. On all but the simplest of sites, the geophysical image will rarely identify the true complexity of the buried archaeology.

Despite this, it should be clear that prospection is an extremely technical activity, in both its theoretical context and application. Consequently, it is unlikely that a simple image of the data would actually represent a satisfactory outcome of many surveys. Schmidt (2003) identified five stages in the process of prospection: measurement and data recording; acquisition procedure; data processing; visualisation; and interpretation. These stages reflect the computational role that is inherent in the endeavour, but it is evident that the *archaeological* output is primarily the interpretation of the data. When one considers the value of ground-based geophysical techniques for archaeological purposes then the interpretation of the data is the most challenging part of the investigation and also the most important. Measured responses are non-unique, that is to say that the same response can originate from different archaeological features, and therefore the interpretation is seldom straightforward and requires a full understanding of the expected archaeology within an area. From an academic perspective it seems that the archaeological community has become too uncritical of the use of such images and this has led to unchallenged and often archaeologically modest output. That is not to say that as a community we are not getting better at collecting and displaying data, but often the image remains the main delivery and very little linkage is made to substantive archaeological problems (Conyers & Leckebusch 2010; Conyers 2010).

Historically, two parallel paths can be charted indicating how geophysical techniques have been used for archaeological purposes. Differences have been observed between those techniques that create area maps and those that image sections through the earth. That difference is not so apparent now as technical developments have allowed rapid and coherent interchange between these outputs and this is most apparent in the application of ground penetrating radar. Perhaps of greater interest to this discussion are the choices relating to manual or time base and automatic GPS location of measurements. There is a logarithmic change in speed and a similar potential transformation in the accuracy of measurement location between these two competing trends. Moreover, while recent technical changes offer great advantages in collecting ground-based geophysical measurements, there must be more cogent reasons for using new and expensive technology than just an increase in speed. Whatever the emerging path for these technologies, we must consider the deliverables in terms of additional understanding of archaeological features, sites and landscapes rather than implementation of new technology for its own sake. The validity of the use of particular techniques lies not in the ability to measure a wanted signal rapidly, but in the understanding of the archaeological significance of that signal.

Specifically we need to consider what new challenges are developing as a result of the increasingly data rich and extensive survey strategies that are being implemented. The most obvious challenge results from the sheer amount of data that can be collected in a short time. If we use conventional strategies for manipulation and interpretation then it is likely that we cannot provide well developed and considered, management focussed outputs in the short time frames that are often required in heritage management applications. A second challenge relates to the interface with other digital technologies and the opportunities that are emerging as a consequence. Geophysical techniques have been used for some time to create a context for excavation, but now we need to consider the context of the survey, as there are often many types of correlated but contrasting digital data sets that form a larger holistic or staged investigation. In particular, we must consider the interplay between remotely sensed data that describe subterranean elements of former landscapes and a wider heritage that is not restricted to the buried zone. This paper will illustrate new or more profitable application areas that are, in part, revealed by novel adaptations of proven techniques and identify those areas that need significant research to counterbalance the recent, radical changes in technology.

What is geophysical survey?

Given the longevity of application of geophysical techniques within archaeology, the requirement to define what constitutes a geophysical survey may seem perverse. However, the need for critical and regular reconsideration of what constitutes the world view of archaeological geophysics is necessary as all techniques, methods and concepts have a shelf life that demotes them to secondary or 'normal' activities. By re-evaluating we re-invigorate and re-affirm their value to the heritage community.

In broad terms one can use the definition suggested by Gaffney & Gater (2003): that geophysical survey is the examination of the earth's physical properties using non-invasive ground based survey techniques to reveal buried archaeological features, sites and landscapes. More recently GeoSIG, the UK based Institute for Archaeologists' special interest group (www.archaeologists.net), has considered a 'standard' for an archaeological geophysical survey and suggested that the definition is a programme of non-intrusive surveys of the physical properties of sediments which determines the presence or absence of anomalies likely to be caused by archaeological features, structures or deposits, as far as reasonably possible, within a specified area or site on land, inter-tidal zone or underwater (IFA 2010). However, this is really only the outer shell of what constitutes such work and a more sophisticated and accurate definition is likely to include a fundamental understanding of how the manner in which we undertake a survey determines the ultimate outcome. Specifically, archaeology is about the life and culture of ancient people and it seems absurd that archaeological geophysicists should choose not to participate in that debate. By acknowledging the limiting nature of the established paradigm we should be able to forge a more thoughtful and rewarding context for archaeological geophysics. By implication this suggests that whilst we understand the limitations of our data, we also accept that this includes the ability to push at the boundaries to reveal greater archaeological 'depth' within our data and hence additional value. In fact, this is not just an optional ability but a requirement if geophysical data are to contribute to the successful management of the archaeological resource. In a world where relevance or 'impact' is becoming increasingly required then advances in the former without the latter are increasingly problematic. By changing and adding to the way in which we undertake ground based survey it is likely that we will be able to contribute to the management and interpretation of significant archaeological areas. This transformation must incorporate a systemic change in the perception of why we undertake ground-based surveys in particular. The dislocation of technical ability from the archaeological agenda at best devalues the work of the archaeological geophysicist and at worst adds serious doubts as to the sustainability of such endeavours. This is particularly true when one considers similar technical advances in other remote sensing areas.

Techniques

The techniques that we use for geophysical survey have not varied much over the last forty or so years, although their implementation has changed dramatically over the last decade (Gaffney 2008). A list of techniques that have been used can be seen in Table 10.1. There are two core techniques that are used throughout the world for management purposes: earth resistance and magnetometry. They are examples of active (anomaly 'induced' by the instrument) and passive (anomaly present without the instrument switched on) techniques and have been successful due to their ability to rapidly map properties that can be related to changes in the subsurface. However, there is a great deal of variation in use of techniques between different countries. For example, magnetometry is the most used technique within UK archaeology (David *et al.* 2008), in France, electromagnetic devices are often to the fore (Aubrey *et al.* 2001), while in the USA ground penetrating radar is commonly used (Conyers 2004). The last technique is increasingly used for archaeological purposes and has benefited from the technical advances linked to utility location (e.g. electricity, water pipes). The variation in use partly reflects the interests of the dominant research institutes in a particular country but also is linked to the type of archaeology that is present and the type of contrasts that are produced. The contrasts are connected to the physical properties of the archaeological strata and the reader is referred elsewhere for further detail (Gaffney 2008; Linford 2006). The principle that underlies archaeological geophysics is that for a feature or stratum to be detected and interpreted it must exhibit some measurable contrast that can be modelled using physics theory. As a result dowsing or divining, although non-invasive 'remote' techniques cannot be included in Table 10.1. It should be understood that each technique depends on a contrast of different properties. As a consequence no single technique can find all 'targets' and the choice of which technique should be used for a particular survey must be carefully considered before

Table 10.1: Geophysical techniques and their frequency of use.

Method	Active or Passive	Frequency of use
Magnetometry	Passive	High
Earth Resistance / Resistivity	Active	High
Ground Penetrating Radar	Active	High / Mid
Electromagnetic	Active	Mid
Magnetic Susceptibility	Active	Mid
Metal Detectors	Active	Low/Mid
Seismic	Active	Low
Induced Polarisation	Active	Low
Microgravity	Passive	Low
Self Potential	Passive	Low
Thermal	Passive	Low

data collecting starts. Some rules of thumb do exist (David *et al.* 2008, Table 3), but matching survey technique to feature type is difficult as local conditions can be highly variable. One of the outcomes of a survey is to find previously unknown features and, therefore, trying to work out which technique is 'best' for a particular survey is problematic. Dependence on a single technique is therefore inherently risky and most management applications require a number of techniques or methods to counteract this problem (Figures 10.1 & 10.2).

Accuracy of location and value added interpretation

While archaeological geophysicists have long worked in the digital domain, adding value to the results by blending different data sets, as opposed to simple display or visualisation, has rarely been achieved in comparison to other areas of archaeology. It has been pointed out elsewhere that the move from geophysical or remotely sensed numeric data to sophisticated interpreted models required for high level archaeological inference is not a well trodden path (Gaffney & Gaffney 2006). However, there is a conclusive benefit associated with such an advance as digital geophysical data becomes directly central to interpretation. As has been stated above one of the recent and major technological changes is the incorporation of GPS into the methodologies for geophysical data capture. Examples of such integration involving all the major techniques can be found in the literature (e.g. GPR (Leckebusch 2005), magnetometry (Gaffney *et al.* 2008) and earth resistance (Dabas 2009)). The use of GPS also, potentially, allows grid-less survey as well as precisely located measurements (Figure 10.3).

Figure 10.1: Fluxgate magnetometer and earth resistance survey at the cemetery for High Royds Chapel, Menston (UK). The cemetery is the resting place for 2,861 unclaimed patients. The burials are believed to be in unmarked paupers' graves and the last burial was in 1969. There is an archive interment plan that shows about 1,000 plots. The magnetic survey (white = -30, black = +30 nT) largely indicates the location of ferrous material. In this case the 'anomalies' are the result of metal grave markers. Evidently, given the variation in markers across the graveyard, it is likely that the identification of individual graves using this technique is sporadic. However, it is apparent that many graves were not unmarked. The markers are, of course, moveable and this distribution can only give an indication of the location of about 200 graves. The earth resistance data (white = 18, black = 21 ohm) are dominated by paths as well as a change in background level that is linked to topography. However, in the north east quadrant can be seen a few, east-west oriented, individual graves.

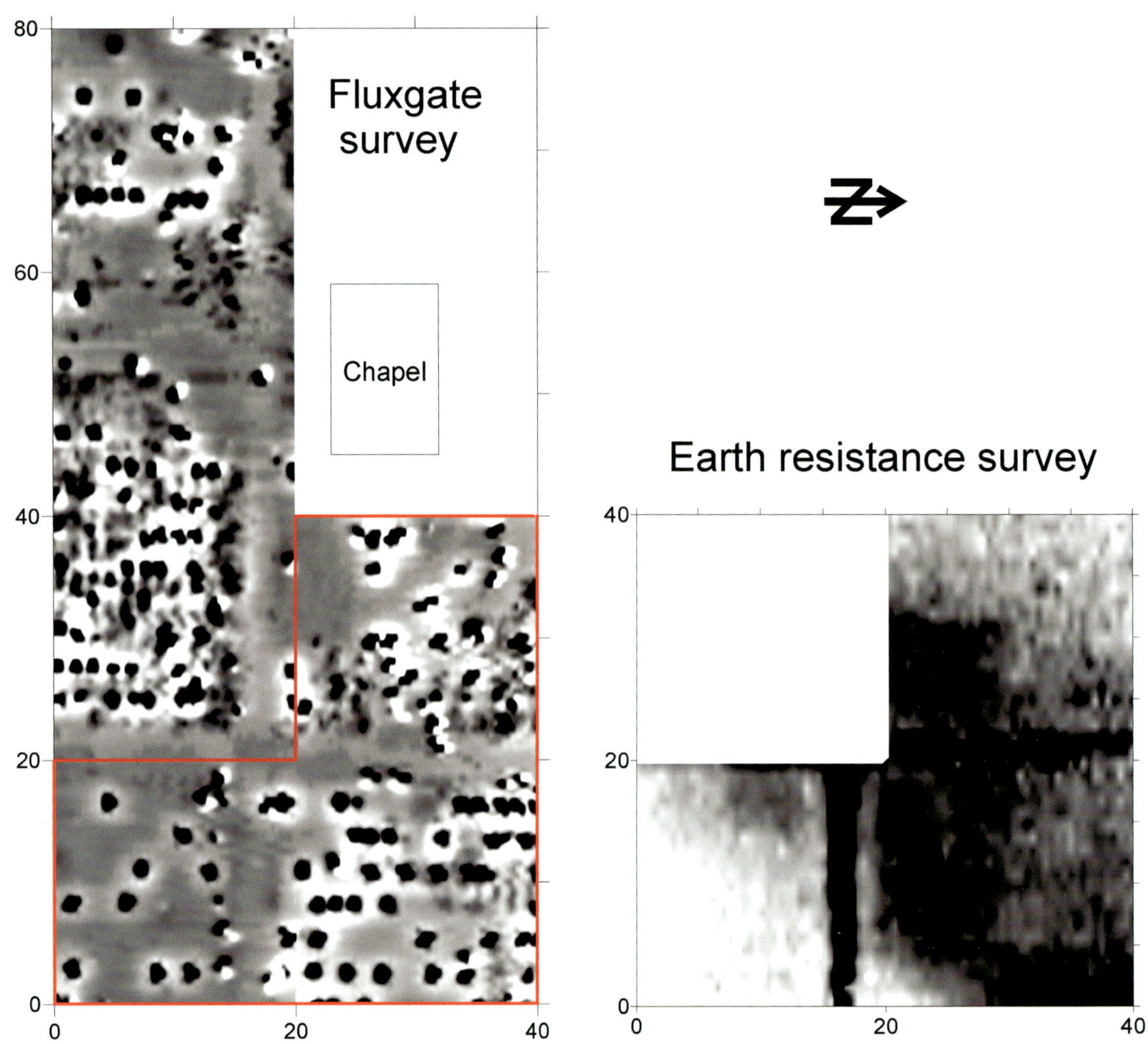

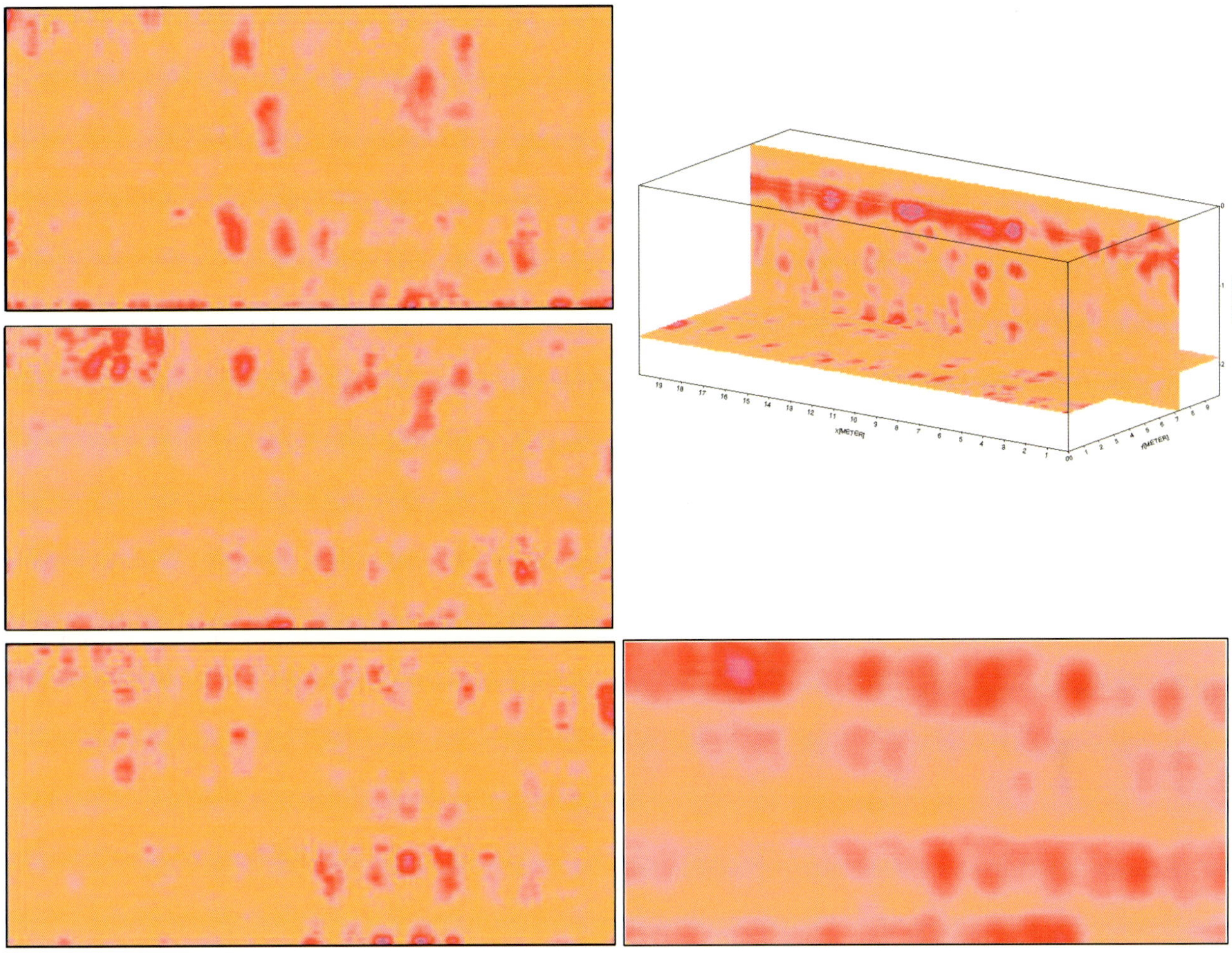

Figure 10.2: Ground Penetrating Radar (GPR) over a small section (20 × 10m) of the High Royds cemetery. Data were collected at 0.25m transects across this area and combined into a data cube. The top right image shows the cube cut both horizontally and vertically. On the left hand part of the diagram thin 'slices' have been generated which show horizontal maps at specific depth. From the top the depths are approximately, 1, 1.5 and 2m. From this information it can be seen that the responses differ significantly with depth and that while only 1,000 grave locations were originally planned for the cemetery is clear that many have more than one interment. In terms of the management of this historic graveyard the bottom right image is important as this is a thick slice that attempts to identify all of the graves within the cube. Combining the three geophysical techniques allows a better understanding of the previously 'unmarked' paupers' graves.

These developments should also be considered with respect to the automation of data collecting. For example, there is a natural progression in the way in which 'Square Array' earth resistance measurements have been collected. We can see that from manual insertion (Aspinall & Saunders 2005), a hand pulled cart system was then developed where the contacts for the measurements were inserted into the wheels (Walker *et al.* 2005). Additionally a trapezoidal array (the ARP© system, Automatic Resistivity Profiling), is now in use where the guidance system is undertaken by GPS (Dabas 2009). The last system is very heavy and was developed to test soils in vineyards. With suitable adaptations the ARP system can cover many hectares per day – a major change from manual data collecting levels. However, the heavy array and the additional GPS means that the system requires a large quad bike to function. As a management tool the rapid collection of many hectares a day is welcome but, on some sites, the use of motorised vehicles may prove impractical or even ethically difficult (see below).

How does geophysical survey fit with the management of heritage?

A number of authors have attempted to map the manner in which geophysical techniques have become 'nested' within archaeological practice (UK: Gaffney & Gater 2006, Clarke 1996; USA: Somers *et al.* 2003). Common to all approaches is a concept of strategies or methodologies varying as a result of the scale of the analysis. While the differences between the strategies are becoming increasingly small due to technological changes, there are still inherently different approaches to heritage management associated with scale. In management terms it has been traditional to use geophysical techniques to indicate presence or absence of archaeology within areas of presumed archaeological significance. However, management at a landscape scale poses additional problems, not least in visualising small scale change, but also identifying and interpreting such change. The latter goal may require terms of reference that are different from interpretation

Figure 10.3: Survey using the gridless Foerster system. The cart is navigated using real time GPS. High data density surveys can be undertaken in areas where conventional data grids are difficult to collect. An additional benefit of systems like this is the topographic data that is collected alongside the geophysical measurements.

at smaller scales as the significance of the mapped anomalies may be unknown, and indeed unknowable, without further, possibly invasive, fieldwork. However, the question of prospecting for archaeology at the landscape level can often be accomplished via simple presence or absence of 'anomalies'.

Site based management

If one were to interrogate the internet geophysics databases of either English Heritage (http://sdb2.eng-h.gov.uk/) or the North American Database of Archaeological Geophysics (http://cast.uark.edu/nadag/), then it would be apparent that, historically, geophysical survey has been used to map individual sites. Many of these single 'site' investigations have been for management purposes and have provided critical information on site extent as well as clarification of the nature of suspected remains (Figure 10.4). While the need to provide context for excavations is important in small scale survey, the ability to produce interpreted maps of the subsurface *without* excavation is an additional key element. For many national bodies across Europe it is likely that geophysical survey of individual sites will remain the most important use of this technology in the immediate future.

Hinterland based studies

Although much survey has been undertaken to inform archaeologists about the nature of archaeological sites it remains true that much work undertaken for general archaeological purposes has actually taken place outside known archaeological sites. This has often occurred within the hinterland of former towns or on land that

Figure 10.4: The World Heritage Site at Cyrene, Libya is a classic problem when considering the protection of unknown elements of a site when standing remains are present. As can be seen from the Quickbird image the site has large areas of upstanding remains and these elements dominate Cyrene. However, there are many open areas that are apparently 'blank'. Although the site conditions are difficult for ground based survey the use of a magnetometer that is guided by real time GPS through a virtual grid circumvents many of the practical issues (Gaffney et al 2008). In the image can be seen some of the additional structures found within the previously empty zones of the site – a management plan that includes both visible and buried archaeology can now be agreed.

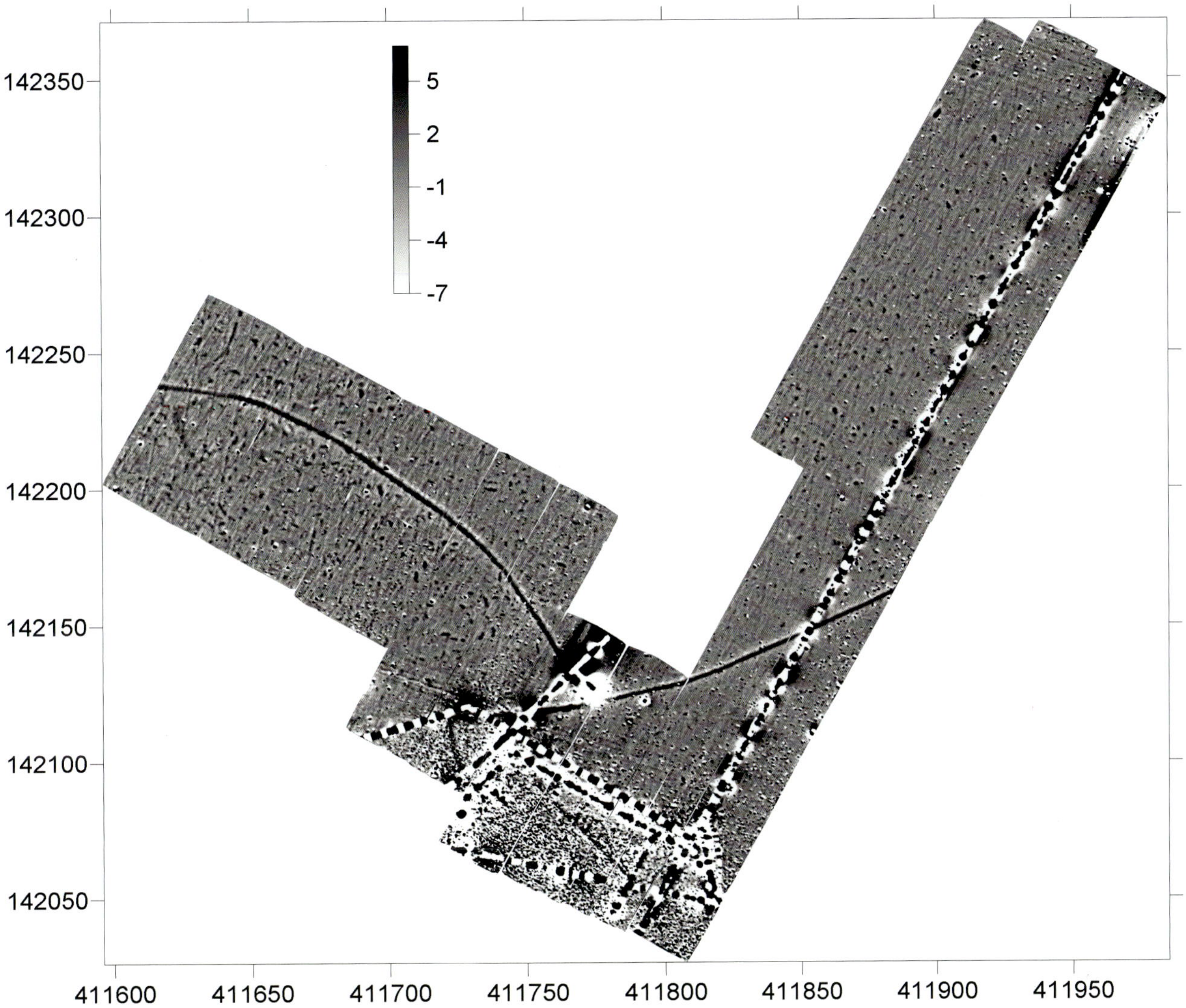

Figure 10.5: Magnetometer data collected using the Foerster gridless navigation system in the field west of Stonehenge. The magnetic data illustrates the variety of interpretive data that comes from such sources. The data are dominated by a modern pipeline and the footprint of the former RAF Stonehenge while the more subtle anomalies from the 'palisade' and smaller scale individual features can be seen.

is believed to have been suitable for settlement in the past. Unfortunately, the majority of such endeavours are bound up in commercial activity and, traditionally, this resource has been difficult to access. Within the UK at least, the inclusion of unpublished survey, evaluation and assessment reports within the Archaeological Data Service (http://ads.ahds.ac.uk/catalogue/library/greylit/) provides access to legacy reports that are of value to the heritage manager. Occasionally, however, these small-scale surveys can add significantly to the understanding of a landscape (Figure 10.5). Surveys within the Stonehenge World Heritage Site, for example, cover more than three square kilometers (David 2005, 14), which is approximately a quarter of the protected area. Significantly, despite the research based opportunities associated with this prestigious landscape, the majority of the geophysical survey in fact stems from commercial activities and chiefly relate to linear projects such as proposed road re-alignments. Under such circumstances the individual surveys contribute to a more meaningful analysis when considered with other similar surveys within the hinterland.

Landscape based studies

True landscape studies using ground based geophysical equipment are very rare. Probably the most complete example is the Heslerton Parish Project (Powlesland this volume). This has been achieved over a number of years, but it is clear that the rapid collecting devices described above place landscape-scale analyses increasingly within reach within short timescales. Such analysis has elements of the traditional site approach, but also *interprets* the context of the sites.

The relevance of rapid and accurately located data collection is evident at the landscape level but it also brings its own challenges. It is important when discussing heritage management, that we acknowledge that some landscapes may contain 'vulnerable' deposits that may be susceptible to erosion or damage. Consequently, there may be restricted access to such landscapes or even contractual agreements reducing farming activity to stabilise the resource. Under these conditions it may not be appropriate to drive tractors and quad bikes across the fields in the manner described above. However, lightweight, human propelled carts have been developed that negate these problems and yet collect data speedily, densely and accurately (e.g. Foerster magnetometers, Gaffney *et al.* 2008). Also some collecting systems such as the Geoscan Research MSP40 (Figures 10.6 & 10.7) allows more than one geophysical measurement to be collected (Walker & Linford 2006) and some systems use (relatively) light-weight quad bikes.

The significance of the new technologies to landscape-scaled studies is not simply their capacity to collect information, but the application of continuous measurement and analysis of space and the extension of the analytical sphere to virtually every part of a landscape. Some years ago the authors considered that the emergent remote sensing technologies were likely to transform our capacity for control of space through the reproduction of the surface of the landscape in an almost seamless fashion (Gaffney & Gaffney 2006; Gaffney 2008). At the time, the consideration of scale in extensive projects such as the North Sea Palaeolandscapes (Fitch *et al.* this volume), drove such an argument. However, whilst the methodological challenges overcome in that project have direct relevance to this discussion (in terms of data resolution, storage, manipulation and visualisation), the scale of such landscape exploration had relatively few counterparts in the terrestrial environment (although see Powlesland this volume).

This situation, however, is changing as our capacity to collect ground-based remote sensed data enhances prospection research agendas. Here we may return to Stonehenge where previous work by Birmingham University provides an example of what might be achieved through digital cartography (Exon *et al.* 2000). This study was aimed at an analysis of funerary and ritual monuments associated with Stonehenge. The study area of approximately 40 by 40km incorporated about 1,200 Neolithic or Bronze Age funerary and ritual monuments, including large cursus and henge enclosures. The monument complexes around Stonehenge often appear unique in their scale and intensity but their analysis is a commonplace problem in terms of landscape. Emerging over millennia the landscape is a 4-dimensional puzzle in which spatial relations of monuments (in Cartesian terms) must be merged with archaeological time (interpretative periods – Neolithic, Bronze Age) and personal time (measured by the scale of monuments). Interpretational significance is situated both in the past through the, presumed, liturgical and social significance of the monuments and the present, where our personal appreciation of the surviving landscape or situation of a monument must be gauged and represented. The archaeological problems of landscape are therefore quantitative, semi-quantitative and qualitative in nature and, in that volume, this was represented by GIS viewshed mapping, bubbleworlds and audio description embedded within a digital landscape.

As the capacity of technology to capture data increases exponentially we face new interpretative challenges. This becomes clear when one considers the scale of the most recent research project- the 'Stonehenge Hidden Landscapes Project'. This aims to integrate earlier larger scale data with work centred on the visual 'territory' of Stonehenge itself. With the development of remote sensing technology this project aims to explore up to 14 square kilometres of landscape with a data density and resolution at a 'human' scale (down to 8cm by volume in some instances), rather than topographic and, as a consequence, has potential reference to liturgical movement within the landscape (Figure 10.8). The outputs of the project are already elaborating on activity around Stonehenge but the output, in such a charged archaeological landscape, requires formats that not only represent interpreted anomalies but also reflects ambient significance. The representation of ring ditches simply as green humps clearly does not adequately reflect the monuments that may have delineated ritual space through their striking visual presence. The project, therefore, aims to re-present this remote sensed data within a gaming environment

Figure 10.6: MSP40 survey within the hinterland of an historic site. This 'low footprint' survey collects both earth resistance and magnetic data at the same time; the earth resistance is measured using contacts attached to the rim of each wheel, while the magnetic data is captured via the vertical white tube located between the cart and the operator.

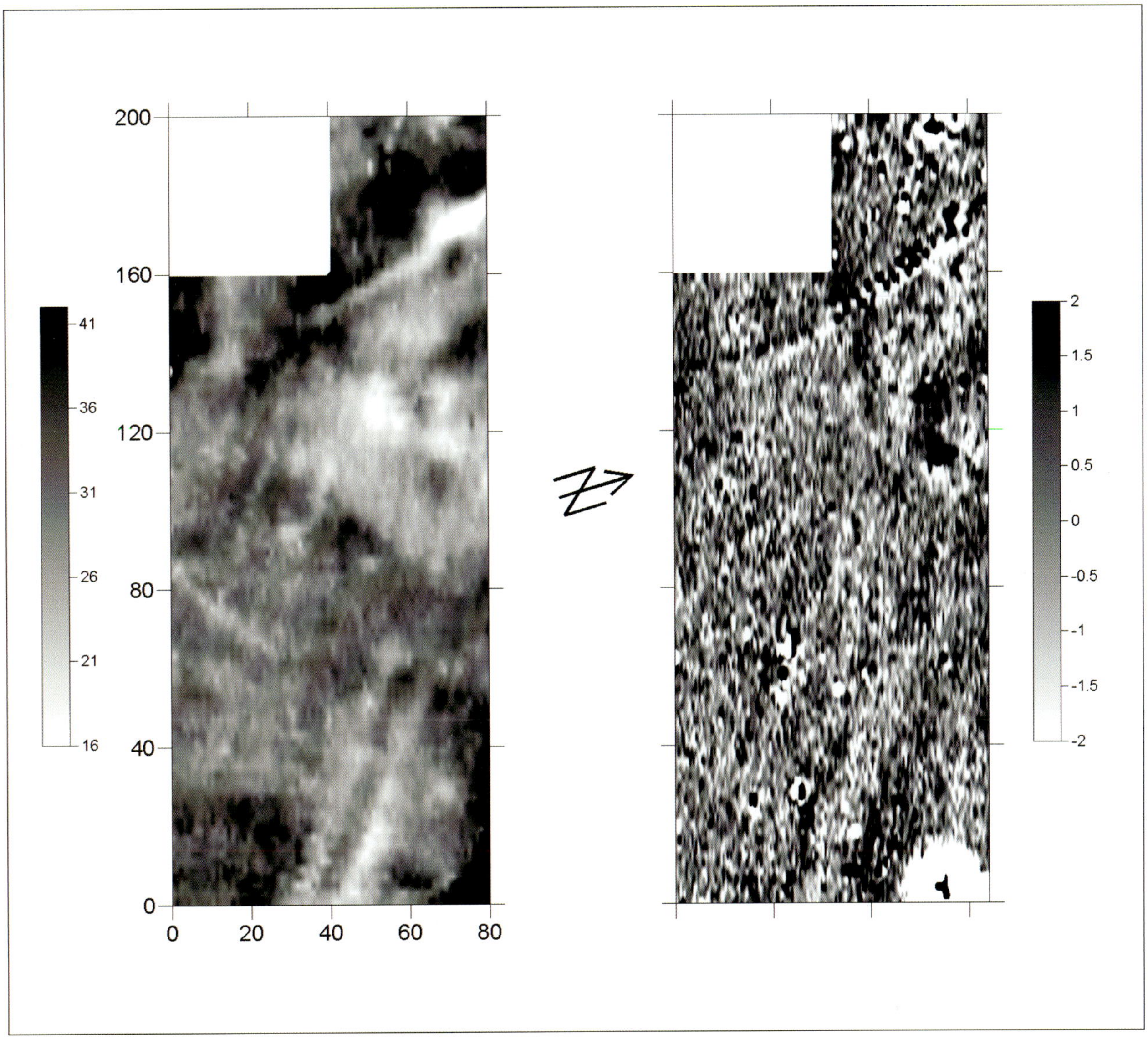

Figure 10.7: Earth resistance (*left*) and fluxgate gradiometer (*right*). Typical data collected using the MSP40 cart around the edge of an historic site. By combining the two data sets greater certainty is gained in the interpretation of this area. This is particularly true in this area where both methods indicate low level responses, probably indicating tracks and garden features, as well as geological and modern responses. As is often the case the latter categories often produce the most apparent anomalies.

to provide the ambient nature of interpretation and support an enhanced phenomenological research context. Such outputs have other qualities, other than simply being alluring. Within interdisciplinary research environments accessible visualisation can help bridge intellectual divides providing accessible interpretation to experts from different disciplines. Such software are also more likely to be understood and appreciated by a visually sophisticated general public, many of whom are eager to engage with archaeology (see also Remondino this volume).

Conclusions

The Stonehenge example demonstrates that, in a shifting technical environment, the key drivers for landscape-scaled prospection are increasing speed, greater data density, larger scale and multiple data types collected in one area. Likewise, key challenges in delivering enhanced management of the buried heritage include low 'footprint' data collecting devices, suitable visualisation and interpretation of small scale anomalies within expansive data cubes. However, in archaeological terms, our capacity to refine data capture and interpretation will, in turn, have to be matched by an academic rigour that supports the direct incorporation of data increasingly collected at a human density and resolution within digital outputs that reflect their archaeological context and interpretation. The future for archaeological geophysics will be a brave new digital world in which data, output and interpretation will merge within software which re-present complex composite data structures as immersive environments which can be interacted with at a personal level

Increasingly, archaeological geophysicists are considering their role in the determination of the value of the archaeological resource and the contribution of the suite of remote sensing techniques that are at

Figure 10.8: Rapid and dense GPR data collection on the 'Stonehenge Hidden Landscapes Project' using a Mala Imaging Radar Array (Mira) 16 channel system which produces a data cube at 8 × 8cm density. The antennas are housed in the yellow box and the data collection is tracked using a Total Station.

their disposal. As a result it is evident that the emerging agenda includes the following aims:

- Relevant for landscapes;
- Flexible for 'vulnerable' sites;
- Contribute to new analytical environments;
- Create real impact on the management and interpretation of the buried heritage;
- Provide not only traditional 'feature' interpretation but also 'broad brush' visualisation;
- Part of the archaeological debate rather than a separate technical discipline.

There is a realisation that to achieve these aims a number of technical and philosophical challenges need to be overcome. Those who work in remote sensing know how rapidly the technical context of their work has changed. It is clearly also true that philosophical context, especially the increasing relevance of interpretation to heritage management is also changing. Accepting these challenges will ensure a significant role for these techniques within the 4-dimensional puzzle that is modern archaeology as well as contributing to the enhanced understanding and ever more effective management of the archaeological resource.

References

Aitken, M.J. 1958: Magnetic prospecting. I. The Water Newton Survey. *Archaeometry* 1, 24–6.

Aspinall, A. & Gaffney, C. 2001: The Schlumberger array – potential and pitfalls in archaeological prospection. *Archaeological Prospection* 8, 199–209.

Aspinall, A. & Saunders, M.K. 2005: Experiments with the square array. *Archaeological Prospection* 12, 115–29.

Atkinson, R.J.C. 1953: *Field archaeology* Methuen, London.

Aubrey, L., Benech, C., Marmet, E. & Hesse, A. 2001: Recent achievements and trends for research for geophysical prospection of archaeological sites. *Journal of Radioanalytical and Nuclear Chemistry* 247, 621–28.

Bevan, B.W. 2000: An early geophysical survey at Williamsburg, USA. *Archaeological Prospection* 7, 51–8.

Clark A. 1996: *Seeing Beneath the Soil*. Batsford, London.

Conyers, L. 2004: *Ground-Penetrating Radar for Archaeology*. AltaMira Press, Walnut Creek.

Conyers, L.B. 2010: Ground-penetrating radar for anthropological research. *Antiquity* 84(323), 175–84.

Conyers, L.B. & Leckebusch, J. 2010: Geophysical archaeology research agendas for the future: some ground-penetrating radar examples. *Archaeological Prospection* 17, 117–23.

Dabas, M., 2009: Theory and practice of the new fast electrical imaging system ARP©. In Campana, S. & Piro, S. (eds): *Seeing the unseen: geophysics and landscape archaeology*. Taylor and Francis Group, London. 105–26.

David, A. 2005: Ground-based geophysical surveys. In Darvill, T. (ed.): *Stonehenge World Heritage Site: an archaeological research framework*. London and Bournemouth: English Heritage and Bournemouth University. 14–15.

David, A., Linford, N. & Linford, P. 2008: *Geophysical survey in archaeological field evaluation*. English Heritage, Swindon.

Exon, S. Gaffney, V. Woodward, A. & Yorston, R. 2000: Stonehenge Landscapes: JourneysThrough Real and Imagined Worlds. Archaeopress, Oxford.

Fassbinder, J.W.E. 2010: Geophysical prospection of the frontiers of the Roman Empire in southern Germany, UNESCO World Heritage Site. *Archaeological Prospection* 17, 129–39.

Gaffney C.F. 2008: Detecting trends in prospecting for the past: fifty years of geophysical survey. *Archaeometry* 50, 313–36.

Gaffney, C., Gaffney, V., Cuttler, R. & Yorston, R. 2008: Initial Results using GPS Navigation with the Foerster Magnetometer System at the World Heritage Site of Cyrene, Libya. *Archaeological Prospection* 15, 151–56.

Gaffney, C. & Gaffney, V. 2006: No further territorial demands: on the importance of scale and visualisation within archaeological remote sensing. In *From Artefacts to Anomalies: Papers inspired by the contribution of Arnold Aspinall University of Bradford 1–2 December 2006*, http://www.brad.ac.uk/archsci/conferences/aspinall/.

Gaffney, C. & Gater J. 2003: *Revealing the Buried Past: Geophysics for Archaeologists*. Tempus, Stroud.

Heron, C.P. & Gaffney, C.F. 1987: Archaeogeophysics and the site: Ohm sweet ohm? In Gaffney, C.F. & Gaffney, V.L. (eds): *Pragmatic Archaeology. Theory in crisis?* BAR 167. 71–81.

Hopewell, D. 2005: Roman Fort Environs in North-West Wales. *Britannia* 36, 225–69.

IFA, 2010: *Draft Standard and Guidance for archaeological geophysical survey*. http://www.archaeologists.net/sites/default/files/node-files/geophysicsSG.pdf.

Leckebusch, J. 2005: Precision real-time positioning for fast geophysical prospection. *Archaeological Prospection* 12, 199–202.

Linford, N.T. 2006: The application of geophysical methods to archaeological prospection. *Reports on Progress in Physics* 69, 2205–57.

Nuzzo, L., Leucci, G. & Negri, S. 2009: GPR, ERT and magnetic investigations inside the Martyrium of St Philip, Hierapolis, Turkey. *Archaeological Prospection* 16, 177–92.

Schmidt, A. 2003: Remote Sensing and Geophysical Prospection. *Internet Archaeology* 15.

Scollar, I., Tabbagh A., Hesse A., & Herzog I. 1990: *Archaeological Prospecting and Remote Sensing*. Cambridge University Press, Cambridge.

Somers, L., Hargraves, M.L. & Simms, J.E. 2003: *Geophysical surveys in archaeology: Guidance for surveyors and sponsors*. U.S. Army Corps of Engineers, Engineering Research and Development Center. ERDC/CERL SR-03-21. www.cecer.army.mil/.

Tsokas, G., Tsourlos, P., Papadopoulos, N., Manidaki, V., Ioannidou, M. & Sarris, A. 2006: Non destructive ERT Survey at the South Wall of the Akropolis of Athens. *Extended Abstracts of Near Surface 2006, Helsinki, Finland, B040*.

Tsokas, G.N., Tsourlos, P.I., Vargemezis, G. & Novack, M. 2008: Non-destructive ERT for Indoor investigation: the case of Kapnikarea church in Athens. *Archaeological Prospection* 15, 47–61.

Walker, R., Gaffney, C., Gater, J. & Wood, E. 2005: Fluxgate gradiometry and square array resistance survey at Drumlanrig, Dumfries and Galloway, Scotland. *Archaeological Prospection* 12, 131–36.

Walker, R. & Linford, P. 2006: Resistance and Magnetic surveying with the MSP40 Mobile Sensor Platform at Kelmarsh Hall. *ISAP News* 9, 3–4.

11 | Marine Geophysics: integrated approaches to sensing the seabed

Antony Firth

Abstract: This paper looks at the use of geophysics and remote sensing in marine archaeology, with the emphasis on their use in field evaluation. Single- and multibeam echosounder, sub-bottom profiler, sidescan, and magnetometer are addressed as geophysical methods; coring, seabed sampling, video and still cameras, Remote Operated Vehicles and divers are all considered as forms of remote sensing. Different motivations, themes, scales and types of operation are discussed, as are their implications for methodological innovation in three areas: position-fixing; event-based recording; and decision-oriented recording.

Introduction

Wessex Archaeology (WA) is a large not-for-profit company principally involved in land-based development-led archaeology. We have become increasingly involved in marine archaeology over the last 15 years and marine geophysics is now a central and substantial part of that work. We also use a variety of techniques that can be described as remote sensing. But rather than looking at individual results, this paper focuses on the integration of archaeological recording.

Our geophysical work usually has a direct consequence for subsequent interventions, whether they are construction works, statutory protection, or further investigations. Whilst the sensors we use – sidescan, sub-bottom profiling, single- and multi-beam echosounder, magnetometer – are common tools in marine surveying, the strength of the link between geophysics and 'what happens next' archaeologically has prompted approaches to recording that are novel. Consequently, this paper shows how integrated approaches to archaeological recording can significantly improve the contribution of marine geophysics to understanding the archaeological heritage of the seabed.

As for many others, much of our work can be characterised with reference to three broad phases of investigation: desk-based assessment; field evaluation; and mitigation. Whilst these phases are normally associated with development-led investigations, this phased approach is common in other scenarios.

Both marine geophysics and remote sensing are especially relevant to the field evaluation phase, that is, to establishing the presence (or absence) of archaeological material on or in the seabed, including its position, extents, period, character and so on. The use of marine geophysics and remote sensing overlaps in the evaluation phase with marine geophysics, providing data and/or imagery directly relating to material on the seabed, but through indirect means (predominantly sound and magnetic disturbance). Remote sensing – as used here – provides physical contact but from a distance, usually a boat on the surface.

Marine geophysics and remote sensing both provide field-based evidence, hence their status as methods of evaluation. The overlap is complementary: the indirect measures applied by marine geophysics are generally effective over a broad area, providing context in terms of a site's overall form and wider setting; remote sensing generates direct physical data but usually in a highly localised manner. Consequently, field evaluation in the marine sphere often comprises a combination of localised physical evidence with more extensive indirect evidence.

Whilst their main application is in field evaluation, marine geophysics and remote sensing are also relevant to desk-based assessment and mitigation. In desk-based assessment it is important to include as sources the results of previous geophysical and remote-sensing surveys, which might involve re-interpreting survey data or core logs not originally obtained for archaeological purposes but which can yield archaeological information nonetheless, as a sort of 'latent' data. The conduct of desk-based assessment should also anticipate marine geophysics and remote sensing in the evaluation phase, identifying key questions and targets for investigation, so that costly field methods are deployed efficiently.

Equally, marine geophysics and remote sensing need to be conducted in a manner that maximises opportunities for field data to inform mitigation and, in some instance, to provide the data that will form the basis of the mitigation itself. For example, high resolution geophysics can in itself serve as a record of a site that is to be destroyed or that needs to be monitored; geophysics data can also be used to provide an outline site plan that can be augmented by limited, targeted diving work, for example. With respect to remote sensing, it is becoming commonplace for mitigation

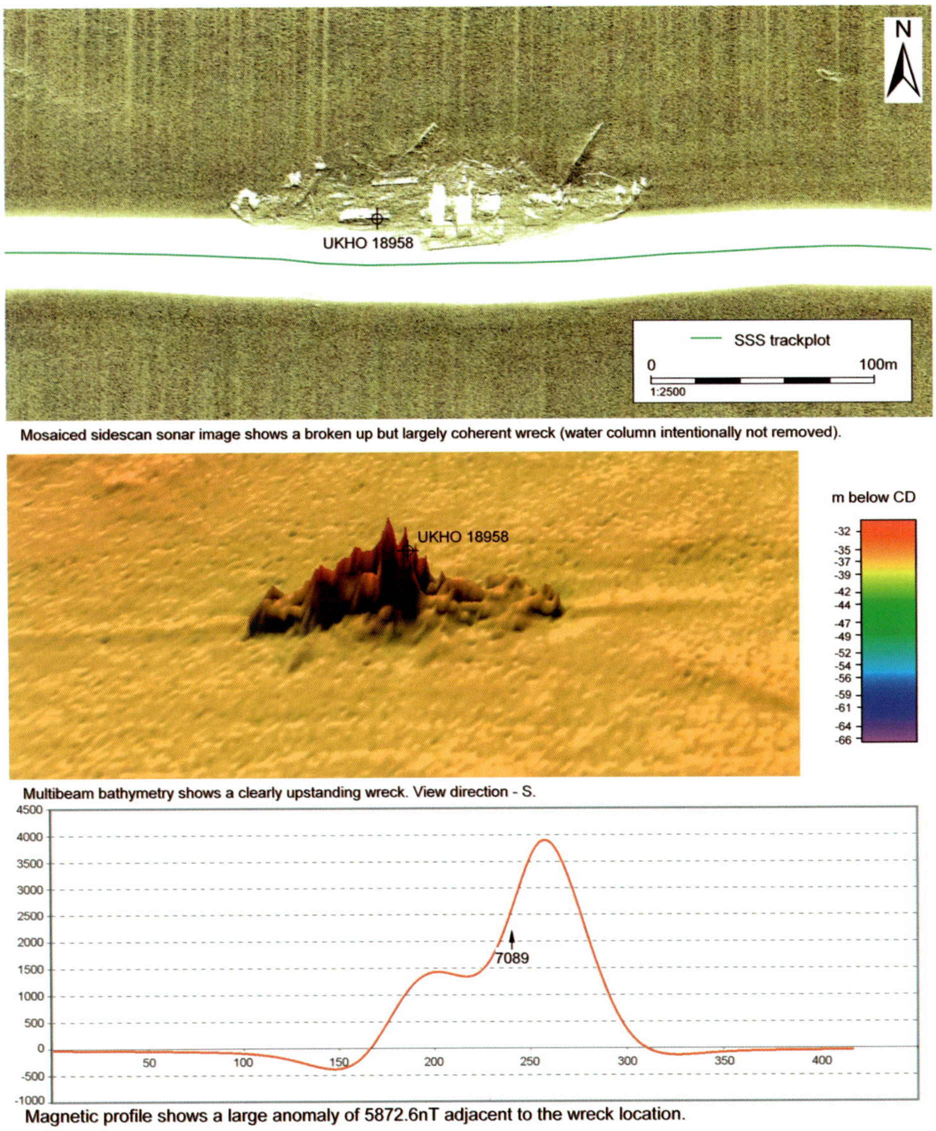

Figure 11.1: Different marine geophysical techniques furnish different types of information about the seabed and are most productive when used in combination rather than in isolation. The data show the wreck of the SS *Mendi*, a troopship sunk by a collision in February 1917 with the loss of over 600 lives. *© Wessex Archaeology, Crown Copyright.*

in the form of detailed palaeo-environmental analysis and scientific dating of key sediment units to be based on sub-samples obtained during the evaluation phase, rather than requiring a further episode of coring.

Geophysics

The principal marine geophysical methods that we use are set out below. A key characteristic of marine geophysics is that these instruments – because they work in different ways – each say something different about the seabed. They are at their most powerful when used in conjunction with each other, to compare and contrast the different datasets for any particular patch of seabed (Figure 11.1). Whilst it was usual to record marine geophysical data on paper even in the recent past, it is now much more usual for data to be recorded digitally, and there are various software packages available to facilitate complex data processing and interpretation.

Single beam echo sounder (SBES)

Single beam echo sounders use a pulse ('*ping*') of sound that rebounds from the seabed. The time that the *ping* takes to return can be calculated to give a value for depth, which can be corrected for tide to give an absolute height. Conventional echosounders provide depth data directly underneath the sensor, which is usually mounted in a vessel's hull or on a pole over the side. By sailing multiple lines, fixed by dGPS, it is possible to build up an elevation model. We usually use survey-specification echosounders in conjunction with sub-bottom profilers to calibrate the 'first return' (see below).

Multibeam echo sounder (MBES)

Multibeam works like a fan of single beam echo sounders to obtain depth data from a swathe of seabed beneath the vessel. This means that each line covers a considerable width of seabed and, by sailing lines that abut each other it is possible to create an elevation model covering 100% of the survey area. The high density of data is very good for imaging complex topographic features such as wreck sites.

Sub-Bottom Profiler (SBP)

Sub-bottom profilers, also referred to as 'shallow seismics', use a pulse of sound like a single beam echo sounder but at a frequency such that some of the energy penetrates the seabed and bounces off the boundaries between layers of different sediment. The returning data provides a vertical section through the sediment, which can be interpreted for different characteristics and for features such as cuts and fills indicating palaeo-landforms. The 'first return' represents the seabed, which can be calibrated with an echo sounder to give good control over the absolute depth of layers. Different sound sources are used, with different frequency and power characteristics. There is generally a trade-off between penetration and resolution: low frequency *'boomers'* provide good penetration of horizons of interest to archaeologists, but at a lower resolution than higher frequency *'pingers'* which do not penetrate as far. *'Chirp'* sub-bottom profilers use a package of frequencies to combine both penetration and resolution. Parametric sonars work on different principles but still use sound to obtain high resolution images of layers of sediment. As sub-bottom profilers only collect data from below the sensor then a series of parallel lines and perpendicular cross-lines are usually acquired in order to build up a picture of the seabed, where the detail of the picture depends on the density of the lines.

Sidescan Sonar

Sidescan also uses sound but projected from either side of a *'fish'* towed close to the seabed from behind the vessel. If the sound is reflected from sediments or objects on the seafloor straight back to the fish then they contrast with areas from which sound is deflected away, or from areas of acoustic shadow behind upstanding items or in depressions. The pattern of bright returns and shadows provides a qualitative but detailed image of the seabed that can be used to identify wrecks, rock outcrops, small anomalies, sediment characteristics and bedforms (such as sand waves and ripples caused by water flow). Again there is trade-off between range and resolution: high frequency sidescan creates a very detailed image but has limited range; lower frequency sidescan has greater range, but may not be capable of resolving the small and ephemeral anomalies of greatest interest to archaeologists. As sidescan sonar obtains data from a swathe each side of the *fish*, then – depending on the spacing of the lines and the range of the side scan – it is possible to build up 100% coverage of the survey area.

Magnetometer

Magnetometers detect minor variations in the Earth's magnetic field using a *fish* towed close to the seabed and far enough behind the vessel to avoid the interference from the vessel itself. Once the magnetic data have been processed it is possible to identify localised variations caused by the presence of ferrous material indicative of wrecks or other debris. The size of anomaly is proportional to the amount of ferrous material and its distance from the sensor: a small anomaly could be something large but deeply buried, for example. Magnetometers provide data for lines rather than swathes, so the detail of a survey is dependent on line density.

Remote sensing

In this paper, remote sensing is used in a relatively broad sense with respect to marine archaeology, encompassing a variety of methods where archaeologists are remote from the seabed. The environmental constraints on human access – needing a boat or breathing apparatus on one hand; weather, tides and poor visibility on the other – are such that all marine archaeology is, in effect, remote sensing. Archaeologists are highly reliant on technology, which mediates their access. Technological dependence combined with the highly changeable character of the environment means that the costs of investigation are high. In turn, investigations have to be of limited duration and high intensity.

The remote sensing methods discussed here are all concerned with achieving a degree of physical access to the seabed from a remote position. The main methods we have employed are as follows.

Coring

To obtain a core of sediment from the seabed to investigate its vertical sequence and the composition of each layer we typically use vibrocores, where a rig is dropped to the seabed on a cable (e.g. Figure 15.30). Within the rig, a vibrating weight drives a core barrel into the seabed. The core is then withdrawn and the rig is recovered to the ship. The core is then removed and cut into 1m sections for examination. Conventional 'shell and auger' coring is also carried out, using a drilling rig on a platform jacked-up on legs above sea level. Whilst vibrocores typically penetrate up to 6–8m into the seabed, conventional cores are often drilled tens of metres into the seabed. In both cases, cores are typically 100mm diameter and are suitable for archaeological recording and for obtaining sub-samples for palaeo-environmental assessment, analysis and scientific dating.

Seabed sampling

Different sorts of grabs and scientific trawls can be used to recover samples from the (near-) surface of the seabed. Typically used for examining flora and fauna that live on the seabed or buried just below its surface, seabed samples can also be used to obtain artefacts and other archaeological indicators. These tools are usually dropped to the seabed on a cable and then recovered to the vessel where the contents are sorted, recorded and sampled, or bagged for processing onshore.

Drop Cameras and Sleds

Video and still cameras can be deployed to the seabed from the surface, either as stationary 'drop' instruments that obtain vertical images of the seabed, or on towed sleds that can be dragged along the seabed with both vertical and oblique cameras. Again, used primarily for ecological purposes, drop cameras and sleds can be used archaeologically to obtain photographs of the seabed.

Remote Operated Vehicles (ROVs)

ROVs are manoeuvrable submersible platforms on which video and still cameras can be mounted. They can be piloted from the surface towards and around archaeological sites without the risks or limitations associated with divers, generating large volumes of imagery and mapping. Larger models have manipulator arms and can be used for complex tasks.

Divers

Although diving does place the archaeologist in direct physical contact with archaeological material on the seabed and might therefore not be regarded as remote, its characteristics as a method are not dissimilar from the other remote sensing methods described here. Generally, diving is constrained by weather, tidal currents, decompression requirements and expense, so that it must be used in a highly targeted way. As well as being brief in duration, the spatial range of a diver is often relatively limited. On the positive side, divers are usually equipped with helmet-mounted video cameras and voice communications so that archaeologists on the surface can see what is present and discuss it with the archaeologist who is diving; in this way the diver serves as a remote sensor for the archaeological team as a whole.

Motivations

As indicated above, the combination of marine geophysics and remote sensing is relevant to field evaluation arising from different motivations.

Development-led

The phasing of investigations in terms of assessment, evaluation and mitigation is associated primarily with investigations prompted by development proposals, where the intent is to establish whether archaeological material is likely to be affected by development and, if significant material is shown to be present, to carry out work to avoid or offset the impacts in a cost-effective manner. Investigations prompted by marine development have strongly influenced our methodological approach, especially with respect to maximising the use of previously-obtained geophysical and geotechnical data, and seeking integration with proposed surveys and their interpretation (Firth 2004; Evans *et al.* 2009).

Strategic research

A major drawback with development-led investigations is that their parameters are fixed by the scope of the development, whilst many of the questions that marine development raises – both methodologically and historically – clearly extend beyond such scope either spatially or in terms of available time or resources. Consequently there are numerous key questions that have to be addressed as a matter of research, albeit research that is closely tied to practical concerns. Our strategic research has been directed towards questions concerning the overall distribution, character and importance of particular types of monument, or of developing methods of data gathering and usage that will enable such questions to be addressed. In such cases we have used both geophysics and remote sensing to acquire data, and experimented with new uses of geophysics and remote sensing techniques (Bickett 2011; Hamel 2011).

Statutory support

Since 2002, a significant element of our work has been to assist the national heritage agencies in England, Wales, Scotland and Northern Ireland in their implementation of statutory heritage protection legislation. Typically our role has been to provide field-based data to establish whether statutory designation is warranted and to monitor sites that are already designated. Whilst greatest emphasis used to be placed on the use of diving archaeologists, in recent years we have increased the use of marine geophysics as a tool for confirming position, establishing extents, guiding the diver to key features and providing the base for site survey.

Multidisciplinary marine science

The high cost of marine survey, for any purpose, places a premium on integration so that the survey can be shared across disciplines. The benefit is all the greater for archaeology as it tends to be the minor partner to engineering, resource mapping, geology and sediment transport, navigation and hazard identification (e.g. ordnance) or ecology (habitats). Whilst we have become adept at the archaeological re-use of geophysical and remote sensing data acquired for other disciplines, there is increasing recognition of the value of building-in archaeological objectives and archaeological staff into multidisciplinary surveys. In particular, the recent and extensive Regional Environmental Characterisation (REC) surveys funded by the UK Aggregates Levy Sustainability Fund (ALSF) have been genuinely multidisciplinary with archaeology taking its place alongside geology and ecology (e.g. EMU Ltd 2009; James *et al.* 2010). It is to be hoped that multidisciplinary surveys will emerge even more strongly as a key driver for the acquisition of marine geophysical and remote sensing data for archaeological use.

Learning and access

A further motivation for using marine geophysics and/or remote sensing is to facilitate learning and access. In our own experience, learning and access has not usually been the main driver for geophysical or remote sensing investigations, but it has been a very valuable outcome, achieved opportunistically or as a planned 'secondary' outcome. Technological advances that enable the integration and scientific interpretation of features on the seabed are also – perhaps incidentally – creating results that are much more readily appreciated by non-specialist audiences. On the one hand, we are acquiring data that is much higher resolution and clearer than previously, whilst on the other we have interpretative tools and delivery mechanisms that allow us to present information more widely and in accessible formats. The contrast could hardly be greater between a few murky slides from underwater and overheads copied from paper rolls of barely discernable sidescan traces, shared with a face-to-face audience, and high quality digital images, clips and visualisations that are shared around the globe using Flickr, YouTube and the like. Technological advances are also creating new forms of public access, whether it is podcasts of diver's underwater descriptions, or learning about geophysical methods by modelling sidescan and multibeam traces using plasticine (Figure 11.2).

Figure 11.2: The increased accessibility of marine geophysical results enables much wider audiences not only to explore monuments on the seabed, but also to learn about the science that underpins the methods. © *Wessex Archaeology.*

Themes

Marine geophysics and remote sensing are directed towards features, sites and their wider setting relating to four major archaeological themes: prehistory; coastal activity; maritime remains (e.g. shipwrecks); and aviation.

Prehistory

The potential presence and significance of prehistoric material from former dry-land sites that are now submerged is currently high on many agendas (see Fitch *et al.* this volume). Sea level change driven by the cycle of glacial and inter-glacial periods that has characterised the UK over the past 700,000 years when our human ancestors have been present, if intermittently, has meant that areas of the Continental Shelf currently lying under tens of metres of sea water distant from today's shore were formerly inhabitable. Artefacts discovered offshore demonstrate that the 'in principle' argument for submerged prehistoric sites is borne out by physical evidence, yet trying to identify the presence of such artefacts beneath the waves is extremely difficult. The approach has, therefore, generally been landscape based – seeking to identify remnants of land surfaces and palaeo-geographic features where palaeo-environmental and scientific dating evidence might survive, and then using such geographical and environmental evidence to develop hypotheses about how the landscape might have been inhabited, including possible 'hot spots' where artefacts might survive. This landscape-based approach undoubtedly reflects the strengths of marine geophysical and remote sensing data and survey tools, as echosounders, sub-bottom profilers, coring and seabed sampling are most suited to area-based investigations.

Coastal activity

Whereas the early prehistoric features we seek to investigate may have been many miles from the contemporary coastline, from later prehistory to the present day there are examples of archaeological sites directly concerned with use or exploitation of the coastal environment and which lie close to the current coast. Coring is a key method of investigating the effects of reclamation on former marine and estuarine areas, for example (Firth 2000), and sidescan can be used to map fishtraps. In some cases, structures have been built in the water whereas in other cases, structures built on land or above water have become submerged as a result of erosion, collapse or subsidence (Figure 11.3).

Maritime

In many respects, maritime sites, especially shipwrecks, are the traditional core theme for marine investigations. The key means of investigation has been by diver, with ROVs increasingly used for investigations in deep water. In terms of geophysics, sidescan and magnetometer have been the main tools, though predisposed towards sites dating to the last 150 years that are likely to have substantial amounts of ferrous (i.e. magnetic) material and be prominent on the seabed on account of the use of metal and industrial construction techniques. In recent years, the use of high resolution multibeam has resulted in eye-catching digital elevation models of wreck sites that can be manipulated and visualised in a very engaging way. Like sidescan and magnetometer though, multibeam – being sensitive to topography – is best suited to prominent and therefore predominantly more modern wrecks. Wrecks that leave only an ephemeral trace on the seabed, and which do not contain large amounts of ferrous material, are difficult to identify using the geophysical tools currently available. High resolution sidescan – used and interpreted by experienced archaeologists – holds the greatest promise for identifying older wreck sites.

Aviation

Aircraft crash sites have become a focus for archaeological interest in the UK only in relatively recent years. Several distinct phases of operations during World War II resulted in numerous losses in the sea around the UK, and these waters have probably the greatest number of aircraft wrecks of any comparably-sized sea area in the world and may hold unique information. Although relatively recent in date it is worth noting that some aircraft types are represented only by their wrecks; there are no surviving examples

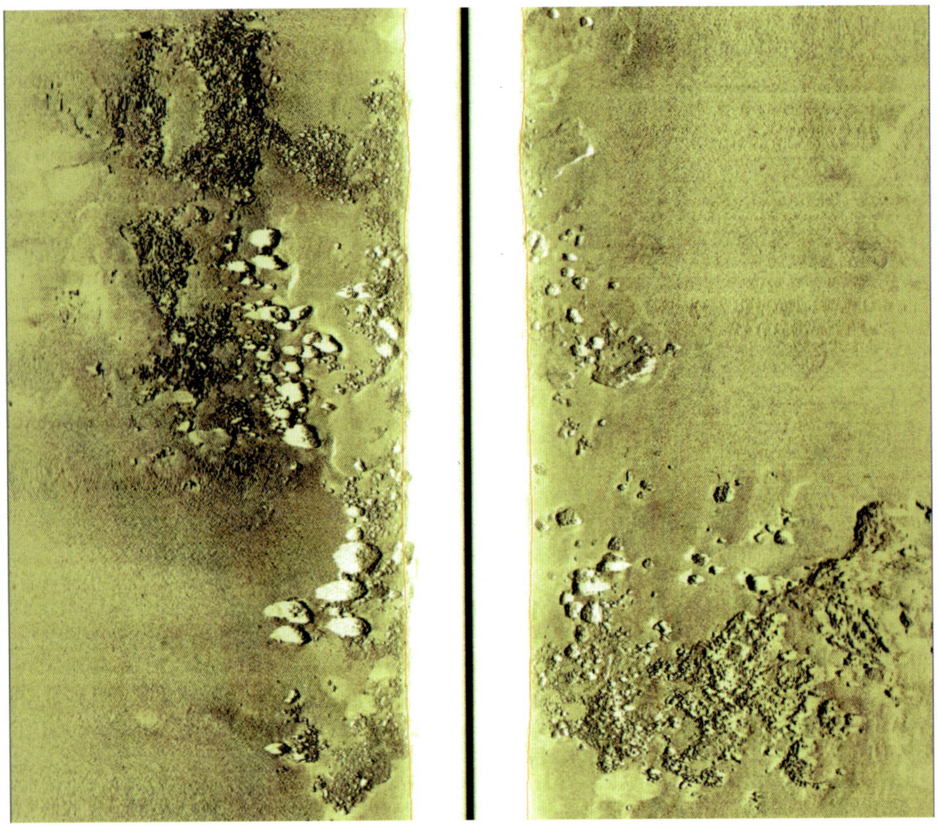

Figure 11.3: As well as wreck sites and remains of prehistoric land surfaces, geophysics can be used to identify the remains of structures that were once at the coast or on land. This sidescan image is thought to show the remains of a church from the once prosperous medieval port of Dunwich in Suffolk, which has largely been lost to the sea. © *Wessex Archaeology, courtesy English Heritage.*

in preservation (Holyoak 2002; English Heritage 2002). And, whilst the fixed infrastructure of 20th century warfare on land has received a fair amount of archaeological attention (e.g. Bacilieri & Thomas 2010), the aircraft crash sites are the principal monuments to combat itself, especially from key campaigns such as the Battle of Britain, the different phases of the Blitz, the Battle of the Atlantic, the Allied strategic bombing offensive, Operation Overlord (D-day) and so on. Military air crash sites are also automatically protected under the Protection of Military Remains Act 1986 (English Heritage 2002). As with shipwrecks, sidescan is the main marine geophysical method for investigating aircraft wrecks, with divers and/or ROVs serving as remote sensing. However, air crash sites are smaller and even more ephemeral than shipwreck sites due to their fragile construction and the often catastrophic nature of their impact with the sea.

Scales

Our involvement in the use of marine geophysics and remote sensing has to accommodate investigations at widely different scales, from broad regional studies to recording and analysis within a site.

Regional

A relatively recent development – at least in terms of archaeological involvement – has been regional surveys encompassing entire sea areas. Following a Government led assessment of aggregate resources and constraints in the Bristol Channel, and anticipating the introduction of Strategic Environmental Assessment (SEA), the marine aggregate industry in the UK carried out its first Regional Environmental Assessment (REA) in the Eastern English Channel, published in 2002 (Posford Haskoning 2002). Although archaeology only featured in the desk-based element of this first REA, subsequent assessments of the Thames, South Coast and Anglian regions have involved archaeological interpretation of newly-acquired data. In a linked but parallel development, ALSF-funded Regional Environmental Characterisation surveys have come to include archaeological objectives – and archaeologists – directly in the acquisition phases. In contrast, Government-led SEAs for the energy industry (oil and gas at first, but later including offshore renewables) have involved only desk-based archaeological work, though even this has been very influential. The approach to licensing offshore wind farm developments has also become regional, with extensive 'zones' subject to assessment, and (at least in some cases) involving archaeologists in survey work. Dealing with such very large areas has required a new approach to the application of marine geophysics and remote sensing, for archaeology as well as other disciplines. The investigations are intended to characterise the sea area, to provide an evidence base for understanding possible cumulative and in-combination effects, and to facilitate more focussed Environmental Statements that will speed the process of seeking consent. As a result of these objectives, surveys can be partial, spatially selective samples rather than seeking 100% coverage. The RECs adopted a 'corridor' approach whereby narrow strips of seabed separated by several kilometres are surveyed. Inferences are then made by using the results from the corridors to 'characterise' the whole area. At least some of the wind farm zones are also adopting a corridor approach, either in acquisition or though selective interpretation of datasets.

Area

The majority of our involvement in development-led archaeology has been at development-scheme scale, i.e. of an area of seabed coinciding with the footprint of the development. The extent and overall form of the survey area varies considerably, reflecting the variety of development schemes. Cables and pipelines, like linear schemes on land, result in a long narrow strip of seabed being investigated. New navigation channels accompanying new ports are also linear, but tend to be quite broad. Aggregate dredging areas are very variable in shape, whilst offshore wind farms usually comprise an irregular but large polygon accompanied by a long corridor to shore for power cables. In each case, the survey area is not driven by what might be most productive or appropriate to understand the archaeology of the area, but by the footprint of the anticipated impact. This is obviously problematic where archaeological questions need answers that lie outside the footprint, hence the value of the regional characterisation approach referred to above. However, the approach to investigation normally encompasses a reasonable margin beyond the immediate footprint, either as a buffer or to enable flexibility and 'micro-siting' (moving elements of the development to reduce impacts) in the detailed design of the development. As well as the form and extent of the survey area being driven by development rather than archaeology, the detailed design of the survey will be optimised for the development, not necessarily for obtaining the best data in archaeological terms. For example, geophysical surveys will usually be carried out in a series of parallel lines spaced at a uniform distance. These lines may not be optimal for imaging wreck sites or palaeo-channels, for example, and there will often be little scope to run additional lines to address new discoveries. Similarly, seabed sampling and coring will be carried out on a grid or will be targeted according to engineering or ecological questions, which may not be optimal for archaeology. Nonetheless, there have been instances when it has been possible for archaeologists to change plans to enable archaeological questions to be addressed alongside the primary concerns. Whilst the majority of our area-based surveys are development-led, we have done some area surveys for primarily archaeological purposes, though generally only where there is an extensive feature (such as a known navigational hazard) where the density of archaeological material is expected to be high.

Site

In contrast to the predominantly development-led area approach to investigation, our archaeologically-led investigations have tended to be site-specific. That is to say, we carry out smaller, localised surveys around the known/presumed position and extent of an archaeological feature in order to obtain the best possible data for archaeological purposes. In the case of shipwrecks, for example, sidescan survey lines will be run along and perpendicular to the apparent long axis of the wreck and slightly offset so that we get the best possible image of the site. Equally, if trying to investigate a palaeo-channel, sub-bottom profiler lines will be run across the channel to obtain a series of cross sections. Seabed sampling may be based on grids or transects if trying to identify broad spatial relationships, or judgement-led where specific elements (e.g. sedimentary horizons) are targeted.

Intrasite

For several decades, geophysical instruments have made it possible to identify the presence and position of an archaeological site such as a wreck. Once located, however, subsequent survey of the site itself had to be based on diver or ROV based methods, such as taped measurements, photomosaics and so on. However, in the last decade or so, major improvements in equipment and positioning have considerably improved the resolution of geophysical surveys, so that it is possible to record the details of a site directly from geophysics. This is most apparent in high resolution multibeam survey, where the digital elevation model is a fully quantified 3D survey from which further measurements and dimensions can be obtained. Although offering qualitative rather than quantitative data, sidescan can offer even better resolution than multibeam, especially of wrecks in deeper water, enabling individual timbers to be discerned, or the position and orientation of individual cannon to be recorded. Although the geo-referencing process can reduce the clarity of sidescan images, it is possible to use the resulting data for obtaining precise positions and dimensions, and for mapping the extent of wreck elements. Notwithstanding, intrasite survey directly from geophysics does need to be supplemented by on-site observations by divers or ROVs because the methods are only capable of recording topographical anomalies; they do not necessarily reveal the archaeological item or feature that caused the topography to be anomalous; or what features or items may be present without causing an anomaly. We have had considerable success, however, in combining geophysics with diver or ROV based investigations, with the general extent and form of a site being derived from the geophysics (thereby limiting the call for time-consuming in-water surveying by divers) and then annotating the basic plan with comments and details based on diver / ROV observations.

To conclude this section, it should be obvious that these different scales of investigation serve different purposes and are complementary rather than alternatives. The greatest advantage and efficiency accrues where there is scope to 'nest' different scales of surveys as appropriate, pinning down regional characterisations with site-specific details, and placing individual aspects of a site within its wider contexts.

Operations

Our approach to marine geophysics and remote sensing has to encompass four phases of operations: data acquisition (i.e. the survey itself); re-use of data (where the data has already been acquired for another purpose); processing and interpretation; and archiving.

Acquisition

In comparison to our interpretation of data acquired by others, new acquisition forms a relatively small proportion of our work. Nonetheless, involvement in acquisition means that we can have direct experience of the entire process and, in particular, has allowed us to explore methodological developments in achieving high resolution datasets and in adapting existing technologies to archaeological objectives. Another aspect of our involvement in acquisition is where we are not carrying out the survey ourselves, but are engaged alongside the survey company, providing a form of watching brief or using the opportunity afforded by the investigation to acquire records and samples for specifically archaeological purposes. Our staff are often called upon, therefore, to attend geophysical or geotechnical surveys, to take part in joint recording and sampling of cores, or to join existing diving teams as 'embedded' archaeologists.

Re-use

The predominant mode of our involvement with geophysical and geotechnical data arising from development-led work is in re-use of data (see also Fitch *et al.* this volume). For the majority of development-led projects, geophysical and geotechnical data is acquired to meet the needs of a wide range of engineering and environmental studies. The timing, objectives and strategy for the survey are not usually determined by archaeological concerns. Effort is required, therefore, to optimise and streamline the processes whereby this non-archaeological data can be used to satisfy archaeological objectives. One methodological consequence is that we carry out a data audit as a first stage of any major project, to ensure that all the relevant data is available in the right format, is suitable for archaeological use and covers the required area. The audit also forms a basis for discussing processing strategies with the client, and to fix estimates of timescale and costs.

Processing and interpretation

Whether we have acquired the data ourselves or are re-using data acquired for other purposes, the subsequent processes are the same. Whatever the type of data, various processes are undertaken to convert, check, collate and optimise data for interpretation. Depending on the data, processing might take place within proprietary software, or through processes we have developed ourselves. Processing needs to be conducted systematically, consistently and conscientiously so that interpretation takes place on a firm foundation. Far-reaching and costly decisions may be made on the basis of our results – relating to a statutory designation or the layout of a wind farm, for example – so our work must be open to scrutiny; a transparent framework that enables our results to be worked-back and contested is one of the important objectives of the processing phase. Interpretation – the conversion of data into meaningful archaeological information – also makes use of a variety of procedures depending on the data type, again ranging from proprietary solutions to in-house mechanisms. Interpretation is predominantly a question of judgement, as there are no practical tools capable of automatically identifying features of archaeological interest. For the reasons outlined above, however, interpretations must be contestable so we have developed working practices to ensure that the exercise of judgement is consistent and can be re-visited in the light of queries or further data becoming available.

Archiving

Archiving marine archaeological data is a burgeoning problem that, in the absence of imminent resolution, results in contractors such as WA becoming *de facto* archives, nominally temporary and without the facilities or procedures of a recognised publicly-accessible archive. Part of the reason for this unfortunate situation is that marine archiving is a complex matter. A project archive may include paper and drawn records, environmental sub-samples and artefacts (which, if waterlogged from the sea, present particular problems), plus digital material ranging from video and still photographs to major datasets. Even the digital datasets may be complex, requiring distinctions to be maintained between 'raw', 'processed' and 'interpreted' data, and making problems for the future because of the need to keep migrating digital archives as software and hardware becomes obsolete or is updated. There are also differences in the anticipated purpose of archiving; in some cases it is intended primarily to provide a secure, permanent record, whereas in others the intention is to enable widespread re-use of the data. Such different approaches have implications for how archive material is best stored and managed. A further level of complexity is added by differences between archiving trends in the marine data and archaeological data communities, exacerbated in some cases by different approaches in each of the UK's home countries. For all of this complexity, archiving is an important operational phase for us to be involved in, because it is vital to anticipate the final form that data will take when considering how best to conduct operations in acquisition, re-use, processing and interpretation.

Implications

Our application of marine geophysical and remote sensing methods, whilst focussed on evaluation, has to accommodate both desk-based assessment and mitigation. As I've shown above, there is very considerable variety in the circumstances we address across motivations, themes, scales and operations. Both our need to cope with such variety, and our experience of it, have brought several methodological concerns to the fore and have propelled us towards some innovative approaches. Three such concerns, which are set out in more detail below, concern position-fixing, event-based recording and decision-orientated recording.

Position-fixing

Our marine work generates numerous datasets. Whilst we can make inferences from each dataset in isolation, using multiple datasets in combination, comparing and contrasting, supports incisive inferences and greater confidence in the results. Accurate horizontal and vertical position-fixing provides a common frame to relate different datasets to each other.

Positioning is also critical as our investigations are often motivated by questions that are themselves spatial. In research, for example, the absolute depth of a horizon identified by sub-bottom profiling or vibrocore helps relate the data to sea level curves. In development, it is critical to know the distance between an archaeological feature and the proposed position of a cable, foundation or dredging area. In statutory protection, areas designated by law must adequately cover the feature they are intended to protect.

Position-fixing to relate datasets to each other and to the real world is vital at sea because, other than a distant coastline or a few navigational buoys, there are no fixed features such as walls, roads or buildings that can be used as a visual reference. Furthermore, the seabed is usually at a distance from the operational platform, separated by a water column that moves laterally and vertically with considerable force and is often so turbid that through-water visibility may be a few metres or zero. As we have to pass equipment or people through this medium, knowing where we are on the surface is a very important aspect of trying to know where we are on the seabed, and where we have been. With visibility so poor, movement so restricted and the costs of investigations so high, achieving an absolute accuracy of 5m is a realistic expectation for equipment that is dropped or towed, with sub-metre accuracy being our aim for divers or ROVs that can be navigated underwater.

In practical terms, dGPS has had a radical impact in marine archaeology. Within Wessex Archaeology, for example, our exploration and take-up of dGPS was much quicker in investigations at the coast and offshore than on land. In the horizontal plane we were quickly confronted by the need to be very careful about geodesy – knowing what projection and co-ordinate system was being used and how transformations were calculated. This need was, of course, driven in part by the increasing inaccuracy of familiar land-based systems as distance offshore increases, and by the variety of systems that are available at sea. Features on the seabed can appear to be displaced by tens of metres by different projections and co-ordinate systems, which was not a concern when position-fixing at sea was itself accurate only to tens of metres. The lack of geodetic information for legacy data that was acquired historically (perhaps only 10 years ago!) is a real limitation on the utility of this information. Having learned the hard way, we have instituted specific protocols for recording geodetic information as metadata and for processing transformations so that they are rigorously checked and are auditable.

Conventional dGPS is not especially strong in the vertical plane, which is challenging because our working level (the sea's surface) changes with short-term impacts (waves and swell) and with longer term tidal cycles. Correcting for these effects is critical in collating data from height-sensitive sensors such as multibeam. Elevation models of the seabed can only be constructed if all of the movement of the multibeam sensor can be cancelled out and its position related to an absolute vertical datum. An array of onboard sensors and fast processing can be used to calculate a 'fixed' position for the multibeam head, which can be related to a vertical datum using tide predictions, very accurate positioning (RTK GPS) or using data from a tide gauge to make corrections afterwards.

Our main methodological development has been in adapting acoustic-tracking technologies for day-to-day use by diving archaeologists or on ROVs. Acoustic tracking is commonplace in offshore engineering, but they are packaged and priced for major industrial applications rather than archaeologists. The advantages it presents over conventional diver-based archaeology are, however, radical. Relating the position of a feature such as a wreck on the seabed to the real world has tended to require extensive relative measurements (themselves often difficult to obtain because of the environment) tied back somewhat imperfectly at a small number of points to a GPS (at best) on the surface. Navigating a diver to such a feature, especially for the first time or, if prospecting across an area of seabed that might, or might not, contain such a feature was even less satisfactory. The diver would start from a nominally fixed point and then work their way from it using a tape measure or offsets from a baseline. In contrast, acoustic tracking uses one or more beacons whose real world position can be calculated to work out the position of a beacon on the diver or ROV. The diver's position can be seen by the surface team on a screen and they can be navigated towards a target. Once at the target, the diver's beacon can be used to build up a network of 3D points delineating the feature. And if no feature is found, there is at least an accurate record of the area that has been searched.

Recognising that we needed to bring the technology in-house if we were to work out how it might best be used, we invested in two different systems and in the software development needed to marry the acoustic tracking outputs to a GIS. This aggregation of two systems was especially important, as it enabled us to bring other georeferenced datasets – charts, previous site plans, and geophysical data – into direct use in the course of diving / ROV operations. That is to say, the diver could be directed around the seabed on the basis of such other data in real time, rather than just using the acoustic tracking data as an overlay post-fieldwork. The integration of acoustic tracking and GIS also enabled the development of a digital recording system for diver- and ROV-based archaeology (below).

Event-based recording

Our approach to recording is informed by the Monument-Event-Source model, with particular emphasis on recording events. In particular we record a variety of sub-events, which are the individual observations that enable archaeologists to assemble an overall 'picture' of a monument. This concern for recording events is directly attributable to the character of the environment and our reliance on geophysical and remote methods. On land, the position, form and character of a monument can be discerned quite readily and there is generally no great difficulty in proceeding directly to make a record of the monument itself. At sea, we tend to get brief glimpses over very short distances, in which much is obscured or unfamiliar. The knowledge gained from these glimpses is quite likely to be challenged or changed by the next observation. Inferences have to be drawn from multiple sources, all of which are partial; none comprehensive. Whilst uncertain, earlier observations may not be 'wrong', so we need to be able to add information without earlier

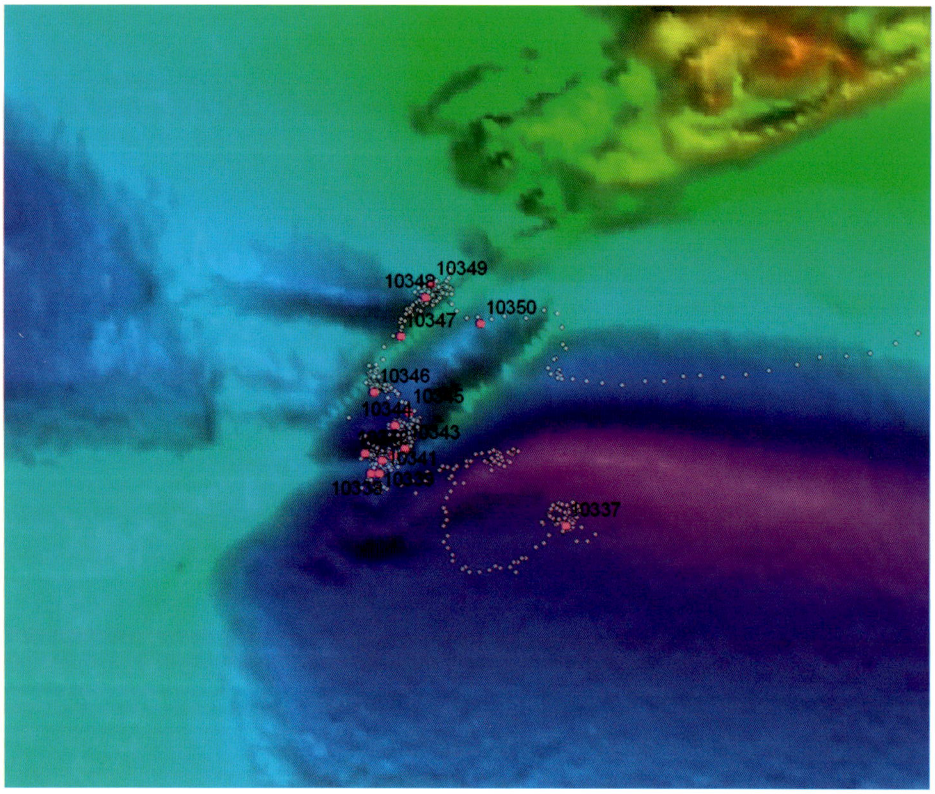

Figure 11.4: Acoustic tracking enables diving archaeologists to be navigated around sites in very low underwater visibility on the basis of geophysical images, and for observations to be recorded spatially. This image shows a diver's track and observations on the site of the *London*, lost in the Thames in 1665. *© Wessex Archaeology, Crown copyright.*

information being deleted or replaced. Further, our recording process has to be transparent, auditable and contestable – by others and indeed by ourselves as new information comes to light. It is this transparency and contestability that renders our observations 'objective' despite their provisional and contingent character. Building up observation on observation, sub-event by sub-event, it can be seen that monument records are an interpretation towards which we progress, not our starting-point.

Two examples show how this works in practice. We have a digital recording system known as Diva that operates through a database attached to the same GIS that we use in conjunction with acoustic tracking, as discussed above. Whilst the diver is being navigated around the geophysics image, they can make a comment on what they find. At the surface, an archaeological recorder using Diva makes a new, uniquely numbered Observation record, which is fixed in space by the acoustic tracking (Figure 11.4). The Diva recording system opens digital pro forma that prompts the recorder for information, including querying the diver on the seabed for further details. If the diver takes photographs or samples, these can be recorded and cross referenced to the Observation. Observations can also be classed so that they are represented in different colours in the GIS as the dive progresses. Where a ROV is being used rather than a diver, then the archaeological recorder at the surface can make Observation records directly in response to their own findings. On completion of the dive, Diva collates the Observations so that they can be reviewed as a whole and in conjunction with other Observations from previous dives. It is on this basis that a Monument record can be developed, still within Diva, in a manner that preserves the link to the original observations.

The second example is our approach to interpreting marine geophysical data. Although not as explicit as in Diva, a similar approach prevails. Each source of information has a unique number block: monument records from desk-based information start at 1,000, sidescan anomalies at 2,000, magnetometer anomalies at 3,000 and so on. Monument records and anomalies might seem to be real 'things' in the first instance, but in conceptual terms they are better regarded as observations. That is to say, a signal returning to an instrument that appears anomalous is really a sub-event of the survey; it only becomes 'an anomaly' through interpretation. Equally, existing records of wrecks and other features on the seabed are often so poor that they are themselves best regarded as time-specific observations, which must be corroborated if they are to afford confidence.

Our way of working is to examine each dataset in its own right and to identify all the anomalies we see in each dataset alone. This is in contrast to approaches that look at datasets in conjunction from the start, implicitly focussing on the more obvious and directly corroborated features rather than on highly ephemeral traces that may be of greatest potential interest to archaeologists. The character of geophysical surveying is that datasets will overlap because different instruments are deployed on the same survey lines and because survey lines run adjacent to each other and are intended to overlap slightly. In addition, there may be crosslines perpendicular to the main survey lines or lines from previous surveys, plus underlying data from monument or wreck records. Using GIS, the records from each data source can be viewed as a whole in order to identify 'groups' which are given their own unique identifiers. A group may be made up of a variety of underlying records – a couple of sidescan anomalies, a magnetic anomaly and an historic reference to somewhere a fisherman snagged his nets, for example – all of which might be some distance apart but, taken as a whole, can be interpreted as a single feature. Importantly, single records are also treated as a 'group' as even uncorroborated records need to be carried forward as possible features pending further information. The grouped anomalies are characterised to arrive at the best interpretation possible taking all the sources into account. Many will be set aside as being unlikely to have an archaeological origin, but nothing is deleted. The next bit of survey work could challenge the earlier interpretation and we need to be able to re-trace our steps, all the way back to the individual observations.

Decision-oriented recording

The difficult environment, uncertainty and cost of investigation all place a premium on maximising the return from every minute of fieldwork; this is no exaggeration. Our efforts – and our clients' cash – have to be focussed precisely on the project aim and objectives, whilst other possible lines of enquiry are ignored. Our fieldwork is thus selective, which is not necessarily different from land-based investigations except that the gradient of selectivity is steeper and more explicit. Strongly selective fieldwork cannot, however, be arbitrary, just a matter of the personal predilection or interest of the principal investigator. Rather, selectivity has to be based robustly on wider considerations about what is important about the historic environment.

The case for a rational system of selective prioritisation of fieldwork makes sense when considering individual sites, but the need is more pressing when there are numerous sites that multiply the pressure on resources and time available, as occurs on major construction projects. Contemplating major fieldwork in connection with the proposed new port at Dibden Bay in Southampton Water, WA drew on recent work on a complex urban scheme in Stonehouse, Plymouth, where conditions requiring extensive building recording were framed around a system of recording levels published by the Royal Commission on the Historical Monuments of England (RCHME). The RCHME levels were adapted to make them applicable to marine sites of all forms, though in practice we have used them predominantly for informing wreck recording.

Our system comprises five principal levels of recording, each of which has an explicit objective tied to decision-making. The basic levels are split into subsidiary levels (see Table 11.1). The levels are not stages; it is not necessary to progress from one to the other, or that the higher levels can only take place once lower levels have been achieved (though the decision to carry out recording to level 3, 4 or 5 may presuppose that information equivalent to level 1 and/or 2 is already available). In practice, investigations are often targeted at a couple of levels (i.e. level 1/2; level 2/3), acknowledging that environmental constraints may limit what can be achieved, and that decisions about deploying and targeting resources may have to be made quickly in the field if the site proves to be more/less interesting than indicated prior to investigation. In this, a key strength is that this system of levels provides an explicit indication of the overall intention of any particular investigation. Staff can make their own decisions on how best to proceed based directly on local circumstances, informed by the overall objective but not tied to a prescriptive list of observations to be made or dimensions obtained. Also, the objectives are clearly related to the decisions that will have to be made by third parties such as developers, regulators or local authorities. Even if they encounter difficulties and recording is partial, field staff can apply the specific objective as a 'test' of what they have achieved, to see whether they have done enough to inform third parties' decision-making.

Although developed with field-based recording in mind, the system of levels can incorporate geophysics and remote-sensing based recording, either as a contributing component that helps meet some of each objective, or as the primary means of investigation, at least for levels 1-3. Practically, geophysics and remote sensing are the predominant means of achieving level 1 recording, often providing the main evidence-base for desk-based assessment and Environmental Impact Assessment. Geophysics and remote sensing also predominate in establishing overall extent and character at level 2; depending on the site, geophysics and remote sensing may also enable generalised

Table 11.1: The five principal levels of recording applied by Wessex Archaeology.

Level	Type	Objective
1	Assessment	A record sufficient to establish the presence, position and type of site.
2	Evaluation	A record that provides sufficient data to establish the extent, character, date and importance of the site.
3	In situ Recording	A record that enables an archaeologist who has not seen the site to comprehend its components, layout and sequences.
4	Removal	A record sufficient to enable analytical reconstruction and/or reinterpretation of the site, its components and its matrix.
5	Inter-site Analysis	A record that places the site in the context of its cultural environment and other comparable sites.

evaluation of date and importance, where the form of a wreck, for example, places it in a broad category of site. Components, layout and sequences (level 3) can also be established using geophysics and remote sensing, especially where high-resolution data can be acquired. We are increasingly using high-resolution geophysics as a basis for in situ recording of extensive sites; and in the case of submerged prehistoric land surfaces and deposits, then geophysics and remote sensing are critical to understanding sequence.

Conclusion

As this paper has demonstrated, geophysics and remote sensing already provide a powerful suite of techniques for evaluating the archaeological heritage on the seabed. Irrespective of motivation, theme, scale or operational phase, effective investigation requires methodological integration. Whilst marine archaeology is demanding and costly, it is also stimulating and engaging, and can – as I hope I have indicated – make a useful contribution to discussion about approaches to geophysics and remote sensing on land as well as offshore.

Acknowledgements

This paper draws on experience gained alongside numerous colleagues at Wessex Archaeology across many projects for a wide range of clients. For the content of this paper I am indebted to my colleagues, past and present, and to the people who commission our investigations, though responsibility for the views expressed must rest with me.

References

Bacilieri, C. & Thomas, R.J.C. 2010: Ephemeral coastal defences of the two world wars in North East England. In Cowley, D.C., Standring, R.A. & Abicht, M.J. (eds): *Landscapes through the Lens Aerial Photographs and Historic Environment*. Oxbow, Oxford. 89–98.

Bickett, A. 2011: *Submerged Prehistory: the MALSF and offshore prehistoric archaeology*. MEPF Project 10/150. Lowestoft: Marine Environment Protection Fund.

EMU Ltd 2009: *The Outer Thames Estuary Regional Environmental Characterisation*. Marine Aggregate Levy Sustainability Fund, ALSF/MEPF (DEFRA). http://www.alsf-mepf.org.uk.

English Heritage, 2002: *Military Aircraft Crash Sites – Archaeological guidance on their significance and future management*. English Heritage.

Evans, A.M., Firth, A. & Staniforth, M. 2009: Old and New Threats to Submerged Cultural Landscapes: fishing, farming and energy development. *Conservation and Management of Archaeological Sites* 11.1, 43–53.

Firth, A. 2000: Development-led Archaeology in Coastal Environments: investigations at Queenborough, Motney Hill and Gravesend in Kent, UK. In Pye, K. & Allen, J.R.L. (eds): *Coastal and Estuarine Environments: sedimentology, geomorphology and geoarchaeology*. Geological Society Special Publication No. 175. London: Geological Society. 403–17.

Firth, A. 2004: Prehistory in the North Sea; questions from development-led archaeology. In Flemming, N.C. (ed.): *Submarine Prehistoric Archaeology of the North Sea: research priorities and collaboration with industry*. CBA Research Report 141. York, Council for British Archaeology. 89–98.

Flatman, J. & Doeser, J. 2010: The International Management of Marine Aggregates and its Relation to Maritime Archaeology. *The Historic Environment: Policy and Practice* 1(2), 160–84.

Hamel, A.T. 2011: *Wrecks at Sea: marine ALSF research in context*. MEPF project 10/152. Lowestoft: Marine Environment Protection Fund.

Holyoak, V. 2002: Out of the Blue: Assessing Military Aircraft Crash Sites in England, 1912–45. *Antiquity* 76(293), 657–63.

James, J.W.C., Pearce, B., Coggan, R.A., Arnott, S.H.L., Clark, R., Plim, J.F., Pinnion, J., Barrio Frójan, C., Gardiner, J.P., Morando, A., Baggaley, P.A., Scott, G. & Bigourdan, N. 2010: *The South Coast Regional Environmental Characterisation*. British Geological Survey Open Report OR/09/51.

Posford Haskoning, 2002: *East Channel: regional environmental assessment*. Unpublished report for East Channel Association. Accessed 18 December 2010 at http://www.eastchannel.info/downloads/environ01.pdf.

PART 3

Exploring the archaeological resource base

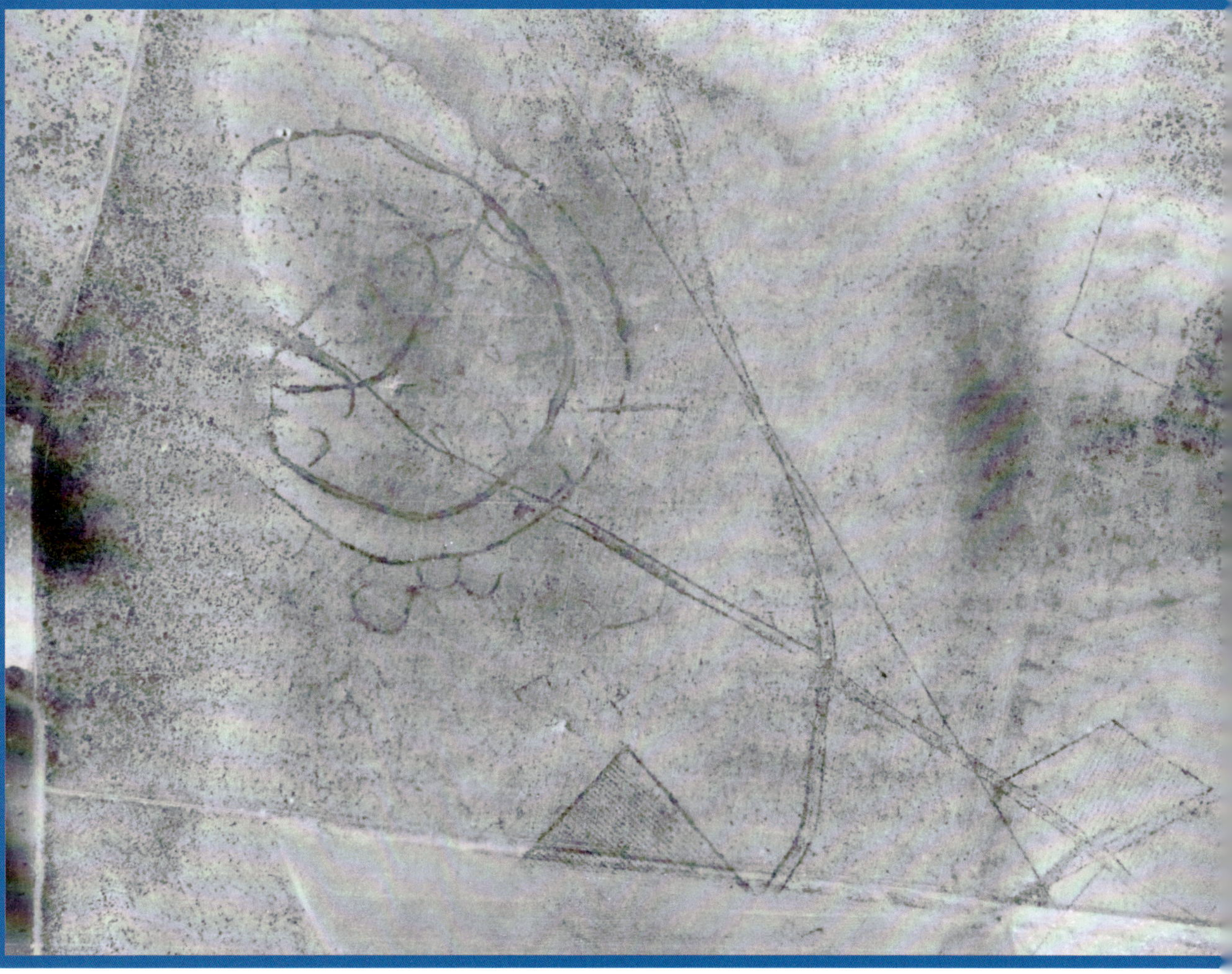

Vegetation marks recorded by chance on a military reconnaissance photograph, taken in 1945, record the prehistoric village, field boundaries and trackways at Masseria Cascavilla, San Giovanni Rotondo in Apulia, southern Italy. Well known sources, such as historic aerial photographs, and cutting edge approaches, such as Airborne Laser Scanning, underpin the creation of reliable archaeological knowledge. *TARA_SJ_682_L21_3678, reproduced under licence. RCAHMS: The National Collection of Aerial Photography.*

12 | The English Heritage National Mapping Programme

Pete Horne

Abstract: The English Heritage National Mapping Programme has now completed archaeological aerial photo interpretation surveys for over 40% of England. It is estimated that the project has recorded over 100,000 archaeological sites, half of which had not previously been recorded. The methodology is always developing as technology and available data sources change, but the underlying principles of maintaining teams of specialists in interpreting aerial imagery and analysing all readily available sources, continues. The programme is a long term one and there are challenges in ensuring the continuity of funding, maintaining access to ever-increasing photo archives and other sources, and increasing the pool of trained interpreters. However, by carefully targeting projects to those areas where the benefit will be greatest, the dramatic results are ensuring that NMP continues to be a key element in managing England's historic environment.

Mapping England's archaeological sites and landscapes

The aim of English Heritage's National Mapping Programme (NMP) is to enhance the understanding of past human settlement by providing primary information and synthesis for all archaeological sites and landscapes visible on aerial photographs or other airborne remote sensed data. The resultant maps and records are intended as the raw material to assist further research, planning and protection of the historic environment. Alongside these primary products, reports provide summary analyses and some themes are taken forward for wider publication and dissemination.

Origins of the National Mapping Programme

The value of mapping from archaeological photographs as an important stage in their interpretation and use for further research goes back to Crawford in the 1920s but took a long time to become fully realised (Barber 2011). The publication of *A Matter of Time* by the Royal Commission on the Historical Monuments of England (RCHME) in 1960 showed how vulnerable extensive archaeological landscapes were when their location coincided with resources needed for a rapidly developing country; the Danebury volume showed how excavation work could be put into a wider context by the analysis of aerial photographs (Palmer 1984).

The aim of NMP was first articulated in 1992/3 following a series of pilot projects (1988–1992), aimed initially at mapping and classifying extensive areas of archaeological cropmarks, and subsequently taking the methodology into an upland area. The pilot projects were initiated to assist the English Heritage Monument Protection Programme (English Heritage 2000) and were developed following the RCHME's experience of two very different aerial photo interpretation and mapping projects in the Yorkshire Wolds and Dartmoor (Stoertz 1997; Soffe 1985). It was clear that the only way to manage buried archaeological remains visible as cropmarks was by gaining a better understanding of them through the interpretation and mapping process. Thus, a project in Kent (Edis 1989) was quickly followed by projects in Hertfordshire (Fenner 1992) and the Thames Valley (Fenner 1994), while the linked Yorkshire Dales Mapping project quickly demonstrated that a similar approach could also be used to rapidly identify and characterize extensive and complex archaeological earthworks and industrial remains (Horne & MacLeod 1995).

Unlocking archives of aerial photographs

The key source materials for NMP are the collections of aerial photographs available to archaeologists and other researchers, particularly those in the National Monuments Record (NMR) and the Cambridge University Collection of Aerial Photographs (CUCAP), now part of the Cambridge University Unit for Landscape Modelling. These collections hold a mixture of oblique and vertical photographs, including those taken for archaeological and non-archaeological purposes. By careful interpretation of the images it is possible to produce synthetic maps and records and so to rapidly enhance our knowledge of the pattern of human settlement in England (Figure 12.1; see also Winton & Horne 2010).

It is perhaps worth stressing that the capture of new aerial photography is not normally considered necessary prior to an NMP project. Because there are so many variables in capturing archaeological information through the medium of aerial photography (Crawford 1929; Wilson 1982; Brophy & Cowley 2005), only rarely will a single flight, or even a single season of flying, reveal a significant proportion of the necessary information.

Furthermore in England, the archives already contain a large number of photographs (both vertical and oblique) taken in various seasonal conditions, and new wide-ranging reconnaissance continues to build on this collection (Grady 2007). The exception is in areas where archaeology is likely to survive as earthworks and it is known little specialist photography exists; a few flights using oblique photography in the right conditions can effectively boost the record for an area and provide illustrative material as was shown by the Malvern Hills project (Winton 2005; Bowden 2005). For NMP the results of such prospective flying just become one source to be used alongside archival sources. The actual number of photographs examined for each project varies considerably but often runs into tens of thousands.

Using the extensive archives rather than commissioning new photographs is effective because NMP tries to record archaeology of all periods visible as structures, earthworks, cropmarks or soilmarks. However, in those landscapes where the survival of remains mainly manifests itself through earthworks, specialist flying and a photogrammetric approach has also proved a very effective means of survey, producing very detailed and accurate mapping as for example in the RCHME Bodmin and Cheviot surveys (Johnson & Rose 1994; Herring *et al.* 2008; Topping 2008; Topping & Pearson 2008), and more recent surveys of Cawthorn Camps and Fylingdales Moor in North Yorkshire using digital photogrammetry (Stone 2003, 2006). This sort of survey tends to take longer than equivalent NMP projects, but the resultant detailed survey, often supported by detailed field work, can be extremely useful for managing specific assets and for gaining a much greater understanding of landscape development (Bowden 1999). The development of LiDAR techniques provides an even better tool for this sort of survey and is producing spectacular results in open and wooded areas (Figure 12.2). And although the aerial photograph archives still have a valuable input to these projects, they become a secondary source, albeit a valuable one, for the historic depth they provide, often showing features that have been degraded, levelled or totally destroyed since the date of photography (Crutchley 2010). Analysis of LiDAR data requires much the same set of skills as for interpreting aerial photographs, and as it becomes more available and affordable it will have a greater impact on NMP. Already standardized images created from LiDAR data by the Environment Agency have become one of the standard sources for NMP, although lacking the flexibility of direct access to the data, this source has the benefit of rapid access and is available through an inter-departmental agreement

Figure 12.1: NMP process of examining photographs, producing rectified orthophotographs, interpreting archaeological information and producing mapping and supporting database records for use within GIS environment. © *English Heritage, Photo NMR 17467/07 NY2637/24 16-Jun-00; Background map © Crown copyright. All rights reserved. English Heritage 100019088.2010.*

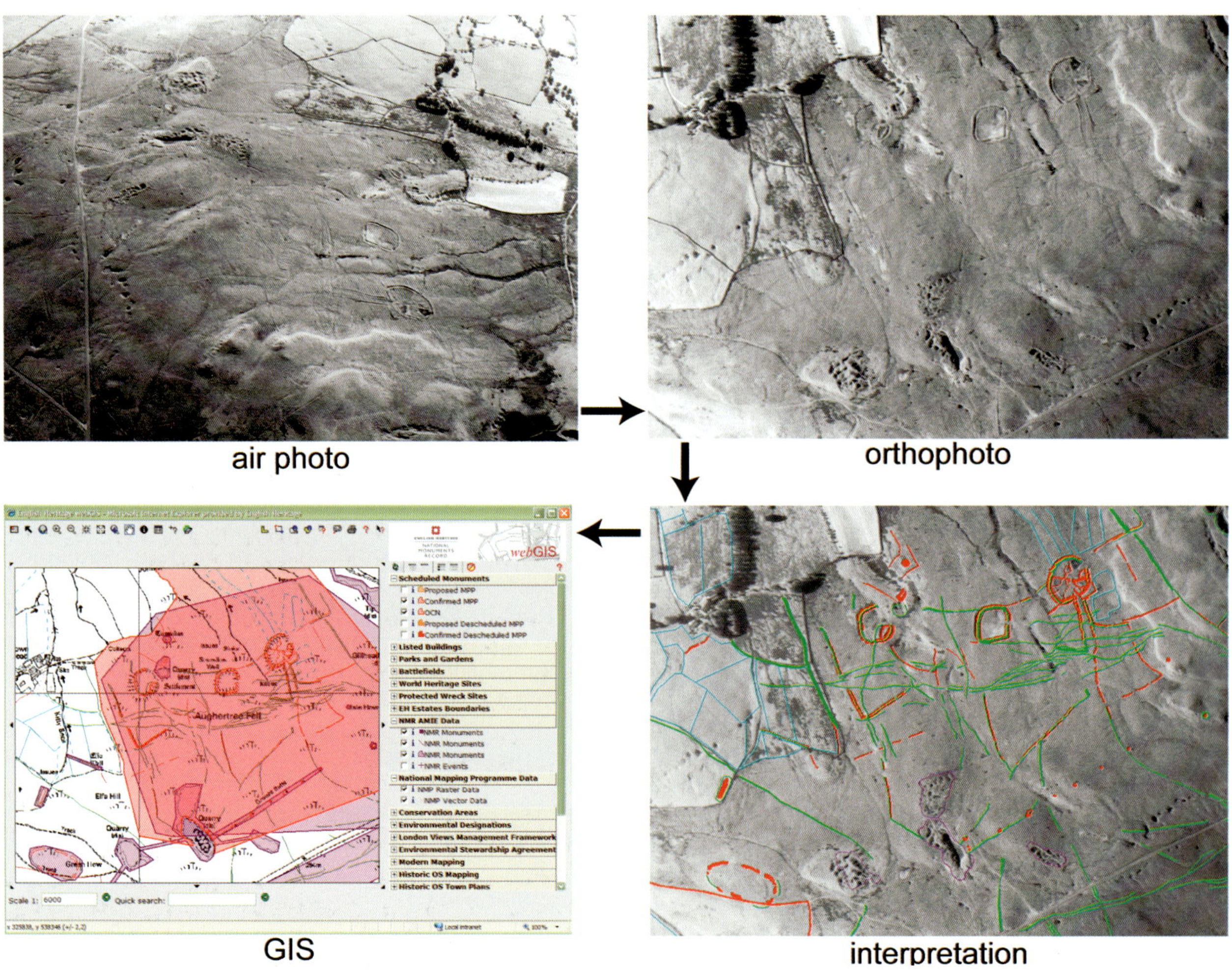

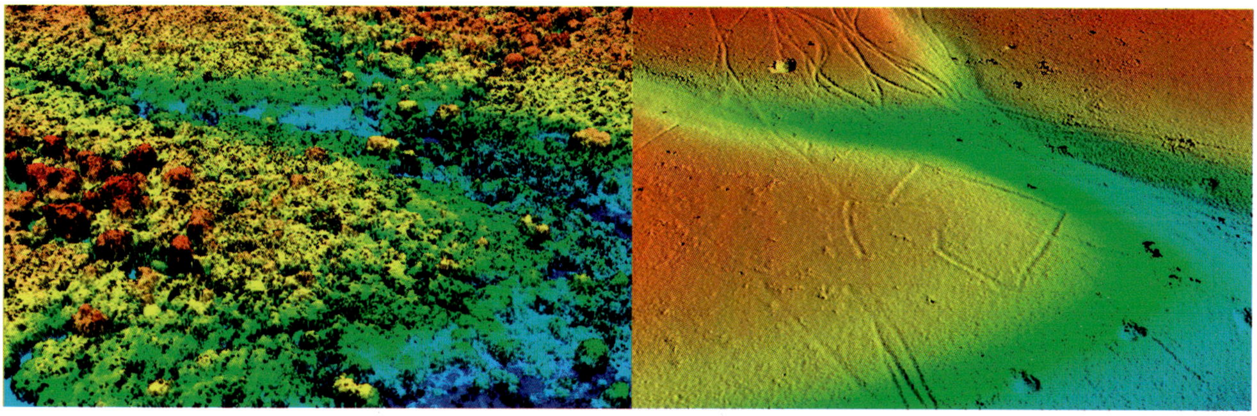

Figure 12.2: LiDAR data showing an area of Savernake Forest showing the tree canopy (*left*) and a prehistoric enclosure and trackways under the tree cover (*right*). © *English Heritage; LiDAR data © Forestry Commission, source, Cambridge University ULM (May 2006).*

at no charge. Other countries have even more complete agreements allowing direct access to data by all government bodies and this is being explored as an important source for archaeological survey (see Bofinger & Hesse this volume). Other techniques such as multi-spectral imaging may well also have their place in future projects (see Beck and Bennett *et al.* this volume).

Interpretation and analysis

Whilst existing archives are vital to NMP, the key to unlocking them are the archaeological air photo interpreters who undertake the work. The reliable interpretation and analysis of archaeological information on aerial photographs is a specialism requiring visual acuity alongside experience and broad archaeological knowledge, an understanding of geological features and farming practices, and an ability to bring all these together consistently and present the data for others to use (see Palmer this volume).

Aerial photo interpretation skills can be applied for the very detailed study of individual sites or can be used to rapidly discover and characterise hitherto poorly known, or undiscovered, sites and whole landscapes at a basic level (e.g. 'here be a Roman fort'), or at a more detailed level that allows for an understanding of relationships between individual features in a spatial and temporal way. Whilst the use of aerial photographs as an adjunct to small scale field survey or excavation is well-established and understood (Bowden 1999), the skills and experience needed for interpretation of sites solely from aerial photographs and existing maps and records are not used widely in archaeology in the UK outside of the NMP programme. NMP attempts to cover the ground as quickly as possible, but in a way that maximizes the flexible further use of the data recorded. The vast majority of the sites and landscapes recorded through NMP will need to be assessed, evaluated and managed without recourse to further analysis (particularly those that are used to inform the extensive agri-environment schemes now being rolled out across England). We need to have confidence in the quality of the NMP product and this means that training is an important part of the programme as a whole. Experience is central to good interpretation and as a general rule we would expect someone new to the job to take about six months to gain the necessary skills and experience to work largely unaided. That said NMP encourages all projects to be run with small teams of 2 – 4 people to allow for experience to be shared and hence ensure the quality of the resultant interpretation. Where projects are run with a single person they are encouraged to frequently consult with other teams or English Heritage staff to the same end.

Maps and records

Initially the mapping part of NMP was done by manual transcription of the interpreted features onto inked overlays supported by a specialist database (Edis *et al.* 1989), but since 1998 digital mapping techniques and GIS compatible recording have become the norm. This has resulted in products from NMP that are much more flexible in use, and has allowed an increase in nominal mapping scale from 1:10,000 to 1:2,500. However, this, along with the increased numbers of photos available for interpretation as the NMR continues to rapidly expand its accessible archive, has had the effect of slowing down the progress of NMP from an average of 15–20 days per 25 sq km to a current rate of around 25 days per 25 sq km.

Although aerial photographs are the main source material for NMP projects the importance of the other sources used is worth remembering. Existing records held in the National Monuments Record and the local Historic Environment Records (HER) are always consulted, are usually the basis for record enhancement, and provide the structure for new records. These sources will contain information about field survey work, finds and possible historic sources and so can inform the interpretation of features seen on the aerial imagery. The history of the development of the NMR and the local HERs means there is a considerable overlap between the two sources, but as they have usually been enhanced separately NMP will usually try and look at both sources. Often a product of an NMP project is the concordance of NMR and HER data, which along with the new NMP data allows for a full up-to-date picture of knowledge for a specific area.

Equally important are the existing maps used by the NMP interpreters. NMP aims to provide an overlay to the currently available Ordnance Survey mapping concentrating on features not already mapped. The AP interpreters have access to digital versions of the OS maps at 1:10,000 or 1:2,500 scale compiled and updated throughout the 20th century, but with the earliest versions usually originating in the 19th century.

Although NMP is national in concept the projects are designed, developed and run in close liaison with local heritage managers and special interest groups and this helps to ensure that local knowledge, priorities and additional sources of information contribute to the work, and the results are more immediately used.

Originally NMP projects were considered largely as standalone products, maps and records being provided in a format that was considered useful for use in a planning environment but allowing for the analysis by specialists to help produce a report on the character of archaeology in an area highlighting specific subjects of interest (Bewley 2001). However, with the increasing agreement of local and national data standards, the importance of NMP linking in with this soon became apparent and so all projects moved on to recording systems that directly input to either the national or local Historic Environment Records. This has the benefit of allowing all dissemination and access to be focussed through these composite resources. This approach seems to be paying dividends as more and more data is being made directly accessible online through initiatives such as the Heritage Gateway (http://www.heritagegateway.org.uk). Currently only the textual database records from more recent NMP projects are available on line, but two projects have trialled different approaches to web mounting the whole NMP dataset of maps and records. The *Flying through Cornwall's Past* web site includes an extensive non technical report on the results of the Cornwall NMP project along with interactive GIS maps (www.historic-cornwall.org.uk/flyingpast), and the results of the Northamptonshire NMP project, with both interactive and downloadable data, are on the Archaeology Data Service web site (http://ads.ahds.ac.uk). Increasingly, NMP summary reports aimed at the professional user are also being made available online as PDF downloads.

The results of NMP surveys

Projects using the NMP methodology have now been completed for 41.6% of England (more than 50,000 sq km) rewriting our understanding of the extent, character and sheer quantity of archaeology visible on aerial photographs (Figures 12.3 & 12.4). Because of the variety of recording systems used as the programme has developed it is difficult to give exact figures for the number of sites recorded, but it is clear that by 2008 more than 100,000 archaeological sites had been mapped directly as a result of NMP. Furthermore more than 50% of the records created by NMP are for sites that have not previously been recorded in either the NMR or local HERs, and for some projects this figure has been over 70%. This means that more than 50,000 archaeological sites have been newly identified, totally transforming our view of the archaeological resource in these areas. This increase is visible even when viewing the data as dots on a map at a national scale: see Figure 12.4 and compare with Figure 12.3 – the densest patches coincide closely with the extent of NMP projects.

This paper will not highlight individual site discoveries, since the real value of NMP is in the new understanding of whole landscapes. However, it is worth stressing that even in England with a long history of field and aerial survey there are still numerous significant discoveries of nationally important sites being made, for example in the recent NMP study of the area around Hadrian's Wall a possible 16 additional Roman camps were identified (Oakey 2009). The great variety of results can probably best be appreciated by browsing the NMP pages on the English Heritage web site (www.english-heritage.org.uk/NMP) and the linked PDF reports of individual projects that provide a glimpse of the sort of insights that are possible using NMP data.

Using NMP results

Although NMP projects provide high quality, and often quite detailed, information, the intention is that this is baseline data for characterisation and strategic management of the overall resource; it is not intended for detailed management of individual sites or mitigation in the event of destruction following evaluation under planning system controls. Indeed, it is inevitable that as a result of individual interpretation, the sheer quantity of aerial photographs now available and the necessary speed of recording for NMP, that sites and nuances will be missed. Quality assurance can help to minimise this but it should always be recognized that the great increase in the number of sites and landscapes identified through NMP are not a definitive statement of what is, or has been there, merely a better guide to it than we have had before. With further work additional sites will be discovered and existing sites will be re-interpreted. The record of the historic environment that NMP contributes to (and is held in the NMR and HERs in England) should be the starting point for more detailed evaluation at a local and regional level and where necessary the stimulus for more detailed evaluation of the original air photos alongside data from other sources.

That NMP data is useful in heritage management is clear, as is shown by the following quotes from those working with heritage management on a day to day level, assessing planning decisions and encouraging local proactive approaches to research and care.

'NMP is a key EH programme delivering high quality information to HERs to be used for managing the historic environment through the planning and other statutory systems but also providing the community with a unique perspective on their environment. In Gloucestershire NMP information has contributed to our research programme in the Forest of Dean, to the RCZA assessment of the Severn Estuary, to the management of the rural landscape through agri-environment schemes and to the accurate assessment of the impact of many development proposals on the archaeological resource.' Jan Wills,

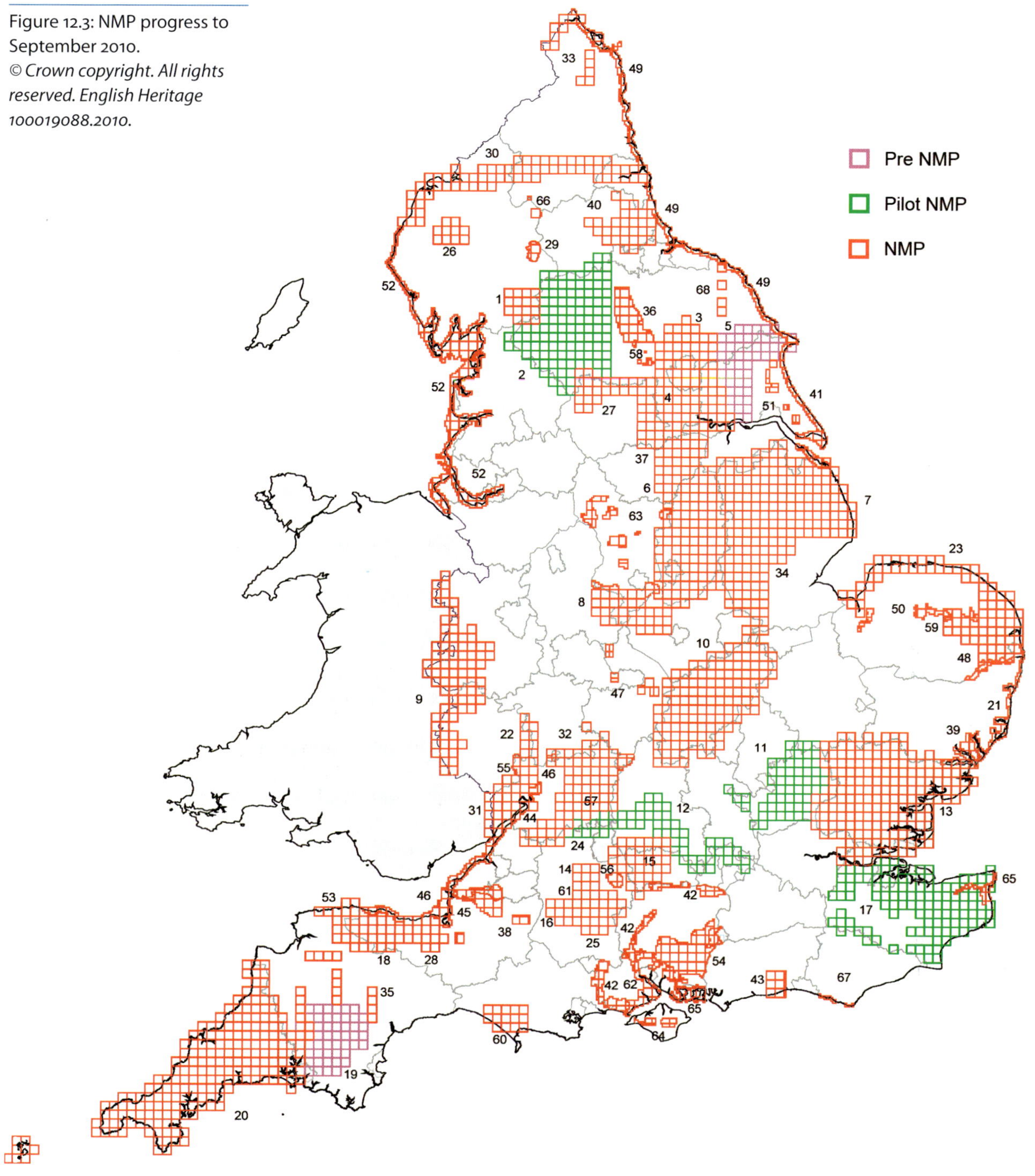

Figure 12.3: NMP progress to September 2010.

County Archaeologist, Gloucestershire County Council Archaeology Service (pers. comm.).

'It [NMP] has been a really useful piece of work that is now used daily in the West Yorkshire HER to help protect the archaeology, to persuade planners and developers of the need for archaeological work, and to inform the wider public, as well as contributing significantly to our research agenda and understanding of West Yorkshire's past.' Ian Sanderson, Principal Archaeologist, West Yorkshire Archaeological Advisory Service (pers. comm.).

The way in which NMP can be used directly for heritage management purposes is seen in English Heritage's approach to the coast. England's coastal heritage has been seen as a priority for enhanced recording to ensure management of a fragile resource (English Heritage 2003). To meet this need projects using NMP methodology accompanied by desk-based record enhancement are recording a narrow coastal strip up to 1km inland as well as the intertidal zone (Horne 2007). These Phase 1 Rapid Coastal Zone Assessment Surveys (RCZAS) are then followed up by Phase 2 field assessments, and the combined results can then feed into the Shoreline Management Plans which bring together the priorities of many factors of which heritage is only one element.

As well dealing with management issues coastal projects have significant research and outreach possibilities. Being an island nation the coast has been a focus for defensive structures, particularly in the 20th

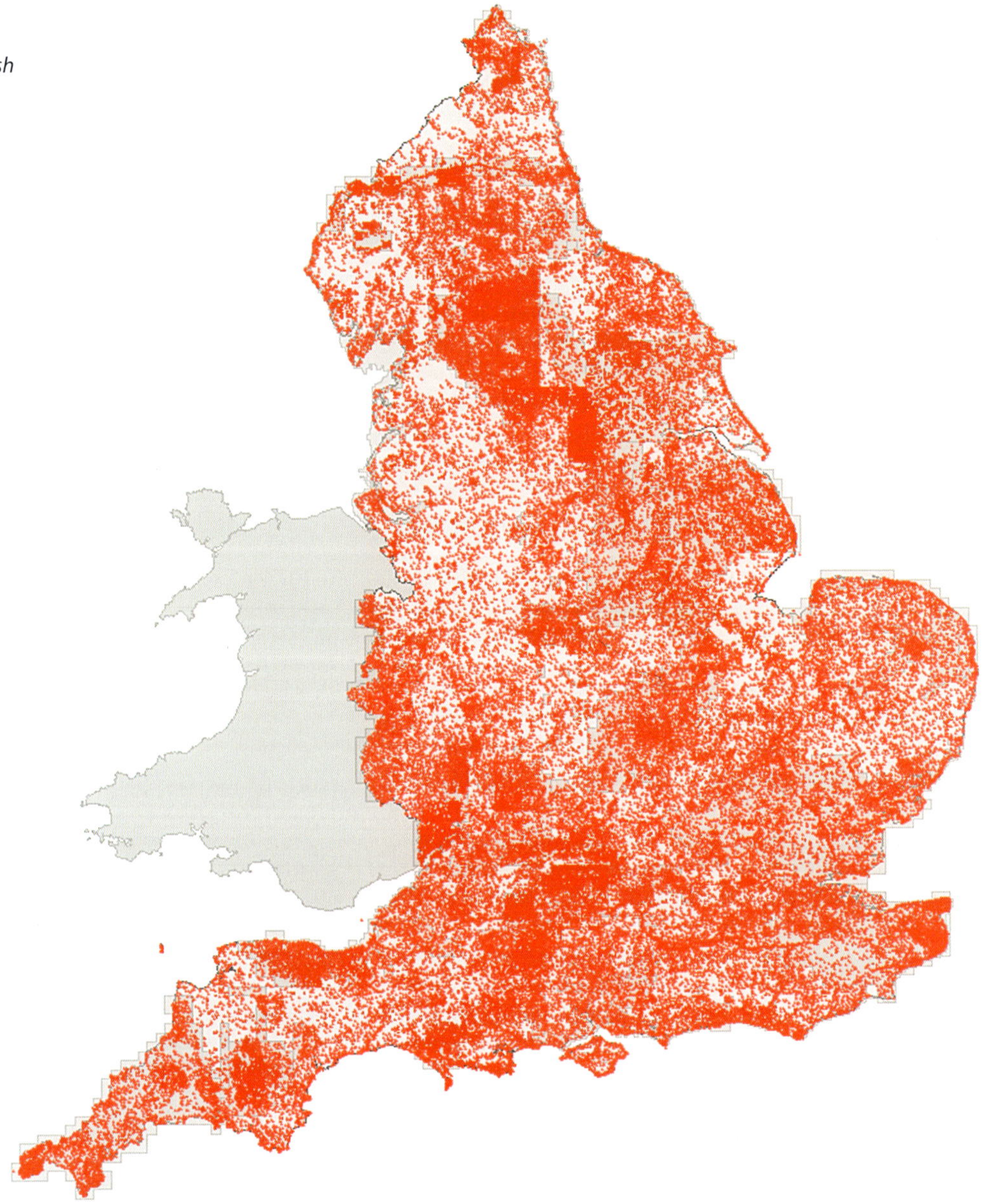

Figure 12.4: Monument records in the National Monuments Record. © *English Heritage; Background map © Crown copyright. All rights reserved. English Heritage 100019088.2010.*

century, and the aerial evidence provides an aid to understanding and putting the surviving structures in context (e.g. Bacilieri & Thomas 2010). The Suffolk RCZAS NMP recorded sites of all periods, but much interest was generated in the identification of 20th century defences along the English coast (Hegarty & Newsome 2005). Therefore as well as providing data for heritage managers to use in planning for change, the results of the Suffolk RCZAS project were developed into a book aimed at a wide audience that has done much to raise the profile of this more recent archaeology and the way in which aerial photos can be used for research (Hegarty & Newsome 2007). It is now available as a print on demand volume or as free PDF downloads chapter by chapter from the English Heritage web site. This model for dissemination has also been followed for other NMP publications such as *Mapping Ancient Landscapes in Northamptonshire* which explores the results through a series of period based landscape essays (Deegan & Foard 2007).

The most recent publication to use the results of NMP as the basis of analysis is *Understanding the Cropmark Landscapes of the Magnesian Limestone* (Roberts *et al.* 2010), providing a much deeper insight into, and context to, the area covered by Derrick Riley's influential *Early Landscape from the Air* (Riley 1980). The discussion is based on the results of four NMP projects, combining them with the results of geophysical survey work and excavations (mainly undertaken as part of the planning process). These form a convincing narrative of the landscape development of a large part of northern England, showing the parcelling of the land for agriculture and settlement from late prehistory through the Roman period, followed by a marked discontinuity in the Anglo-Saxon period. Whilst the process of NMP literally puts sites on the map with a record that ensures they are considered as part of heritage management for the landscape, this sort of wider analysis allows better definition of priorities for designation or further research purposes.

Looking forward

NMP has now been running for many years and methodologies have matured and developed, although the principles underlying it remain the same. Funding streams have always been a major issue but the value of the NMP approach has been appreciated, for example fitting in well with the needs of the RCZAS and the Aggregates Levy Sustainability Fund (ALSF) that have provided a major focus for the programme over recent years. That said there are always new issues that arise and the NMP manual is in a constant draft state, updated regularly. At the outset each NMP project must define its own specific aims and lay out the project structure according to the English Heritage Management of Research Projects in the Historic Environment (MoRPHE) guidelines (http://www.english-heritage.org.uk/morphe).

Amongst new techniques, LiDAR has attracted much attention over the last decade and as the quality of the data and the flexibility of software has improved it has been possible to look at it as a possible additional source for NMP (Holden *et al.* 2002; Crutchley 2008). LiDAR was first used alongside aerial photographs from the outset of the Savernake Forest NMP, although previous projects had tested its value (Crutchley 2010). A preceding project on the Forest of Dean had, surprising to some, shown the benefits of aerial photo analysis even in a heavily wooded environment (Small & Stoertz 2006), but subsequent LiDAR survey had shown even more potential. The Savernake project demonstrated that LiDAR could be the primary data source for survey in wooded environments, but that aerial photos were also important for the wooded area. The importance of using aerial photos alongside LiDAR was demonstrated even more clearly for the gaps between the trees and beyond the wooded areas where sites were only visible as cropmarks or soilmarks and so would not normally be recorded by the laser scanning process (Crutchley *et al.* 2009).

It is clear that future NMP projects will require careful assessment of the cost-effectiveness of using LiDAR data to meet NMP aims; a multi-disciplinary project now underway for an upland area of northern England will help inform that debate (Ainsworth 2009).

It is not just the possibilities of new sources of data such as LiDAR and multi-spectral techniques that have to be considered. Ever more aerial photographs are becoming available for analysis, most importantly through online resources such as Google Earth™ and Bing and through the accessioning and cataloguing of historic archives. There is a natural inquisitive desire to understand landscapes more deeply and to examine all possible sources of data to inform that process. Often the aerial photographic evidence will enable more detailed work than is necessary to meet the aims of NMP and although the use of additional sources (airborne, ground-based or documentary) will often allow a much fuller picture to be developed the value of this work needs to be balanced against the ability to cover the ground quickly. One of the key challenges in many countries is getting access to the aerial imagery to allow this sort of work. In England we are in the envious position that in some areas we need to prioritize which sources are used (Winton & Horne 2010). We have to accept that the importance of NMP is increased by ensuring coverage of as large an area as possible even though being selective in the sources and methods used will inevitably result in some sites escaping notice.

NMP in the long-term

The aim of NMP is still to complete the interpretation and mapping of archaeological sites and landscapes across England through the analysis of readily available aerial photographs and other related sources. This is a long-term programme that with current resources and methodology will take at least a further twenty years. The NMP Strategy describes the planned way forward through a continuous series of projects of varying size that will target those areas where the results will be of most use and also that will allow for partnership, the development of new techniques and help to ensure a sustainable pool of skilled professionals capable of undertaking this work (Horne 2009).

Keeping the results of completed NMP projects up to date becomes an increasing consideration as the area covered increases. New material becomes available either as a result of ongoing specialist aerial reconnaissance, increasingly targeted by the use of copies of NMP maps in the air by airborne archaeologists, or by the addition of new material to accessible archives. Where this material is the result of selective photography to record sites seen from the air, or is accessible online through a single source such as GoogleEarth™ this can be a relatively straightforward exercise. English Heritage has a programme called 'Recce Recording' that ensures the most important discoveries in areas already covered by NMP are recorded and mapped to a similar standard and made available through the National Monuments Record (Winton & Horne 2010). Projects have also been funded to see how such approaches can be undertaken at a local level and collections of historic photographs are increasingly valued and being made available (e.g. the Aerofilms collection for the United Kingdom and the TARA collection for Europe and beyond see Ferguson this volume).

New technologies continue to provide further possibilities and provide the opportunity for much more detailed and comprehensive landscape survey than is achievable just from an NMP approach (e.g. Powlesland this volume), and it is likely in the future that some of these techniques may provide a better first source of information than aerial photography in some landscapes (e.g. LiDAR described above). However, in areas where sites are revealed as cropmarks, soilmarks or earthworks aerial photography is an effective tool for site identification and the NMP approach is likely to remain the most cost-effective way of increasing our understanding of broad landscape interpretation for the immediate future. The results so far suggest that most of England will benefit from this approach and it is likely that this is the case in much of the rest of the world. But in differing landscapes, with differing histories and differing access to a range of aerial source material the approach will need to be carefully considered and developed to meet local circumstances.

Conclusion

If we want to manage the heritage in a manner that provides us and our descendants with a real insight on how our ancestors have laid the foundations of what is around us (both physical and conceptual), their successes and their failures, their innovations and their conservatism, their needs and their beliefs, then we need to take a comprehensive look at the totality of the landscape all around us.

Resources will never allow every detail to be managed and preserved, but by using approaches that allow us to identify what is 'normal' and what is 'exceptional', what makes up the overall pattern and where that pattern changes, we can identify where we need to target our efforts to protect or record before loss, and for this programmes like the National Mapping Programme are needed. The techniques used must be flexible, applying traditional and new techniques appropriately and in a way that is sensitive to local environments. The approach can be different for urban areas, rural areas, coastal areas, upland areas, wooded areas, and may be targeted to those areas with a perceived high risk or high productivity – but without a programme that provides a baseline dataset we cannot prioritize resources in the rigorous way that is needed to gain greatest benefit in the face of major challenges.

References

Ainsworth, S. 2009: Miner-Farmer landscapes of the North Pennines Area of Outstanding Natural Beauty. *Research News* (Newsletter of the English Heritage Research Department) 11, 12–6.

Barber, M. 2011: *A History of Aerial Photography and Archaeology. Mata Hari's Glass Eye and other stories.* English Heritage.

Bacilieri, C. & Thomas, R. J. C. 2010: Ephemeral coastal defences of the two World Wars in North East England. In Cowley, D.C., Standring, R.A. & Abicht, M.J. (eds): *Landscapes through the lens: aerial photographs and historic environment.* Oxbow, 89–98.

Bewley, B. 2001: Understanding England's Historic Landscapes: An Aerial Perspective. *Landscapes* (2001) 1, 74–84.

Bowden, M. (ed.) 1999: *Unravelling the landscape: an inquisitive approach to Archaeology.* Tempus, Stroud.

Bowden, M. 2005: *The Malvern Hills, An Ancient Landscape.* English Heritage, London.

Brophy, K. & Cowley, D. (eds) 2005: *From the air: understanding aerial archaeology.* Tempus, Stroud.

Crawford, O.G.S. 1929: *Air-Photography for Archaeologists.* Ordnance Survey Professional Papers No. 12, London, HMSO.

Crutchley, S. 2008: Shedding light on the past: using airborne LiDAR to understand ancient landscapes. In Vassilopoulos, A., Evelpidou, N., Bender, O. & Krek, A. (eds): *Geoinformation Technologies for Geocultural Landscape Analysis: European Perspectives.* CRC Press, 87–103.

Crutchley, C. 2010: *The Light Fantastic. Using airborne LiDAR in archaeological survey.* English Heritage.

Crutchley, C., Small, F. & Bowden, M. 2009: *Savernake Forest. A report for the National Mapping Programme.* Research Department Report Series 29–2009, English Heritage.

Deegan, A. & Foard, G. 2007: *Mapping Ancient Landscapes in Northamptonshire.* English Heritage, Swindon.

Edis, J. 1989: *The Classification of Cropmarks in Kent.* RCHME unpublished report.

Edis, J., MacLeod, D. & Bewley, R. 1989: An archaeologist's guide to classification of cropmarks and soilmarks. *Antiquity* 63, 112–26.

English Heritage 2000: *MPP 2000. A review of the Monuments Protection Programme, 1986–2000.* English Heritage Policy Document.

English Heritage 2003: Coastal *Defence and the Historic Environment.* English Heritage Guidance (available as a pdf from http://www.english-heritage.org.uk/upload/pdf/CoastalDefenceEH.pdf). A summary is available also http://www.english-heritage.org.uk/upload/pdf/coastal_defence.pdf.

Fenner, V. 1992: *Cropmarks in Hertfordshire: A report for the National Mapping Programme.* RCHME unpublished report.

Fenner, V. 1994: *The Thames Valley Project: A report for the National Mapping Programme.* RCHME unpublished report.

Grady, D. 2007: Cropmarks on Clay, Getting the Timing Right. In Mills, J. & Palmer, R. (eds): *Populating Clay Landscapes.* Tempus, Stroud.

Hegarty, C. & Newsome, S. 2005: *The Archaeology of the Suffolk Coast and Inter-tidal Zone: A report for the National Mapping Programme.* Suffolk County Council and English Heritage internal report.

Hegarty, C. & Newsome, S. 2007: *Suffolk's defended shore: coastal fortifications from the air.* English Heritage, Swindon.

Herring, P., Sharpe, A., Smith, J. R. & Giles, C. 2008: *Bodmin Moor – An Archaeological Survey: Volume 2 The Industrial And Post-Medieval Landscapes.* English Heritage.

Holden, N., Horne, P. & Bewley, R.H. 2002: High-resolution digital airborne mapping and Archaeology. In Bewley, R. & Rączkowski, W. (eds): *Aerial Archaeology: developing future practice.* Proceedings of the NATO Advanced Research Workshop on Aerial Archaeology, 15–17 November 2000, Nato Life Sciences, Series 337, 173 –80.

Horne, P. 2007: Air, Land and Sea: an integrated approach to England's coastal heritage. In Jarockis, R., Kraujalis, R. & Musson, C. (eds): *Past from the air. Aerial archaeology and landscape studies in Northern Europe*. Department of Cultural Heritage, Lithuania, 43–56.

Horne, P.D. 2009: *A Strategy for the National Mapping Programme*. English Heritage.

Horne, P.D. & Macleod, D. 1995: *The Yorkshire Dales Mapping Project. A report for the National Mapping Programme*. RCHME internal report.

Johnson, N. & Rose, P. 1994: *Bodmin Moor: An Archaeological Survey Volume 1: The Human Landscape to c 1800*. English Heritage/Royal Commission on the Historic Monuments of England, London (reprinted 2008).

Oakey, M. 2009: *Hadrian's Wall World Heritage Site National Mapping Programme Project Summary Report*. Research Department Report Series no, 73–2009, English Heritage.

Palmer, R. 1984: *Danebury, an Iron Age Hillfort in Hampshire: an aerial photographic interpretation of its environs*. RCHME Supplementary Series Vol. 6, London.

Riley, D.N. 1980: *Early Landscapes from the Air: studies of crop marks in Nottinghamshire and South Yorkshire*. Sheffield.

Roberts, I., with Deegan, A. & Berg, D. 2010: *Understanding the Cropmark Landscapes of the Magnesian Limestone*. Archaeological Services WYAS, Leeds.

RCHME 1960: *A Matter of Time: an archaeological survey of the river gravels of England*. RCHME, London.

Small, F. & Stoertz, C. (eds) 2006: *Gloucestershire Forest of Dean NMP, National Mapping Programme Report*. Research Department Report Series 28–2006 English Heritage.

Soffe, G. 1985: *The Archaeology of Dartmoor: an air photographic survey*. RCHME unpublished report.

Stoertz, C. 1997: *Ancient Landscapes of the Yorkshire Wolds: aerial photographic transcription and analysis*. RCHME, Swindon.

Stone, J. 2003: Cawthorn Camps, North Yorkshire – a Photogrammetric Approach. *Archaeological Prospection* 10(2), 153–8.

Stone, J. 2006: *Fylingdales Moor, North Yorkshire, Photogrammetric Work*. Research Department Report Series No. 76/2006, English Heritage.

Topping, P. 2008: Landscape Narratives: the South East Cheviots Project. *Proceedings of the Prehistoric Society* 74, 323–64.

Topping, P. & Pearson, T. 2008: *The South-East Cheviots Project: a descriptive account of the Prehistoric Landscape*. Research Department Report Series No. 83/2008, English Heritage.

Wilson, D.R. 1982: *Air photo interpretation for archaeologists*. Batsford, London. Second edition 2000: Tempus, Stroud.

Winton, H. 2005: *The Malvern Hills, Report for the National Mapping Programme*. Aerial Survey Report Series, English Heritage internal report.

Winton, H. & Horne, P. 2010: National archives for national survey programmes: NMP and the English Heritage aerial photograph collection. In Cowley, D.C., Standring, R.A. & Abicht, M.J. (eds): *Landscapes through the lens: aerial photographs and historic environment*. Oxbow. 7–18.

13 | Integrating survey data – the Polish AZP and beyond

Włodek Rączkowski

Abstract: The Polish AZP has been already presented many times. It is treated as a basic, objective database of archaeological sites, and since in some regions the project is completed already the question has been asked: what next? The new concept of so called AZP_2 project is being discussed. Non-invasive methods have already been accepted as integral to the project. Up till now in some pilot projects four different categories of information generated by applied methods are gathered: field-walking data (as previous AZP), aerial photographs, geophysical data and test trenches. There is a need to integrate all information within a relational database. The discrepancies between data sources become an important pretext for further thought on the reasons behind the differences. The concept of AZP_2 generates at least two other problems. It is not enough to integrate data and methods. Both heritage management officers and field archaeologists have to understand the new concept and be able to interpret a variety of integrated data. Both practice and mentality need to develop, and all groups should learn about new technology (including GIS) and complex issues related to integration of data.

Introduction

The Polish Archaeological Record (AZP) based on the field-walking method has been presented and discussed many times (Jaskanis 1996; Prinke 2002; Rączkowski 2005). It was founded on guidelines formulated in the 1970s based on the theoretical thinking that was current at the time. Polish archaeologists then focused mainly on settlement pattern studies within the framework of a micro-regional approach (Hensel 1968). As a result, data on the geographical context of the location of a site, its function and chronology was essential in this field of study. Empiricism, therefore, played a dominant role. Another important aspect relating to the introduction of the AZP, the collection of information on archaeological heritage, was connected to the need for more effective protection of archaeological sites during the intense period of economic expansion Poland experienced in the 1970s. The category of 'archaeological site' was of utmost significance for both aspects of the programme in the archaeological thinking of that time. What is more, the past was regarded through the prism of the 'archaeological site'.

The AZP programme began in 1978 (Kobyliński 1999) and is, with various levels of intensity, still under way today. Close to 90% of the area of Poland has been surveyed and in some regions the programme has been completed. The experience so far and awareness of the benefits and limitations of the programme (and the results achieved) have given rise to the need for a broader discussion on its future.

Figure 13.1: Location of the Wielkopolska region in Poland.

The Wielkopolska Region – looking into future?

In the Wielkopolska region (Figure 13.1) the AZP was completed in 2003. Even in its final stages of completion the discussion began – what next? The important issue here was that the discussion began amongst conservators, but was not taken up by the academic milieu. Therefore, for a number of years the future of the AZP was discussed only from the perspective of the identification and management of archaeological heritage. The result of this was a series of articles published in 2006 presenting the continuation of the AZP programme from the perspective of conservation service, though based on new assumptions and with newly formulated aims. Four basic areas of action were identified (Starzyński 2006):

1) so-called settlement pattern studies, the aim of which is to verify the AZP data so far by carrying out additional field-walking survey (completing any missing information, verifying the existence of sites and their character);
2) the verification of fortified settlements from the Bronze Age, Iron Age and early Middle Ages, medieval and modern fortifications, by carrying out a detailed survey and assessment of their current state of preservation;
3) the on-site verification of megalithic burial mounds following earlier analysis of archival data, including detailed survey and assessment of current preservation conditions;
4) rescue excavations of sites under threat.

As this list shows, the conservation activities focus mainly on the identification (gathering information about sites) of archaeological heritage resources, which should permit the formulation of guidelines on their protection and management. Archaeological heritage resources are understood here as the sum of the archaeological sites with an emphasis on earthworks. The term 'settlement pattern studies' is very misleading as the activities involved are mostly limited to fieldwork and has nothing in common with either the concept of micro-regional or regional studies (e.g. Hensel 1968; Jankhun 1977) or spatial archaeology (e.g. Clarke 1977).

At the same time the academic community in Poznań undertook various initiatives relating to the introduction of non-invasive methods into Polish archaeology. First of all, the focus was on widening the use of aerial photographs (e.g. Nowakowski *et al.* 2005), and with time other projects emerged connected to the application of geophysical methods (Kiszkowski & Wrzesiński 1996; Ducke & Müller 2004). As a result, initiatives arose for using non-invasive methods in relation to the aforementioned aims of AZP within the discussions taking place in conservation too. In particular, this concerned the 'settlement pattern studies' project, otherwise known as AZP_2.

AZP_2 – an escape from the dogma of field-survey?

The result was the start of the AZP_2 programme's pilot projects. The choice of the areas selected for the trials was governed by two basic criteria – firstly the identification of current threats and secondly the cognitive value based on surveys so far. The first pilot projects were limited to carrying out further field-walking and the selection of sites for trial trenches (Starzyński & Dernoga 2006). Geophysical surveys were also carried out on particularly valuable sites (identified as such from field-walking results). Since 2003 aerial survey has been regularly carried out (by J. Nowakowski and W. Rączkowski) in order to either investigate a region or to search for certain type of site (i.e. barrows or megalithic graves). Therefore the AZP_2 project today comprises the following actions:

1) field-walking,
2) aerial reconnaissance and mapping archaeological features (mostly pits),
3) geophysical survey (on selected sites),
4) test trenches (on selected sites but mostly those recognised from the air but not during field-walking).

This collection of methods selected for the AZP_2 certainly does not make full use either of the potential that exists today or the potential that archaeologists can actually utilise. In my opinion this is due to two factors: the tradition of archaeological research (and as a result of the mentality of archaeologists) and the interest in non-invasive surveys in Wielkopolska in the academic community.

Why field-walking?

It is quite obvious why field-walking has been selected as the main method in the AZP_2 project, since there is a long tradition of such research. Józef Kostrzewski transformed field-walking from simple field trips into a significant field survey method in the early 20th century. The fact that it was recognised as an effective tool for increasing information on the presence of traces of humans in the past increased its popularity (especially in the 1960s and 1970s and particularly amongst archaeologists concerned with settlement pattern studies). The many years of field-walking practice shaped and consolidated the thinking of Polish archaeologists about archaeological sites. In this context an archaeological site is understood as (Mazurowski 1980, 19):

... a section of space in which archaeological material is grouped together with a context explaining them, it has the characteristic of being separated from other similar spaces in which there are no archaeological remains....

The result of such thinking is the belief that only the presence of fragments of artefacts is proof of past human activity on a given site. This has far reaching consequences in practice, as no other data (e.g. cropmarks) can be used as a basis for defining a particular archaeological site and recording it on the register. Therefore, the cropmarks, for example, visible on aerial photographs MUST be verified by the presence of archaeological material attained as a result of field-walking or trial trenches.

Many years of field-walking surveys according to the AZP model have led to a situation where many archaeologists find it impossible to imagine other prospecting methods. Also they are not able to think about sites in categories other than those included in the Archaeological Site Record Card in the late 1970s. This means that environmental factors are mostly recorded as the most important and treated as real factors which influenced human decisions on the location of their settlements or other form of activities. The effect of this is rather strange, as the information recorded on the Card is the subject of analysis allowing conclusions to be made about the past (e.g. Rybicka 1995; Pelisiak 2003). However, such behaviour systematically strengthens the opinion that field-walking is the best prospecting method. This belief is not altered by the fact that the preparation stage of many recent infrastructure projects reveals the many shortcomings of field-walking and the imprecision of the information gained.

Another negative consequence of field-walking in the AZP model is the huge reduction in analytical categories, as archaeologists reduce the collected material to primarily four basic types: settlement, cemetery, camp-site, non-defined trace. Other categories of remains occur sporadically. The variety of human activity already disappears from the archaeologists' field of interest at the stage when archaeological material is gathered as many may not produce distinctive material on the surface (e.g. pit alignments).

Despite its numerous limitations, field-walking remains the basic method in Polish archaeology, which is empirically and objectively oriented. The presence of ceramic fragments or flint tools is unequivocal proof of the presence of humans in a given place and such thinking is transferred into research and conservation practice. This is why field-walking had to remain the foundation of the AZP_2 project.

Why aerial reconnaissance?

In this instance the answer is not at all obvious. Despite the presence of aerial photographs in Polish archaeology since the 1920s and 1930s, it is still not present in the consciousness of archaeologists as an effective method of uncovering the past. Very often the role of aerial photographs is simplified to an illustration of the location or presentation of the geographic terrain. The shortcomings and limitations of such an approach were highlighted as far back as the 1930s by Konrad Jażdżewski (1938). Even today, not much has changed on this issue. As a result, even though the AZP programme foresaw the use of aerial photographs this has never actually happened in practice (Kobyliński 1999).

The end of the 20th century witnessed a whole range of new initiatives connected to further attempts showing how useful aerial photographs in Poland can be (e.g. Kobyliński 2005; Nowakowski *et al.* 2005). The consciousness of archaeologists finally began to awaken with the particularly spectacular discoveries of archaeological sites dating from the Neolithic (e.g. many settlements such as Brześć Kujawski type – Figure 13.2, *Kreisgrabenanlage* near Biskupin – Figure 13.3, the megalithic long barrows in Bielejewo), through to the Bronze and Iron Age (e.g. the Early Iron Age fortified settlement in Jurkowo – Figure 13.4) to the Middle Ages (e.g. the medieval town of Szamotuły – Figure 13.5). There was a marked increase in interest in aerial reconnaissance expressed by archaeologists, though this still concerns only a few regions.

The effectiveness of aerial survey, especially in Wielkopolska, and the role of the courses in how to apply aerial photographs held at the University in Poznań, had an undeniable influence on the inclusion of this method in the AZP_2 project.

Why geophysical survey?

The decision to introduce geophysical methods into the AZP_2 project seems rather surprising. There was a long-held belief that these methods were not useful under Polish conditions (a similar situation to aerial photographs). The positive results achieved by Tomasz Herbich (e.g. 2003; Herbich & Peeters 2006) and Krzysztof Misiewicz (e.g. 1993, 1998) were largely related to archaeological sites in other countries and were better known internationally than in Poland. Only recently has this perception slowly changed (though there is yet a long way to go to convince people of how useful these methods really are). I believe that one of the factors affecting the gradual acceptance of geophysical methods is the more persuasive presentation of the results of the surveys (a more visual approach stirring the imagination of archaeologists). Furthermore, a number of projects on significant sites have been realised in a short period of time which have shown the potential of these methods in Poland (e.g. Ducke & Müller 2004; Dernoga *et al.* 2007; Harding & Rączkowski 2010).

For many years the Institute of Archaeology and Ethnology PAS in Warsaw (T. Herbich and K. Misiewicz) was the leading centre concerned with geophysical methods. Today, a greater number of academic centres have undertaken this type of initiative and private companies are being set up offering services in this field.

Figure 13.2: Kaczkowo, Kujawy Region. An example of Brześć Kujawski type settlement. Photo W. Rączkowski, 25.05.2009. © *IP UAM.*

Figure 13.3: *Kreisgrabenanlage* discovered from the air near Biskupin. Photo W. Rączkowski, 30.06.2008. *© Archaeological Museum in Biskupin.*

Why trial trenches?

Despite the fact that non-invasive methods in archaeological prospecting have been accepted (i.e. aerial photographs and geophysics), the belief exists amongst Polish archaeologists that only material left behind after human activity is evidence of the existence of an archaeological site. Therefore, no image – whether in the form of an aerial photograph of cropmarks or the visualisation of geophysical survey results – is sufficient indication to be able to enter the site on the record and conservation register. It is necessary to find archaeological material (ceramic fragments, flint tools etc.) or determine the presence of a cultural layer. If such material is not found during field-walking then it is necessary to carry out trial trenches.

An additional role of trial trenches is to indicate how well preserved the archaeological features beneath the topsoil are. Polish archaeologists believe that only in this way it is possible to ascertain the cognitive value and how well the archaeological features have been preserved.

Why not other methods?

In Polish archaeology there is no tradition of a search for new technological solutions and the application of new methods. The age of excavations in Biskupin in the 1930s, during which many attempts to use various methods were made (Grossman & Piotrowski 2005), has long since past. There is no sense of a need to experiment or to apply modern technology. Traditional thinking is unusually strong and reinforces the standards worked out many years ago. It uses two basic arguments against innovation – high cost and incompatibility with Polish conditions. The first argument shows that archaeologists prefer to spend money digging as excavations are still treated as the only method allowing discovery the past. The second view is astonishing – something has not yet been tested but it has already been judged to be useless. This concerns not only non-invasive methods but also methods used in documentation or analysis (e.g. GIS).

It is not surprising therefore that Polish archaeologists have not even heard of many methods (as they do

Figure 13.4: Jurkowo, Wielkopolska Region. Biskupin type fortified settlement discovered in 1999. Photo W. Rączkowski, 11.03.2010. *© IP UAM.*

Figure 13.5: Mutowo, Wielkopolska Region. The first identification of the medieval town at Szamotuły, photographed in 2006. Photo W. Rączkowski, 3.07.2006. © *IP UAM.*

not want to hear and as a result do not look for such information). Satellite images are known only via Google Earth™ and are not deemed to be especially useful. Up till now LiDAR has been used only once – in 2008[1]. It is easy enough to imagine its influence on what we know about archaeological heritage in a country where nearly 30% of the land surface is covered by forests. Any other prospecting methods simply do not exist in archaeologists' consciousness.

AZP_2 – a summary

Summarising the decisions relating to the form of the AZP_2 project, it is clear that it has been heavily influenced by tradition in Polish archaeology. The result is a belief in the effectiveness of field-walking and the necessity of using trial trenches in order to confirm the cognitive value of an archaeological site.

From this perspective, applications of non-invasive methods such as aerial photographs and geophysics are evidence of conservation bodies/authorities slowly opening up to new ideas, the gradual acceptance of the ideas that emerged from the Malta Convention and their introduction into practice. It is a long, drawn-out process as it is not easy to change the mentality of archaeologists and those responsible for the shape of conservation politics. In this situation the fact that at least somewhere in Poland such initiatives are being undertaken is a positive note, even though they certainly are not fully satisfactory.

Tradition still in power

The realisation of a programme with four elements as described above generates further problems. Up till now the AZP programme was restricted to the identification of archaeological sites, completing the Archaeological Site Record Card and the gradual development of a database of the identified sites. A computerised database (AZP_Max) has been introduced which is mandatory in some regions (Prinke 1992), and a spatial database (mAZePa) also exists as a module of the MapInfo programme (not widely used as basically the conservation services do not have the required programming available) (Prinke 2002). Thus the single 'modern' tool used in the management of information on archaeological sites is the AZP_Max.

The AZP_Max is a textual database, based on the Archaeological Site Record Card questionnaire, which was developed in the mid 1980s and updated in 1990s. Part of the data contained in the database can be simply analysed, though much of the information is held only in the form of a description and therefore cannot be interrogated automatically. There is no possibility of extending the programme further, and as a result the database does not fulfil the requirements necessary today in order to realise the AZP_2 project.

The consequence of such a situation is that in the realisation of the AZP_2 project the conservation service in Poznań receives results from those carrying out the survey which are not in any way integrated. The results of field-walking are contained in the Archaeological Site Record Card, 1:10,000 maps (on paper) and in the AZP_Max database (independent from the earlier AZP). The results of aerial survey are in digital format and, as well as digital photographs, they contain flight paths registered via GPS and a database of images with the georeferences of each one. All data is prepared using the MapInfo standard. Mapping of selected cropmarks is additionally carried out, which are also georeferenced. Geophysical survey provides the visualisation of the sites (with georeferences) and a commentary with interpretation. Finally, the results of the trial trenches are returned on paper (drawings) together with a written report.

Data on the sites gained in such a way does not permit full integration of the results from the various methods applied and due to this the material is difficult to apply and use practically. Certain reports are held separately and it is not easy to control all the activities which were

1 Jacek Nowakowski presented preliminary results at a conference *Sezon archeologiczny 2008-2009. Konferencja sprawozdawcza* in Poznań in April 2010.

Figure 13.6: Parchanki, Kujawy Region. An example of sand/gravel extraction without any control from the archaeological heritage services. Photo W. Rączkowski, 10.05.2006. © *IP UAM*.

carried out, nor the results attained. This gives the impression that the application of various methods serves rather to show off to others – 'Look what we're doing!' – than used effectively. When whatever decisions regarding the protection of archaeological sites are taken, only the field-walking results are used, as the rest are still considered to be rather difficult to apply. An additional difficulty is the fact that these methods operate on different spatial levels – from a regional dimension (aerial survey) to a fragment of the site (trial trenches).

Towards a model of relational database(s) to integrate archaeological data

There is nothing surprising in the statement that in the current situation only the development of a textual and spatial relational database can solve the existing problems. Obviously, it is necessary to take into account the special characteristics of the data recovered using various methods. It is not therefore possible to create one general database. It seems that a better direction to aim for would be the construction of a specific database for the data generated via particular methods, but with a unified system to record the spatial information. Only information integrated in such a way would permit the comparison of data achieved via various methods and critical reflection on the results attained (Rączkowski 2006).

It is also standard practice and rather obvious that during the AZP_2 data is collected on archaeological information (location, chronology, function, character of the feature, and the presence of cultural layers). This becomes the basis for the evaluation of the archaeological sites (past landscapes are rarely considered). Such an evaluation should be the foundation for the formulation of aims and forms of protection (see above – second basic aim of the AZP_2 project). I believe that this is insufficient information to permit such important decisions to be made. The methods used within the framework of the AZP_2 project also allow the collection of information on the dangers threatening the site and the landscape. The identification of various threats (cultural or natural) could have a significant influence on the shaping of archaeological heritage protection politics and the actions then undertaken.

The methods selected for the realisation of the AZP_2 project, especially field-walking and aerial survey, as well as existing remote sensing data (such as *Web Map Services* resources), permit the data on processes taking place in the landscape, urban development, infrastructure projects etc., to be continually updated. Observations recorded during field-walking or aerial survey can add to existing information on threats, in particular those beyond the control of the conservation services. This particularly concerns individual activities undertaken without permission (Figure 13.6). The data collected via such methods can be introduced into a separate module of the database and allow a swift reaction to threats faced.

The proposed relational database model should consist of three basic parts (Figure 13.7): a database of archaeological record, a database of threats and a database of public spatial data (e.g. WMS). The first database will comprise various modules containing data and their structure which are characteristic of the data generated within the frameworks of particular methods. Extension of the database should also be possible depending on requirements as should the introduction of new prospecting methods. The assumption is that all the data will have unified georeferencing.

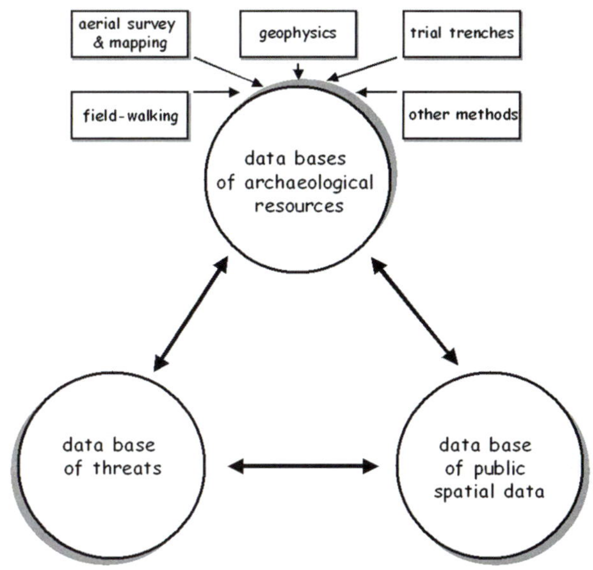

Figure 13.7: Scheme presenting the general modules of the proposed database.

Relational database as a challenge

As described above, the form of the AZP_2 in the current pilot projects is a compromise between tradition and new possibilities. Tradition plays a key role and at the data collection and collation stage it certainly dominates. There exists therefore a need for change and the introduction of a new model. The consequences are far-reaching for all project participants (Figure 13.8).

A change in data preparation practice for the conservation service is fundamental and is due to changes in technology which allow the use of a relational database and the resulting integration of various data necessary for archaeological interpretation. With this new technology comes a change in procedures during field surveys as well as in the preparation of survey results. This means that there is also a need to become familiar with using the new technology and the computer programmes that accompany it. For those in field-walking (AZP) or excavations this may well be a major psychological barrier. The same goes for archaeological heritage management officers who for years have used only the AZP_Max and paper maps. Who knows, perhaps persuading people to accept these new needs and to learn how to use new tools in their work will be a greater problem than resolving the structural problems of a relational database.

A separate issue is the interpretation of data already in the database. The information from surveys already carried out in these methods cannot simply be reduced to a statement on the presence or lack of archaeological remains. In particular, the differences in results attained from the various methods should generate deeper, critical thought on the reasons why such a situation occurred (Rączkowski 2006). Only then will a database such as this be a useful tool in the evaluation and management of archaeological heritage. This means that archaeological heritage management officers will need to have greater knowledge of the possibilities and limitations of the applied methods.

Final remarks

The concept and practice of the AZP project has, to a great extent, shaped the thinking of Polish archaeologists on the past, on ways of carrying out fieldwork and the concepts of protection of archaeological heritage. Perhaps due to this, and the dominant positivist approach to archaeology, it is difficult to introduce changes to the existing action plan. The set organisational structures and attitude to archaeology certainly make the introduction of new solutions, methods and concepts more difficult. It is not surprising therefore that the need to continue the current AZP in some form gave rise to a new concept (the AZP_2) which to a large extent refers to tradition. Even the small number of new elements, the non-invasive methods (geophysical and aerial survey), are treated in a traditional way. The resulting effort, though it can in certain cases yield excellent results (e.g. the integrated surveys of the medieval town of Szamotuły), does not create new values and does not offer valuable change.

Polish archaeologists must change their mentality both on a theoretical level (departing from a narrowly understood positivism) and to the consequences of being more open to 'new' suggestions for methodology and methods, including the application of new research tools (e.g. Gołembnik 2009). The lack of a breakthrough in the (mainly psychological) barriers present in Polish archaeology may shortly lead to the loss of both the great potential the AZP offers and the chance to introduce non-invasive methods to a wider audience. If the current state of affairs continues (as is happening in the majority of regions in Poland) this may lead to archaeology being compromised and seen as an anachronistic discipline, ill adapted to the contemporary world.

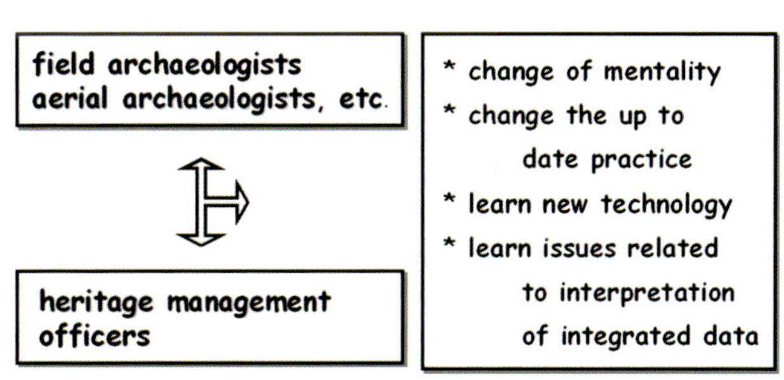

Figure 13.8: Database as a challenge for heritage management officers and archaeologists doing a variety of surveys and fieldworks.

References

Clarke, D.L. 1977: Spatial Information in Archaeology. In Clarke, D.L. (ed.): *Spatial Archaeology*. Academic Press, London, 1–32.

Dernoga, M., Herbich, T., Pietrzak, R., Rączkowski, W. & Święch, D. 2007: Discovering medieval Szamotuły (Wielkopolska region): a multidisciplinary approach. *Studijné Zvesti* 41, 131–4.

Ducke, B. & Müller, J. 2004: Die geomagnetische prospektion. In Czebreszuk, J. & Müller, J., (eds): *Bruszczewo: Ausgrabungen und Forschungen in einer prähistorischen Siedlungskammer Großpolens* 1. Leidorf, Poznań, Kiel, Rahden (Westf.), 61–8.

Gołembnik, A. 2009: *Podstawy metodyczne zastosowania nowoczesnych technologii w archeologii*. Paper presented at the conference *Cyfrowa przeszłość. Metody komputerowe w archeologii* in Warsaw in May 2009.

Grossman, A. & Piotrowski, W. (eds) 2005: *Badacze Biskupina*. Muzeum Archeologiczne w Biskupinie, Biskupin.

Harding, A. & Rączkowski, W. 2010: Living on the lake in the Iron Age: new results from aerial photographs, geophysical survey and dendrochronology on sites of Biskupin type. *Antiquity* 84, 386–404.

Hensel, W. 1968: W sprawie metody mikrogeograficznej w archeologii. *Sprawozdania z prac naukowych Wydziału Nauk Społecznych PAN* 3, 27–9.

Herbich, T. 2003: Archaeological geophysics in Egypt: the Polish contribution. *Archaeologia Polona* 41, 13–55.

Herbich, T. & Peeters, C. 2006: Results of the magnetic survey at Deir al-Barsha, Egypt, *Archaeological Prospection* 13, 11–24.

Jankhun, H. 1977: *Einfuhrung in die Siedlungsarchaologie*. Walter de Gryter, Berlin, New York.

Jaskanis, D. 1996: Próba oceny metody Archeologicznego Zdjęcia Polski na podstawie doświadczeń ogólnokrajowego koordynatora. In Jaskanis, D. (ed.): *Archeologiczne Zdjęcie Polski – metoda i doświadczenia. Próba oceny*, MKiS, Warszawa, 9–38.

Jażdżewski, K. 1938: Lotnictwo na usługach prehistorii. *Z Otchłani Wieków* 13(3–4), 33–41.

Kiszkowski, P. & Wrzesiński, J. 1996: Prospekcja elektrooporowa Ostrowa Lednickiego. *Studia Lednickie* 4, 167–95.

Kobyliński, Z. 1999: Siedemdziesiąt lat archeologii lotniczej w Polsce. *Światowit* 1 (nowa seria), fasc. B, 112–22.

Kobyliński, Z. 2005: *Archeologia lotnicza w Polsce. Osiem dekad wzlotów i upadków*. Państwowe Muzeum Archeologiczne, Instytut Archeologii i Etnologii PAN, Warszawa.

Mazurowski, R. 1980. *Metodyka archeologicznych badań powierzchniowych*. PWN, Warszawa, Poznań.

Misiewicz, K. 1993: Resistivity survey in prospection of settlement sites. *Archaeologia Polona* 31, 105–15.

Misiewicz, K. 1998: *Metody geofizyczne w planowaniu badań wykopaliskowych*. Instytut Archeologii i Etnologii PAN, Warszawa.

Nowakowski, J., Prinke, A., & Rączkowski, W. 2005: Latać, latać i... interpretować: problemy i perspektywy polskiej archeologii lotniczej. In Nowakowski, J., Prinke, A. & Rączkowski, W. (eds): *Biskupin... i co dalej? Zdjęcia lotnicze w polskiej archeologii*. IP UAM, OODA, MAB, PTP, Poznań, 11–23.

Pelisiak, A. 2003: *Osadnictwo. Gospodarka. Społeczeństwo. Studia nad kulturą pucharów lejkowatych na Niżu Polskim*. Wydawnictwo Uniwersytetu Rzeszowskiego, Rzeszów.

Prinke, A. 1992: The Polish National Record of Archaeological Sites: A Computerization. In Larsen, C.U. (ed.): *Sites & Monuments. National Archaeological Records*. Nationalmuseet, Copenhagen, 89–93.

Prinke, A. 2002: Introducing Information Technology to Archaeological Resource Management: Towards GIS-Based SMR of Mid-Western Poland. In García Sanjuán, L. & Wheatley, D.W. (eds): *Mapping the Future of the Past. Managing the Spatial Dimension of the European Archaeological Resource*. Universidad de Sevilla, Sevilla, 85–96.

Rączkowski, W. 2005: Tradition in power: vicious circle(s) of aerial survey in Poland. In Brophy, K. & Cowley, D. (eds): *From the Air. Understanding Aerial Archaeology*. Tempus, Stroud, 151–67.

Rączkowski, W. 2006: Towards integration: two prospection methods and some thoughts. In, Campana, S. & Forte, M. (eds): *From Space to Place. 2nd International Conference on Remote Sensing in Archaeology*. Archaeopress, BAR International Series 1568, Oxford, 203–6.

Rybicka, M. 1995: *Przemiany kulturowe i osadnicze w III tys. przed Chr. na Kujawach: Kultura pucharów lejkowatych i amfor kulistych na Pagórach Radziejowskich*. Muzeum Archeologiczne i Etnograficzne, Łódź.

Starzyński, A. 2006: Założenia do programu ochrony zabytków archeologicznych na terenie województwa wielkopolskiego. *Wielkopolski Biuletyn Konserwatorski* 3(1), 29–31.

Starzyński, A. & Dernoga, M. 2006: Badania osadnicze na terenie Wielkopolski. *Wielkopolski Biuletyn Konserwatorski* 3(1), 32–44.

14 | As far as the laser can reach... Laminar analysis of LiDAR detected structures as a powerful instrument for archaeological heritage management in Baden-Württemberg, Germany

Jörg Bofinger and *Ralf Hesse*

Abstract: A new project has been initiated by the State Office for Cultural Heritage Management Baden-Württemberg for the laminar analysis of LiDAR data in Baden-Württemberg. This project is part of the archaeological prospection system in Baden-Württemberg. New techniques for detection of archaeological structures using special data processing methods on LiDAR data were developed. The focus of this project is on the totality of visible structures presumed to be archaeological sites or at least elements of the historic landscape in LiDAR imagery in the whole state of Baden-Württemberg. Based on the complete high resolution LiDAR data set of the surface of Baden-Württemberg a comprehensive mapping of all potential sites in an area of 35,751km^2 is the aim of the three year project.

The story so far – first use of LiDAR imagery in Baden-Württemberg

For several years the Department of Archaeology of the State Office for Cultural Heritage Management in Baden-Württemberg has been using LiDAR (Light Detection And Ranging), or Airborne Laser Scanning (ALS), as a tool for the detection and visualization of archaeological sites.

The first steps in ALS were made in Baden-Württemberg in 2003. In preparation for extended field investigations in the surroundings of the Heuneburg early Iron Age hill fort in Upper Swabia, there was the opportunity to scan an area of 20km^2 of the Danube valley around this important site (Bofinger *et al.* 2006, 2007). This survey was contracted to a specialized company who conducted the survey and data collection in October 2003, surveying from an altitude of 500m with a raster width of 0.5m, allowing a highly accurate, geo-referenced digital terrain model to be created (Figure 14.1). After a very dry summer old meanders and water courses, especially on the Danube floodplain, were clearly apparent in excellent conditions for detailed survey and mapping. The modern method of LiDAR terrain mapping was useful in making these recordings. Even in this first project high precision LiDAR imagery registration was specified, so that even inconspicuous details such as the slight bank of an ancient rampart system in the fields northwest of the Heuneburg appears very distinctly in the shaded relief. This hook-shaped rampart is hardly discernible in the present landscape; only the ditch in front of the rampart is clearly visible on aerial photographs. This example clearly illustrates the ways in which the results of conventional aerial archaeology can be improved significantly upon (Figure 14.2).

Whilst the Heuneburg study area was scanned at very high resolution (point spacing 0.5m) on special demand, the participation in the Culture 2000 program 'European Landscapes – past, present, future' offered the possibility to extend the LiDAR work. This provided the possibility for comparing data sets with different resolution. The State Topography Authority of Baden-Württemberg (Landesamt für Geoinformation und Landentwicklung Baden Württemberg, LGL) provides DTM-data at a grid width of 1m and an elevation accuracy of ± 0.15m. On the other hand, for the Heuneburg two datasets were available with a much higher resolution provided by specialized enterprises: one with a grid width of 0.5m and an elevation accuracy of ± 0.10m and another with a grid width of 0.25m and an elevation accuracy of ± 0.15m (Figure 14.3).

A detailed study of these three LiDAR data sets with different grid widths indicated that the cheaper solution of the State Record would normally be sufficient for the identification and visualization of archaeological structures. Thus we could see that the State Topography Authority data clearly provided an excellent basis for further investigations. Beyond this, it is especially useful to employ the high resolution LiDAR data only in the case of extremely detailed and sensitive structures such as small walls or wooden elements. Exceptions to these are very prominent impacts on the landscape due to current changes, such as flooding or landslides, when contemporary recording of 'updated' data by special LiDAR flights is recommended.

From single points of interest to country-wide analysis

In the first years of LiDAR use in Baden-Württemberg the focus was only on discrete localities, such as the Heuneburg and the tumuli in its environs. The upper

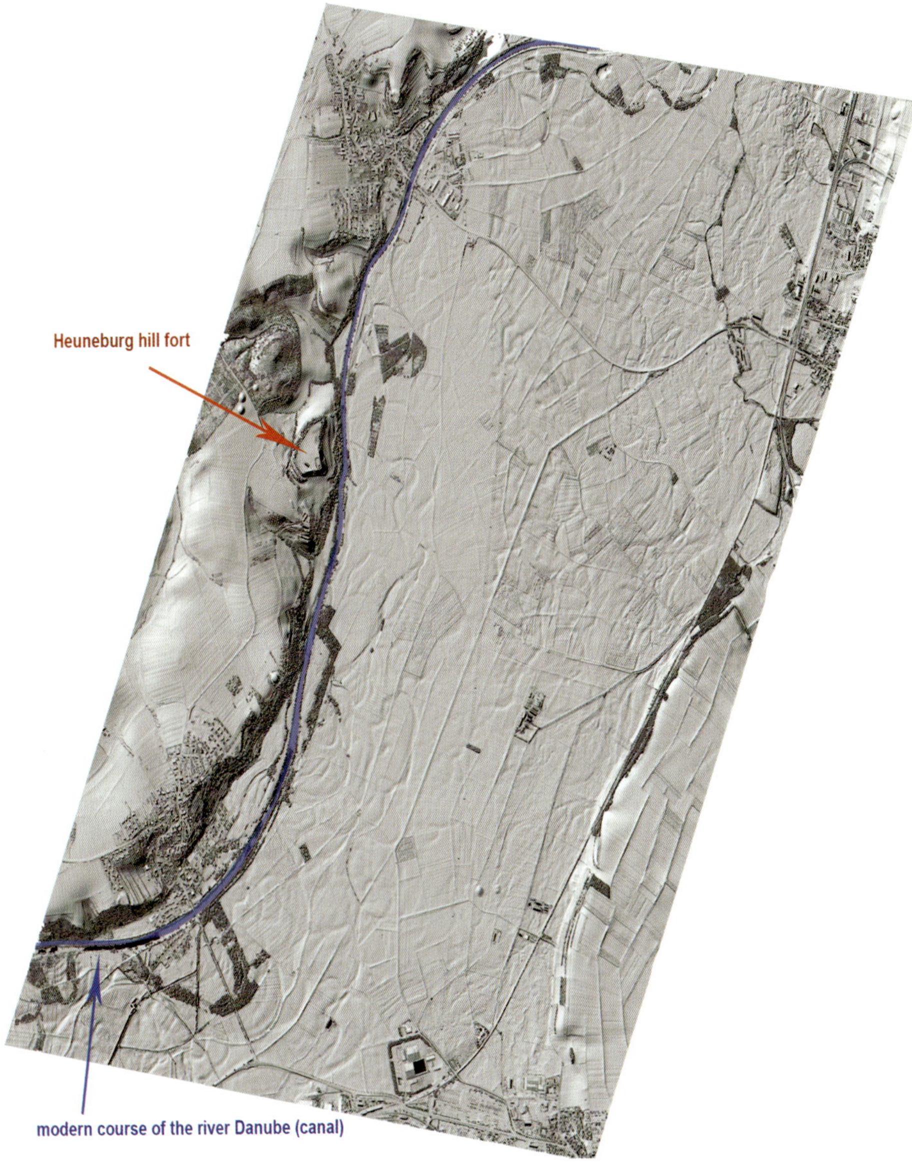

Figure 14.1: LiDAR scan of the landscape around the Heuneburg in the Danube valley. A 20km² area of this region saw the first implementation of this new technology on archaeological research activities. *© LAD/Toposys Ltd.*

Germanic Limes is another example for such detailed on-site study of LiDAR data (Bender 2009). Only shaded relief maps were used as a basis for interpretation, and data were only produced for specific projects and needs. In contrast to these first steps, a new project of LiDAR analysis is focused on the totality of visible structures presumed to be archaeological sites or at least elements of the historic landscape in LiDAR imagery in the whole state of Baden-Württemberg.

The Federal State of Baden-Württemberg in south-western Germany is rich in archaeological heritage from the Palaeolithic onwards (cf. LAD, 2009, for an overview of recent archaeological research in Baden-Württemberg). In 2009, the State Office for Cultural Heritage Management Baden-Württemberg (Landesamt für Denkmalpflege, LAD) launched a three-year project aimed at the complete archaeological mapping of Baden-Württemberg. For the laminar analysis of the complete pre-existing Baden-Württemberg LiDAR record, a plan for the examination and interpretation of this record was developed, with a pilot phase running between May 2009 and summer 2010. This study is based on ALS of the whole of Baden-Württemberg (35,751km²) by the State Topography Authority undertaken in 2000–5, and fortunately available cost free to the archaeological heritage service for scientific and heritage management purposes. The project aim is to complete the data base for cultural heritage management, building from the 72,000 sites already listed (including those detected by aerial archaeology), by locating all potential sites in the LiDAR data record which could be elements of the historic landscape.

However, it is unknown to what extent the current state of knowledge approximates the actual number of sites. This is particularly relevant given that 39% of the state is covered by forest which renders large areas as blank spaces for archaeological prospection by aerial photography. And the need for an updated inventory is pressing as residential and industrial sprawl, construction of roads, railway lines and pipelines, mechanised agriculture and forestry practices, as well as looting, pose serious threats to known and unknown archaeological sites. Against this urgent need for spatially extensive archaeological prospection, LiDAR data are being developed as a valuable archaeological

Figure 14.2: The synergy of LiDAR imagery and 'conventional' aerial archaeology is shown in the case of the ramparts in the Heuneburg's outer settlement. The ridge can be clearly seen in the LiDAR, while details such as old trenches can only be seen in the aerial photograph. © *LAD/Toposys Ltd, photo O. Braasch.*

tool. Colour-coded maps of Local Relief Models (LRM) are found to be a valuable tool for archaeological prospection. First results of the project confirm the feasibility of using LiDAR-based data for the archaeological mapping of very large areas.

Methodology for laminar LiDAR analysis

High-resolution Digital Elevation Models (DEM) based on airborne LiDAR have in recent years become an important data source for the prospection, mapping and monitoring of archaeological sites. Such data are now becoming increasingly available on a regional or even national scale. In most archaeological applications, LiDAR is applied to relatively small areas (up to a few square kilometres). LiDAR DEM are mostly visualised as shaded relief images which allow viewing of the land surface under different simulated lighting conditions (elevation and azimuth) as well as vertical exaggeration (e.g. Harmon *et al.* 2006; Bofinger *et al.* 2006; Boos *et al.* 2008; Risbøl *et al.* 2006). While the

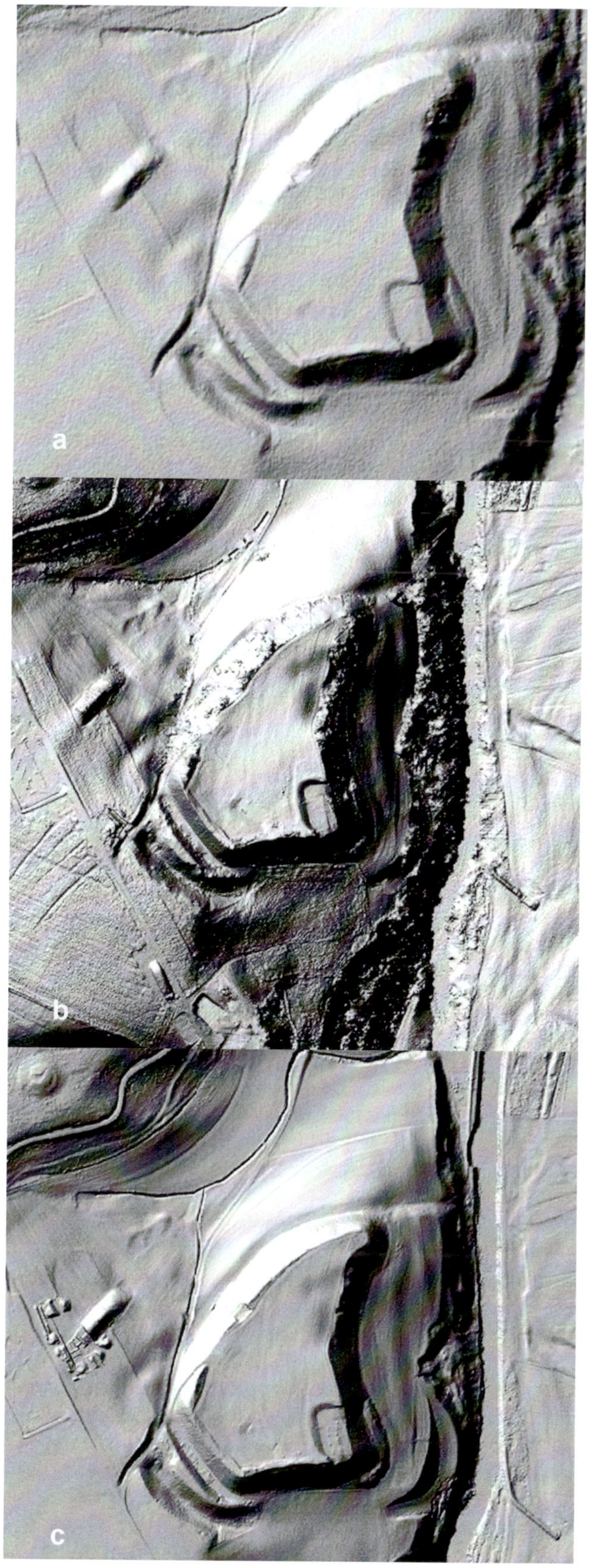

Figure 14.3: Comparison of three LiDAR records with different resolution. a) transformed raw data of the national record Baden-Württemberg (grid width 1m); b) high resolution LiDAR (grid width 0.5m); c) very high resolution LiDAR (grid width 0.25m). © *LAD/Toposys Ltd/Arctron GmbH/LGL www.lgl-bw.de.*

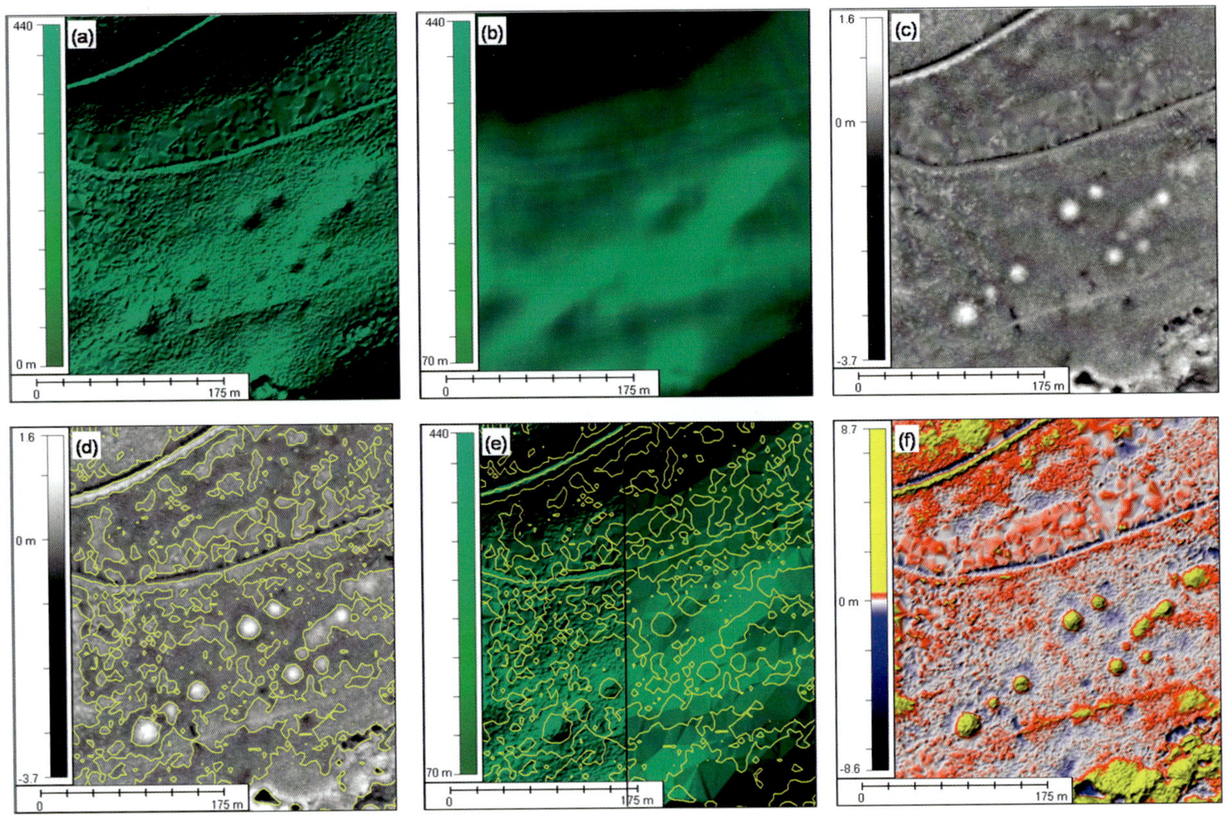

Figure 14.4: Data processing steps for the generation of the Local Relief Model (LRM). a) DEM generated from point cloud data; b) low-pass filtered DEM; c) difference map between DEM and smoothed DEM; d) zero-contour lines in difference map; e) extraction of point elevations and interpolation of purged DEM; f) colour-coded visualisation of the LRM. © *LAD/LGL www.lgl-bw.de.*

experimental manipulation of lighting conditions allows a visual optimisation of individual features, it is a time-consuming process as the visibility of potential archaeological features depends to a large degree on the chosen illumination angles (e.g. Devereux *et al.* 2005). For a spatially extensive prospection project covering thousands of square kilometres, like the present project in Baden-Württemberg, time is an important constraint. Therefore, and to enhance the reliability of the prospection results, improved visualisation techniques are required in archaeological applications of LiDAR. Experimentation with approaches for improved visualisation from other disciplines (Hiller & Smith 2008; Loisios *et al.* 2007; Rusinkiewicz *et al.* 2006) did not deliver satisfactory result for archaeological prospection.

In the first phase of the present project, a new approach was therefore developed and implemented (Hesse 2010). It is based on the observation that archaeologically relevant structures are usually characterised by very low relief relative to the elevation range of the surrounding landscape. Because they thus often only appear as subtle features in the conventional shaded relief visualisation, one goal of LiDAR data processing for archaeological prospection was identified as the problem of extracting local small-scale, low relief features from the DEM and eliminate as far as possible the large-scale landscape forms from the data.

Several data processing steps (Figure 14.4) have to be applied to extract small-scale (detail) topographic features for archaeological interpretation:

(a) A DEM is produced from the vegetation-filtered LiDAR point cloud data (in the present case with a pixel size of 1 × 1m).
(b) A low pass filter is applied to the DEM. This smoothed elevation model represents a first approximation of the large-scale landscape forms. The kernel size of the low pass filter determines the spatial scale of features which will be captured in the LRM. In the present case, a kernel size of 25m is used for the low pass filter. This size was found experimentally to result in a good representation of many previously known archaeological features and is therefore assumed to work well for the detection of previously unknown features. Features much larger in diameter or cross-section are uncommon; furthermore, they would be conspicuous in conventional shaded relief images of the DEM. Degraded representation of features much smaller than the kernel size may become a serious issue if they are underlain by strongly convex or concave terrain (e.g. hilltops or ridges, valley bottoms), but is less pronounced on smooth slopes.
(c) By subtracting this smoothed elevation model from the DEM, a first approximation of the local relief is achieved: only small-scale topographic features are preserved in the model while the

large-scale landscape forms are eliminated. However, because small-scale features are smoothed rather than eliminated by the low pass filter, the model derived by this approach is biased towards small features, i.e. the local relief elevations are progressively underestimated as spatial extent of the features increases.

(d) The zero-metre contour lines in the difference map are the limits between positive and negative topographic anomalies.

(e) The DEM point elevations along these zero-metre contour lines are extracted and a purged DEM is created from these elevations. This purged DEM represents the large-scale landscape forms after cutting out rather than smoothing small-scale features.

(f) Subtraction of this purged DEM from the original DEM results in the final LRM which reflects less biased elevation information of small-scale features relative to the landscape at large. The visual interpretation is facilitated by colour-coding (e.g. warm colours for positive and cold colours for negative anomalies); such colour-coded LRM images can also be draped over conventional shaded relief DEM (see Figures 14.7, 14.9, 14.11 & 14.12).

In comparison to using a simple difference map between the DEM and its low pass or median filtered derivate (Doneus & Briese 2006; Hiller & Smith 2008), the LRM derived using this approach results in a less biased representation of small-scale topographic features which reflects more truthfully the elevations of these features relative to the surrounding landscape and thus allows the direct measurement of feature volumes heights. On the other hand, it is less computationally expensive and easier to implement than the kriging based filtering suggested by Humme *et al.* (2006) if an efficient workflow for data processing is required.

Data management and application of the LRM workflow

The vegetation-filtered LiDAR point cloud data supplied by the State Topography Authority of Baden-Württemberg for the area of 35,751km² amounts to around 1 Terabyte in size, consisting of 160,000 single files of filtered last pulse data. The enormous amount of data that had to be processed required the acquisition of capable hardware (8-core Xeron with 16 GB RAM and 4.5 TB hard disc) and software (ENVI).

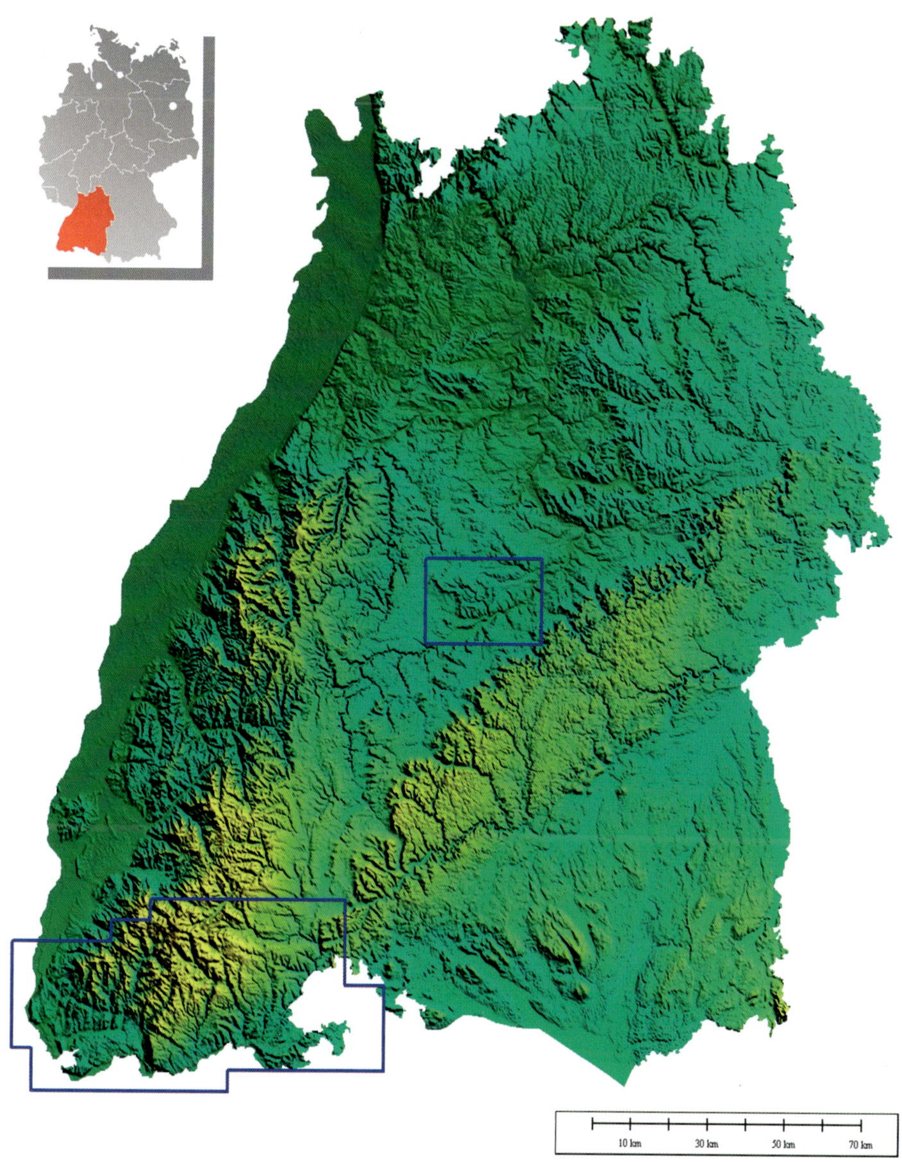

Figure 14.5: Map of Baden-Württemberg showing the current status of the project (summer 2010). Prospection has been completed in the outlined regions. © *LAD/LGL www.lgl-bw.de.*

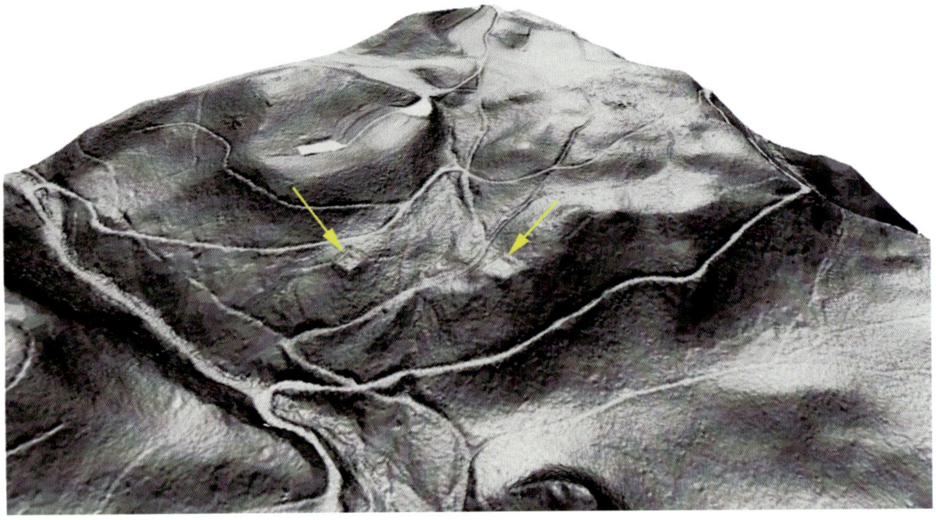

Figure 14.6: Fortifications flanking a sunken road, Bad Säckingen (3D view of shaded relief DEM; yellow arrows mark fortifications). Similar features in the region date to the 16th/17th century. © *LAD/LGL www.lgl-bw.de.*

Dedicated software for the efficient management of the data and the semi-automatic implementation of the LRM workflow was not available at the outset of the project. Therefore, two graphical user interfaces were developed using Visual Basic for Applications (VBA). The first user interface allows the efficient interactive management of all raw, semi-processed and processed data. The current data processing and archaeological prospection status is documented for data segments of 10 × 10km. Several data processing steps implemented in VBA can be interactively executed for these data segments. For data processing steps that are executed in ENVI the processing status is also documented in the user interface. Furthermore, relevant metadata like the number of known archaeological sites in each data segment or the average LiDAR raw data quality are displayed in the user interface, preview maps of the data segments can be displayed and short notes can be stored for each data segment.

Finally, data segments can be selected and all relevant data opened for visualisation and mapping. This includes:

- the LiDAR-derived DEM, DSM and LRM,
- raster maps showing the LiDAR raw data quality in terms of point density and data gaps,
- raster maps showing surface depressions (e.g. dolines) derived from the LiDAR DEM,
- topographic and geological maps,
- aerial photographs,
- vector data of present settlements, roads etc.,
- vector data of known archaeological sites and find spots.

The prospection of the data and mapping of sites is based on the visual interpretation of a combination of all relevant data. Known archaeological sites and find points are used as reference for the qualitative and quantitative properties of archaeological features.

A second graphical user interface serves as a toolbox for actual prospection. It allows orientation within the 10 × 10km segment and documentation of the mapping status on the scale of one square kilometre as well as the interactive manipulation of the illumination in the shaded DEM visualisation. It also provides templates for the creation of new vector objects for the mapping of potential archaeological features.

First results: overview

After development and implementation of methodology, workflow and data management, the processing of the LiDAR point cloud data, the DEM and LRM for the whole of Baden-Württemberg were generated. Further LiDAR-derived raster maps showing raw data quality, data gaps and surface depressions were generated. Data processing was largely finished in January 2010. The subsequent archaeological prospection then concentrated on two large test regions, the forest region Schönbuch in the centre of Baden-Württemberg and the region southern Black Forest and Upper Rhine (Figure 14.5).

In the Schönbuch region (600km²) 2,513 potential archaeological sites have been identified by prospection of the LiDAR data compared with 1966 previously known sites and find spots. In the southern Black Forest and Upper Rhine region prospection of an area of 2,700km² resulted in 57,936 potential sites compared with 3726 previously known sites and find spots. Most features mapped as potential archaeological sites can be related to historic or prehistoric resource use. For example, terraced slopes and ridge-and-furrow document agricultural use, while mining traces, slag heaps and thousands of kiln stances allow new insights into spatial patterns of mining, ore processing and related fuel supply. Furthermore, a large number of potential burial mounds and several previously unknown fortifications have been detected. Hollow trackways and relict field parcel boundaries have also been mapped in large numbers and they should allow inferences regarding the location of settlements.

First results: examples

The following examples will give an impression of the diversity of archaeological sites detected in the first phase of the project. Two fortifications flanking a hollow trackway were identified in the vicinity of Bad

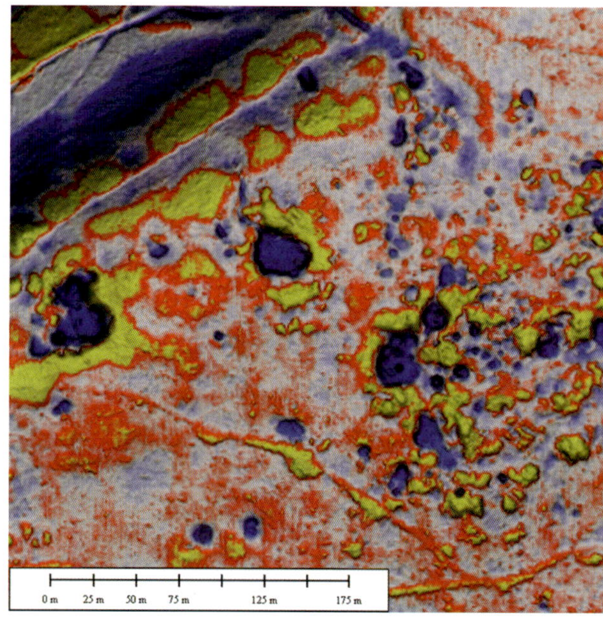

Figure 14.7: Mining traces, Albführen (colour-coded LRM draped over shaded relief DEM). © *LAD/LGL www.lgl-bw.de.*

Säckingen. Based on their similarity with known sites in the region, they probably date to the 16th or 17th century AD (Figure 14.6). In comparison to such characteristic structures, it is much more difficult to suggest date ranges for other detected features. This is certainly the case of mining traces, such as those near Albführen, which cover approximately 50 hectares (Figure 14.7) and whose date is still a matter of conjecture.

A particularly conspicuous phenomenon in the southern Black Forest are large numbers of kiln stances. The archaeological analysis of individual kiln stances probably provides only limited and likely redundant insights. As the LiDAR-based prospection allows their distribution to be mapped largely independent of the present vegetation cover, spatial patterns of resource use can be reconstructed (Figure 14.8).

The traces of former agricultural activities, such as terraced slopes or ridge-and-furrow, are elements of the cultural landscape rather than archaeological sites in a traditional sense, but they provide valuable information on land use and can help to locate former settlements. Ridge-and-furrow fields in a presently forested area in the Schönbuch region (Figure 14.9), for example, are unlikely to be identifiable during a field visit given their low relief (approximately ±10cm).

A high number of potential burial mounds were detected in the southern Black Forest and Upper Rhine region. Because of the similarity of burial mounds and other features such as small natural mounds, mining-related waste heaps or slag heaps from ore processing in the LiDAR data, the confidence in the interpretation of positive relief anomalies as burial mounds is generally quite low if examined individually. However, the frequently observed arrangement of mounds in discrete clusters and the lack of spatial relationships with mining or ore processing can reduce this uncertainty. It is expected that limited on-site examinations will provide sufficient information to characterise entire groups of mounds.

On-site verification and comparison with known archaeological sites

The verification of the results of the archaeological prospection is beyond the scope of the present project. At the present stage, a full verification of all identified potential archaeological sites is not envisaged. Rather, the prospection results are used in archaeological investigations accompanying construction projects like highways, railways, pipelines or industrial areas. This in turn provides valuable feedback, eventually allowing a quality assessment of the prospection.

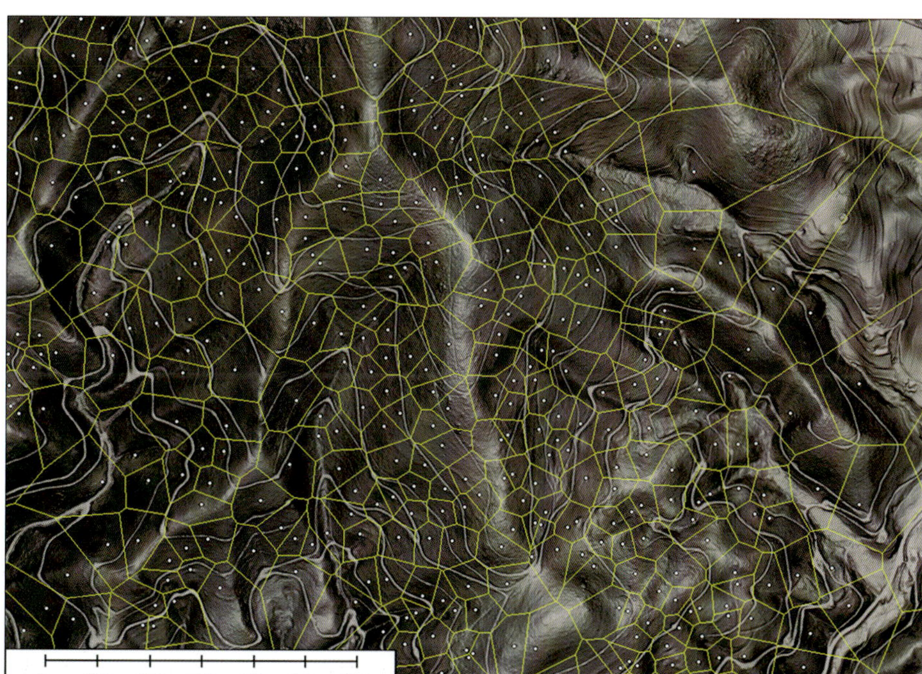

Figure 14.8: Spatial distribution of kiln podia, southern Black Forest (vertically illuminated shaded relief DEM; white dots: positions of kiln podia; yellow lines: corresponding Thiessen polygons). © *LAD/LGL www.lgl-bw.de.*

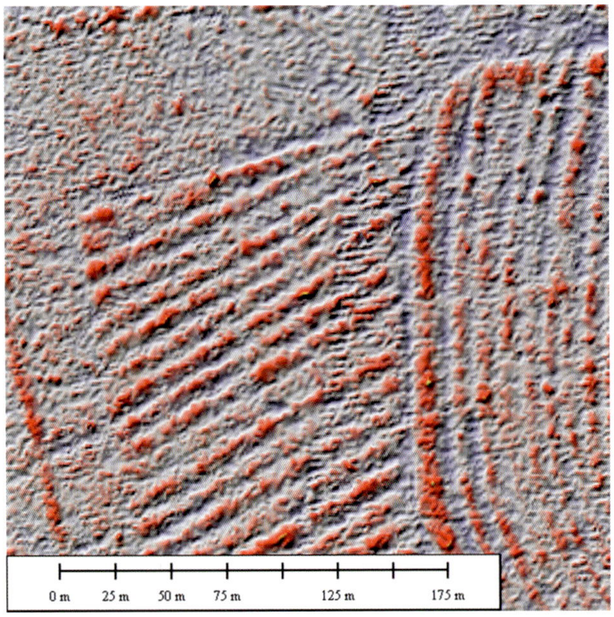

Figure 14.9: Ridge and furrow, Schönbuch (*left* – aerial photograph; *right* – colour-coded LRM draped over shaded relief DEM). These structures show that fields existed in areas that are today covered by forest. © *LAD/LGL www.lgl-bw.de.*

However, some verification has been undertaken. A first verification of a slight positive relief anomaly of 20cm with a diameter of 20m as a potential burial mound was made possible in summer 2009 in the context of archaeological investigations along the course of a future pipeline (Figure 14.10; Bofinger *et al.* 2010). One of the newly discovered sites confirmed during a site visit is a fortification consisting of an 80m long rampart and ditch near Mühlhausen im Täle (Figure 14.11). It is probably medieval in date and a previously unknown part of a series of fortifications along the Fils valley.

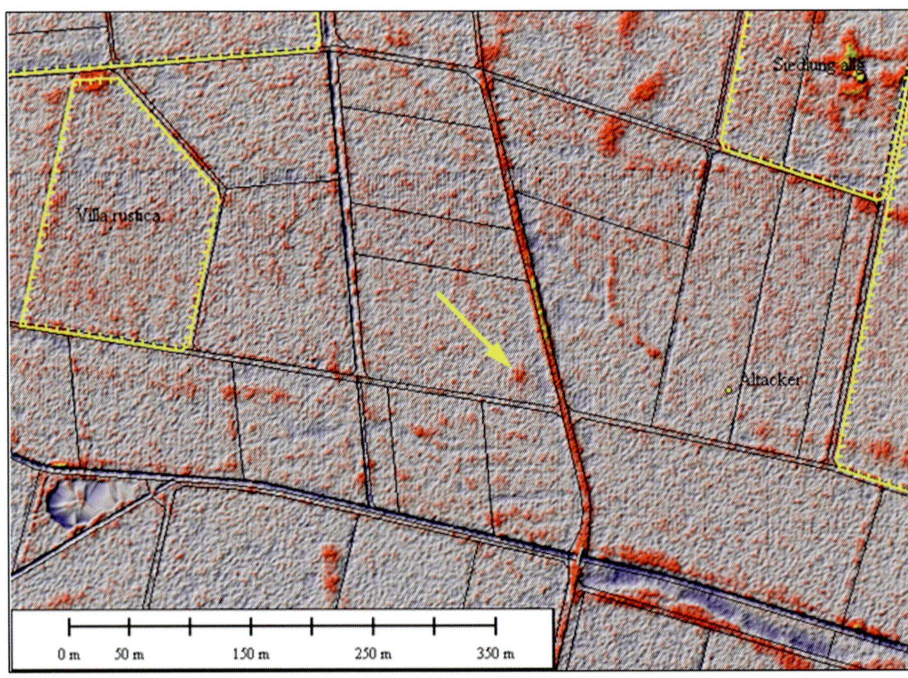

Figure 14.10: Iron Age burial mound, Goldburghausen (*top* – colour-coded LRM – *bottom*: excavation of stone circle lining the perimeter of the burial mound – photograph: M. Meyer). © *LAD/LGL www.lgl-bw.de.*

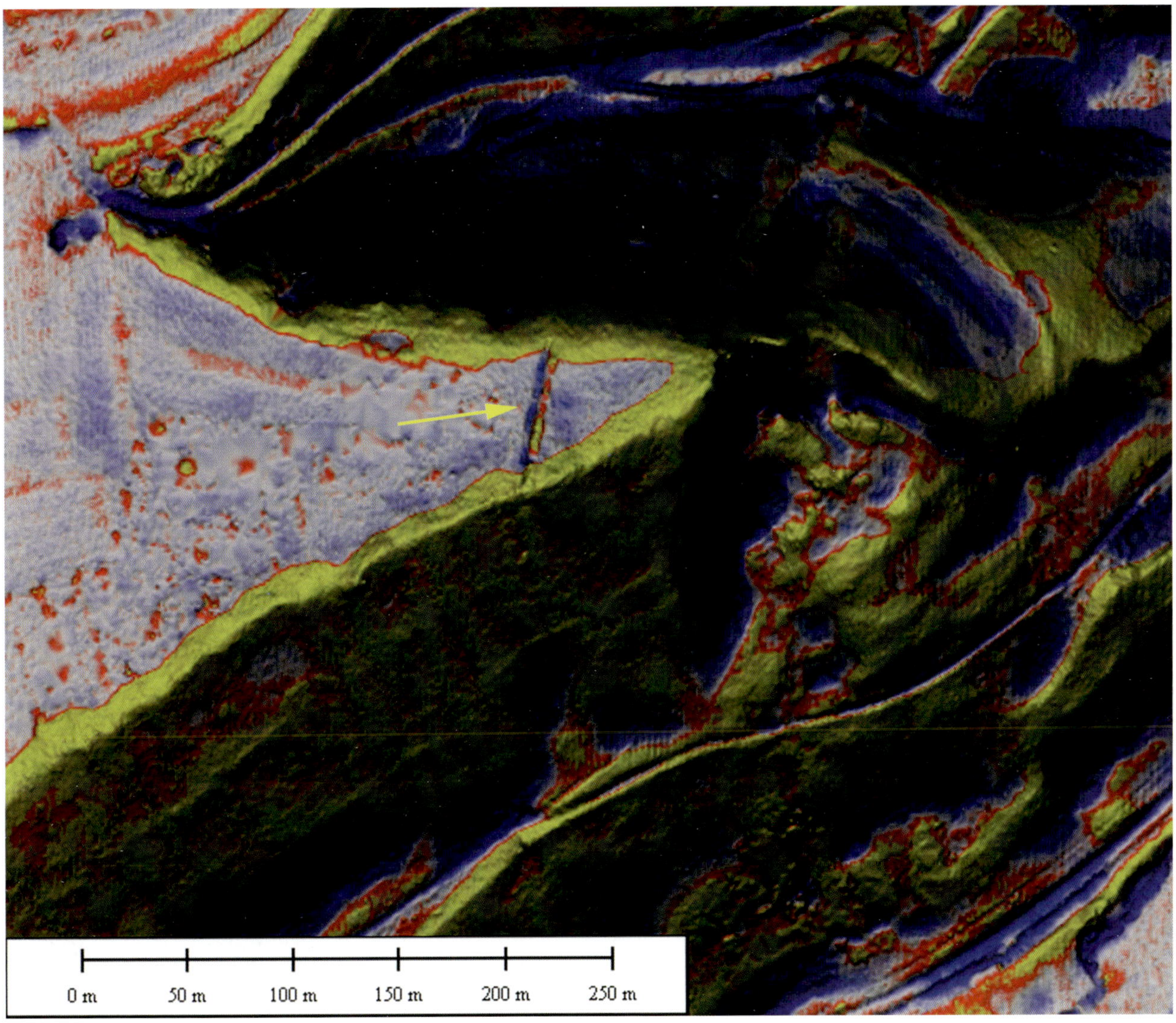

Figure 14.11: Fortification on a mountain spur, Mühlhausen im Täle (colour-coded LRM draped over shaded relief DEM; yellow arrow points to ditch and rampart). Note small mounds and subtle agricultural structures west of the fortification. © *LAD/LGL www.lgl-bw.de.*

In the vicinity of Gottmadingen, a prehistoric circular earthwork or ring fort on a hilltop (Figure 14.12) clearly shows the potential of LiDAR-based mapping for the revision and expansion of the knowledge for known archaeological sites. The area protected as a prehistoric monument actually lies outside the site which is easily recognisable in the LiDAR DEM. Such inaccuracy in the localisation of protected monument areas is probably due to the problem that older site descriptions often lack accurate coordinates. Using LiDAR data, corrections can be made. In many cases, this also leads to new discoveries or research questions, such as the date and associations of several terraces and hollow roads visible on the north-eastern to southern flanks of the hill. Given the arrangement of some of these terraces, they may represent an outer rampart of the ring fort.

Outlook: potential for further applications and developments

LiDAR data may even be a valuable tool for research on time periods from which we cannot expect any remaining topographic surface anomalies: the Palaeolithic. The mountain range Swabian Alb is well-known for caves with abundant and high-quality Palaeolithic artefacts such as mammoth ivory carvings or swan bone flutes. Today the surrounding landscape is farmed and forested. One is tempted to imagine the Ice Age landscape of the Palaeolithic simply by replacing fields and forests with arctic tundra. This, however, ignores the fact that the Swabian Alb is a karst landscape with abundant topographic depressions (dolines). These dolines today drain to an underground system of caves which under glacial conditions was sealed by permafrost. A simulated flooding of these topographic depressions reveals that the Palaeolithic inhabitants of this landscape faced a complex mosaic of different habitats, at least in summer (Figure 14.13). Furthermore, such landscape reconstructions may in the future help to locate Palaeolithic hunting and occupation sites beyond the well-known caves.

As for the development of data processing and interpretation, future work will concentrate on (a) the statistical assessment of size and spatial distribution of sites, (b) the combined processing and interpretation of raster and vector data (i.e. taking into account shape

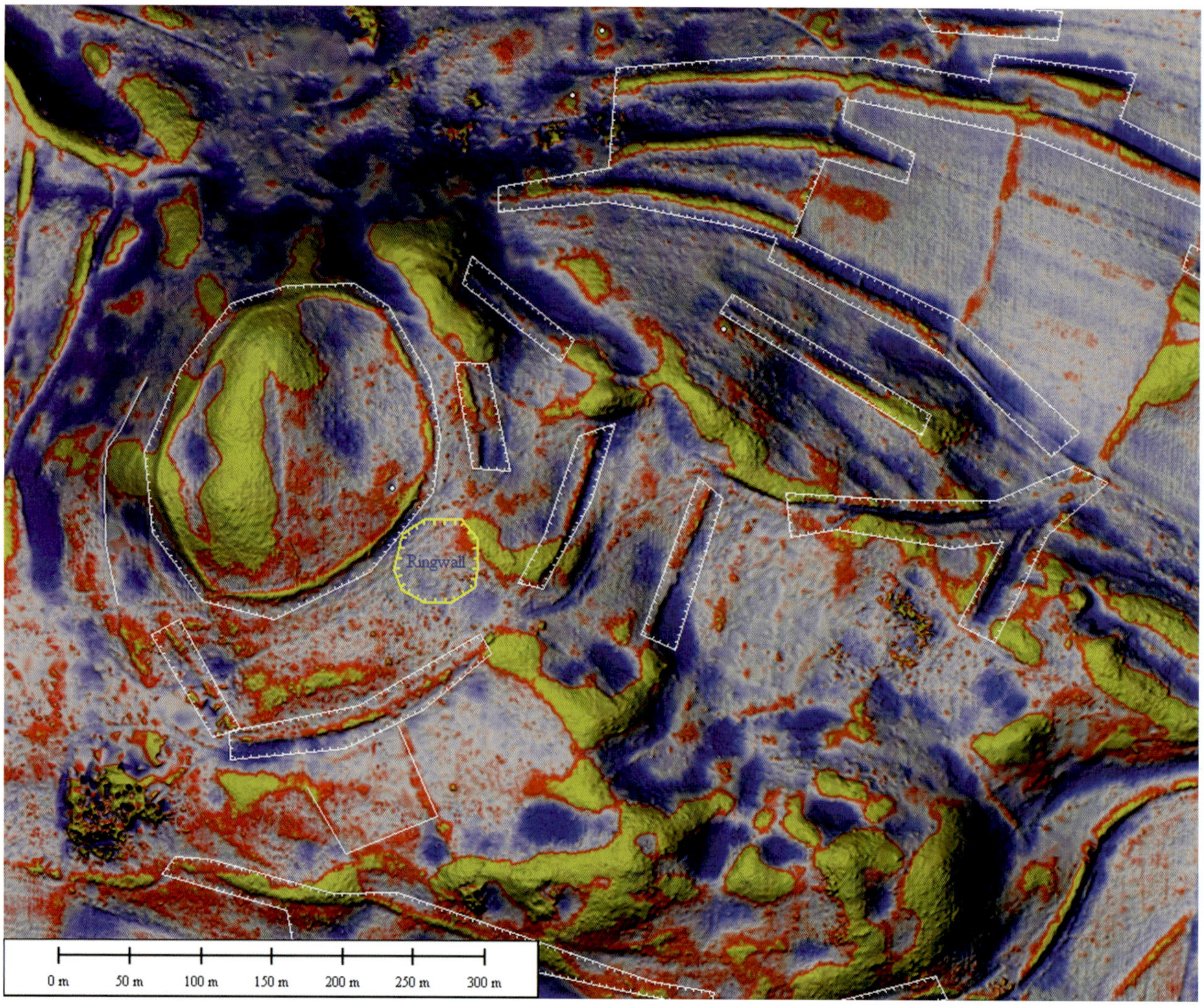

Figure 14.12: Prehistoric hill fort, Gottmadingen (colour-coded LRM draped over shaded relief DEM; yellow: erroneous delineation of earthwork, white: features mapped on the basis of LiDAR data). Note the alignment of terrace steps that may represent an outer rampart belonging to the hill fort. © *LAD/LGL www.lgl-bw.de.*

and volume of relief anomalies). The aim of this work is (c) the (semi-) automated detection of selected features. Progress to date has demonstrated that such data are a very powerful tool for the extensive and comprehensive mapping of earthwork archaeological sites and monuments, even those characterised by very slight relief, and in a wide variety of environments. LiDAR prospection is especially effective in forested areas where other prospection techniques have limited efficiency. By providing a landscape-scale view of archaeological sites including fortifications, settlements, resource exploitation and transport networks, LiDAR prospection has the potential to greatly improve our understanding of sites and their interrelationships in their cultural environment.

References

Bender, S. 2009: Spuren von Kleinkastellen beim Gleichener See? *Der Limes* 3, 10–11.

Bofinger J., Kurz, S. & Schmidt, S. 2006: Ancient maps – modern data sets: different investigative techniques in the landscape of the Early Iron Age princely hill fort Heuneburg, Baden-Württemberg. In Campana, S. & Forte, M. (eds): *From Space to Place. 2nd International Conference on Remote Sensing in Archaeology*. BAR International Series 1568. Archaeopress, Oxford, 87–92.

Bofinger, J., Kurz, S. & Schmidt, S. 2007: LiDAR – High Resolution Raster Data as a survey tool. In Figueiredo, A. & Velho, G.L. (eds): *The world is in your eyes. Proceedings of the XXXIII Computer Applications and Quantitative Methods in Archaeology Conference, Tomar, Portugal 2005*, Lisbon 2007. 255–60.

Bofinger J., Merkl, M.B. & Meyer, M.G. 2010: Untersuchungen eines neu entdeckten hallstattzeitlichen Grabhügels und in einem Rechteckhof bei Riesbürg-Goldburghausen, Ostalbkreis. *Archäologische Ausgrabungen in Baden-Württemberg* 2009, 96–100.

Boos, S., Müller, H., Hornung, S. & Jung P. 2008: GIS based processing of multiple source prospection data in landscape archaeology. In Lasaponara, R. & Masini, N. (eds): *Advances on Remote Sensing for Archaeology and Cultural Heritage*

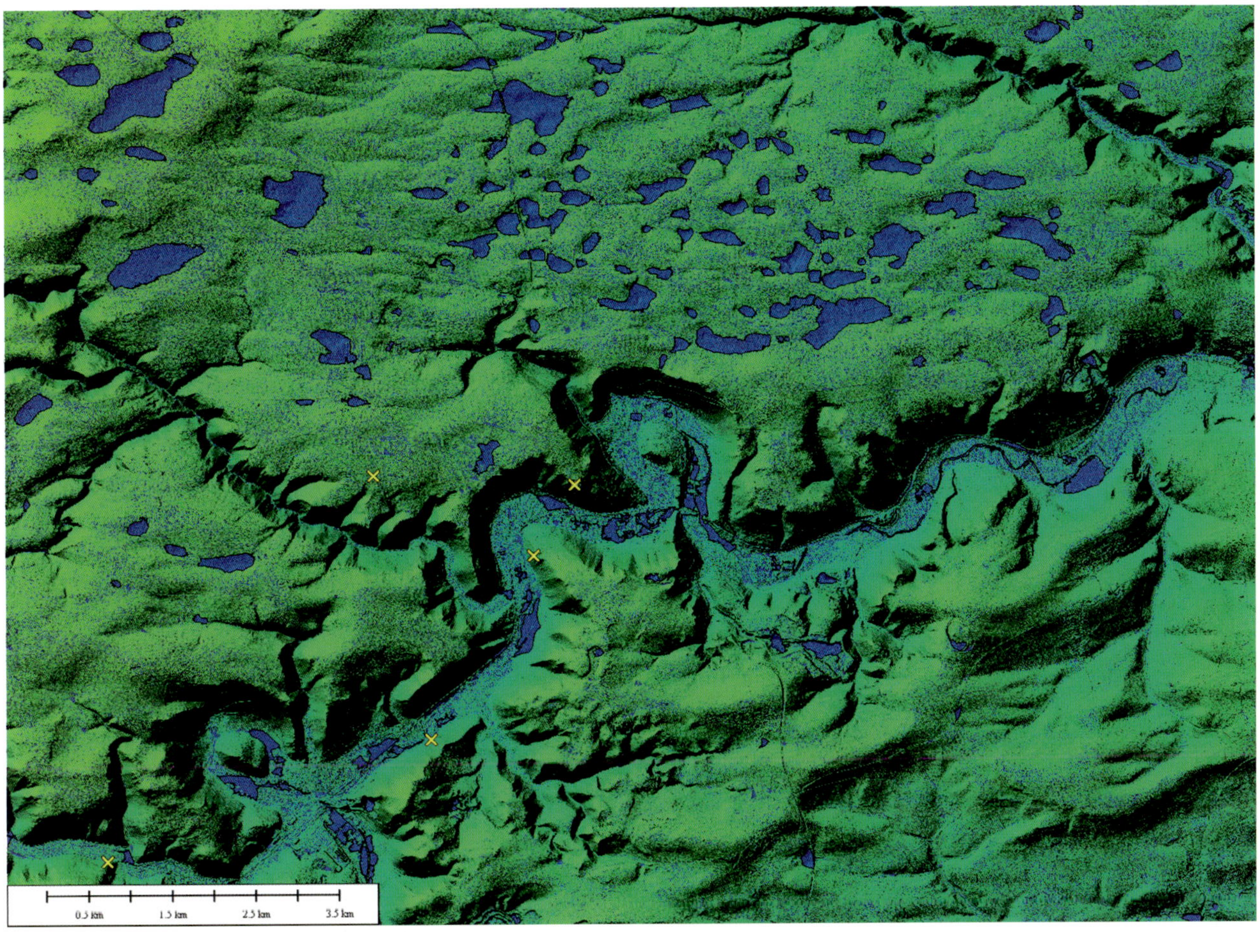

Figure 14.13: Palaeolithic landscape reconstruction, Swabian Alb (shaded relief DEM with closed topographic depressions shown in blue). In the summer during the Glacial period, the landscape was likely structured by meltwater lakes in permafrost-sealed dolines. © *LAD/LGL www.lgl-bw.de.*

Management. Proceedings of the 1st International Workshop on Advances in Remote Sensing for Archaeology and Cultural Heritage Management – Rome, 30 September – 4 October, 2008. 113–7.

Devereux, B.J., Amable, G.S., Crow, P. & Cliff, A.D. 2005: The potential of airborne LiDAR for detection of archaeological features under woodland canopy. *Antiquity* 79, 648–60.

Doneus, M. & Briese, C. 2006: Full-waveform airborne laser scanning as a tool for archaeological reconnaissance. In Campana, S. & Forte, M. (eds): *From Space to Place. 2nd International Conference on Remote Sensing in Archaeology*. BAR International Series 1568, Archaeopress, Oxford, 99–105.

Harmon, J.M., Leone, M.P., Prince, S.D. & Snyder, M. 2006: LiDAR for archaeological landscape analysis: a case study of two eighteenth-century Maryland plantation sites. *Antiquity* 71, 649–70.

Hesse, R. 2010: LiDAR-derived Local Relief Models – a new tool for archaeological prospection. *Archaeological Prospection* 17(2), 67–72. DOI: 10.1002/arp.374.

Hiller, J.K. & Smith, M. 2008: Residual relief separation: digital elevation model enhancement for geomorphological mapping. *Earth Surface Processes and Landforms* 33, 2266–76.

Humme, A., Lindenbergh, R. & Sueur, C. 2006: Revealing Celtic fields from LiDAR data using kriging based filtering. In Maas, H.-G. & Schneider, D. (eds): *Image Engineering and Vision Metrology. Proceedings of the ISPRS Commission V Symposium, Dresden, 25–27 September 2006*. XXXVI, part 5.

LAD 2009: *Archäologische Ausgrabungen in Baden-Württemberg 2009*. Theiss, Stuttgart.

Loisios, D., Tzelepis, N. & Nakos, B. 2007: A methodology for creating analytical hill-shading by combining different lighting directions. *Proceedings of the International Cartographic Conference, ICC, 2007 Moscow*. Published on CD.

Risbøl, O., Gjertsen, A.K. & Skare, K. 2006: Airborne laser scanning of cultural remains in forests: some preliminary results from a Norwegian project. In Campana, S. & Forte, M. (eds): *From Space to Place. 2nd International Conference on Remote Sensing in Archaeology. 2nd International Conference on Remote Sensing in Archaeology*. BAR International Series 1568, Archaeopress, Oxford, 107-12

Rusinkiewicz, S., Burns, M. & DeCarlo, D. 2006: Exaggerated shading for depicting shape and detail. *ACM Transactions on Graphics (Proceedings SIGGRAPH)*, 25(3). 1199-205.

15 | Between the Lines – enhancing methodologies for the exploration of extensive, inundated palaeolandscapes

Simon Fitch, Vincent Gaffney, Benjamin Gearey and *Eleanor Ramsey*

Abstract: In recent years there has been an increasing appreciation of the archaeological potential of European coastal shelves. This interest has not, however, simply been associated with the progress of traditional maritime history but rather the development of a wider marine archaeology aimed at exploring the settlement and cultural sequences of the immense, prehistoric landscapes that lie off some of our coasts. Inaccessible until quite recently, the development of a variety of remote sensing technologies has made exploration of these inhospitable landscapes a real possibility. The information that is being provided from such work is fundamentally changing our perception of the archaeology of the Palaeolithic and Mesolithic. This paper discusses the application of legacy seismic data to map these palaeolandscapes and contrasts the value of 2D and 3D data for these purposes. The results of new research in the North Sea and off the west coast of the United Kingdom are presented and used to test wide sampling strategies that have the potential to explore areas of the sea that have not been mapped using 3D technologies. The paper argues that a range of prospection strategies should be employed within these environments and that they should be optimised to support specific research goals. In the light of marine development and current economic uncertainty, the paper argues that heritage curators and researchers must take full advantage of data sets that have cost billions to capture and that provide unparalleled opportunities for research and management.

Introduction

In recent years there has been a burgeoning appreciation of the archaeological potential of the European coastal shelves. This interest has not, however, simply been associated with the progress of traditional maritime history but rather the development of a wider marine archaeology aimed at exploring the settlement and cultural sequences of the immense, prehistoric landscapes that lie off some of our coasts (Bailey *et al.* 2010; Coles 1998; Peeters *et al.* 2009). Inundated following the last glacial period, it is increasingly understood that many aspects of early prehistoric settlement evidence across the north western European littoral are incomplete and, perhaps, not truly comprehensible without an adequate understanding of these enigmatic and largely unexplored regions. It may therefore be surprising that, prior to the last decade, the landscape potential of the marine archaeological record was rarely appreciated – even within the discipline (Oxley & O'Regan 2001; Roberts & Trow 2002; Fleming 2004).

The opportunity for marine sediments to contribute significantly to our knowledge of more than c. 900,000 years of intermittent hominin occupation of northwest Europe should, perhaps, have been self evident[1]. For much of this time the area that would become mainland Britain was not actually an island but a peninsula of the contiguous continental land mass (Stringer 2006). The dramatic topographic changes indicated by the evidence for inundation and periodic emergence were largely the result of a series of glaciations during which global temperatures declined, ice sheets expanded and sea levels fell, at times by as much as 125m (Wenban-Smith 2002). Reconstruction of the nature and pace of change is, however, complex and there are a number of competing models frequently cited within the literature. These include Jelgersma (1979), Lambeck (1995), Peltier *et al.* (2002) and Shennan *et al.* (2000, 2006). However, these studies tend to generate models that are relatively coarse and concerns have been expressed as to whether their output is adequate for the requirements of archaeological landscape reconstruction (Ward & Larcombe 2008). Archaeologically, there are equally difficult interpretational issues relating to these emergent plains. In most instances, the lack of evidence for settlement, and the unlikely success of exploration, has led to the marginalisation of the area within the literature or, if considered, the relegation of inundated regions such as that within the North Sea to the status of a land bridge supporting colonisation routes into the landscape that would eventually become Britain. There was relatively little appreciation that such areas could

[1] As this paper was concluded finds from Happisburgh in Norfolk suggested that hominin occupation of the United Kingdom might be extended as par back as 900,000 (letter to Nature. Parfitt *et al.* (2010) Early Pleistocene human occupation at the edge of the boreal zone in northwest Europe. *Nature* 466, 229-33, 8 July 2010.

support a vast, resource-rich countryside that was available for settlement and, indeed, may often have been more attractive for occupation than the terrestrial zones that now provide much of our information on early prehistory. For the period under consideration here, the Holocene, the largest of the inundated areas within the North Sea basin has been christened Doggerland (Coles 1998; Gaffney *et al.* 2007). This was a vast plain, bisected by large rivers that, after the end of the Devensian glaciation was briefly available for settlement and then gradually covered by the rising sea until about 7,000 years ago, when the current shorelines around the North Sea were established.

Understanding the archaeology of this period is a major challenge even though available evidence hints that substantial deposits and evidence may be preserved within the marine environment (Bailey *et al.* 2010; Dix & Westley 2006; Fleming 2004). There are individual archaeological finds from as far north as the Viking Banks (Long *et al.* 1986), significant concentrations of archaeological and palaeoenvironmental finds across the southern North Sea (Reed 1913; Louwe Kooijmans 1970) and records of submerged sites including that at Bouldnor Cliff in the Solent (Momber 2000). Indeed, the relict landscapes identified in the North Sea, off the south coast of England and within the Irish Sea, are so large that it cannot be certain that the societies that inhabited these regions are directly comparable to those attested within terrestrial contexts (Gaffney *et al.* 2009). Marine landscapes offer new and intriguing opportunities for exploration and discovery not previously faced by terrestrial archaeologists. Moreover, development and dewatering on land threatens those terrestrial ecosystems that retain substantial palaeoenvironmental deposits essential to our understanding of the environments that supported early prehistoric societies. Consequently, the increasing evidence for extensive caches of palaeo-sediments within marine environments achieves huge importance. However, the areas under investigation are supra-national in scale and may be masked by tens of metres of water or sediment. This exceptional archaeological resource provides those archaeologists who wish to explore the landscape, and the heritage managers who seek to preserve them, with a unique set of legal, technical and methodological challenges[2].

It is no wonder that one reviewer recently suggested that exploration of the inundated landscapes that exist across the globe may well prove to be one of the last great challenges for archaeology (Bailey 2010).

A step change in our attitude towards these landscapes has been the consequence of technological development (COWRIE 2010; see Firth this volume). Marine archaeology, of course, has had recourse to a variety of data sources when exploring marine environments and these may include excavation, where conditions permit, seabed sampling, shallow coring, bathymetric survey, and a variety of remote sensing technologies including seismic reflection profiling. Some of these datasets may have been acquired for a variety of non-archaeological purposes and possess differing characteristics and utility for archaeological application. For example, surveys collecting seabed samples or involving shallow coring can provide detailed chronological, sedimentological and environmental data but frequently have a relatively poor spatial framework. High-resolution bathymetry can provide excellent images of the seabed topography but cannot represent submerged features that lack a bathymetric expression. For early prehistory, the requirement for regionally extensive data across the entire area of the continental shelf is such that, aside from precision and contiguity, issues of scale and resolution are also of considerable importance. Currently, only seismic reflection datasets are likely to provide maps for inundated later prehistoric landscape features at a regional level.

However, marine seismic acquisition is undertaken for a variety of purposes and involves varying data densities, coverage, depths of penetration and resolution. Consequently, there is often a choice to be made when using such data and it is entirely possible that individual surveys may not be appropriate for use by archaeologists with specific research agendas. The decision to use such information will therefore depend upon archaeological requirement and the fit of available data on the grounds of resolution or scale of survey. In many ways this position is not so different to that experienced by terrestrial archaeologists who often have valid reasons to choose spatially extensive, low-resolution sensors in preference to high-resolution sensors (Gaffney & Gater 2003). The latter technologies may often only operate at site level and have little relevance to research that is concerned with the investigation of geomorphology or behaviour at landscape level. Within a marine context, extensive datasets, often characterised by their low resolution, may not initially appear to support the requirements of detailed archaeological investigation. However, they can provide an invaluable topographic framework to guide detailed work or into which higher resolution survey, shallow boreholes, seabed samples and bathymetric data can be integrated (Gaffney *et al.* 2007, 2009). These extensive datasets may also be used within modelling programmes which may not be supported by less extensive datasets. We can explore some of these issues by considering the nature of these data sets and some examples of their recent use (see also Firth this volume for a discussion of survey techniques).

[2] The wider legal issues associated with management of cultural resources are complex and will not be discussed here. For Britain these have recently been summarised in a forthcoming guidance note produced by COWRIE (Collaborative Offshore Wind Research into the Environment) dated 17/May/2010, and also extensively in a succession of Strategic Environmental Assessment (SEA) reports which can be consulted and downloaded at http://www.offshore-sea.org.uk/site/scripts/sea_archive.php. Whilst much discussion in the past has tended to equate marine archaeology with the study of wrecks, recent guidance has increasingly emphasised marine environments as landscapes in their own right and the relevance of SEA directives and the Council of Europe *European Landscape Convention* to promote the protection, management and planning of European marine landscapes is increasingly evident (Roberts & Trow 2002).

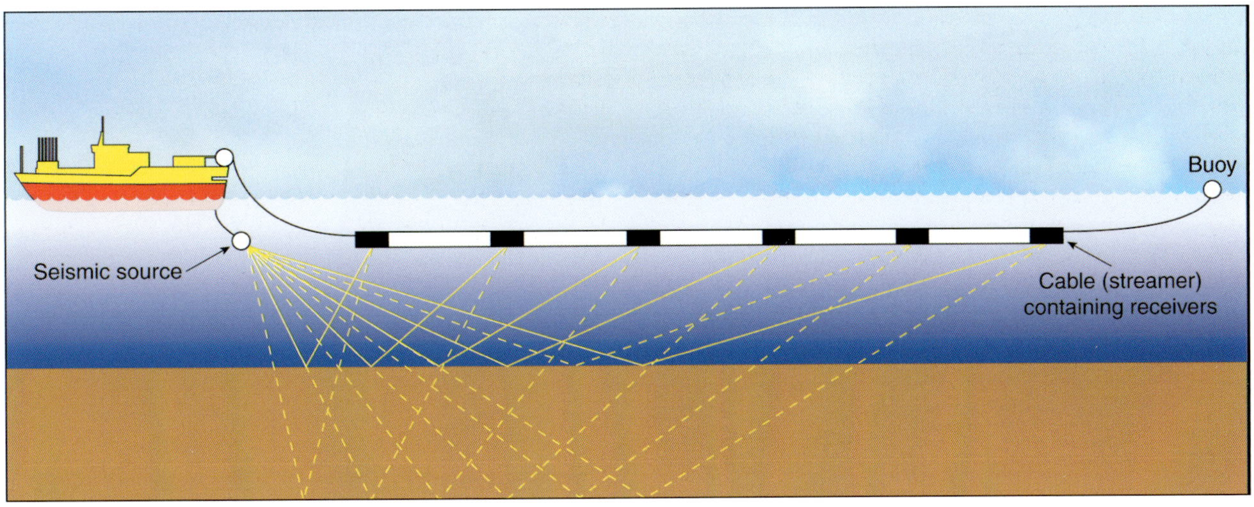

Figure 15.1: Typical marine seismic reflection acquisition. *From Gaffney* et al. *2009, figure 3.4.*

Seismic reflection survey

Seismic reflection surveying involves the transmission of acoustic energy into the subsurface and recording the energy reflected from acoustic impendence contrasts. The reflections produced as acoustic impendence contrasts are predominantly the product of changes in lithology. With appropriate processing this allows the production of pseudo-depth sections of the subsurface structure with the vertical axis being the two-way travel time to the reflector. Although the basics of this technique are common, the details vary according to a range of applications including the investigation of deep crustal structure, hydrocarbon exploration and near seabed sediment structure (e.g. Salomonsen & Jensen, 1994; Velegrakis *et al.*, 1999; Praeg 2003; Bulat 2005). These diverse applications dictate different acquisition parameters that in turn determine the resolution and depth of penetration of the survey as well as the costs involved in acquiring the data. Consequently, the relative merits of a range of available seismic reflection data types needs to be assessed when considering the investigation of submerged, and partially buried features.

Standard marine acquisition involves towing an energy source and a cable (streamer), containing pressure sensitive receivers, to record the reflections from the underlying strata (Figure 15.1). In single fold data, only one reflection is received from any point in the subsurface. However, many seismic profiles are multi-fold and reflections can then be summed in order to increase the signal-to-noise ratio of the seismic profile.

Traditional seismic reflection data is generally referred to as 2D as it is acquired as a series of discrete vertical profiles using a single streamer towed behind the vessel (Figure 15.1). In contrast, 3D reflection seismic data involves the towing of multiple streamers which support the rapid collection of a series of closely spaced lines (Figure 15.2). This survey configuration provides significant advantages. Seismic response

Figure 15.2: Typical 3D marine seismic reflection acquisition. *From Gaffney* et al. *2009, figure 3.5.*

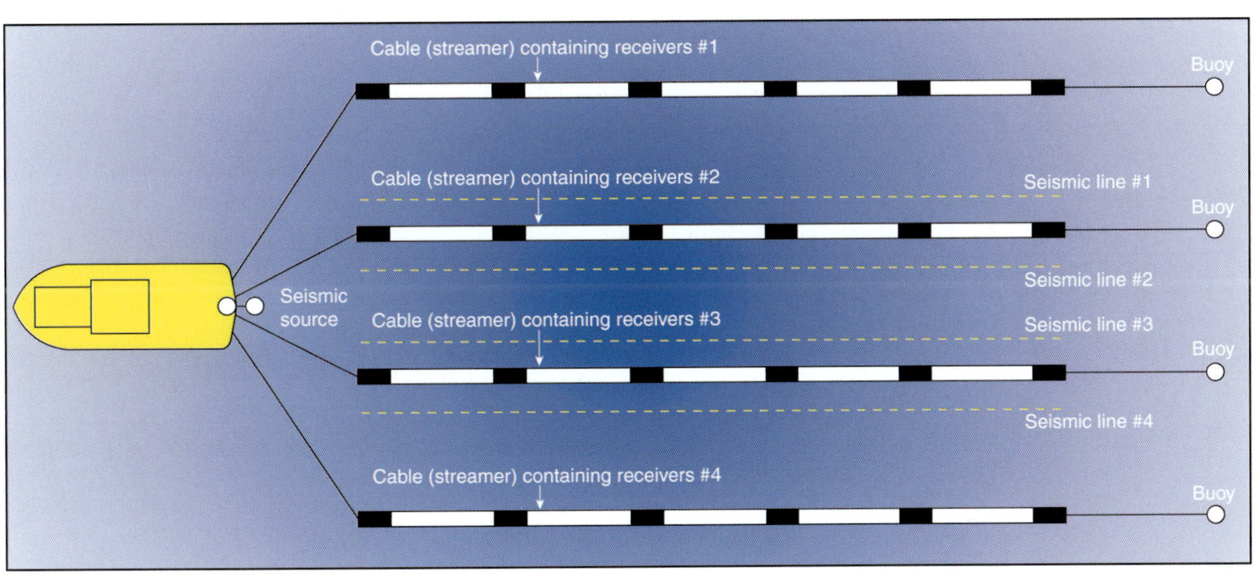

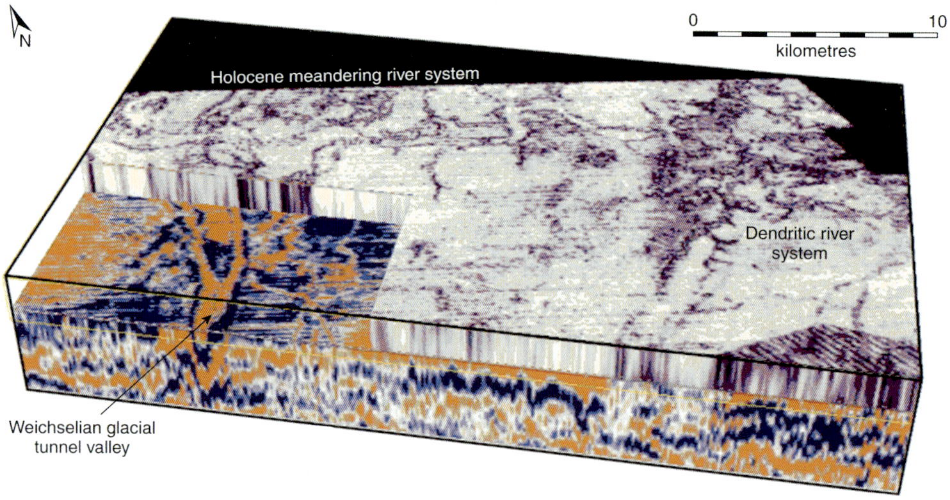

Figure 15. 3: A cube of conventional 3D data with a section removed to demonstrate the ability to slice the data in the horizontal and vertical dimensions and to reveal landscapes of differing age. *From Gaffney* et al. 2009, *figure 5.14*.

is correctly positioned in space and, in the case of data acquired for hydrocarbon exploration, is 'binned' within data volumes with a resolution of 12.5m × 12.5m × 4 milliseconds, or multiples thereof. Once treated in this manner a feature can be mapped from bin to bin, removing the potential errors involved in the interpretation of 2D data. Moreover, instead of relying on vertical profiles, the volume can be sliced in any direction. Of particular importance to the investigation of relatively shallow, and flat, landscape features is the ability to produce a horizontal slice (time slice) through the data as this can, in many cases, be interpreted as a map showing a range of sedimentary features (Figure 15.3). The interpretation of 3D seismic data has improved significantly in recent years due to the development of a range of new techniques originally designed to improve geological interpretation for hydrocarbon exploration and production. Once a stratigraphic marker of interest has been identified, it can be mapped across the 3D seismic volume to produce a horizon that may have a geomorphological or chrono-stratigraphic value and, in some cases, the output can approximate the original land surface itself. The value of such data for the interpretation and analysis of inundated landscapes and modelling past settlement or land use should be clear.

Another advance in 3D seismic interpretation has been the development of opacity rendering techniques (Kidd 1999). Following conversion of conventional 3D seismic data into a voxel (3d pixel) volume, each voxel contains information from the original portion of the 3D seismic volume that it occupies together with an additional user-defined variable that controls

Figure 15. 4: A volume model of the data illustrated in Figure 15.3. A Holocene river channel (blue) overlies an older tunnel valley (gold and infilling sediments in purple). *From Gaffney* et al. 2009, *figure 3.9*.

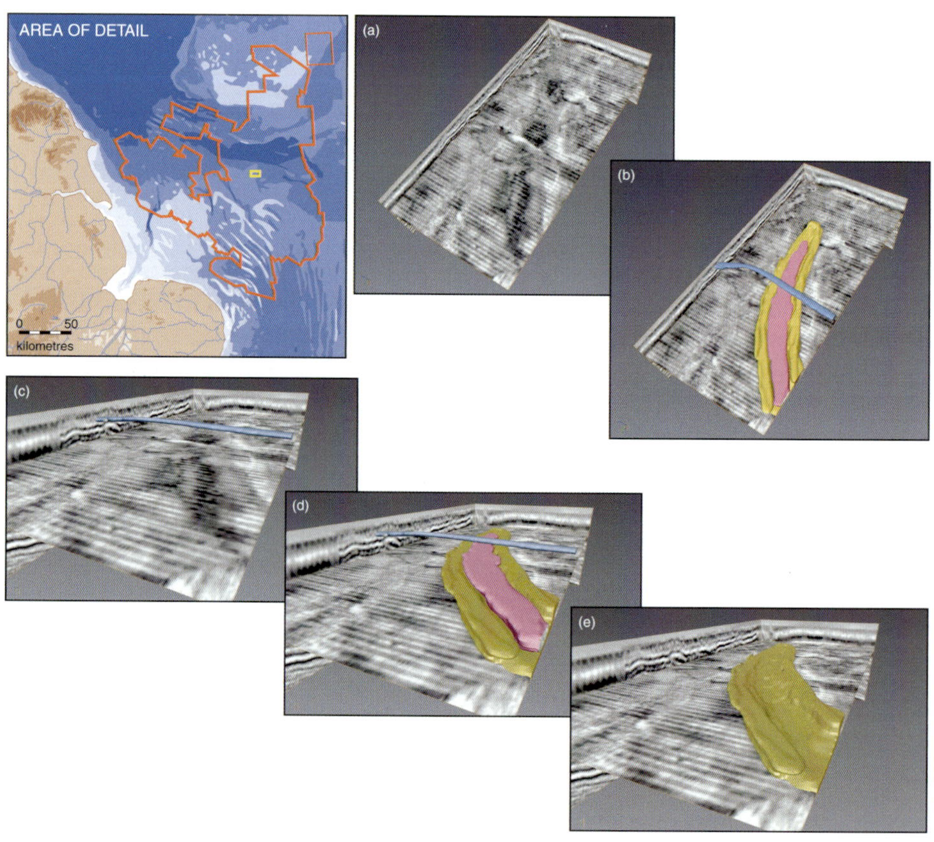

Figure 15. 5: Study areas referred to within the text 1) North Sea Palaeolandscape Project 2) Humber REC 3-4) West Coast Palaeolandscape Project. *ASTER DEM is a product of METI and NASA. ETOPO2v2 is the property of the National Geophysical Data Centre, NOAA, US Dept of Commerce.*

its opacity. The opacity of individual voxels can then be varied as a function of their seismic amplitude (or any other attribute), allowing the user to examine only those voxels that fall within the particular amplitude (or attribute) range of interest. By using appropriate opacity filters it is possible to image depositional systems such as buried fluvial channels. This exploits seismic characteristics, which are in part lithologically dependent, and different from the surrounding materials, thus permitting the surrounding strata to be made transparent whilst preserving all but the smallest channels as opaque features (Fitch *et al.* 2005). In archaeological terms such processing also provides further insight into the stratigraphic relationship of features identified and, through their volume and sedimentary characteristics, the opportunity to assess whether such features have the potential for preservation of archaeological or environmental data (Figure 15.4).

Generally, the ideal dataset for the investigation of submerged prehistoric landscapes within the region would be high-resolution (>100Hz) 3D seismic data with appropriate borehole control. Such a dataset would provide high (metre or less) vertical and lateral resolution and a laterally continuous coverage. Unfortunately, such systems involve slower survey rates, higher costs and do not usually provide the extensive output required to explore landscapes at a supra-national scale. Commercial 3D seismic datasets, which possess a significantly coarser resolution, may appear to be less suitable for archaeological exploration but even these can provide maps containing important information from shallow deposits. Consequently, even with a bin spacing of 50m, the spatial coverage of such datasets and published outputs demonstrate that these data have the potential to provide an extensive reconnaissance tool for the investigation of submerged landscapes. The application of such technologies, within the Holocene at least, has been amply demonstrated by a number of projects carried out within British waters (Figure 15.5).

Regional survey within British maritime waters

Much recent research on British marine palaeolandscapes, beginning with the North Sea Palaeolandscapes Project (NSPP), has been carried out by the authors, and through the IBM Visual and Spatial Technology Centre (VISTA) and Birmingham ArchaeoEnvironmental (BAE) laboratory at the

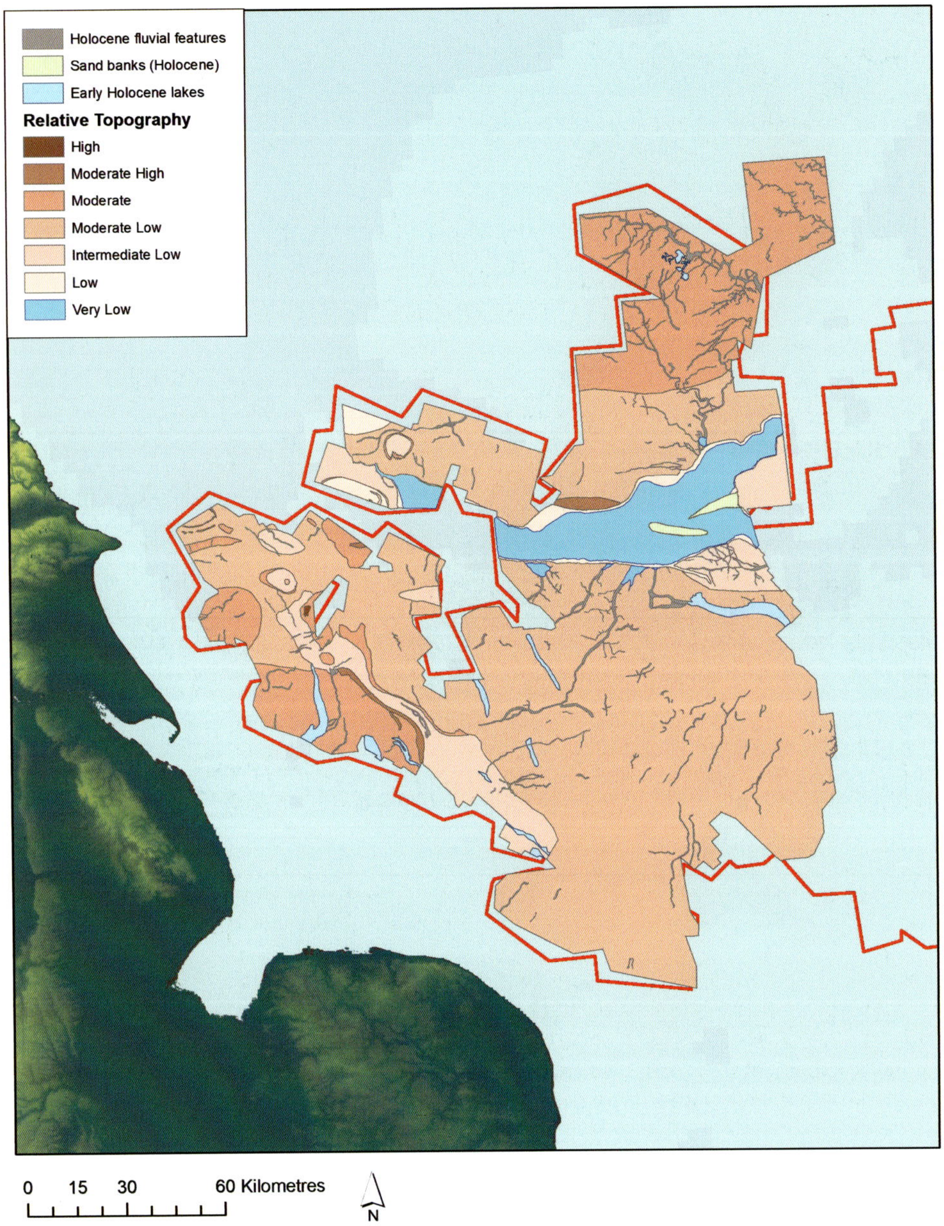

Figure 15.6: General plan of topographic features identified through the North Sea Palaeolandscapes Project (Gaffney et al. 2007, 2009). *ASTER DEM is a product of METI and NASA. ETOPO2v2 is the property of the National Geophysical Data Centre, NOAA, US Dept of Commerce.*

University of Birmingham. Much of the detail of this research has been published and need not be repeated here but the NSPP research team undertook mapping of the submerged European Mesolithic landscape known as Doggerland over an area of about 23,000km^2 of the English sector of the North Sea (Gaffney *et al.* 2007, 2009). The opportunity for landscape investigation in such unprecedented detail was facilitated by access to 3D seismic data collected mainly for use by the petroleum industry. The primary dataset for the NSPP was provided by PGS Ltd and was represented by the merged 3D seismic dataset known as the 'Southern North Sea Megamerge' (PGS 2005). In archaeological terms, the data has a relatively coarse resolution (between 12.5m and 50m), although the intrinsic 3D nature of the data, and its landscape scale, facilitates the production of maps containing information from several metres of Holocene strata. Consequently, the data demonstrated that the Dogger Bank formed an emergent plain during the Holocene with complex meandering river systems and associated tributary channels and lakes dominating the region. This project is ongoing. The National Atmospheric and Oceanographic Administration (NOAA) recently agreed to support further work focusing on the Dutch Sector, with the aim of producing a unified map of the landscape across the southern North Sea for the early Mesolithic period (Figure 15.6).

There can be few who would argue that the results of the NSPP have been little less than revolutionary to our understanding of these landscapes, although developments in petroleum industry data collection suggest that the current situation can be further improved through the implementation of extensive, high definition 3D (HD3D) survey. These surveys potentially offer greatly improved vertical resolution and feature definition (Figure 15.7; Long 2003). This has been achieved by technology that supports acquisition of data through a denser 3D spatial sampling grid and improved frequency bandwidth recovery than was available to traditional 3D seismic reflection surveys. Although not widely accessible at present, archaeological research undertaken at the University

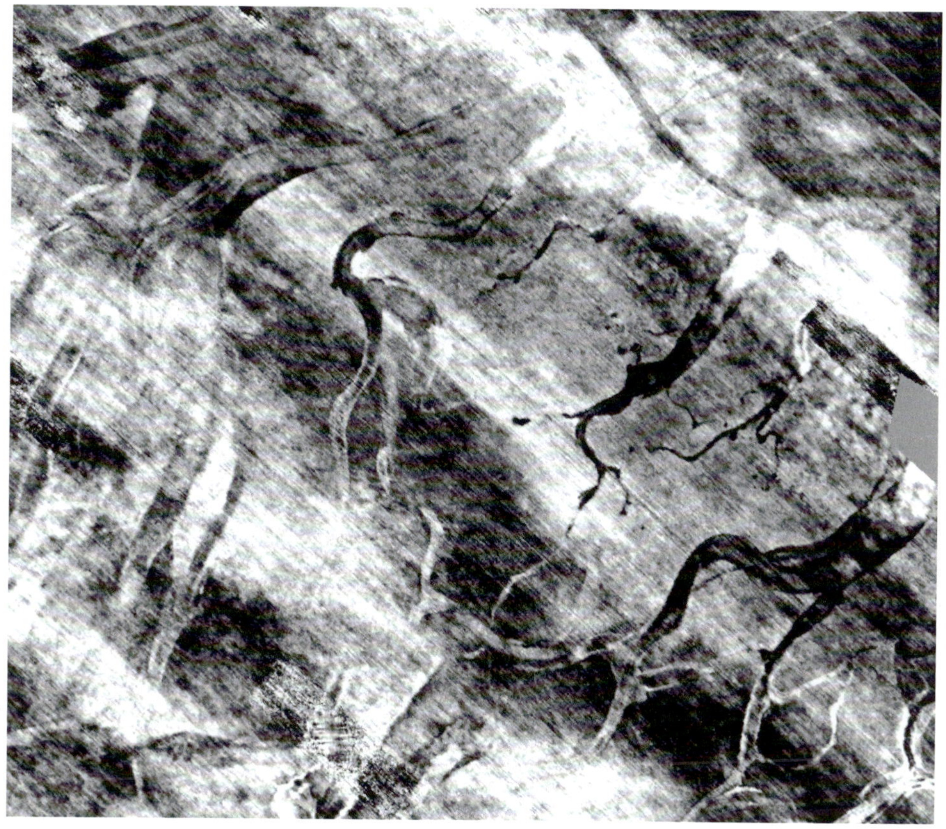

Figure 15.7: A Shallow time slice from HD3D data in Qatar. The resolution of the complex meandering channel systems is outstanding, and signal-to-noise quality is a step improvement over historical seismic data in such shallow waters. *Data courtesy of Maersk Petroleum (Qatar) Ltd.*

of Birmingham on HD3D data from the Arabian Gulf suggests that this data may be eminently suitable for the exploration of areas where noise or water depth are issues (Cuttler *et al.* 2010; Mueller *et al.* 2006). The acquisition of such data is becoming more frequent within the mineral sector and, as access to surveys is improved, future archaeological applications are likely to result in finer resolution, broad area, palaeo-landscape investigations.

Unfortunately, extensive HD3D datasets are only rarely available for archaeological study, whilst contiguous, traditional 3D surveys are not universal in their coverage or always accessible to archaeologists for a variety of commercial or logistical reasons. These are important points. We can be confident that past, habitable environments on the European coastal shelves are considerably more extensive than the available 3D data sets and in the absence of such information we remain essentially uninformed of the conditions or significance of these wider regions. Consequently, there is an imperative not simply to refine our methodologies relating to 3D seismic interpretation, but to explore and implement novel methodologies for areas where 3D data does not exist or is restricted in access. If this is not attempted we risk missing data critical to our understanding of earlier prehistoric settlement. Heritage managers, unable to manage a resource that is defined essentially by what we do not know, will be prevented from protecting what may be a one of the largest and best-preserved cultural resources in Europe, or indeed the world.

Researchers at Birmingham have begun to explore this problem using 2D surveys, as these have a greater geographical spread than 3D surveys and therefore, potentially, can be used to supplement contiguous surveys or provide data in their absence (Figure 15.8). With respect to the use of 2D data there are clearly two issues that deserve mention. Initially, it is generally true that 2D surveys have a greater vertical resolution than extensive 3D surveys. This suggests that these data can be used to clarify and improve the interpretation of 3D data where the datasets, and features of interest, are congruent. Unfortunately, when comparing the interpretation of extensive horizontal 3D data with 2D surveys, the acquisition pattern for 2D data often appears problematic. This frequently involves the collection of multiple profiles with a spacing that may be several orders of magnitude greater than the trace spacing (i.e. the horizontal sampling interval along the profile). This method of acquisition has significant disadvantages for archaeological prospection. Firstly, the reflected seismic energy is assumed to have originated from a point directly beneath the profile even though it could have originated from a point laterally offset from it. Secondly, the spacing between lines may be so wide that it can be difficult to map the position of a feature across the region of interest. Figure 15.9 demonstrates how wide line spacing can lead to several equally valid interpretations.

There are therefore a number of issues with the use of 2D data as an exploratory landscape tool and these include:

- whether 2D data can be used to provide landscape interpretations in areas where 3D data is either unavailable or absent and
- whether 2D data can be used to refine interpretation based on traditional 3D survey.

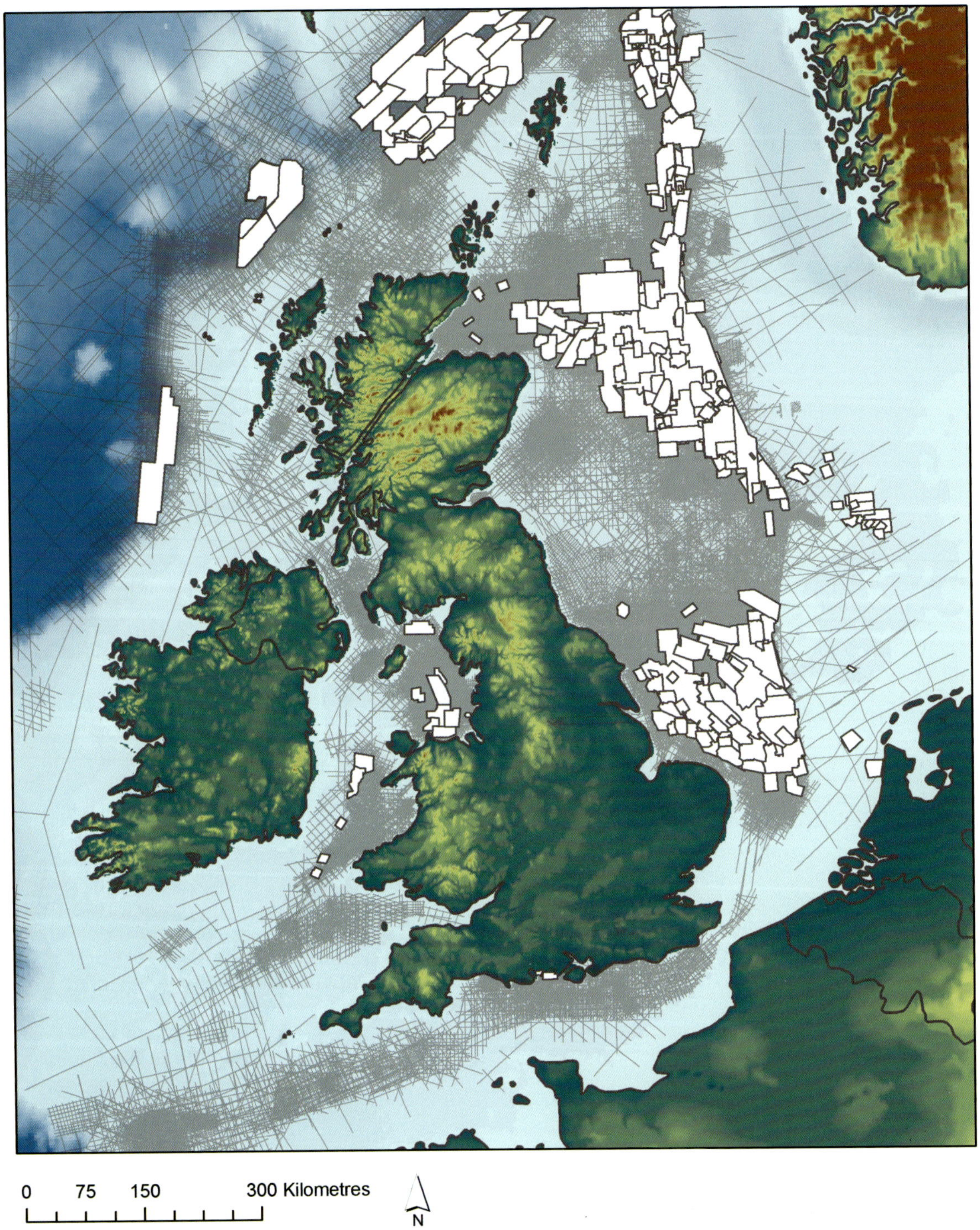

Figure 15.8: 2D lines (grey) and 3D survey (polygons) availability around the mainland UK. *ASTER DEM is a product of METI and NASA. ETOPO2v2 is the property of the National Geophysical Data Centre, NOAA, US Dept of Commerce.*

Figure 15.9: (a-d) Four possible interpretations of a channel morphology based on a coarse 2D seismic grid. *From Gaffney* et al. *figure 3.6.*

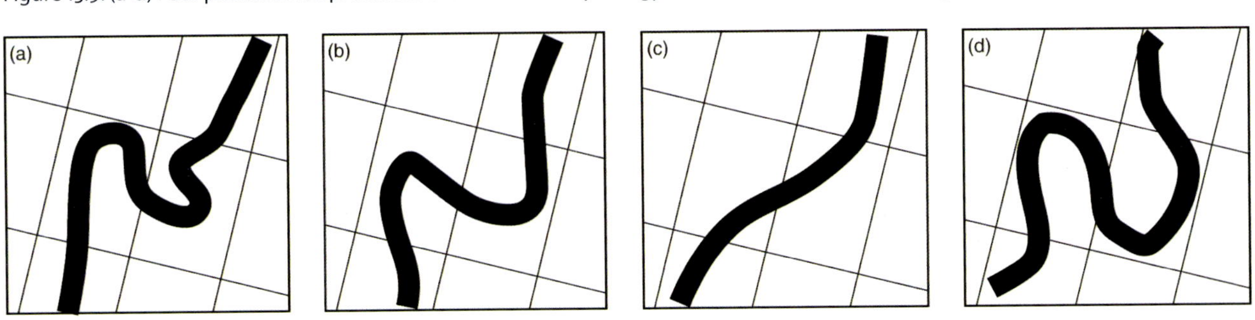

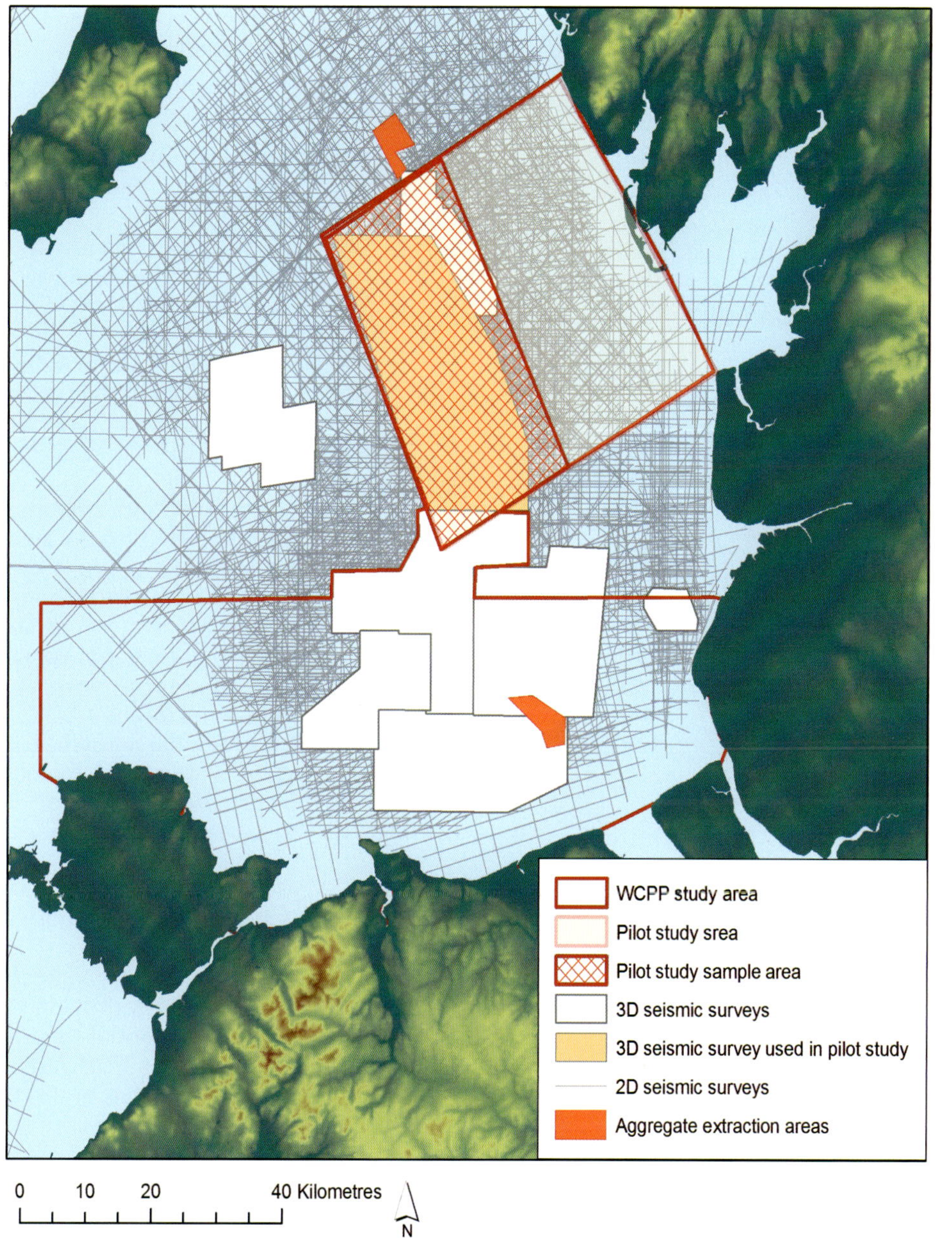

Figure 15.10: The red outline is the WCPP main study area outline. The pilot study area is shaded yellow,, shown in relation to the coast of England and Wales. *ASTER DEM is a product of METI and NASA. ETOPO2v2 is the property of the National Geophysical Data Centre, NOAA, US Dept of Commerce.*

Using 2D data

A specific output of the NSPP of particularly significance to this paper is the data audit carried out to identify the sources of remote sensed data potentially available for the study of inundated landscapes around the English coastline (Bunch *et al.* 2010). This study identified several areas around the United Kingdom where sufficient data, often in the form of older 2D survey, might support studies similar to the NSPP and which could contribute to our understanding of the development of the Mesolithic and, potentially, Palaeolithic periods in England and Wales, Ireland, Scotland and the Isle of Man. The west coast of Britain was identified as a key area that possessed a significant data resource and, through the work of Professor Martin Bell, had also been subject to a significant regional research programme on early Holocene archaeology (Bell 2008). In 2009 English Heritage commissioned a pilot project in the Irish Sea through the Marine Aggregates Levy Sustainability Fund (Figure 15.10). The aim of this work was to investigate the utility of available 2D survey data for extensive mapping and this was anticipated as a precursor to a larger project off the west coast of the United Kingdom.

The situation of the Irish Sea is comparable to that of the North Sea in the existence of a significant coverage of traditional 2D and high-resolution 2D seismic lines, but differs with respect of the lesser availability of 3D data. It therefore provides an ideal test bed to develop a methodology for investigating coastal areas with variable data availability (Figure 15.10). The overall aim of the pilot study, known as the West Coast Palaeolandscapes Pilot (WCPP), was therefore to develop a methodology that might be utilised in areas where existing 3D seismic data coverage was limited or absent. Such a methodology had to be able to provide baseline data to facilitate future management of the submerged prehistoric resource, through the limited mapping and identification of the Late Palaeolithic and Mesolithic landscapes.

Berr/Dti	Object of interest	Available geomatics sensors and data
CS9NAME	Line Name	Length
HY752D1002	H751-11	14.1
HY752D1002	H751-13	25.9
HY752D1002	H751-15	25.7
HY752D1002	H751-28N	31.7
HY752D1002	H751-30N	22.12
HY752D1002	H751-32N	20.4
HY752D1002	H751-17	20.84
HY752D1002	H751-34	18.8
HY752D1002	H751-36	20.6

UKOGL	Object of interest	Available geomatics sensors and data
Operator	Line Name	Length
SHELL	SW81-720(2)	4.04
SHELL	SW81-720(3)	4.06
SHELL	SW81-720(4)	10.12
SHELL	SW81-721	13.0
SHELL	SW81-722	19.4
SHELL	SW81-723	15.8
ULTRAMAR	U038-87-10A	4.31
ULTRAMAR	U038-87-11	9.33
ULTRAMAR	U038-87-12	2.1
ULTRAMAR	U038-87-12A/1	3.69
ULTRAMAR	U038-87-13	2.58

BGS		
Survey	Line Name	Length
BGS	Line 11 1972	51
BGS	Line 3 1972	33.5
BGS	1968	34.5

Table 15.1: 2D lines acquired for use within the West Coast Palaeolandscapes Pilot.

After consideration, the WCPP team adopted the following methodology.

1) A standard 3D seismic dataset within the pilot study area was utilised to identify archaeological features within the study area.
2) A small number of 2D seismic datasets over and around this area was obtained for comparative purposes.
3) The 2D datasets were then investigated to determine if they contained the features identified within the 3D dataset and the intersections recorded.
4) The spatial configuration of all existing 2D datasets within the region were then assessed against known features to ascertain their capacity to provide adequate landscape data across the pilot study area.

The selected 3D dataset used within the project consisted of a single standard seismic survey covering an extensive area of the west. This survey, known as 'Morecambe and Satellites', is a 3D seismic reflection survey acquired using standard technology by the contractor Centrica, and provided to the University of Birmingham for research purposes. The resulting digital survey has a bin spacing of 12.5m. Quality of the data was good and it responded well to serial time slicing. The data also proved to be of suitable for full processing and archaeological interpretation. Initial results suggest that the main limitation of this dataset for archaeological research is due to the relatively ephemeral characteristics of the Holocene deposits within the area. Information was confined to a relatively small number of slices, and thus a small vertical resolution. This was, however, an issue of the prevailing geology, rather than a feature of the dataset itself.

The earliest landscape features identified by this project were the two broad areas of Late Upper Palaeolithic fluvio-glacial plains (Figures 15.11 & 15.12). These were present beneath palaeo-periglacial outwash and are significant as they were formed by drainage from nearby glaciers, possibly in the English Lake District. In landscape terms, they potentially acted as a barrier to human movement and, as a consequence, suggest a low potential for archaeological remains. However, the identification of a mammoth tusk during monitoring in the Humber region (Wessex Archaeology report to BMAPA 2006) suggests that these areas are not necessarily devoid of archaeological potential. The Early Mesolithic was represented within the study area by two large river systems (Figure 15.13), along with several other smaller river systems. These have been identified as braided or anastomosed rivers (Rosgen 1994). These types of rivers provide rich and varied environments with the potential to sustain a wide variety of animal and plant resources. As a consequence, these represent highly attractive areas for Mesolithic activity. Other features identified within the data include several areas of higher ground (Figure 15.13). These may well have acted as focal points in the wider landscape or provided opportunities to observe game. Such upstanding features are also of importance as they that would have formed islands during inundation and represent the final possibly habitable areas within the region. The largest of these upstanding features in Figure 15.13 would also have been attractive to human populations due to the proximity of two large river systems, one to the south of the area of higher ground, and another cutting through it.

2D datasets acquired for use by the project derived from three sources (Table 15.1 & Figure 15.14). The first, obtained by the British Geological Survey between 1968 and 1972, comprised three shallow seismic surveys which ran across extensive sections of the study area. Data from these surveys were derived from a combination of common seismic sources (sparker) and (pinger). The seismic data were made available in the form of scanned paper rolls in TIFF format and the corresponding survey track plots available on DVD. There was some variability in quality within these surveys. The selected sparker datasets were adequate for full processing and archaeological interpretation based on the frequency, range and filter settings chosen during acquisition. The pinger datasets visualise a very

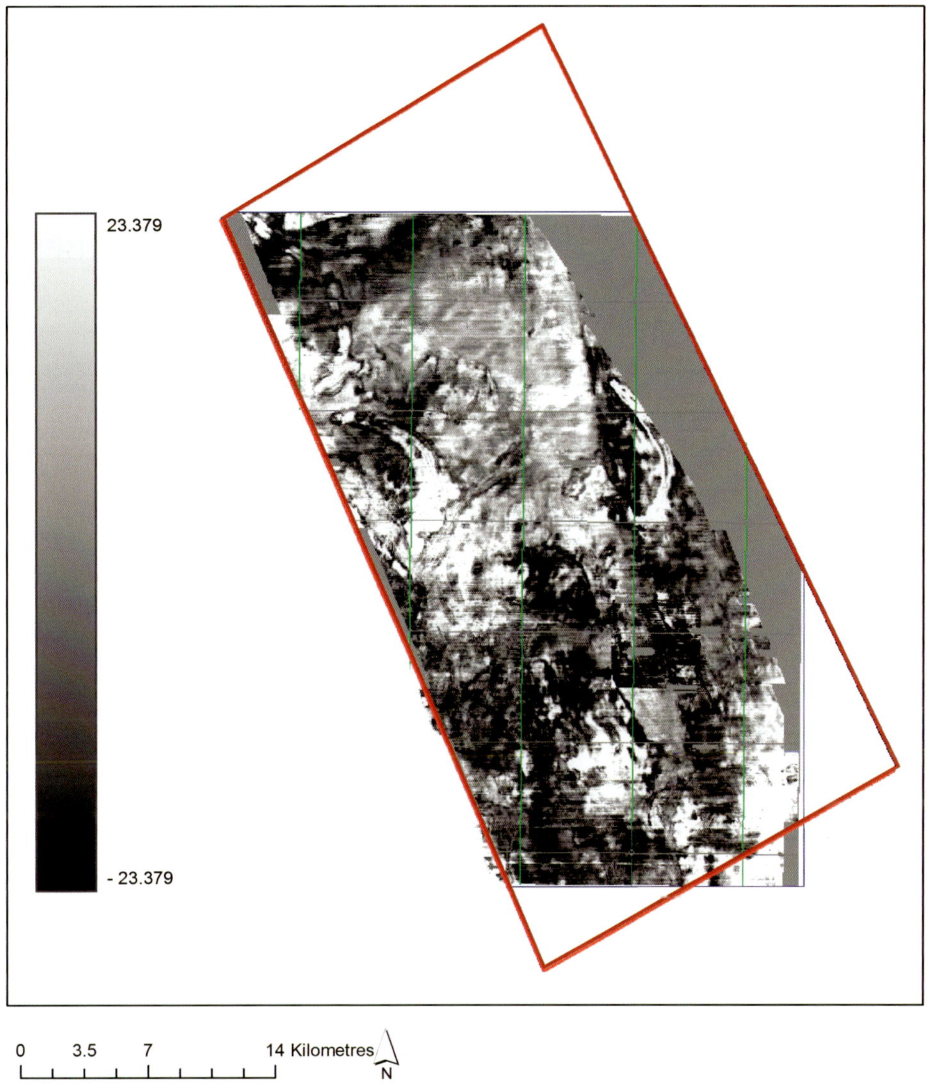

Figure 15.11: Timeslice image of the 3D dataset at 0.076s. *Data courtesy of Centrica UK Ltd.*

shallow section of the seabed and the indistinct images were less reliable for archaeological interpretation. A further selection of traditional petroleum industry 2D seismic surveys was guided by the requirement for data that would intersect with several of the most significant features identified from the 3D analysis. These data were originally obtained by S&A Geophysical Ltd for Hydrocarbon Resources Ltd in 1975, and released to the British Department of Trade and Industry (DTI) in 1980. The original data were obtained from BERR/ Phoenix (representing the DTI's data store), which holds a significant number of surveys from this area. The data is stored on original paper survey rolls which were subsequently scanned by Phoenix Data Solutions Ltd for conversion to standard SEG-Y files. The rolls were also scanned and provided in TIFF format. However, data compression techniques used on the images prevented these images from being displayed reliably. Although the data was of variable quality, even when the selected lines originated from the same survey, the digital format allowed application of a range of processing techniques to optimise the data for interpretation. Unfortunately, some lines remained unsuitable for analysis and it is acknowledged that the scanning/conversion of the paper records into digital formats also potentially introduced an unknown element of error into the process of interpretation.

The final 2D dataset consisted of two UKOGL shallow seismic surveys comprising 11 survey lines covering the onshore sections of the study area. These were obtained during 1981 and 1987 respectively. The 1981 survey was undertaken by Prakla Seismos for Shell UK Ltd and transcribed for the UKOGL Veritas data services UK Ltd. The survey data was acquired utilising a standard airgun as a source and recorded directly to tape as SEG-Y 32 bit floating-point data. The 1987 survey was undertaken by Horizon Ltd for Ultramar Exploration and processed at Horizon Exploration Ltd. Data from both surveys was provided to Birmingham for research purposes by UKOGL.

Initial examination of the 2D seismic lines suggested that a combination of standard interpretation procedures coupled with associated GIS recording should be employed during analysis. Initially, the data was examined utilising standard seismic-stratigraphic procedures (Mitchum *et al.* 1977). Digital 2D data were imported into SMT Kingdom 8.2 (64bit) seismic analysis software and a seismic attribute analysis performed. Generation of this information, however, failed to identify any new features and there was only minor refinement of the identified features. Once completed, interpretation of anomalies was undertaken directly within the seismic analysis software and recorded as a

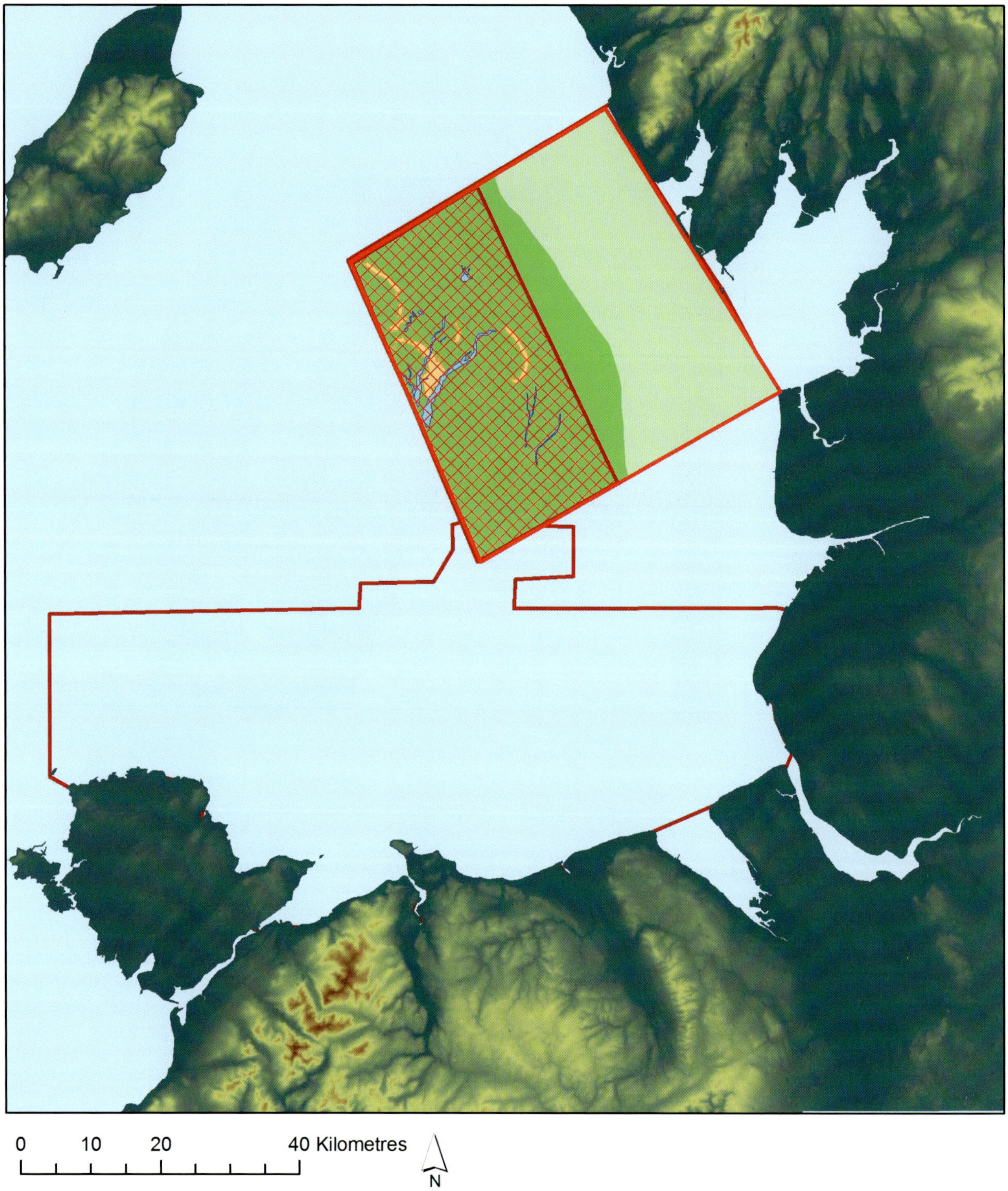

Figure 15.12: The Early Holocene landscape within the WCPP study area. The light green zone reflects near shore areas which would have represented higher ground, whilst the dark green areas for a lower lying plain. *ASTER DEM is a product of METI and NASA. ETOPO2v2 is the property of the National Geophysical Data Centre, NOAA, US Dept of Commerce.*

series of 'culture' files. This information was exported directly into the project GIS for further processing.

The scanned analogue (paper) data were converted into digital SEG-Y files using the Chesapeake ImagetoSEGY software. As the corresponding survey track log data had been provided this was added to the image file during conversion. Once this had been achieved, processing followed the standard digital 2D interpretation process with individual incised features, and possible landscape features, being recorded. It is important to note that this method does not provide as precise a location for these features as might be achieved with the original survey data. Despite this, the error margin associated with the locations is in the range of metres and the accuracy achieved is sufficient to permit future surveys to target features of interest with relative ease.

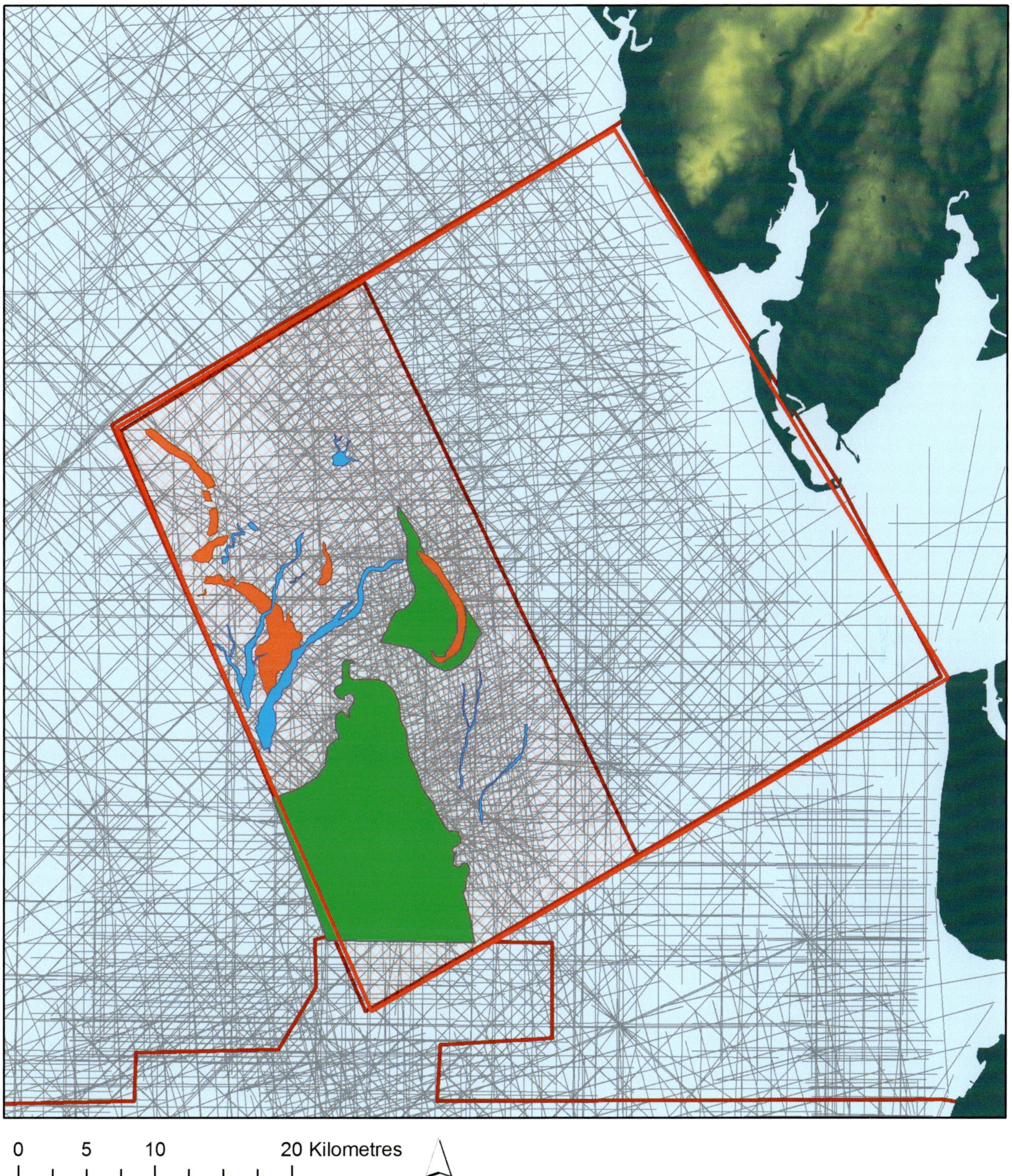

Figure 15.13: Map of palaeolandscape features identified within the 3D seismic data set against Key: *Blue* = Probable Holocene fluvial channels and related features: *Red* = Geological features forming regional highs: *Green* = Late Upper Palaeolithic fluvioglaical floodplains. *ASTER DEM is a product of METI and NASA. ETOPO2v2 is the property of the National Geophysical Data Centre, NOAA, US Dept of Commerce.*

Examination of the 2D datasets chosen for detailed study demonstrated their capacity to identify features of interest (Figure 15.15). Whilst a number of deeper features were not observed within the BGS datasets, the fluvioglacial plain and several of the target palaeochannels, which clustered in the deep-water sections of the study area, were identified during analysis. The results, however, were undoubtedly constrained by the availability of data for analysis, which was restricted to only a small number of lines for exploratory purposes and also by data quality in some areas. The assessment suggested that results were not strongly dependent upon the age of the data. Analysis of the earliest survey line available, from 1968, provided some of the best results obtained. Those from 1972 contained some noise and discontinuous reflectors. Furthermore, it was observed that a pinger dataset from 1968, although

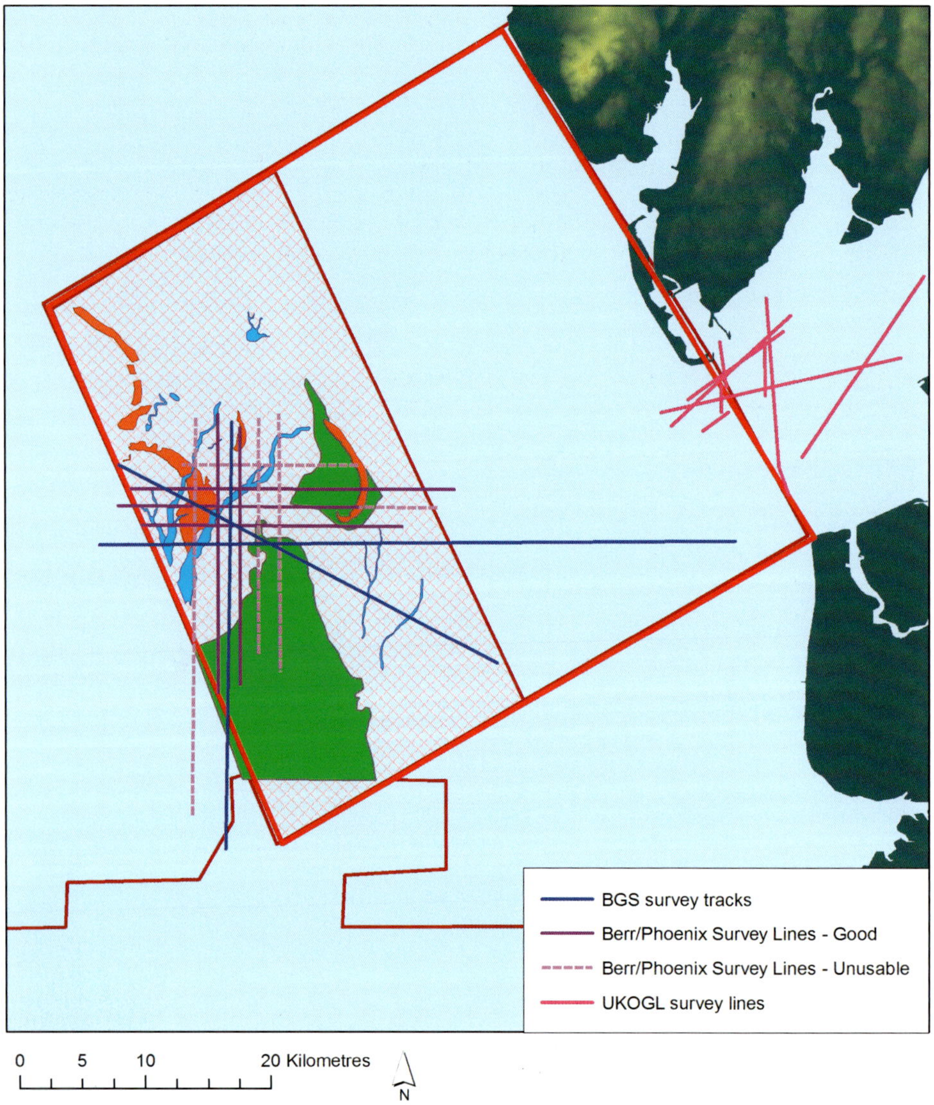

Figure 15.14: 2D lines used within the WCPP study area. *ASTER DEM is a product of METI and NASA. ETOPO2v2 is the property of the National Geophysical Data Centre, NOAA, US Dept of Commerce.*

characterized by poor penetration, was still able to provide information on features of interest (Figure 15.16). Despite this, the legacy data did, initially, provide a worryingly high failure rate in terms of lines that might not be susceptible for analysis. Initially, this was assessed as high as 40% from the small sample of lines selected for analysis as part of the pilot project. However, these lines were themselves from a limited number of surveys and later work suggests that the overall failure rate is considerably lower when data is available from multiple surveys. Figure 15.17 illustrates the line failures in a larger sample of data within the Irish Sea and currently being analyzed as part West Coast Palaeolandscapes Main Project[3]. This includes data on 239 lines in total, 46 of which were corrupted, or partially corrupted, and unusable. This represented a 19.25% failure rate overall. However, there is a lower failure rate in terms of line length which amounted to 14.75% in total or 797km from a total of 5,404km.

Examination of the SeaZone Solutions bathymetric data[4] as a supporting data source suggested that the only major deep recognisable within the dataset, and which may have had palaeogeographic significance, was one associated with Morecambe Bay, and this was confirmed by the UKOGL 2D datasets which crossed this area. Consequently, whilst bathymetric data acted as a valuable background for the work, it only possessed a minor capacity to identify features of archaeological significance outside the shallow marine zone.

The results indicated that it was possible to identify landscape features within the available legacy dataset. The primary features within the 3D data sets were identified and, with a degree of caution, it is likely that the overall trend of the main fluvial system in this area could be identified. What was also apparent was that the fluvioglacial plains were less well resolved within the 2D data, primarily because of their extensive nature. This may, in part, also be due to the issues of noise and

3 The West Coast Palaeolandscapes Main Project is funded by the Aggregates Levy Sustainability Fund through English Heritage and the Welsh Assembly. Partners include the IBM Visual and Spatial Technology Centre (University of Birmingham), Dyfed Archaeological Trust, and the Royal Commission on the Ancient and Historical Monuments of Wales.

4 SeaZone data licence No. 022010.003.

poorer spatial coverage across these areas. However, the results overall suggested that the most important issue relating to the use of 2D data sets for analysis is not generally their capacity to provide information so much as their overall density and availability. Consequently, it seemed reasonable to assess what the impact of varying 2D data availability might be in relation to the known results within the study area. To support this it was necessary to generate a map of the intersections derived from a range of potentially available 2D line configurations within the area.

This required the following steps.

- Digital GIS layers containing the locations of all 2D surveys within the area were imported, irrespective of whether these had been analysed or not, along with associated track log information.

Figure 15.15: Observed intersections between features observed within the 3D dataset and the 2D seismic datasets. Fluvial features are given in blue. The fluvioglacial plain is shown in green. *ASTER DEM is a product of METI and NASA. ETOPO2v2 is the property of the National Geophysical Data Centre, NOAA, US Dept of Commerce.*

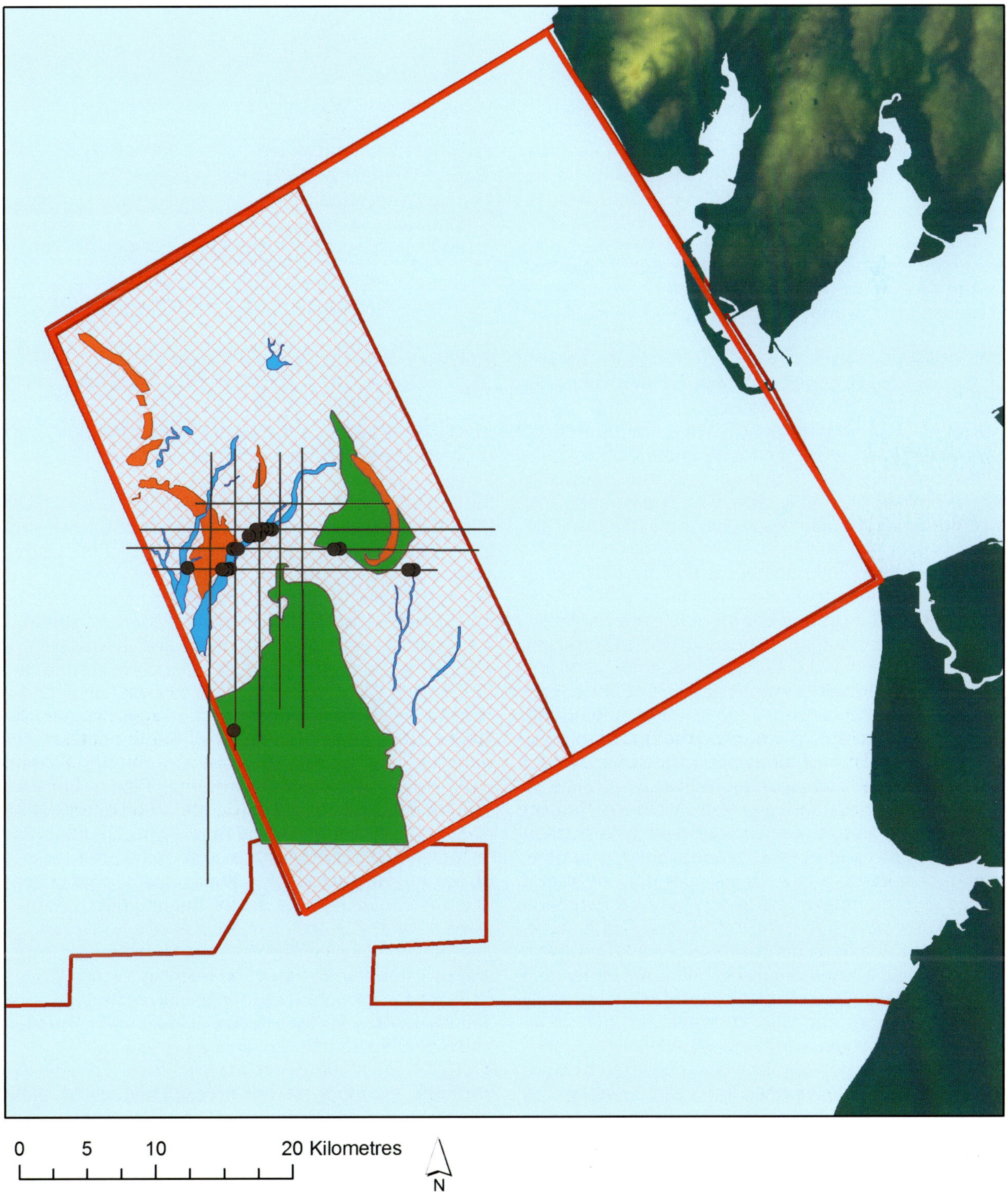

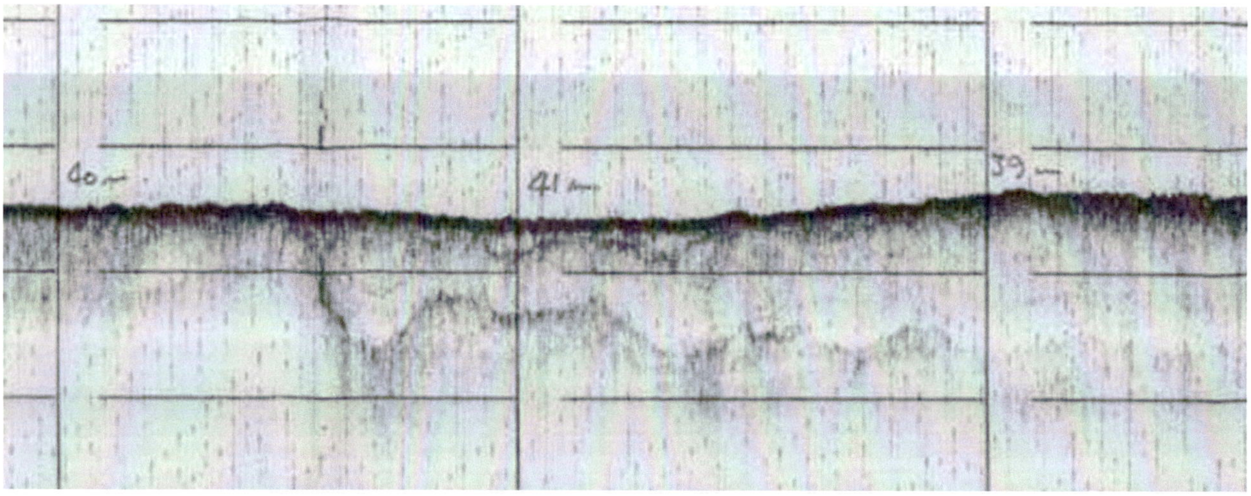

Figure 15.16: Palaeochannel feature located with the 1968 pinger data. *Data courtesy of BGS.*

- A line shapefile was generated of all of the 2D lines that intersected the features identified within the 3D seismic data. Points were created along each of the cropped lines at no less than 50m intervals, a resolution comparable to most 3D datasets.
- Where the points overlay a feature identified from the 3D data the point was assigned an attribute relating to the underlying feature (Figure 15.18).

This procedure generated an 'ideal' feature dataset of the maximum number of features that might be recorded using the existing 2D dataset. The results indicated, not surprisingly, that the numerous intersections of the 2D data with underlying features suggest, visually at least, that full access to all available 2D data would provide an adequate approximation of the underlying archaeological landscapes (Figure 15.18).

This dataset was investigated further by decimating the available lines to explore the impact of varying availability of data. A numeric (double) attribute field – RandomS – was created for lines, and filled with random values between 0 and 1. Sub-samples of 10%, 20% and 50%, designed to reflect various data availabilities, were then generated by selecting by attribute at RandomS <= 0.1, 0.2, and 0.5. As during the previous procedure, points at no less than 50m were again generated for each of these sub-samples (Figures 15.19 & 15.20). Visual inspection of the data suggested that the definition of underlying features varied considerably in relation to line sample size. It was determined that at 10%, although features were clearly present, definition of size and alignment was poor. However, definition was considerably improved by a 20% data sample, and proved nearly as good as extensive coverage at 50%. It should be noted, however, that the study area has a very dense line coverage that is not necessarily representative of the line coverage for the wider area covered by the whole of the 2D line shapefile. In addition, the area within which the features were identified was particularly dense. It can be suggested that the definition of features is therefore related to the relative 2D seismic line density of a particular area. The line density (determined by [Sum line length/ area of study area]*100), for the entire pilot study area was

Sample % of lines within Study Area	Line Density (%)
Whole Area	0.90
Targeted over features	1.23
50%	0.47
20%	0.18
10%	0.09

Table 15.2: Line density within sample area.

0.9%. The density of sub-samples within the study area is presented in the Table 15.2.

Several other factors are important in discussing these data. It is clear that the availability and configuration of 2D lines is not likely to be uniform around the United Kingdom coast and that some assessment of the impact of varying configurations of 2D lines would be invaluable. Consequently, it was decided to resample the pilot area 2D data set using line configurations from a series of sample areas selected around the English and Welsh coasts in order to ascertain what we might expect if we were to undertake a comparative survey in these areas using all available data. For this purpose, four additional areas around the western coast were selected for further analysis. Choice of sample areas was based solely on visual identification of concentrations of lines (Figure 15.21). The coverage of the lines within these areas ranged from relatively uniform to clustered. A polygon shapefile was created over the known features within the original study area (Area 1) and this used as a mask over selected locations. The lines within each of these areas were then exported and placed over the identified features (Figure 15.22, A to D), and a 50m point shapefile was created for each. The line density for the additional areas was calculated and is presented in the Table 15.3.

With the exception of the Bristol Channel, all these sample configurations exceed line densities of 20% for the sample survey area and therefore approximate the minimum suggested line density. All the images suggest that an approximation of the landscape data

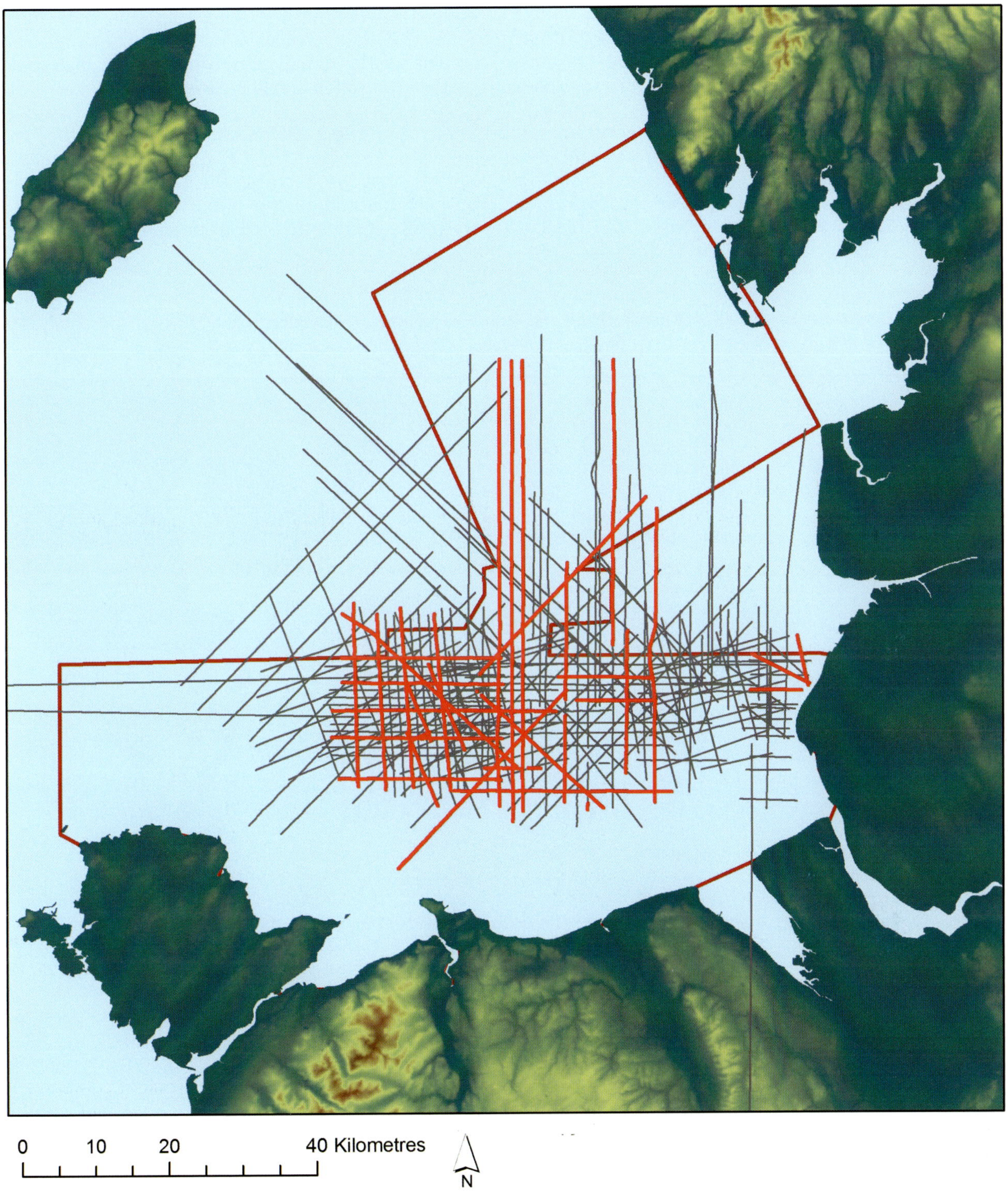

Figure 15.17: Line failures (in red) recorded for data currently processed as part of the West Coast Palaeolandscapes Main Project. *ASTER DEM is a product of METI and NASA. ETOPO2v2 is the property of the National Geophysical Data Centre, NOAA, US Dept of Commerce.*

Area Number	Line Density (%)
1 (Part of original pilot study area)	1.16
2 (Welsh sector of the Irish Sea)	0.95
3 (Welsh sector – Cardigan Bay))	0.21
4 (Scottish sector of the Irish Sea)	0.23
5 (Bristol Channel)	0.18

Table 15.3: Additional areas with line density.

generated by the 3D study might be acquired if the 2D line data was available within the configurations used. Qualitative variation was significant. Not surprisingly the Welsh sector of the Irish Sea, which has the highest line density, provides a visually coherent picture. The gridded configuration in the Scottish sector of the North Sea (Figure 15.21, area 4), although second lowest in terms of line density, provides a reasonable approximation of the extensive fluvioglacial plains in the south. Elsewhere, in Cardigan Bay and the Bristol Channel (Figure 15.21, areas 3 & 5), line densities and configurations are such that the derived image would have value but would not be comprehensive in output. Clearly, lower line densities would be less likely to provide images that were sufficient for extensive landscape interpretation, although they might still locate individual features, including palaeochannels, and these would still have considerable interpretative value for other archaeological activities including sediment sampling. In most cases, however, supporting data sets might be required. For example, ALSF funded Regional Environmental Characterisation data (held by the BGS) is available in the Bristol Channel. This contains further digital 2D seismic profiles that could assist interpretation and line density in this area

If these semi-quantitative outputs have value then the experience provided by the West Coast pilot study suggests that we can use the procedures outlined above to investigate the opportunities for detailed survey elsewhere in UK or European territorial waters.

Figure 15.18: Results of the intersection of all of the 2D lines that intersected the features identified within the 3D seismic data. *ASTER DEM is a product of METI and NASA. ETOPO2v2 is the property of the National Geophysical Data Centre, NOAA, US Dept of Commerce.*

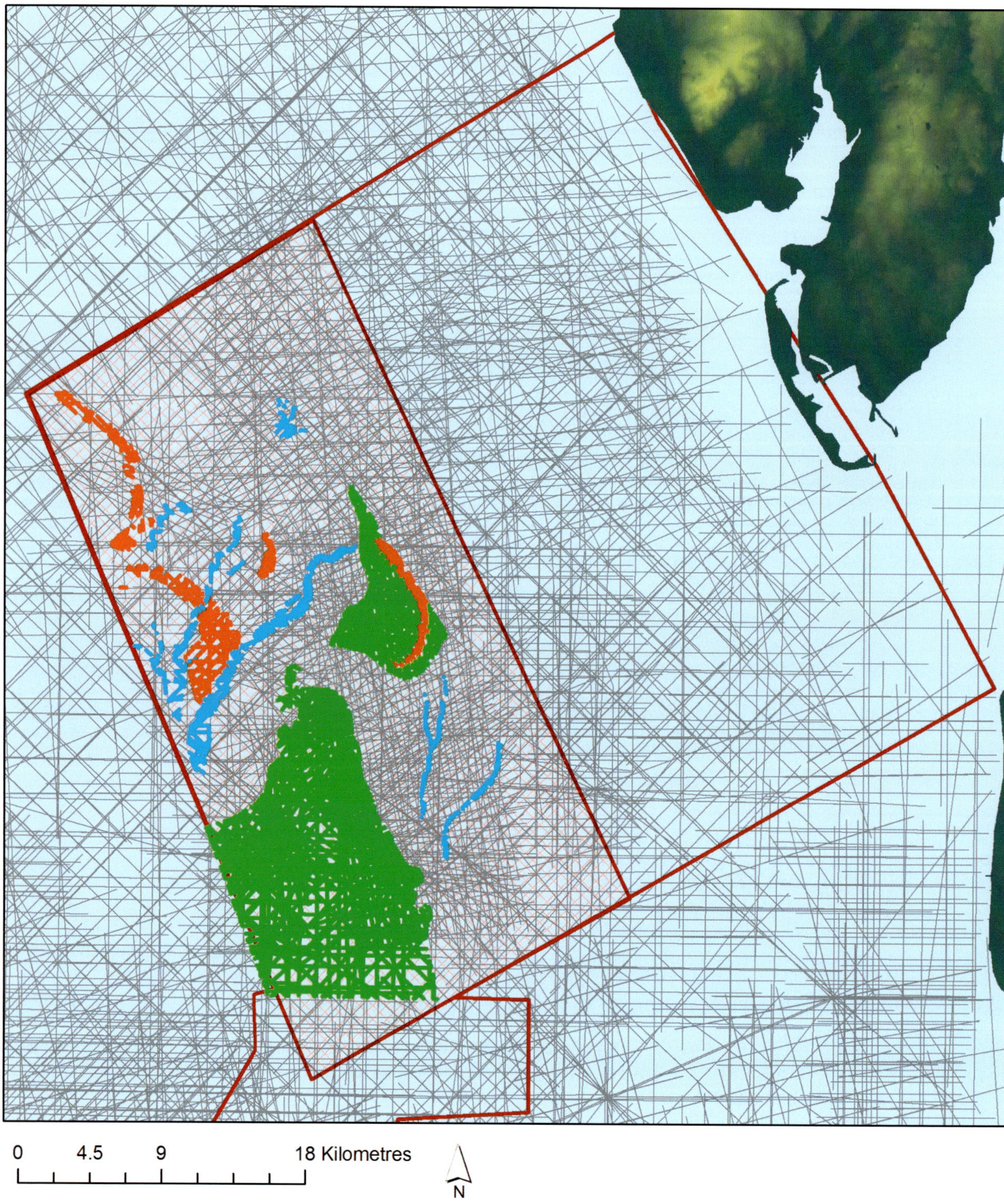

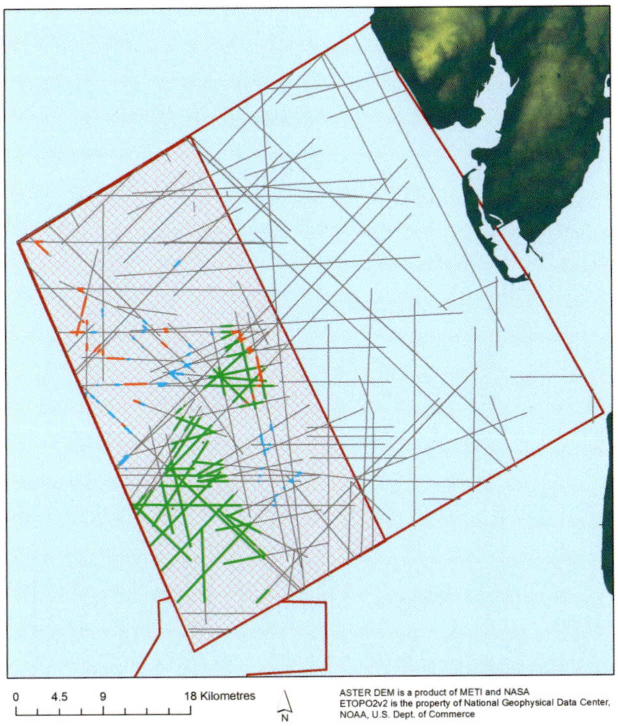

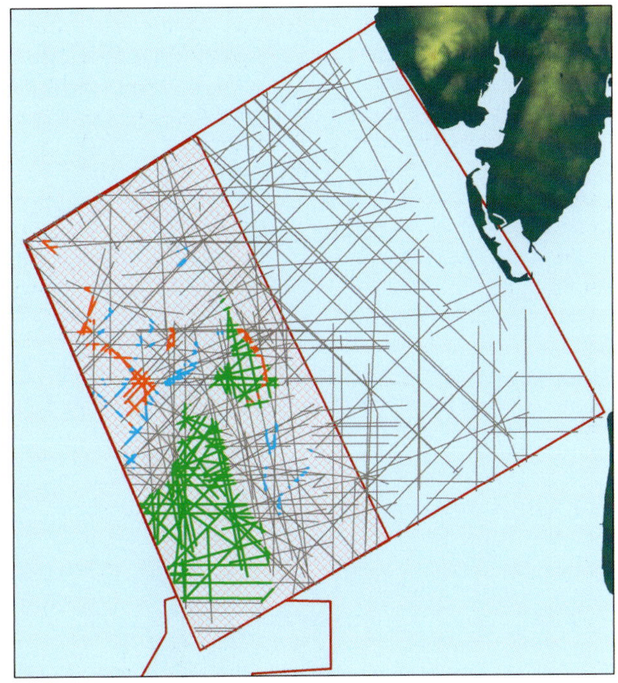

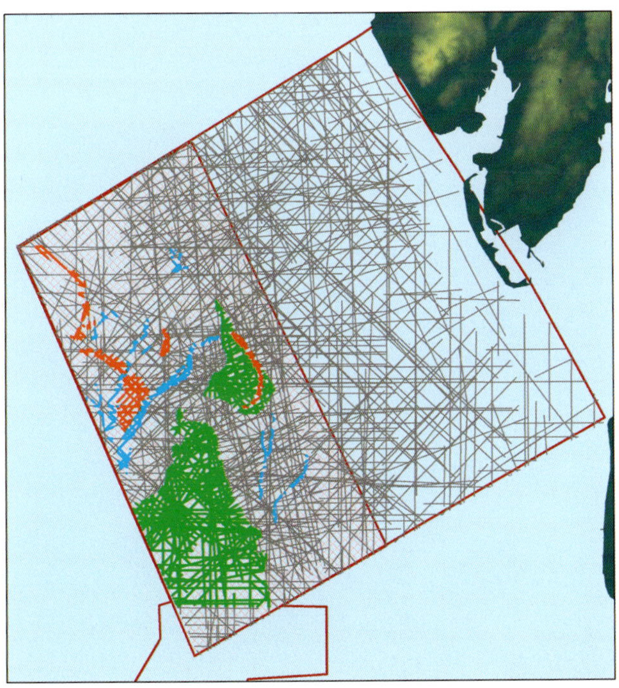

Figure 15.19: Intersection of features using the sub-samples at 10% of lines (*top left*), 25% of lines (*top right*) and 50% of lines (*bottom left*). *ASTER GDEM is a product of METI and NASA' ETOPO2v2 Global 2 Minute Database, National Geophysical Data Center, National Oceanic and Atmospheric Administration, US Dept of Commerce.*

It seems reasonable to suggest that areas with line densities of 20%, but preferably 50%, or more, of the densities in the original study area, are required to provide data that might support extensive landscape interpretation. Appropriate areas can be located using the distribution of 2D line data held in the Birmingham GIS. Figure 15.8 illustrates the density of 2D line coverage around the United Kingdom in relation to primary 3D surveys. This data supported the creation of a grid of 10km × 10km squares generated around the UK coast which could be used to map areas where the line density was potentially suitable for sub-sampling and also the proportion of available lines that should be incorporated into any analysis.

The line density within each grid square was calculated as a percentage (line length to area), and added to the grid square polygon as an attribute (Figure 15.23). Images were created to show the percentage of lines in each grid square needed to be included in any analysis to ensure a maximum line density of 0.5% and 0.2% respectively (Figure 15.23, A & C). The impact of failures of data has already been stressed. Consequently, data was also mapped assuming a worst-case scenario where only 80% of the lines might prove amenable or available for analysis (Figures 15.23, B & D). This suggests that if a research project were solely reliant upon 2D data for mapping purposes then, in many circumstances, it might be necessary to acquire all available datasets to achieve a reasonable level of line coverage that would guarantee results comparable to those provided by the West Coast pilot project

In considering Figures 15.8 and 15.23, it is not surprising that the distribution of both 2D and 3D data largely reflects the interest of energy companies and the intensity of energy exploitation in the southern and northern North Sea. This means that although extensive areas around the coast may always require significant acquisition of new data, specific, archaeologically sensitive areas are likely to be accessible for study because there is sufficient seismic coverage. This includes, for instance, the site of the Viking Bank lithic find and the northern shore of Europe during the maximum extent of exposed land c. 18k BP (Long *et al.* 1986; Fitch *et al.* 2007, Figure 9.2). Moreover, extensive data-rich

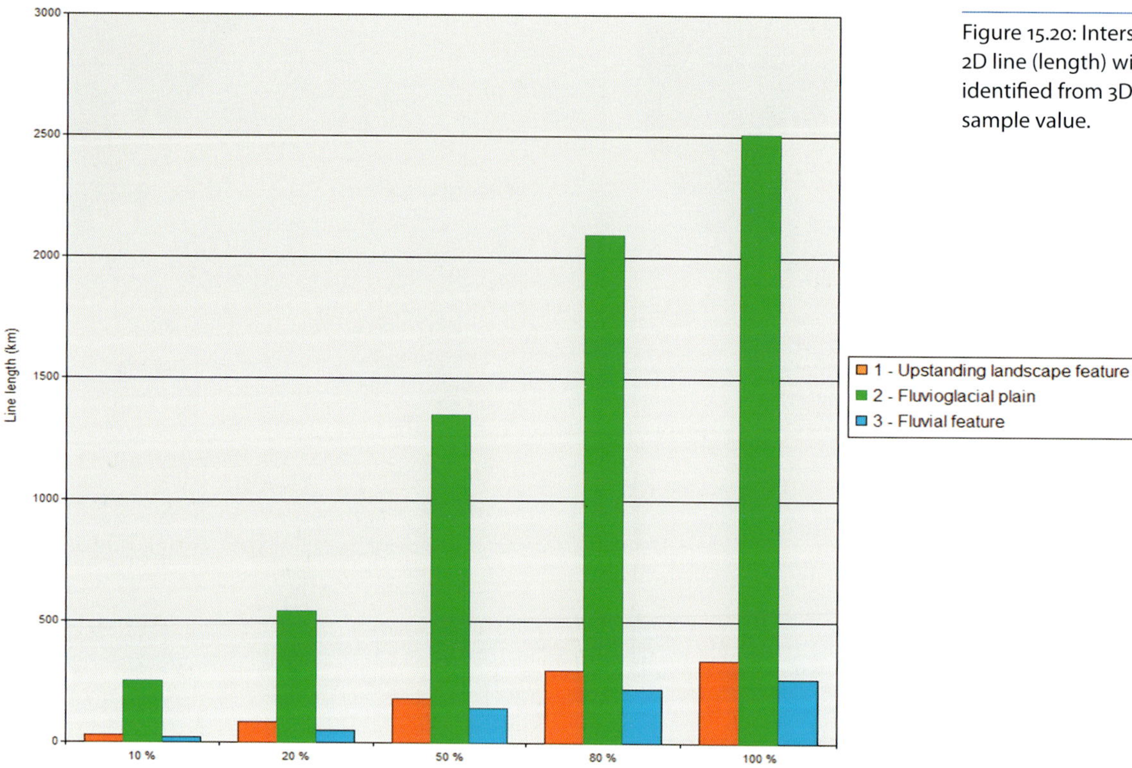

Figure 15.20: Intersection of 2D line (length) with features identified from 3D data by sample value.

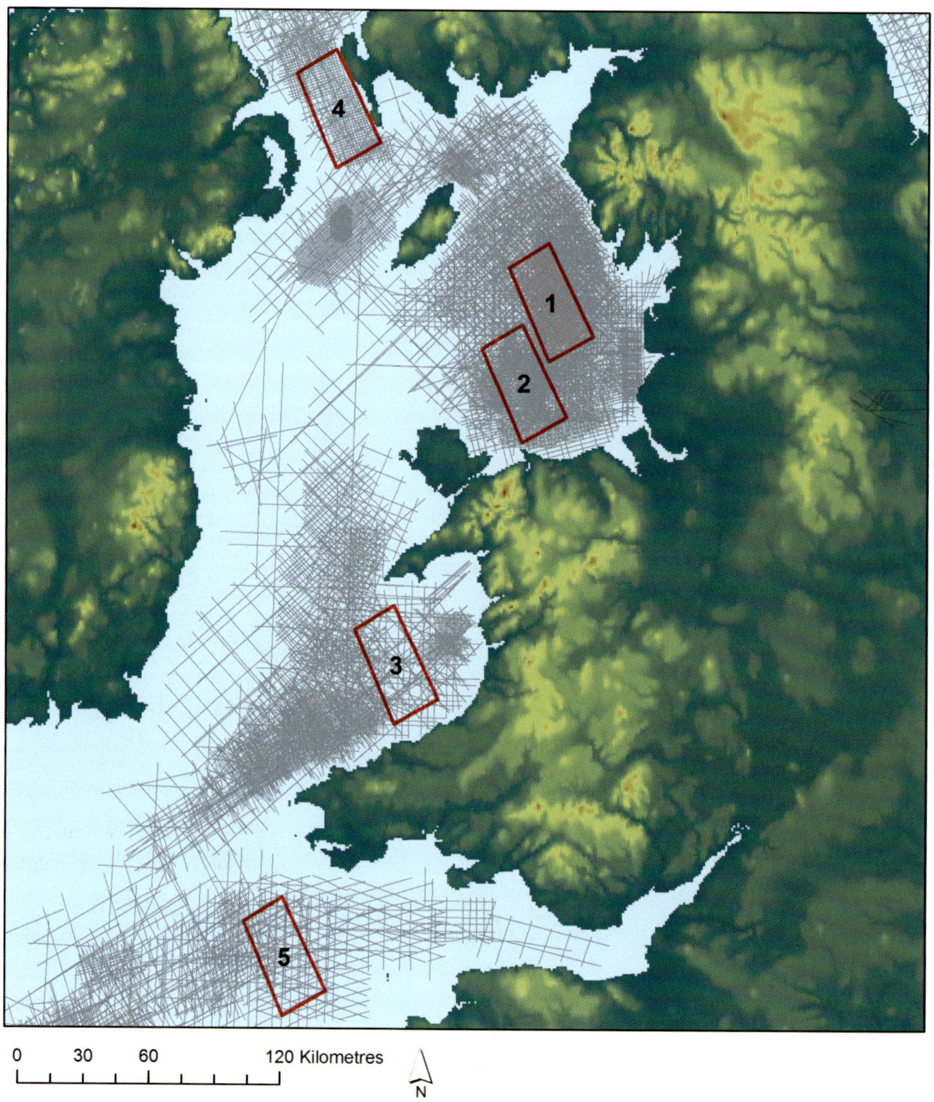

Figure 15.21: Locations of the selected sample areas of line concentrations around the western coast. *ASTER DEM is a product of METI and NASA. ETOPO2v2 is the property of the National Geophysical Data Centre, NOAA, US Dept of Commerce.*

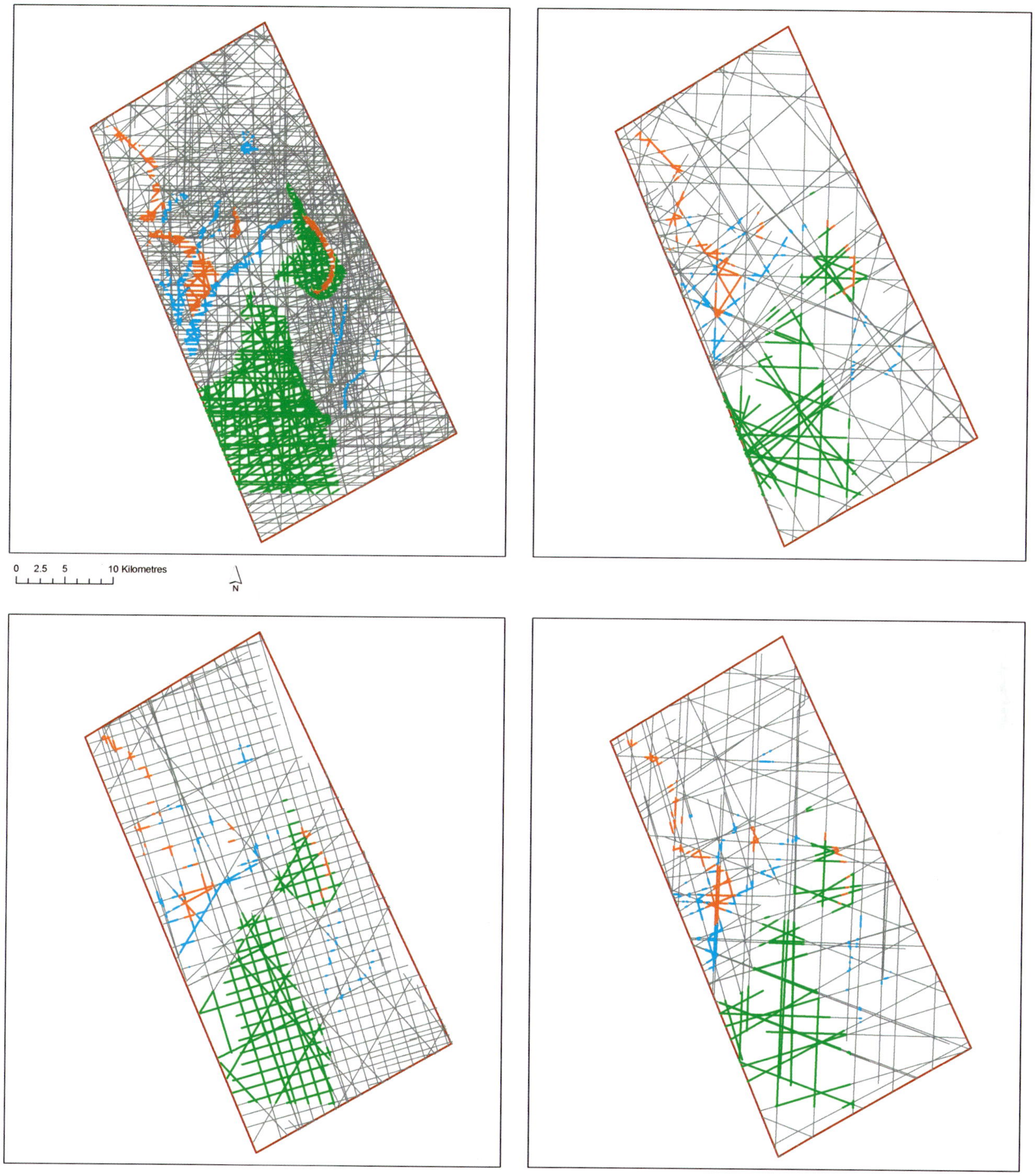

Figure 15.22: West Coast Palaeolandscapes pilot study area sampled using schemes derived from the Welsh sector of the Irish Sea (top left, Area 2), Cardigan Bay in the Welsh sector (top right, Area 3), the Scottish sector of the Irish Sea (bottom right, Area 4) and the Bristol Channel (bottom right, Area 5). *ASTER DEM is a product of METI and NASA. ETOPO2v2 is the property of the National Geophysical Data Centre, NOAA, US Dept of Commerce.*

areas off the Moray Firth provide real opportunity for exploring the landscapes associated with Scotland's currently contentious later Palaeolithic history (Anon 2007). Despite this, it remains true that data availability is limited or absent over much of the territorial waters of the United Kingdom and that the majority of these areas may contain extensive palaeolandscapes and this is likely to be true of comparable areas elsewhere within Europe, including the Baltic and Adriatic. Specific, sensitive areas, including the sea area off the north east coast of England, which is associated with important early Holocene settlement at Howick, remains to be surveyed to the level required for detail landscape survey (Waddington 2007; Waddington & Pederson 2007).

The results of this limited retrospective sampling indicate that extensive projects based upon 2D data are indeed viable in areas where data is available and conducive to analysis. Where these conditions are

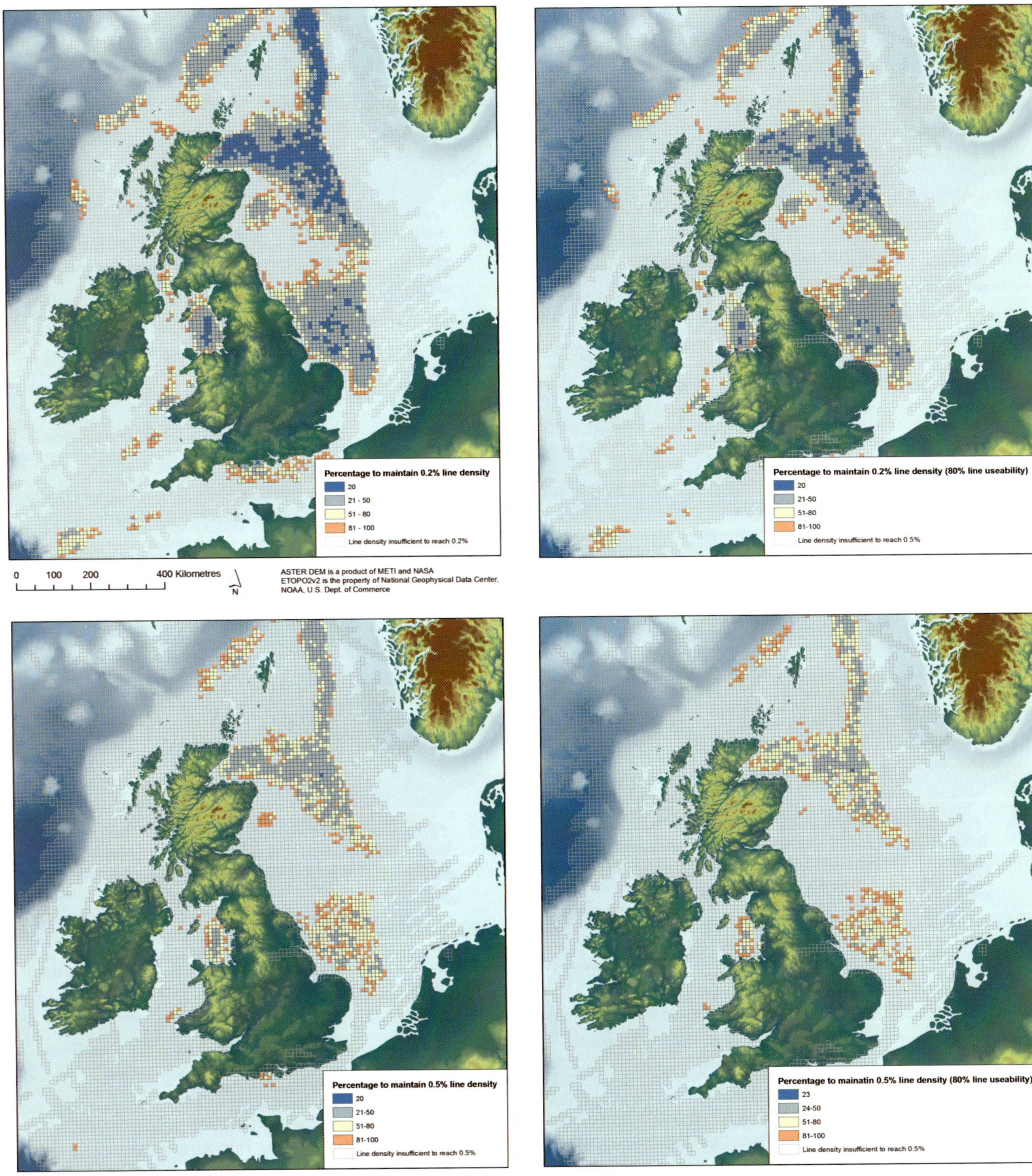

Figure 15.23: Data availability for marine prospection within UK territorial waters based upon line percentage requirements and line failure rates. *Top left*: The percentage of lines in each grid square needed to be included in any analysis to ensure a maximum line density of 0.2% assuming that all lines are useable. *Top right*: The percentage of lines in each grid square needed to be included in any analysis to ensure a maximum line density of 0.2% assuming that only 80% of the lines may be useable. *Bottom right*: The percentage of lines in each grid square needed to be included in any analysis to ensure a maximum line density of 0.5% assuming that all lines are useable. *Bottom right*: The percentage of lines in each grid square needed to be included in any analysis to ensure a maximum line density of 0.5% assuming that only 80% of the lines may be useable. *ASTER DEM is a product of METI and NASA. ETOPO2v2 is the property of the National Geophysical Data Centre, NOAA, US Dept of Commerce.*

met primary outlines of landscape features, required by archaeologists and heritage managers, can be identified. However, it is equally clear that, given the variation in data availability, then a pragmatic approach may be required in some areas to acquire sufficient 2D data to achieve a positive outcome. In support, it is generally true that our territorial waters have often been surveyed for a variety of reasons and that complimentary supporting data from a variety of public and private sources may be available within many of the areas identified for future research. It is also important to stress that, even if access to data becomes

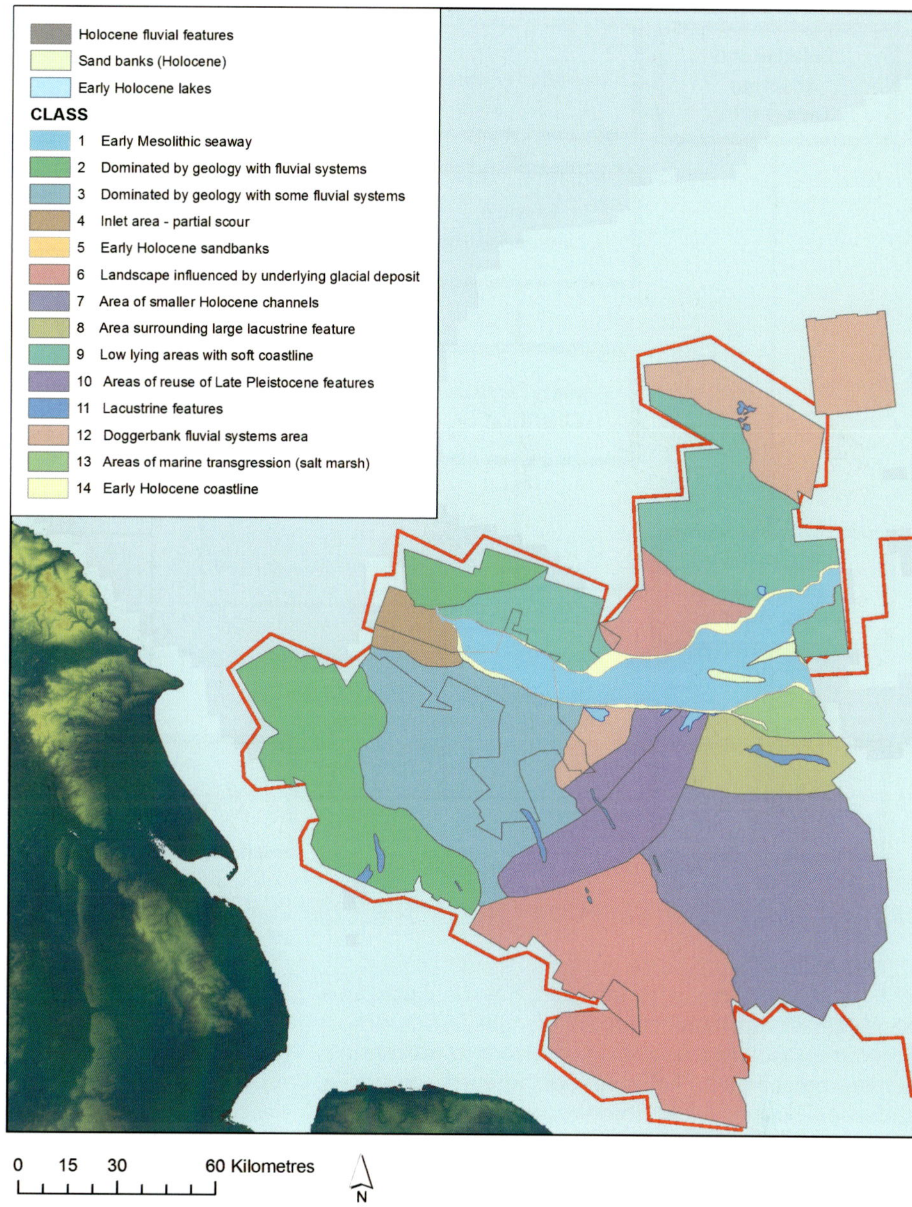

Figure 15.24: HLC areas defined within the NSPP research area. *ASTER DEM is a product of METI and NASA. ETOPO2v2 is the property of the National Geophysical Data Centre, NOAA, US Dept of Commerce.*

an issue in respect of the scale of future research, it is a fact that little or no archaeological baseline data exists for the submerged prehistoric resource across most of the offshore areas surrounding the United Kingdom and Europe more generally. There is, therefore, a real value in almost any information that can be acquired. Certainly, the results presented here suggest that there is no reason why mapping should not be attempted in the majority of marine areas within European territorial waters.

Refining 3D interpretation using 2D data: Historic Landscape Characterisation

Although lacking the full and comprehensive framework associated with 3D/HD3D survey, the fusion of 2D with 3D seismic data has the benefit of being able to combine legacy datasets to produce a higher definition interpretation, maximising the information value of known data assets and, potentially, reducing the need for re-survey. The option to use 2D data to refine relatively low resolution 3D data was highlighted in the North Sea Palaeolandscapes study (Gaffney *et al.* 2007). There, supplementary detail of erosion surfaces from 2D lines across the Outer Silver Pit was used to suggest that not only had that feature been transformed from a relatively placid lake environment to a fast flowing estuary, but that these currents had also removed the basin's potential for sediment sampling (Briggs *et al.* 2007). Then value of such an observation in terms of management or planning for coring should be clear. However, the potential for refinement of interpretation can also be demonstrated at a landscape level through the use of the North Sea data as part of a historic landscape characterisation (HLC) programme (Figure 15.24).

HLC programmes are now an established management tool within British terrestrial archaeology and, increasingly, in marine foreshore or shallow marine contexts. There is now a considerable literature relating to HLC methodologies, although here it is worth stressing that the primary characteristic of

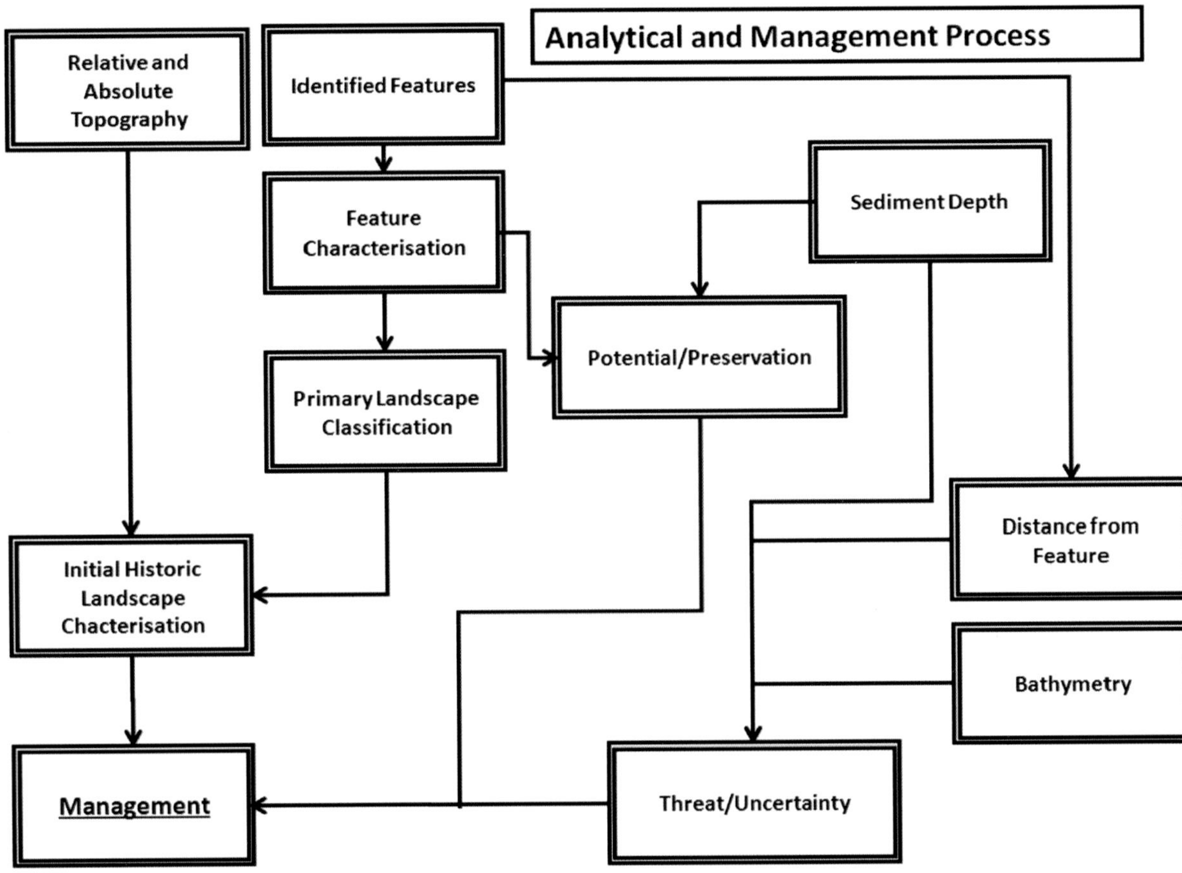

Figure 15.25: The analytical and management process within the NSPP data flow. *Gaffney & Thomson 2007, figure 9.4.*

an HLC programme is to support overall landscape management and the curation of contiguous areas, rather than isolated points or polygons representing archaeological sites or areas of interest within a landscape (Aldred & Fairclough 2003; Clark *et al.* 2004; Fairclough 2001; Fairclough & Rippon 2002). Broad characterisation zones, on land, may incorporate many contemporary factors including landscape aesthetics as well as archaeological or historic data (Barratt *et al.* 2007). In curatorial terms, the output from such work assists managers in assessing the overall impact of change across the entire region rather than privileging artificial constructs such as sites or conservation zones. In the case of the North Sea, the relative inaccessibility of the landscape demanded modification of the application – if not the underlying philosophy or methodology of HLC. The results of the HLC analysis of the NSPP data contrast with terrestrial HLC analyses. Whilst the latter reflect a historical palimpsest of features overlain on a topographic backdrop, the North Sea HLC zones were obtained primarily from topographic and morphological data and the interpretation of these zones derives, in part, from their significance as part of a presumed hunter-gatherer economy. Examples of such a monolithic land use are rare in terrestrial contexts but they are not unknown (i.e. military landscapes in North America which have proven susceptible to HLC analysis – Barratt *et al.* 2007).

Within the North Sea, the landscape was classified into fourteen broad areas based upon their depositional history and major landscape features (Fitch *et al.* 2007). The dividing lines for many of the landscape zones observed coincided broadly with known watersheds between observed fluvial features. This data probably represents the best general zoning in terms of potential Holocene land use achievable using the available data, but also, to the extent that it may correlate with broad economic activity, the data may carry considerable potential to act as the basis for more detailed behavioural modelling. In managerial terms this data was further refined using the workflow outlined in Figure 15.25 and the detail of the outputs has already been published (Fitch *et al.* 2007, 110–8.)

However, the HLC data from the NSPP has recently been used and refined by another survey project off the Humber estuary. This provides a novel example where legacy data has been used to guide new data capture and, in a cost-effective manner, where the new 2D data was able to refine the earlier 3D interpretation (Fitch *et al.* 2010; Benike *et al.* 2004; Fitch *et al* 2010; Novak & Bjorck 2002).

The Humber Regional Environmental Characterisation (REC) project was funded by the Marine Environment Protection Fund under the wider Aggregate Levy Sustainability Fund as administered by DEFRA (http://www.alsf-mepf.org.uk/). A consortium, including the British Geological Survey (BGS), Birmingham University's Institute of Archaeology and Antiquity, Marine Ecological Surveys (MES) and Gardline Environmental, was created to collect data on seabed habitats, species and features of archaeological interest that exist at

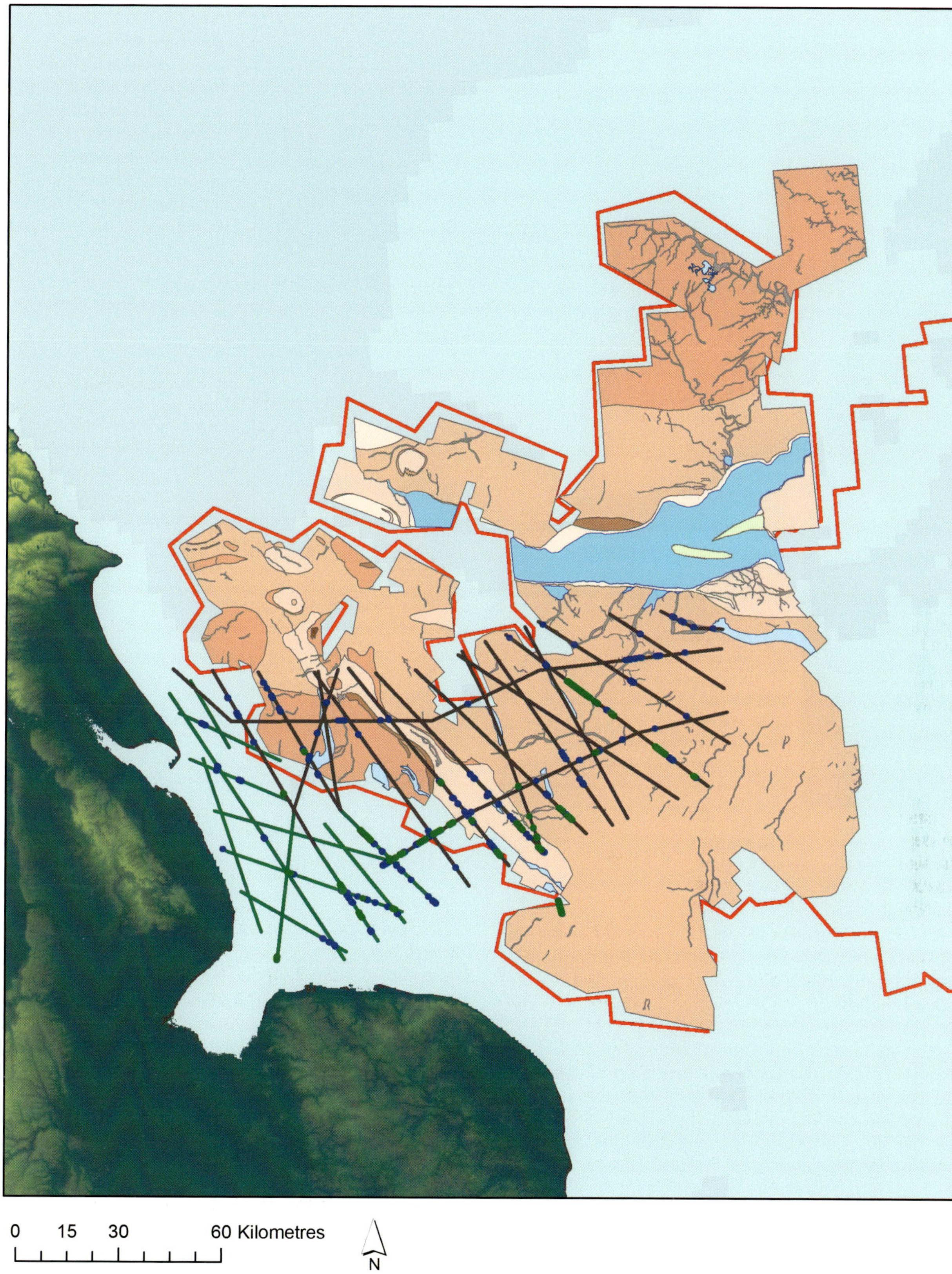

Figure 15.26: The location of fluvial (blue dots) and Holocene (green dots) landscape feature identified within the Humber REC survey lines. *ASTER DEM is a product of METI and NASA. ETOPO2v2 is the property of the National Geophysical Data Centre, NOAA, US Dept of Commerce.*

a regional scale (Figure 15.26). The scope of work for this project called for the acquisition of geophysical and acoustic data amounting to about 3,000km of line data within an area of 11,221km². The data consists mainly of widely spaced lines covering the greater part of the survey area, and four detailed survey areas over known or potential archaeological and biological sites of interest. The shallow seismic geophysical dataset provided for the Humber REC consisted of 2D Boomer lines collected by Gardline Surveys Ltd and provided

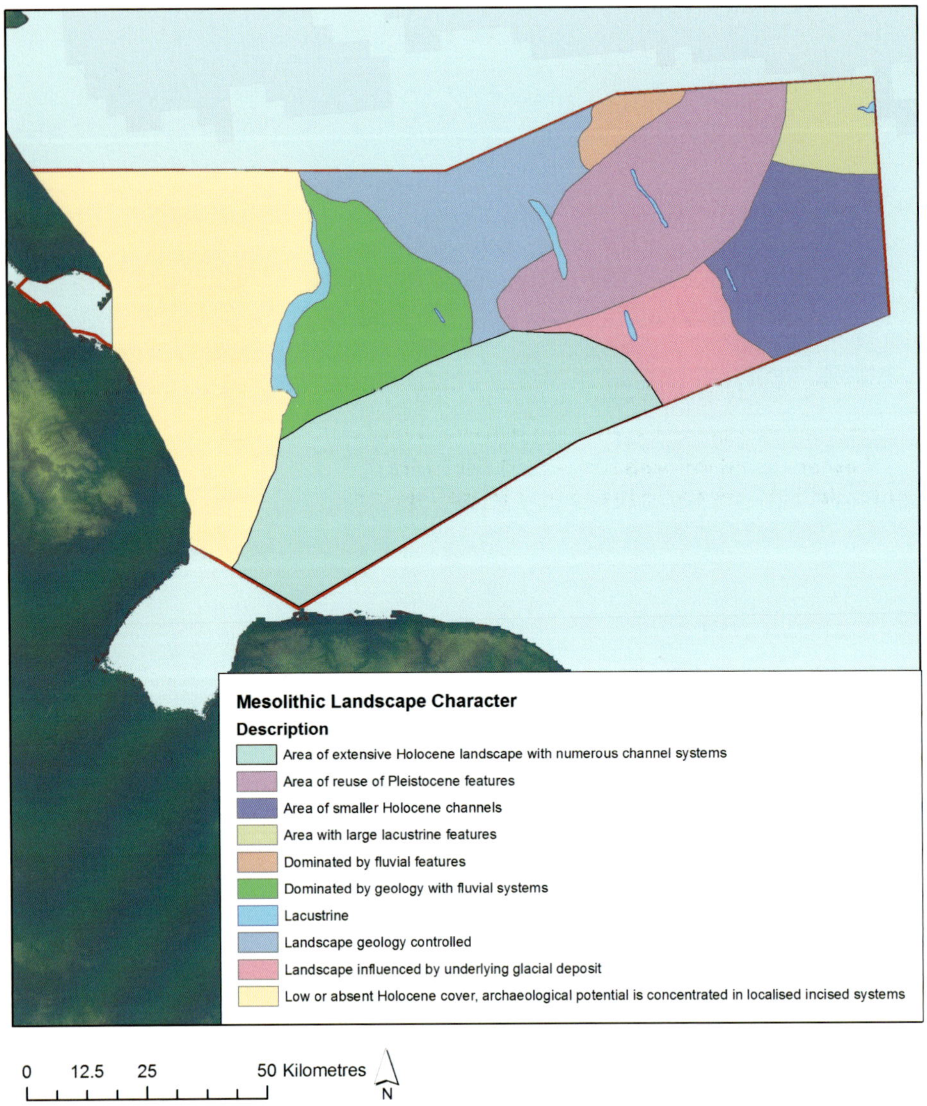

Figure 15.27: Submerged prehistoric landscape characterisation resulting from the use of Humber REC Survey datasets. *ASTER DEM is a product of METI and NASA. ETOPO2v2 is the property of the National Geophysical Data Centre, NOAA, US Dept of Commerce.*

to Birmingham University in standard SEG-Y format. The data was inspected in detail at Birmingham to determine the depth and extent of identified features and anomalies.

For this paper, the important point of this work is that the survey area overlaps with the area of the NSPP and the results of the NSPP provided the basis for interpreting the marine landscape in advance of the Humber REC fieldwork. However, the high resolution survey over the landscape features recorded by the NSPP in turn represents an important validation of previous work and allows for the expansion of the submerged landscape record to the west of the original NSPP study area.

With respect of the HLC outputs to the Humber REC it is hardly surprising that that the new study area can be characterised into broadly similar areas to that produced by the NSPP (Figure 15.27). However, notable modifications to the zones have been made in the near shore areas to reflect the additional, detailed information provided by the REC assessment (Figure 15.28 & Table 15.4). Although the Humber REC used eight of the existing 14 categories defined by the NSPP, two additional categories or landscape zones were created. Within the areas already characterised by the NSPP, some 103 Holocene channels and 37 areas relating to Holocene land surfaces were located by the Humber REC project. These were utilised to corroborate the existing characterisation, and where necessary a reclassification of HLC zones was produced. In the areas not covered by NSPP data some 120 areas containing archaeologically significant landscape features and 47 areas of Holocene landscape were identified and utilised to assist the classification of new areas.

The proportional relationship between the distribution of archaeologically significant Holocene features across character zones is shown in Table 15.5. By utilising the area of survey covered by the geophysical lines it becomes possible to approximate the density of archaeologically significant landscape features within each of the character zones. The sum of the survey lines (not taking into consideration multiple passes) was approximately 1,300km for the offshore survey area, and 530km for the near shore survey area. In numeric terms, the Humber REC led to a reclassification of the NSPP data which amounted to about 1,150km² of about 6,548km² of the area characterised by the NSPP, or about

Table 15.4: Landscape characterisation as defined through the Humber REC.

Landscape characterisation as defined by the Humber REC		
Description	**Used in NSPP**	**Area km²**
Area of extensive Holocene landscape with numerous channel systems	NEW	2,390
Area of reuse of Pleistocene features	YES	1,570
Area of smaller Holocene Channels	YES	1,070
Area with large lacustrine features	YES	350
Dominated by fluvial features	YES	170
Dominated by geology with fluvial systems	YES	1,080
Lacustrine	YES	160
Geology controlled landscape	YES	1,070
Landscape influenced by underlying glacial deposit	YES	590
Low or absent Holocene cover, archaeological potential is concentrated in localised incised systems	NEW	2,590

Table 15.5: Length of Survey Lines for Humber REC character areas.

Length of Survey Lines for Humber REC character areas						
Description	**Area km²**	**Total Of LENGTH Outer Survey Lines**	**Length of Survey Lines for Humber REC character areas** **Total Of LENGTH Inner Survey Lines**	**Number of Features /landscapes Identified along lines**	**Length of Features/ landscapes along lines**	**Projected % of character area containing Holocene features/landscape (km²)**
Area of extensive Holocene landscape with numerous channel systems	2392	143	210	172	46.1	13% (312km²)
Area of reuse of Pleistocene features	1567	270		34	11	4% (64km²)
Area of smaller Holocene Channels	1074	148		13	4.6	3% (33km²)
Area with Large Lacustrine features	356	53		9	3	5% (20.15km²)
Dominated by Fluvial	167	25	23	1	0.06	1% (2km²)
Dominated with Geology with Fluvial systems	1077	118		9	1	0.8% (9km²)
Lacustrine	158	76	4	5	0.1	0.1% (0.2km²)
Landscape Geology Controlled	1074	203		5	0.9	0.4% (5km²)
Landscape influenced by underlying glacial deposit	591	120		22	5.8	5% (29km²)
Low or absent Holocene cover, archaeological potential is concentrated in localised incised systems	258	136	291	35	8.5	2% (5km²)

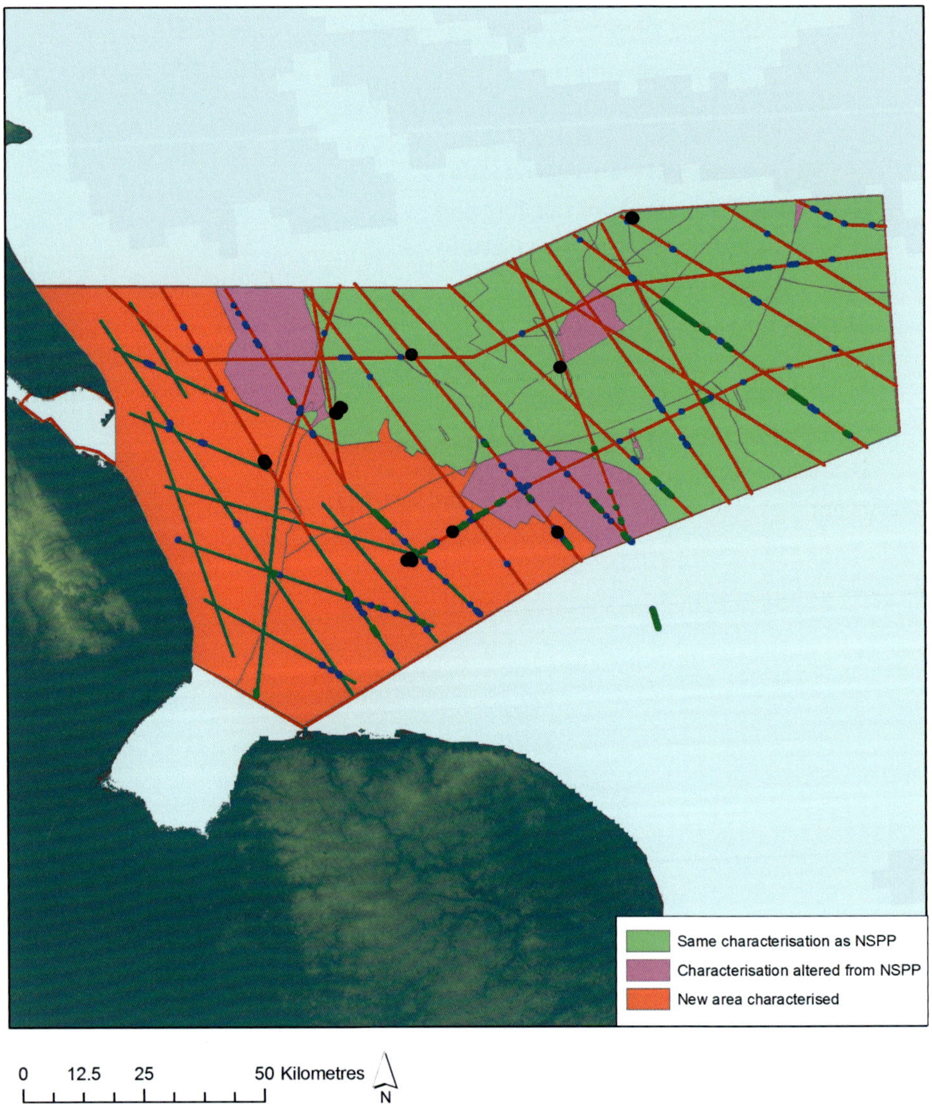

Figure 15.28: Refinement of HLC zones based on 2D data. Existing (unaltered) NSPP character areas are shown in green, whilst re-characterised NSPP areas are shown in pink. The newly characterised are is shown in red. *ASTER DEM is a product of METI and NASA. ETOPO2v2 is the property of the National Geophysical Data Centre, NOAA, US Dept of Commerce.*

17.5% of the total area (Figure 15.28). Clearly, a significant gain in precision can be achieved by integrating 2D and 3D data in this manner.

A final aspect of the use of 2D data provided by the Humber REC relates to its detailed use to guide 'ground observation' of features. A specific example of such an activity relates to the sediment cores recovered from a transect across palaeochannel features some 50km offshore from Titchwell in the study area (survey number: 18_100N). This was carried out using a 5m vibrocore rig operated from the Gardline vessel *Sea Profiler*. The cores were recovered and transported to the University of Birmingham for stratigraphic recording and sub-sampling (Figure 15.30). The suite of sediments from the palaeochannel fill consists of basal clays overlain by humified peats sealed by organic silts, which are in turn overlain by marine sands of the current seabed. These deposits infill a channel incised into the basal boulder clay and reach a maximum thickness of over 3m. Analysis of pollen, plant macrofossil, beetles and ostracod/forams is currently in progress, but initial results demonstrate excellent preservation of these sub-fossil remains. Radiocarbon and Optically Stimulated Luminescence dating provides a chronological framework for the palaeoenvironmental records and demonstrate that the basal peats were accumulating during the 7th–8th millennium BC. The pollen, macrofossil and beetle records are beginning to shed valuable light on the landscape of Doggerland during this period, whilst the ostracod/forams are providing information regarding the timing and nature of the marine transgression that resulted in the eventual inundation of the area.

This work is significant for two reasons: firstly, the acquisition of stratigraphic information allows further 'fine tuning' of geophysical data in terms of the character of the submerged deposits represented by this information. For example, in the case of the palaeochannel discussed above the stratigraphic boundary between the basal boulder clay which forms the predominant drift deposit in the REC study area and the overlying silts and clays is apparent as a distinct positive reflector. The correlation of geophysical and associated stratigraphic records in this manner should permit more robust interpretation of seismic data and gives us greater confidence in the characterisation of the features identified using such survey. Given the logistical difficulties and cost implications of offshore work, as well as likely future development pressures on the marine zone, the ability to reliably identify

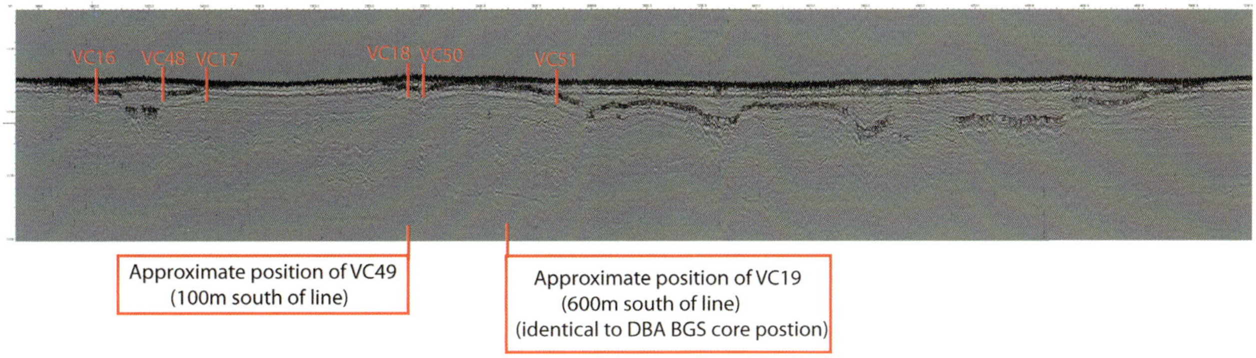

Figure 15.29: Seismic profile showing palaeochannels and core locations within the Humber REC. *Image Copyright: The University of Birmingham.*

locations of high palaeoenvironmental potential is highly significant.

The capacity to accurately identify *in situ* deposits of palaeoenvironmental potential cannot be overemphasised. However, an outstanding question is how future work might locate and investigate any archaeological remains preserved in or sealed beneath deposits of this kind. Currently, the majority of palaeoenvironmental sequences that provide information regarding changes in relative sea level have been sampled from coastal and near shore environments (e.g. Shennan & Horton 2002). Few palaeoenvironmental sequences are available from distances far off shore (up to 50km) or in relatively deep water (25m+). Understanding the character of the environment of Doggerland itself as well as the pattern and process of the early Holocene inundation requires datasets from across as wide a geographical area of the landscape as possible. This may become more important as preliminary results from the REC analysis suggest that currently established rates of early Holocene relative sea level change may not apply especially well to it, perhaps raising questions regarding the utility of regional sea level curves (e.g. Kiden *et al.* 2002). This is critical in terms of understanding human activity and for assessing cultural responses to the flooding and ultimate loss of Doggerland.

Conclusions

It should be clear that the benefits of a strategy that can exploit a greater range of legacy data than has previously been used are considerable. At a purely practical level, archaeologists and heritage managers could never hope to replace or replicate this resource if it had not been provided for other purposes. Indeed, the investment required to acquire data for the area of the North Sea Palaeolandscapes project alone has been estimated as equivalent to a century of heritage funding for a national service equivalent to English Heritage (Powlesland 2010). We are unlikely, therefore, to have any alternative to legacy data sources to explore the majority of areas associated with prehistoric habitation and marine inundation. Although there is an emerging consensus that the coastal shelf may be key to understanding the process of prehistoric colonisation of the region, and to contextualise and interpret the settlement pattern within the terrestrial record of all the surrounding countries, our current knowledge of the majority of the area remains intensely speculative. Consequently, any opportunity to explore those marine areas previously available for human habitation and associated extensive 2 or 3D data should be accepted with some alacrity (Figures 15.8 & 15.23).

In such circumstances, remote sensing, in one form or another, is likely to remain the only practical route towards the investigation of the majority of the inundated landscapes surrounding our coasts. However, it should be emphasised that there is no single data source or methodology that will satisfy the requirements of all archaeologists or heritage managers. The methodologies and technologies

Figure 15.30: Collecting sediment samples using a vibrocorer during the Humber REC. © *Dr Ben Gearey.*

chosen for research and management will always depend upon the nature of the archaeological questions being posed. Some projects, including those which seek detailed sediment sampling, proxy or even direct evidence for settlement or land use, may well require high resolution survey and demand the acquisition of new data in areas which have not previously attracted survey. In other circumstances, the availability of the extraordinarily large, pre-existing data sets that have been acquired around our coasts for other purposes have the capacity to inform and guide the development of European research agendas in their own right. In the case of research involving supra-national behavioural or settlement modelling, the relatively coarse 2D and 3D data sets that are available may well be adequate for such purposes. Together with improved data on sea level rise and geomorphological change, these data have the potential to provide dramatic, new insights into landscapes which may be key to our regional models.

It should also be noted that as national governments become more conscious of their responsibility to preserve the immense heritage landscapes that exist within national waters, there will be a greater requirement for data that is fit for management purposes and, one suspects, can be defended legally. Europe's oceans have always been contested areas and their continued development as strategic economic zones (driven by fishing, energy, aggregates, telecommunication and shipping) will demand ever more accurate mapping. Detailed information on the distribution of heritage resources will increasingly be required by planners from countries with marine possessions. However, it remains true that much current research in marine contexts is made possible by the generosity of commercial institutions that often provide data at no cost and governmental agencies which fund processing or data acquisition. As the European economy falters during 2010, such support will become even more important. Unfortunately, current financial stringencies may force some data suppliers to take a harder position on charging researchers for data. In the future, costs may become a limiting factor for research within UK waters at least. At the moment that archaeologists had begun to overcome the issues of exploring the vast, inundated prehistoric landscapes preserved within the region, it would be tragic if access to data prevented European heritage managers realising their cultural potential. Consequently, the development of cost-effective methodologies is of considerable importance. We must promote methodologies that integrate existing, extensive and relatively coarse data with spatially restricted but higher resolution data and use these to formulate future directed survey and prospection. We hope that the results presented within this paper provide one opportunity to support a strategic shift in marine heritage management and prospection.

Acknowledgements

We would like to thank the following companies, institutions and individuals for their support for the projects whose results have contributed to this publication. English Heritage, the Marine Aggregates Levy Sustainability Fund, the University of Birmingham, PGS, Centrica UK Ltd, DEFRA, British Geological Survey, MES, Gardline, Maersk Oil (Qatar), Chesapeake Technology Inc., Seismic Micro-Technology Ltd (SMT). Pheonix, SeaZone. IBM, VSG and Mechdyne, the Qatar Museums Authority, Prof. Dave Tappin (BGS), Dr Bryony Pearce (MES), Vince Grove (Gardline), Dr Andrew Bellamy (BMAPA, UMD), Dr Huw Edwards, Dr Tim Bird and Dr Chris Loader (PGS), Dr Kath Buxton (English Heritage), Dr Magnus Alexander (English Heritage), Dr Nic Flemming (University of Southampton), Mr Paul Hatton (Information Services, University of Birmingham), Ms Valerie Scadeng (VSG), Dr Jim Roche (IBM), Mr Richard Cashmore (Mechdyne), Esbern Hoch (Maersk Oil Qatar), Faisal Abdulla Al-Naimi (QMA), His Excellency Sheikh Hasan Bin Mohammed Al-Thani (QMA).

ETOPO2v2 Global 2 Minute Database, National Geophysical Data Center, National Oceanic and Atmospheric Administration, U.S. Dept. of Commerce, http://www.ngdc.noaa.gov/mgg/global/etopo2.html The original data of ASTER GDEM is the property of METI and NASA http://www.gdem.orsdac.or.jp/GDEM/E/index.html

References

Aldred, O. & Fairclough, G. 2003: *Historic Landscape Characterisation: taking stock of the method*. English Heritage and Somerset County Council, London.

Anon 2007: Orkney finds confirm early Scottish colonisation. *British Archaeology* 97. http://www.britarch.ac.uk/ba/ba97/news.shtml.

Bailey, G. 2010: Review of *Europe's Lost World: the Rediscovery of Doggerland*, by Vince Gaffney, Simon Fitch & David Smith, 2009. York: Council for British Archaeology. (CBA Research Report 160). *Cambridge Archaeological Journal* 20, 145–6.

Bailey, G., Westley, K., Davies, W., Gaffney, V., Gibbard, P., Firth, A., Flemming, N., Schreve, D., Hosfield, R., White, M. & Bates, M. 2010: *Maritime and Marine Historic Environment Research Framework: Draft Palaeolithic Resource Assessment*. http://www.southampton.ac.uk/archaeology/research/projects/maritime_research_framework.html.

Barratt, G., Dingwall, L., Gaffney, V., Fitch, S., Huckerby, C. & Maguire, T. 2007: *Heritage management at Fort Hood, Texas: Experiments in Historic Landscape Characterisation*. Archaeopress, Oxford.

Bennike, O., Jensen, J. B., Lemke, W., Kuijpers, A. & Lomholt, S. 2004: Late- and postglacial history of the Great Belt, Denmark. *Boreas* 33, 18–33.

Briggs, K., Thomson, K. & Gaffney, V. 2007: A Geomorphological Investigation of Submerged Depositional Features within the Outer Silver Pit, Southern North Sea. In Gaffney, V., Thomson, K. & Fitch, S. (eds): *Mapping Doggerland. Mapping Doggerland: The Mesolithic Landscapes of the Southern North Sea*. Archaeopress. Oxford. 43–60.

Bulat, J. 2005: Some considerations on the interpretation of seabed images based on commercial 3D seismic in the Faroe-Shetland Channel. *Basin Research* 17, 21–42.

Bunch, M., Gaffney, V. & Thomson, K. 2007: Coordinating marine data sources. In Gaffney, V., Thomson, K. & Fitch, S. (eds): *Mapping Doggerland. Mapping Doggerland: The Mesolithic Landscapes of the Southern North Sea*. Archaeopress. Oxford. 11–21.

Clark, J.G.D. 1936: *The Mesolithic Settlement of Europe*. Cambridge University Press, Cambridge.

Clark, J., Darlington, J. & Fairclough, G. 2004: *Using Historic landscape characterisation: English Heritage's review of HLC Applications 2002 – 03*. English Heritage and Lancashire County Council.

Coles, B.J. 1998: Doggerland: a speculative survey. *Proceedings of the Prehistoric Society* 64: 45–81.

COWRIE, 2010: *Offshore Geotechnical Investigations and Historic Environment Analysis: Guidance for the Renewable Energy Sector*. Draft document (July 2010).

Cuttler, R., Fitch, S. & Al-Naimi, F.A. 2010: Considering the 'Terra Incognita' and the implications for the Cultural Resource Management of the Arabian Gulf Palaeolandscape. In Potts, D., Al Naboodah, H. & Hellyer, P. (eds): *Archaeology of the United Arab Emirates: Proceedings of the Second Conference on the Archaeology of the UAE*. Trident press, London.

Dix, J., Quinn, R., & Westley, K. 2004: *A Re-assessment of the Archaeological Potential of Continental Shelves*. Located at: http://www.arch.soton.ac.uk/Research/Aggregates/shelve-report.htm.

Fairclough, G.J. 2001: Boundless Horizons – Historic Landscape Characterisation. *English Heritage Conservation Bulletin* 40(March 2001), 23–6.

Fairclough, G.J. & Rippon, S.J. (eds) 2002: *Europe's cultural landscape: archaeologists and the management of change*. EAC Occasional Paper no 2, Europae Archaeologiae Consilium and English Heritage, Brussels and London.

Fitch, S., Thomson, K., & Gaffney, V.L. 2005: Late Pleistocene and Holocene depositional systems and the palaeogeography of the Dogger Bank, North Sea. *Quaternary Research* 64, 185–96.

Fitch S., Gaffney, V. & Thomson K. 2007: Heritage management and the North Sea Palaeolandscapes. In Gaffney, V., Thomson K. & Fitch S. (eds): *Mapping Doggerland: The Mesolithic Landscapes of the Southern North Sea*. Archaeopress, Oxford. 105–8.

Fitch, S., Gaffney, V., & Ramsey, E. 2010: West Coast Palaeolandscape Pilot Project. Unpublished manuscript. Aggregates levy Sustainability Fund/ English Heritage.

Fitch *et al.* 2010: http://www.barch.bham.ac.uk/projects/pdfs/Humber%20REC.pdf.

Flemming, N.C. 2004: *Submarine Prehistoric Archaeology of the North Sea*. CBA Research Report 141.

Gaffney, C. & Gater, J. 2003: *Revealing the Buried Past: Geophysics for Archaeologists*. Stroud, United Kingdom: Tempus.

Gaffney, V., Fitch, S. & Smith, D. 2009: *Europe's Lost World: the rediscovery of Doggerland*. CBA Research Report. No. 160.

Gaffney, V., Thomson, K. & Fitch, S. (eds) 2007: *Mapping Doggerland: The Mesolithic Landscapes of the Southern North Sea*. Archaeopress. Oxford.

Jelgersma, S. 1979: Sea-level changes in the North Sea basin. *Acta Universitatis Upsaliensis, Symposia Universitatis Upsaliensis Annum Quingentesimum Celebrantis*, 2, 233–48.

Lambeck, K. 1995: Late Devensian and Holocene shorelines of the British Isles and North Sea from models of glacio-hydro-isostatic rebound. *Journal of the Geological Society, London*, 152, 437–48.

Louwe Kooijmans, L.P. 1970: *Mesolithic bone and antler implements from the North Sea and from the Netherlands*, Ber. Rijksdienst oudheidk. bodemonderz. 20/21, 27–73.

Kidd, G.D. 1999: Fundamentals of 3D seismic volume visualization. *The Leading Edge* 18, 702–9.

Kiden, P., Denys, L. & Johnston, P. 2002: Late Quaternary sea-level change and isostatic and tectonic land movements along the Belgian-Dutch North Sea coast: geological data and model results. *Journal of Quaternary Science* 17, 535–46.

Long, A. 2003: Removing seismic noise and increasing resolution in the East Java Sea: High Density 3D acquisition and processing. *PetroMin* 29(8), 26–9.

Long, D, Wickham-Jones, C.R. & Ruckley, N.A. 1986: A flint artefact from the northern North Sea. In Roe, D. (ed.): *Studies in the Upper Palaeolithic of Britain and North Western Europe*, BAR International Series, 296, 55–62.

Mueller, C., Luebke, H., Woelz, S., Jokisch, T., Wendt, G, & Rabbel, W. 2006: Marine 3-D seismic investigation of a late Ertebølle settlement in Wismar Bay (SEAMAP-3D case study). *Geophysical Research Abstracts*, 8, 08447, SRef-ID: 1607-7962/gra/EGU06-A-08447.

Novak, B. & Bjorck, S. 2002: Late Pleistocene–early Holocene fluvial facies and depositional processes in the Fehmarn Belt, between Germany and Denmark, revealed by high-resolution seismic and lithofacies analysis. *Sedimentology* 49, 451–65.

Oxley, I. & O'Regan, D. 2001: *The Marine Archaeological Resource*. Institute of Field Archaeologists Paper No. 4.

PGS, 2005: *Visualise the Big Picture with PGS MegaSurveys*. http://www.pgs.com/ upload/31163/data.pdf.

Peeters, H., Murphy, P.L. & Flemming, N. 2009: *North Sea Prehistory Research and Management Framework 2009*. Amersfoort: NSPRMF.

Peltier, W.R., Shennan, I., Drummond, R., & Horton, B. 2002: On the postglacial isostatic adjustment of the British Isles and the shallow viscoelastic structure of the Earth. *Geophysical Journal International* 148(3), 443–75.

Powlesland, D. 2010: *Review of Gaffney V., Fitch S. & Smith D. 2009: Europe's Lost World: the rediscovery of Doggerland*. CBA Research Report. No. 160.

Praeg, D. 2003: Seismic imaging of mid-Pleistocene tunnel-valleys in the North Sea Basin – high resolution from low frequencies. *Journal of Applied Geophysics* 53, 273–98.

Reid, C. 1913: *Submerged Forests*. Oxford.

Roberts, P. & Trow, S. 2002: *Taking to the Water: English Heritage's Initial Policy for The Management of Maritime Archaeology in England*. Unpublished policy statement, English Heritage, (http://www.englishheritage. org.uk/server/show/nav.8381).

Salomonsen, I. & Jensen, K.A. 1994: Quaternary erosional surfaces in the Danish North Sea. *Boreas* 23, 244–53.

Shennan, I., Bradley, S., Milne, G.A., Brooks, A., Basset, S.E. & Hamilton, S. 2006: Relative sea-level changes, glacial isostatic modelling and ice-sheet reconstructions from the British Isles since the Last Glacial Maximum. *Journal of Quaternary Science* 21(6), 585–99.

Shennan, I & Horton, B. 2002: Holocene land- and sea-level changes in Great Britain. *Journal of Quaternary Science* 17, 511–26.

Shennan, I., Lambeck, K., Flather, R., Horton, B., McArthur, J., Innes, J., Lloyd, J., Rutherford, R., & Wingfield, R. 2000: Modelling western North Sea palaeogeographies and tidal changes during the Holocene. In Shennan, I. & Andrews, J. (eds): *Holocene Land-Ocean Interaction and Environmental Change around the North Sea*. London: Geological Society, 299–319.

Stringer, C. 2006: *Homo Brittanicus: The Incredible Story of Human Life in Britain*, Penguin Books, London.

Velegrakis, A.F., Dix, J.K. & Collins, M.B. 1999: Late Quaternary evolution of the upper reaches of the Solent River, Southern England, based upon marine geophysical evidence. *Journal of the Geological Society, London* 156, 73–87.

Ward, I. & Larcombe, P. 2008: Determining the preservation rating of submerged archaeology in the postglacial southern North Sea: a first-order geomorphological approach. *Environmental Archaeology* 13(1), 59–83.

Waddington, C. (ed.) 2007: *Mesolithic Settlement in the North Sea Basin: A Case Study from Howick, North-East England*. Oxford, Oxbow.

Waddington, C. & Pedersen, K. (eds) 2007: *Mesolithic Studies in the North Sea Basin and Beyond: Proceedings of a Conference Held at Newcastle in 2003*. Oxford, Oxbow.

Wenban-Smith, F. 2002: *Palaeolithic and Mesolithic Archaeology on the Sea-bed: Marine Aggregate Dredging and the Historic Environment*. Wessex Archaeology, http://www.wessexarch.co.uk/files/projects/BMAPA-Protocol/mads2.pdf.

16 | Aerial archives for archaeological heritage management: The Aerial Reconnaissance Archives – a shared European resource

Lesley Ferguson

Abstract: Now based in the Royal Commission on the Ancient and Historical Monuments of Scotland (RCAHMS) in Edinburgh, the military aerial reconnaissance imagery contained within The Aerial Reconnaissance Archives (TARA) provides an exceptional record of landscapes and of past human activity captured from the 1940s to the 1990s across mainland Europe. This paper provides a brief overview of the content of TARA from an archaeological perspective and highlights some of the discoveries and achievements since its transfer to RCAHMS in 2008. The contribution will also look forward and explore some of the possibilities and partnerships which are essential for future development, as well as highlighting some of the challenges to be faced in order to open up access for discovery and exploration to the archaeological community and beyond.

Introduction

Aerial photographs are the original form of archaeological remote sensing, unless one includes the elevated perspective afforded by a horse as used by William Stukeley (1687–1765) on his extensive antiquarian tours of discovery (Piggott 1976). The comprehensiveness of the viewpoint has meant that for more than a century the aerial perspective has established itself as a most effective means of recording monuments and documenting landscape, though much of the emphasis of this work has been on observer-directed aerial reconnaissance in light aircraft (e.g. British Academy 2001) or, more recently, on satellite imagery (e.g. Parcak 2009). Over the last two decades, however, there has been a growing recognition of the importance of historic aerial photographs which span the 20th century (e.g. papers in Bewley & Rączkowski 2002; Cowley *et al.* 2010a). Although still a largely unexplored resource, this imagery has enormous potential for the primary discovery and recording of previously unrecognised archaeological sites, for the documentation of already known sites and for the characterisation of the wider cultural landscape. This paper presents an outline of the contents and potential of The Aerial Reconnaissance Archives (TARA) for cultural resource management across Europe. This imagery, which has its origin as a weapon of war in a continent torn apart during World War II, is a remarkable resource for the study of European landscapes (and beyond).

The Aerial Reconnaissance Archives

In 2008 the world's second largest collection of historic aerial imagery was transferred to the care of the Royal Commission on the Ancient and Historical Monuments of Scotland (RCAHMS) in Edinburgh in a joint initiative with The National Archives (UK) and Keele University to secure its long-term future. Known as TARA (The Aerial Reconnaissance Archives), the collection contains millions of images taken primarily for military intelligence purposes dating from the earliest days of World War II (Figure 16.1) through to recently declassified UK Government imagery up to the late 1980s. This integration into RCAHMS, a heritage organisation with considerable experience in using and generating aerial imagery, provides an opportunity to widen access of the collection to a pan-European audience of fellow professionals and to explore avenues for cooperation to exploit this epic archive. This paper will summarise the origins and content of the collection and through examples, assess some possibilities for its use within heritage management.

Background to TARA

The quantity of aerial reconnaissance images taken, especially during World War II, is astonishing and although there are in excess of 11 million images surviving in TARA, many millions have been destroyed (see Cowley *et al.* 2010b, 3–4 for discussion of patterns of discard). Individual campaigns or monitoring of a specific place could generate tens of thousands of images, as at Brest in the autumn and winter of 1941 when the monitoring of ship movements was particularly critical (Babington Smith 1957). By 1945 the average daily intake of new negatives and prints was 85,000, and by the end of the war, millions of photographic images of the world had been acquired by the air forces of the UK, Canada, South Africa and the US at the home of the Allied Central Interpretation Unit (ACIU) at Medmenham, Buckinghamshire, England. This was where all wartime allied intelligence was centralised, and where the imagery analysis informed the planning or assessed the impact of most military operations for the duration of the war. As noted by the photographer, Patricia Macdonald (1991), *'the history of aerial photography is in some senses basically a history of knowledge as power'*. Some 5.5 million images survive

Figure 16.1: Reconnaissance flight, Garigliano Plain, Italy, 1940s. From an album of aerial photographs of Italy and North Africa compiled by Dr Kenneth Steer, later Chief Executive, RCAHMS. *TARA STEER 040, reproduced under licence. RCAHMS: The National Collection of Aerial Photography.*

from these wartime accumulations at ACIU and these images are at the heart of TARA, representing the most accessible and instantly informative of all the distinct collections. As well as providing a window onto key moments in 20th century history (Figure 16.2), the images and accompanying sortie plots document in detail where and when military reconnaissance sorties were undertaken from 1939 to 1945. Once vitally important images illustrative of a current or recent situation, it is their historic value that is now of crucial significance to archaeologists and historians. In the words of John Bradford *'it should give sober satisfaction that photographs for which lives were risked, and lost, continue to be of value many years afterwards, making fresh contributions to knowledge'* (Bradford 1957).

It was also at the Interpretation Unit at Medmenham that many scholars and academics honed their observation and interpretation skills, and were later influential in their post-war careers in developing the use of aerial photography. Kenneth St Joseph of Cambridge University, for example, was one of those who eloquently demonstrated the potential of aerial imagery for peacetime research. Indeed, in 1947, Kenneth Steer of RCAHMS, a former intelligence officer, was commenting on the *'epoch-making importance for archaeological research'* of the military National Air Survey of Great Britain due for completion that year (Steer 1947). Following the conclusion of the war in 1945, many of these former wartime photographic interpreters, then in academic or professional positions, advocated the wider use of the aerial photographs for teaching and research purposes. However, it was not until the early 1960s that the surviving ACIU images were transferred to the Geography Department of the College of North Staffordshire, later Keele University, where Professor Stanley Beaver intended their use for educational purposes.

Keele University was established as the official place of deposit for all government non-UK aerial imagery and over the next four decades, three further collections were deposited: the Mediterranean Allied Photo Reconnaissance Wing (MAPRW) archive of prints of southern and central Europe taken between 1939 and 1945 arrived in the 1980s; captured Luftwaffe reconnaissance photographs of mainly eastern Europe were declassified and released in the early 1990s; and imagery dating from 1938 to the late 1980s from the Joint Air Reconnaissance Intelligence Centre (JARIC), the main provider of photographic intelligence to the UK Ministry of Defence and wider intelligence community, was deposited between 2004 and 2008. Full details of the distinct collections are available at aerial.rcahms.gov.uk.

Current uses of TARA

Through the years extensive work has been undertaken to create and computerise the limited indices to the imagery and facilitate the use of the archives. Bomb disposal companies are currently the largest users of the imagery, identifying quantities of unexploded wartime ordnance in the historic aerial photographs prior to modern construction or development across many areas of Europe. Tragically large quantities of unexploded ordnance remain one of the legacies of World War II with a number of deaths each year from previously unlocated bombs. Beyond this there has only been limited use by a wide and varied number of professionals, such as historians, archaeologists, geographers, environmentalists, lawyers and computer games programmers. Much remains to be done to open up access to this incredible information resource to the academic community as well as to those researching family history, teachers in a range of curriculum subjects, military historians, town and country planners, geologists and conservationists across Europe and indeed, many other parts of the world. The potential of the collection for use in all these areas is completely undeveloped.

TARA provides an unprecedented record of whole areas of Europe prior to changes in agricultural practices, post-war reconstruction and patterns of industrialisation, and adds an historic and landscape dimension to the documentation of the heritage.

Various recent papers highlight the potential of historic aerial collections including TARA (Cowley & Ferguson 2010) and one of the objectives for RCAHMS is to widen access and expand the use of the historic imagery within the heritage sector across Europe.

Opening up access to TARA

As part of a programme to improve access to TARA by reaching a worldwide audience a new website has been created (aerial.rcahms.gov.uk) to provide detailed descriptions of the component archives, summaries of the available services, and access to a steadily growing number of downloadable geo-referenced images. To date, some 12,000 images are visible, free to the researcher at postcard size but for a small subscription, at a larger scale. Cataloguing to international standards for archival description (ISAD(G)) and digitisation is ongoing and new imagery is added to the website on a daily basis. On the day of the launch of the new website, articles appeared in newspapers across the globe and some 16 million hits were made on the site, illustrating the interest. At the same time as the website was launched, a new public research facility was opened in the RCAHMS premises in Edinburgh where, by appointment, visitors can freely access all the flight plots and images in the ACIU archive. A paid search service operates for researchers who cannot travel to Edinburgh but it is hoped that finding aids will be available on-line in due course.

There are however many barriers to the use of the historic aerial imagery and one of these is the scale of the collection itself. It has taken in excess of ten years to make the ACIU archive accessible and it will be decades before the MAPRW, Luftwaffe and JARIC imagery are similarly available. With no comprehensive finding aids or basic geographical location information, identifying imagery can only be achieved once flight plots have been catalogued. To help accelerate this, RCAHMS must develop partnership projects with colleagues across central and eastern Europe, in particular, to undertake research and help to create the indices so essential for wider appreciation and use of the collection.

Solutions for the plotting and public dissemination of UK photographs cannot readily be extended to the worldwide imagery. Cataloguing of each and every image is just not feasible, nor is the identification on an individual basis of places or features. The geographical centre-point and the automatic calculation of an approximate extent of each image is achievable and will be the standard level of indexing applied to the collection in the future. But with developments to the website it is hoped that additional keywords or descriptions can be contributed by researchers worldwide to improve the searchable information available about individual images.

Using TARA for heritage management

The surviving imagery provides high quality geospatial data of whole areas of Europe from predominantly vertical coverage and was generally a critical

Figure 16.2: Caen, France. The historic town of Caen was devastated following the D-day landings. *TARA ACIU RAF 106G/3170/11 2 October 1944, reproduced under licence. RCAHMS: The National Collection of Aerial Photography (ACIU Archive).*

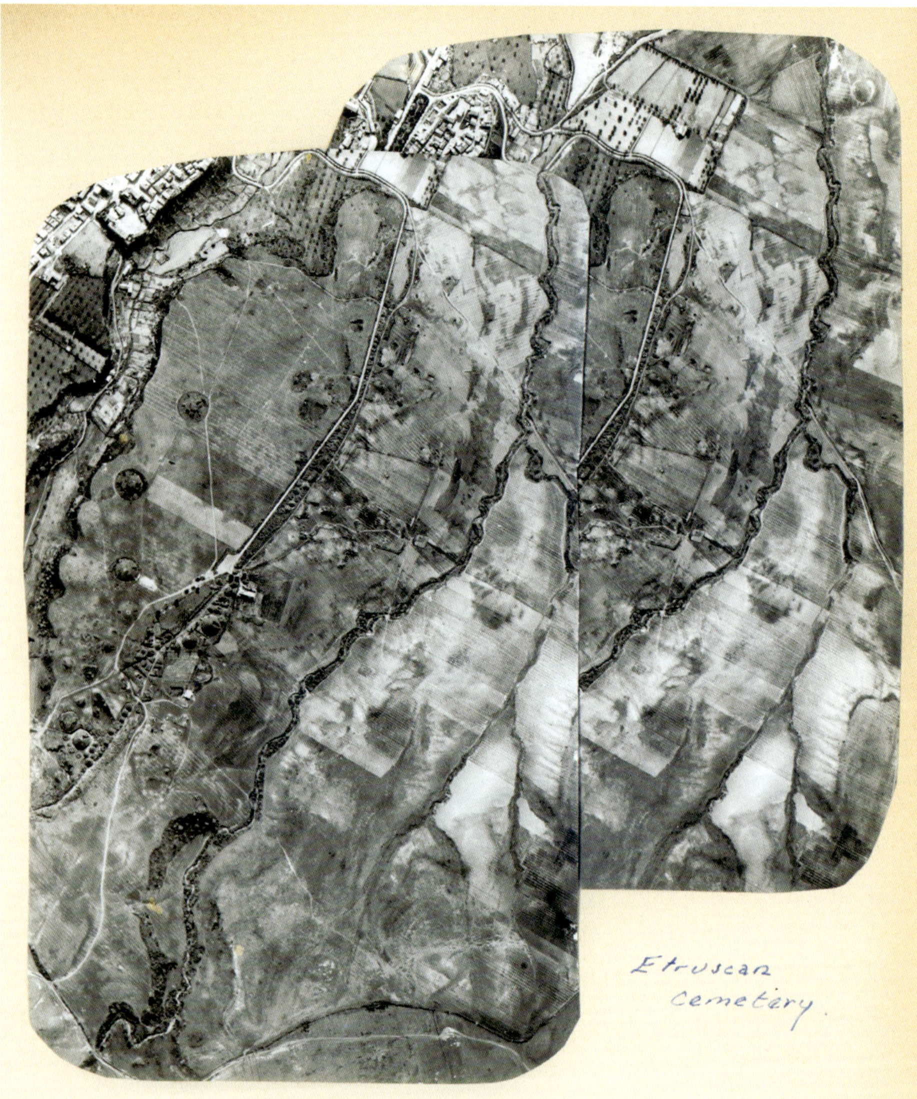

Figure 16.3: Caere, Banditacchia, Italy. Two photographs have been positioned together for stereoscopic viewing of the extensive relict landscape of upstanding and ploughed out Etruscan tumuli. From an album of aerial photographs of Italy and North Africa compiled by Dr Kenneth Steer, later Chief Executive, RCAHMS. *TARA STEER 061, reproduced under licence. RCAHMS: The National Collection of Aerial Photography.*

component of the military intelligence gathering process, such as determining the positions of potential targets or providing the most up-to-date information about a particular place. Despite the operational difficulties inherent in undertaking clandestine aerial photography in wartime, the basic requirement for resulting images was quality, to allow interpretation, and scale, to facilitate the accurate plotting and preparation of scaled plans. New wartime interpreters were taught that a vertical air photograph was not a picture but a precise mathematical document from which maps and plans could be made or details analysed, attributes which are equally applicable and relevant today to the professional archaeologist, for ease of integration with other data to map the archaeological and historic landscape (Figure 16.3). Of course, some countries have their own large collections of aerial photographs, of which Italy is an excellent example (Cerraudo & Shepherd 2010), but for many countries TARA holdings are likely to provide the earliest aerial record.

Snapshots in time

In some cases there is blanket coverage of whole countries or regions where outdated maps meant that for planning purposes the military required an accurate understanding of the nature of the landscape, the location and extent of settlements, industrial plant or the road and railway network. One such example was Belgium where in anticipation of the invasion by Germany, comprehensive aerial reconnaissance was undertaken of the entire country in a series of covert operations (Babington Smith 1957). Other sorties, sometimes repeated, targeted the coastline, river valleys, key settlements or strategic positions and these locations, crucial in 20th century warfare, were sometimes just as important in historic or even prehistoric times (Figure 16.4).

Often the earliest aerial images of areas, the photographs provide a 'snapshot in time' of whole landscapes, long-recognised monuments and historic towns (Figures 16.5 & 16.6), providing a dated benchmark from which change can be documented or previously unrecognised sites recorded (Figure 16.7). In the decades following the conclusion of World War II, wholesale 'improvements' to agricultural practices in some countries resulted in the widespread destruction of historic landscapes and levelling of monuments. Although aerial reconnaissance programmes targeted the recording of individual sites, as for example in France (Agache 1975), it is the historic vertical aerial photographs which provide the detail necessary to

Figure 16.4: Due to its location to the north of Dieppe, the oppidum of Bracquemont was photographed 'by chance' in this US sortie. TARA ACIU US7GP 2196 3068 5 July 1944, reproduced under licence. *RCAHMS: The National Collection of Aerial Photography (ACIU Archive).*

appreciate and map the landscape changes, crucial information that is not available from any other source. For post-war expansion of towns and cities or urban redevelopment, the aerial perspective provides a visual statement of fact, often documenting building or industries which have been constructed and demolished within the space of fifty or sixty years. For larger upstanding sites, linear monuments, such as old roads or defences, geographically spread complexes, and designed gardens or landscapes where planting layouts can have a relatively short lifespan (Figure 16.8), the historic vertical imagery can be especially valuable, providing a more comprehensive overview of the features than can be obtained through modern oblique photographic survey alone.

Archaeological prospection and documentation

With numerous sorties taken at different times of the year and at varying heights, it is not surprising that archaeological sites feature in many of the photographs, including previously unidentified cropmark sites (Bradford 1957; Oltean & Hanson 2007) enabling, in some cases, the reassessment of the archaeology of whole areas. Although the suitability and quality of the photographs varies, it is likely that sites still await discovery, many which since their capture on film may already have been destroyed but whose existence is still key to understanding the archaeological landscape.

An inventory or detailed list of all historic monuments, destroyed or extant, is an essential requirement for all aspects of heritage management, from the basic knowledge base through to preservation. Many countries, such as France or Denmark, had started the compilation of a national inventory of archaeological and historical monuments in the 19th century but for others, basic information gathering is still in its infancy and as with any academic research, historic aerial photographic collections are just one tool available to archaeologists alongside an analysis of documentary sources, maps, fieldwork and excavation. The sheer size of the collections and wide geographical spread precludes any kind of rapid assessment but key to discovery within the collection is a detailed knowledge and understanding of each country or area, its terrain, the current state of the archaeological record and for cropmarks, an appreciation of any regional nuances in their formation, as demonstrated through integrated research projects such as the English Heritage National Mapping programme which utilise the historic imagery to provide detail of surviving or destroyed sites and landscapes (Horne 2009; Winton & Horne 2010; Horne this volume).

Characterising landscape

Beyond this inventory of monuments where sites are usually depicted as dots or lines on a map, aerial collections provide the wider geographical context essential for the appreciation and analysis of the chronological depth exhibited in the landscape. With increased awareness of this historically significant cultural landscape and its exclusion from consideration in the decision making process during development or planning, new methodologies were developed in Britain to analyse the historic content of the wider

Figure 16.5: Brielle, Netherlands, the historic town is contained within massive ramparts and water defences. *TARA ACIU RAF 106G/2785/3091 11 September 1944, reproduced under licence. RCAHMS: The National Collection of Aerial Photography (ACIU Archive).*

Figure 16.6: Mostar, Bosnia and Herzegovina. An oblique view of the 16th century bridge which was destroyed by Bosnian Croat artillery in November 1993. The bridge has now been fully reconstructed and reopened in 2004. *TARA ACIU RAF 235/279/518/74 21 January 1944, reproduced under licence. RCAHMS: The National Collection of Aerial Photography (ACIU Archive).*

landscape and assess how people exploited and changed their physical environment. Unimaginable until relatively recent years, modern technology and GIS systems enable vast quantities of geographical information to be combined to provide an overview of the historic use of the land. Aerial photographs are a crucial tool for such landscape characterisation research projects and have become one of the standard resources used in the last two decades (Herring 1998; RCAHMS 2000, 2001). Modern aerial imagery can provide a benchmark against which surviving historic elements can be mapped and preservation pressures identified, while historic imagery can add an additional chronological dimension and identify elements lost through fifty years of modern development. Now a standard analysis tool in many areas, historic landscape characterisation has been successfully adapted to suit different places and available sources of information, as recently demonstrated by Turner and Crow in two different study areas in Thrace and Naxos (2010). These approaches are important in the development of broader landscape-scale thinking by archaeologists, which are required for the implementation of the European Landscape Convention (ELC).

Documenting what has been lost

Beyond archaeological uses, monuments and their accompanying communities which bear witness to a glorious industrial past have often been destroyed without any recognition of their heritage value. Historic aerial photographs can capture some of this, the remnants of rural industries, such as windmills, or the extensive industrial complexes depicted in

Figure 16.7: Lucera, Italy, Neolithic settlement, recorded by chance on an aerial photograph of the aerodrome. *RAF 680/234 4057 21 May 1945, reproduced under licence. RCAHMS: The National Collection of Aerial Photography (St Joseph Collection).*

Figure 16.8: Lustheim, Bavaria, April 1945. Adjacent to an airfield, the designed landscape at the palace at Lustheim has been spared the destruction of the bombs. It is one of the rare preserved baroque gardens in Germany and a comparison with modern photographs illustrates the changes that have taken place in the garden design. *TARA ACIU RAF 683/1149/4070 25 April 1945, reproduced under licence. RCAHMS: The National Collection of Aerial Photography (ACIU Archive).*

wartime which have now long gone or changed beyond recognition. To the industrial historian it is not just the physical structures that are of interest but the format and shape of the surrounding architecture can reveal information about the manufacturing processes undertaken within.

Conflict archaeology and event history

As a resource for military historians, aerial photographs can be dynamic documents and illustrations of modern history in the making as, for example, it is possible to see the crowds lining the streets to welcome the liberating forces into Copenhagen in May 1945 (Figure 16.9). As one of the functions of aerial reconnaissance was to determine whether military activity had been successful or had failed, imagery often illustrates events as they happened (e.g. Montecassino – Cerraudo & Shepherd 2010). Events that were to become defining moments in world history, such as the D-day landings, emerge stage by stage in the aerial photographs as 'Operation Overlord' happened, and combined with oral and documented history, add a landscape dimension not possible from other sources. There is also a human aspect to the imagery, not necessarily immediately apparent visually, but represented indirectly in the images which are capable of interrogation by historians, as demonstrated to great effect by analysis of the photographs of the Auschwitz concentration and extermination camps (Uziel 2010). Wider uses include the location of crashed wartime aircraft, locating war graves (Abicht 2010) or adding to the information and material assembled by family historians, eager to supplement their historical and chronological knowledge of places into a geographical context to understand more of their ancestors life or death. With recognition in recent years that the remnants of the two World Wars merits analysis and protection, various studies have been undertaken to document surviving relics (Barclay 2005; Bacilieri & Thomas 2010). As part of the research ground survey, analysis of contemporary military documents and some oral history has been examined but close examination of the historic aerial photographs has often provided the most comprehensive visual record of the sites at a time they were still in operation or shortly after their abandonment.

Conclusion

Historic aerial photographic collections have much to offer the modern heritage professional, providing in many cases the earliest aerial documentation of the European landscape from World War II onwards before the post-war modernisation of agricultural practice, urban growth and modern development. They provide a benchmark against which change can be measured, studied and understood, and in some areas, archaeological sites identified, all within a

Figure 16.9: Copenhagen, May 1945. History happening before your eyes: crowds line the street to welcome the allies to liberate the city. *TARA ACIU RAF 106G/LIB17/4103 12 May 1945, reproduced under licence. RCAHMS: The National Collection of Aerial Photography (ACIU Archive).*

geographical and landscape context. As well as offering considerable potential as an information resource for heritage management, we have to also remember that these aerial archives are themselves actually also one of Europe's cultural assets. Much of the archive is now 70 years old and after many years of use, requires active conservation and ongoing protection from the ravages of time. With only half of the collection accessible and the content of the other half not fully understood, there are decades of research, cataloguing and digitisation ahead to realise the full potential of TARA.

References

Abicht, M.J. 2010: Using Wartime aerial photographs to locate lost grave sites. In Cowley, D.C., Standring, R.A. & Abicht, M.J. (eds): *Landscapes through the Lens. Aerial Photographs and Historic Environment.* Oxbow, Oxford. 263–5.

Agache, R 1975: Aerial reconnaissance in northern France. In Wilson, D.R. (ed.) 1975: In *Aerial reconnaissance for archaeology.* CBA Research Report No. 12. 70–80.

Babington Smith, C. 1957: *Evidence in Camera the Story of Photographic Intelligence in World War II.* London, Chatto and Windus.

Bacilieri, C. & Thomas, R.J.C. 2010: Ephemeral coastal defences of the two world wars in North East England. In Cowley, D.C., Standring, R.A. & Abicht, M.J. (eds): *Landscapes through the Lens. Aerial Photographs and Historic Environment.* Oxbow, Oxford. 89–98.

Barclay, G.J. 2005: The Cowie Line: a Second World War 'stop line' west of Stonehaven, Aberdeenshire. *Proceedings of the Society of Antiquaries of Scotland* 135 (2005), 119–61.

Bewley, R. & Rączkowski, W. (eds) 2002: *Aerial Archaeology. Developing Future Practice.* NATO Science Series I: Life and Behavioural Sciences 337. Amsterdam.

Bradford, J. 1957: *Ancient Landscapes. Studies in Field Archaeology.* London.

British Academy, 2001: *Aerial Survey for Archaeology. Report of a British Academy Working Party 1999.* Compiled by Robert Bewley, London, British Academy.

Cerraudo, G. & Shepherd, E.J. 2010: Italian Aerial Archives: Case Study. In Cowley, D.C., Standring, R. A. & Abicht, M. J. (eds): *Landscapes through the Lens. Aerial Photographs and Historic Environment.* Oxbow, Oxford. 237–46.

Cosgrove, D. & Fox, W.L. 2010: *Photography and Flight.* London, Reaktion Books.

Cowley, D.C. & Ferguson, L.M. 2010: Historic aerial photographs for archaeology and heritage management. In Campana, S., Forte, M. & Liuzza, C. (eds): *Space, Time, Place. Third International Conference on Remote Sensing in Archaeology, 17th–21st August 2009, Tiruchirappalli, Tamil Nadu, India.* BAR -S2118, 97–104.

Cowley, D.C., Standring, R.A. & Abicht, M.J. (eds) 2010a: *Landscapes through the lens. Aerial Photographs and Historic Environment.* Oxbow, Oxford.

Cowley, D.C., Standring, R.A. & Abicht, M.J. 2010b: Landscapes through the lens: an introduction. In Cowley, D.C., Standring, R.A. & Abicht, M.J. (eds) 2010a: *Landscapes through the lens. Aerial Photographs and Historic Environment.* Oxbow, Oxford. 1–6.

Herring, P. 1998: *Cornwall's historic landscape: presenting a method of historic landscape assessment.* Truro, Cornwall County Council.

Horne, P.D. 2009: *A Strategy for the National Mapping Programme.* English Heritage.

Macdonald, P. 1991: Aerial Perspectives. *Photoresearcher,* No. 3, European Society for the History of Photography.

Oltean, I. & Hanson, W. 2007: Reconstructing the archaeological landscape of Southern Dobrogea: integrating imagery. In Ehlers, M. & Michel, U. (eds): *Remote Sensing for Environmental Monitoring, GIS Applications, and Geology VII, Proceedings of SPIE* 6749, No. 674906.

Parcak, S.H. 2009: *Satellite Remote Sensing for Archaeology.* London, Routledge.

Piggott, S. 1976: *Ruins in a Landscape. Essays in Antiquarianism.* Edinburgh.

RCAHMS, 2000: *The Historic Landscape of Loch Lomond and the Trossachs.* Edinburgh.

RCAHMS, 2001: *The Historic Landscape of the Cairngorms.* Edinburgh.

Steer, K. 1947: Archaeology and the National Air-photograph Survey. *Antiquity* 81(xxi), March 1947, 50–3.

Turner, S. & Crow, J. 2010: Unlocking historic landscapes in the Eastern Mediterranean: two pilot studies using Historic Landscape Characterisation. *Antiquity* 84, 216–29.

Uziel, D. 2010: The Holocaust from above: Auschwitz imagery and beyond. In Cowley, D.C., Standring, R.A. & Abicht, M.J. (eds): *Landscapes through the Lens. Aerial Photographs and Historic Environment.* Oxbow, Oxford. 267–79.

Winton, H. & Horne, P. 2010: National archives for national survey programmes: NMP and the English Heritage aerial photograph collection. In Cowley, D.C., Standring, R.A. & Abicht, M.J. (eds): *Landscapes through the Lens. Aerial Photographs and Historic Environment.* Oxbow, Oxford. 7–18.

PART 4

Using remote sensed data: interpretation and understanding

Placing monuments such as these plough-levelled Iron Age forts at Aytonlaw in the Scottish Borders on record is only the start of a process of interpretation leading to understanding and knowledge that can support effective management. *SC773000, reproduced under licence. RCAHMS: The National Collection of Aerial Photography.*

17 | Remote sensing for the integrated study and management of sites and monuments – a Central European perspective and Czech case study

Martin Gojda

Abstract: Recent innovations in remote sensing techniques have been profoundly changing the possibilities of what is traditionally termed aerial archaeology. These changes are discussed and reflected not only in the West but also in Central European countries, most of which entered aerial archaeology as late as the 1990s after the pan-European collapse of communist regimes. Legal and administrative barriers on flying, taking and publishing aerial photographs for whatever purpose in most of them were responsible for the delay in launching continuous aerial survey programmes in that part of Europe. In spite of this delay activities in remote sensing during the last two decades have enormously enriched the cognitive and methodological capacity of archaeology in the study of past human settlements and landscapes, and also helped to open discussion on theoretical issues. This contribution illustrates how an ongoing aerial archaeology programme in Bohemia has influenced ideas on past settlement forms and dynamics, and how important aerial photography is for the monitoring and documentation of Czech cultural heritage. For the future in Bohemia testing of ALS potential for a large-scale mapping of archaeological landscapes will be of major importance (state-funded pilot project 2010–11) and subsequent acquisition of LiDAR of the whole Czech Republic.

Introduction

Current archaeological practice in Central European countries of the former Soviet bloc has been influenced by progressive acceleration of social processes caused by the collapse of communism. One of the most important consequences of this is large-scale impact on both urban environments and rural landscapes through a boom in construction activity, which represents an unparalleled threat to archaeological heritage. As a result developments in archaeology in the heart of Europe at the end of the 20th century have been largely driven by factors external to the professional community, and in many cases beyond their control. As a result archaeologists are forced to choose research themes, approaches and strategies with account to these external factors – not only in field-project strategies but also in data management and storage, their use, analysis, interpretation, and, last but not least, in making them available by publications to the wider public.

In the face of the continuing large-scale threats to the archaeological heritage since the 1970s methodologies have had to develop in a dynamic way. Problems concerning the strategy of rescue projects and approaches to the most effective evaluation of limited time and budget were particularly pressing. Increasingly, field projects (excavations) of threatened sites have applied probability and sampling strategies. The necessity of implementing sampling strategies in excavations has been greatly assisted by increasing support of non-invasive methods. Although the application of non-destructive methods for data collection is widely understood as currently the most effective means of generating archaeological heritage protection policy, as well as a useful tool for research in some sub-disciplines (such as spatial- and landscape archaeology), it is excavation which continues to dominate archaeological fieldwork in the post-communist era throughout former Soviet bloc countries.

At the same time large-scale improvements in science and technology have influenced cognitive process and archaeological methods, especially in survey techniques. In the last two decades a huge increase in the quantity and quality of data for archaeological study of the human past is largely due to progress in what is generally termed *remote sensing (of the Earth)*. As a result one of the most effective survey disciplines ever applied in archaeology – aerial survey (reconnaissance) and photography, most commonly aerial archaeology – has been influenced dramatically, up to the point that even the term is now considered inappropriate by some scholars.

Thus, the questions to be answered in this paper are: how has the application of aerial survey in Central Europe, specifically in Bohemia, contributed to the current view of ancient settlement dynamics on the one hand and, how has this discipline supported the process of monitoring and documentation of sites and monuments? To begin answering these questions, this paper will open with a brief overview of development from traditional aerial archaeology to more complex remote sensing applications in current archaeology and past landscape studies.

Early developments: who set the agenda for aerial archaeology?

In addressing the issue of the origins of aerial archaeology, it is appropriate to ask three questions:

1. Which part of the world saw the first aerial survey campaign aimed at identifying unknown buried archaeological sites?
2. Who most markedly shaped the content of the field in its beginnings and who influenced most seriously the form of aerial archaeology from the theoretical and methodological point of view?
3. Who, using aerial survey, collected information that had a fundamental impact on the archaeological knowledge of a historical landscape and peoples living there in the past?

The answers to these questions establish that, in its beginnings, the field was formed by several personalities whose general contribution to aerial archaeology was, besides their own abilities and knowledge, influenced significantly by the technical potential of their equipment (especially aircraft and cameras) as well as by geographical characteristics of the landscapes in which they worked. Thus, two very different and distant geographical areas became the cradle of aerial archaeology: the desert and dry steppe landscapes of the Near East (especially East Mediterranean, Sinai, Levant, Mesopotamia), later also the more distant areas of the Middle East (Iran) and North Africa on the one hand, and Western Europe (England) on the other (Bewley 2005; Deuel 1969; Downey 1980; Rączkowski 2002, 28–42; Musson 2005).

A figure with undeniable primacy in the practice of aerial archaeology was O. G. S. Crawford, the founder of *Antiquity*. He was the first to publish his discoveries and to define through them the principles underlying the identification of archaeological sites and features in the field. Simultaneously, he introduced to specialized literature the procedures of gaining and processing field data (especially Crawford 1924; Crawford & Keiller 1928; for analytical evaluation of Crawford by a non-British scholar see Rączkowski 2002, 42–61). However, this happened about twenty years after aerial imaging had begun to take an important part in the discovery and photographic documentation of architectural and archaeological monuments. The development of methodological owes much to the French scholar P.A. Poidebard. The first chapter of his 1934 work is vitally important as it is devoted to the methods of aerial survey in archaeology as he developed them in the specific environment of the arid desert in which he had been working (e.g. the technique of backlight imaging from low altitudes or ground survey of sites immediately after their identification, Poidebard 1934). Crawford himself considered Poidebard the first to have made a creative contribution to the general development of the methodology of aerial archaeology (Kennedy & Riley 1990, 51–63).

In Central European countries (such as Hungary, Poland, Germany and former Czechoslovakia – Czech and Slovak Republics since 1993) aerial archaeology in the period between World War I and World War II was limited to aerial photography of prehistoric earthworks (such as hillforts), medieval standing and ruined buildings, and sites under excavation (i.e. Biskupin in Poland, which was photographed regularly from balloons in 1935–9). In Germany systematic aerial photography of sites and monuments had been taken since 1928 by the Hansa Luftbild company and since 1935 by the Luftwaffe (Braasch 1997; Kobylinski 2005; Krasnodębski 2005).

A promising turning point for large-scale inclusion of aerial survey into central European archaeologies could have been an invitation in 1938 to Crawford by the German Lilienthal Company to Berlin. His lecture became a basis for a book *Luftbild und Vorgeschichte* in which Crawford's achievements (and early German photographs) had the potential to greatly influence scholars in Central Europe who, at that time, were much more aware of German journals and books rather than English. Unfortunately, the start of World War II terminated any potential in this area. In Austria too attempts were made in the 1920s and 1930s to photograph and interpret aerial images and transform sites documented on them into maps (Doneus *et al.* 2001, 12–3). The very first aerial photos taken by Czech archaeologists are of sites excavated by themselves and by American expedition between 1929 and 1932. Although there were a few attempts to undertake airborne archaeological prospection it took six decades until Soviet bloc countries were able to embark on regular aerial survey.

The last two decades: new challenges

Central Europe witnessed the beginnings of systematic long-term aerial reconnaissance and photography for archaeology in Germany (I. Scollar) and in Austria (G. Spitzer and H. Friesinger) in the early 1960s (Deuel 1969; Doneus *et al.* 2001). A few attempts to organize survey flights are detectable also in former Czechoslovakia, but pilot projects – one in Moravia and one in Slovakia – were not launched until as late as mid-1980s (Bálek 1995; Kovárník 1995; Kuzma *et al.* 1996; Visy 1997).

When the Soviet bloc collapsed in 1989/90 a new era in the history of aerial archaeology started. Large territories became the target of local scholars who wanted to explore potential of these 'virgin' areas for the identification of buried sites and landscapes by means of aerial prospection. Some of them soon invited specialists from western countries (UK, Germany, France) in which the discipline had a long tradition to assist in this development. Of special importance for Central European beginners was the assistance of former military pilot and aerial prospector/photographer Otto Braasch from Germany (winner of the 2004 EAA's annual Heritage Prize). Communication between scholars from all over Europe has proved extremely fruitful (Gojda 1997), and there has been an almost continuous chain of international projects supported mostly by EU programmes (such as RAPHAEL and Culture 2000) over the last fifteen years. Summer courses in aerial archaeology, seminars, workshops and exhibitions have been regularly organised, publications produced (Bewley & Rączkowski 2002; Burgeois &

Meganck 2005; Gojda 2007; Nowakowski *et al.* 2005) and a documentary film produced (http: //www.kar.zcu.cz/videoarchiv.php). These have all helped to spread aerial archaeology among professionals, university students and the wider public.

Also significant for the development and spread of aerial archaeology in Europe during this period have been efforts to encourage (through the Aerial Archaeology Research Group (AARG) and European Association of Archaeologists (EAA) the lifting of restrictions on flying and aerial photography (Braasch 2002).

Apart from the many aerial archaeological projects in Central European countries focused on the identification of unknown features, sites and landscapes, on increasing their number in national sites and monuments records, and on detecting new site types (for a high-standard project of this type see Schwarz 2003), a few more complex projects were carried out. These have testified the potential of combining various non-destructive and traditional field methods, including small-scale excavations of sites detected via aerial survey, and study of vertical aerial/satellite photographs. This methodology has usually been applied to projects studying regional settlement history, structure and dynamics (for Bohemia see next section). Finally, theoretical issues of aerial archaeology cannot be omitted. In the past this was an under explored area, but currently, theory is being discussed more frequently. This is most significantly due to Polish scholars from the University of Poznań, especially W. Rączkowski, author of the only monograph on methodology and theory in aerial archaeology (Rączkowski 2002, see also 2005; Zuk 2005; Palmer 1989; Brophy 2005; Brophy & Cowley 2005).

Thus, over the last two decades the hopes of professional communities in the wake of the fall of the Iron Curtain seem to have been fulfilled. In the meantime new possibilities for the discipline arrived. They relate to improved technology (availability of satellite data with very high – sub meter – resolution, mostly from private satellite systems, such as IKONOS and QuickBird; Airborne Laser Scanning (ALS/LiDAR – see Doneus & Briese this volume); airborne thermal infrared scanning) and with current possibilities to study satellite and aerial images freely on internet servers (such as Google Earth™) where they are usually presented in the form of (geo-referenced) orthophotomaps. Also of importance are Global Positioning Systems (GPS), which since early the 1990s have helped in low altitude aerial survey by enabling digital recording of photographed sites into a GPS station, to record flight tracks, and not to loose time (very expensive in flying aircraft) by paper map navigation. Finally, aerial/satellite data processing has also been very much improved, especially through the introduction of GIS into past landscape and environmental studies. Digital photography, another innovation of the last two decades, allows quick and effective processing of aerial photographs immediately after landing. It offers also other advantages, such as quick copying of images in original quality and the storage of photographs in large numbers in digital databases and archives.

Aerial photographs and other remotely sensed data

During the last twenty years or so European archaeology has been strongly linked with natural sciences and technology innovations. GIS, GPS, high resolution satellite images, hyperspectral scans, aerial orthophotos and LiDAR/ALS data (to name only those linked to the detection and record of data from remote distance) are just the most important tools and products devised for use in disciplines other than archaeology, but applied extensively in current archaeological practice (most recently e.g. Parcak 2009; Lasaponara & Masini 2008). Apart from archaeology there is hardly any discipline among human sciences that cooperates so widely with natural sciences, and this is a factor obviously in the favour of archaeology amongst both research communities and the wider public.

However, the extremely dynamic development of sophisticated instruments, operating today preferably in a digital environment, is not free of danger. Metal detector heritage looters, able to identify sites with buried artefacts quickly even from rough locations, represent just one side of the problem. The other side is our incapacity to protect heritage effectively. Moreover, a strict demand to publish excavated and surveyed sites and features with high spatial accuracy, so that GIS spatial analyses, mathematical and statistical procedures can be applied in data processing, is counter-productive as well.

Since the early 1970s a new kind of dataset started to be accessible for the study of the Earth's surface, including archaeology. These are images captured by the first satellite systems operating for civilian purposes (Fowler 2010). In 1960 a term *remote sensing* was used for this kind of continuous photography and scanning of the complete surface of Earth (Hnojil 2005). This term was later introduced also to archaeological terminology as more or less equivalent to *satellite archaeology* (Parcak 2009). Later, since the 1990s the term remote sensing started to be used in a wider sense, to include both images sensed from a great distance (space) and photographs taken from aircraft flying at high altitude. Recently, the term is being used even more broadly to refer to all techniques for archaeological prospection where there is no physical contact with sensed (measured) archaeological situations (features, artefacts, layers etc.). Consequently, geophysical survey is included into remote sensing to name the most frequently used method (see Gaffney & Gaffney this volume).

In fact, the difference between various sensors which record land surface as image data sets is just technical and for archaeology most products of remote sensing are photographs and panchromatic images which are to be analysed and interpreted visually (but see Beck and Bennett *et al.* this volume, and Hanson 2008). This raises two issues to be addressed. Firstly, comparison of the value of vertical images and oblique photographs in terms of their spatial accuracy and the transcription of interpreted data into plans/maps, and secondly the methodological problems associated with different archaeological approaches to the land surface. These can be divided between approaches where specialist aerial reconnaissance in a low altitude aircraft produces hand-held photographs only of those sites/areas/features which the observer considers important – an approach that is biased by time pressures, changing light conditions during flight and also by the personal interests/experience of the prospector (see Palmer 2005 for a discussion of these issues). On the other hand there is the interpreter working on the ground with images – both vertical and oblique, aerial or satellite – which were taken for many reasons, rarely specifically for archaeology and ancient landscape study. While there is no space here to list advantages and limitations of the two approaches, they certainly both are valuable in their own way and it would be unreasonable to exclude one in favour of the other. They can be combined in a useful way and, at the same time, they are each specific enough so that one cannot replace fully the other. For example, oblique photography of historical monuments, ruined or semi-ruined architectural remains and earthworks carried out in very specific winter conditions (i.e. late afternoon long shadows, slight snow cover, trees free of leaves) has no equivalent in high altitude vertical photographs taken with no respect to the specific season and time of day necessary for achieving the desired result.

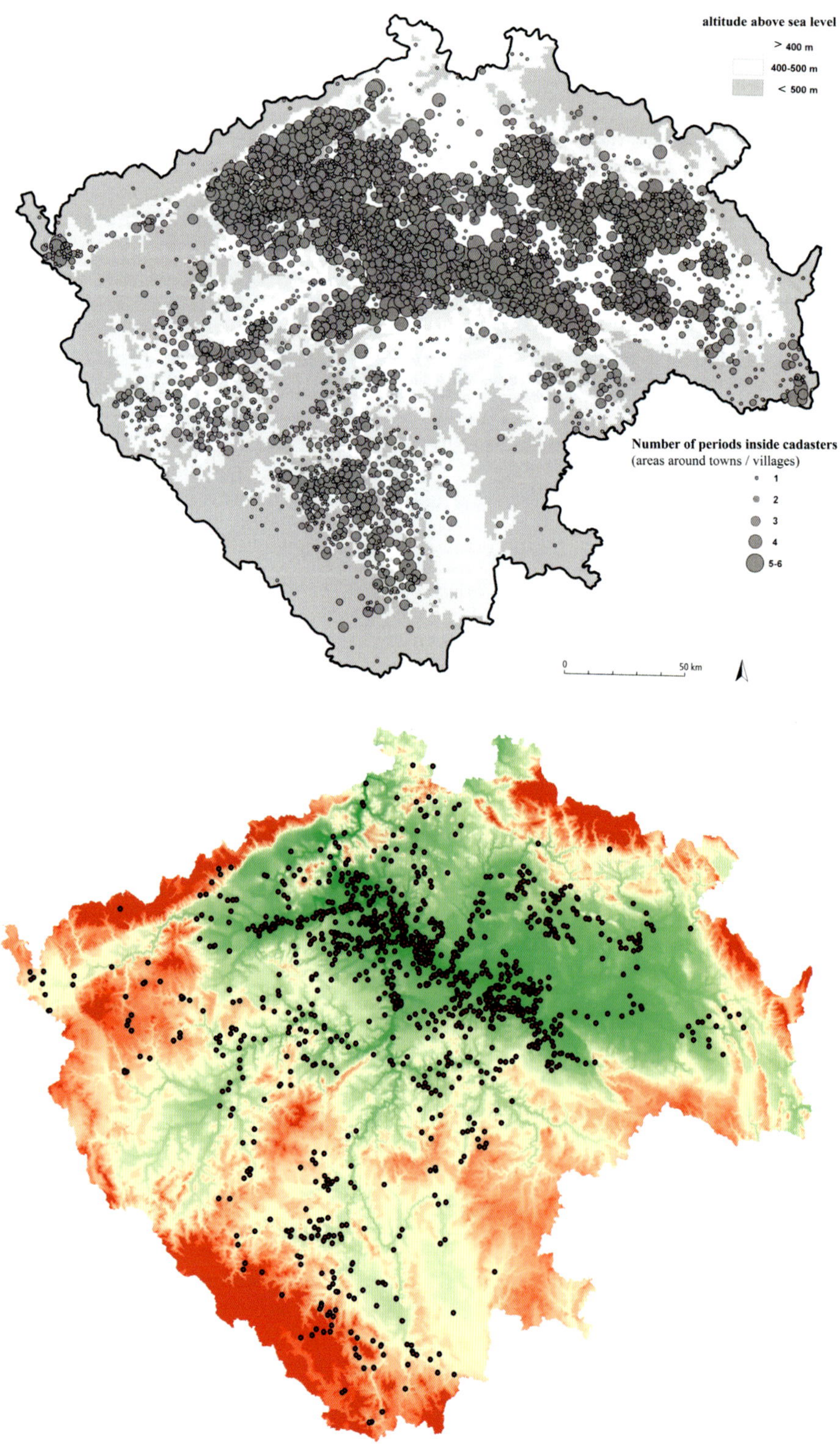

Figure 17.1: Map of prehistoric sites in Bohemia, western part of the Czech Republic. The distribution reflects both settlement activities traditionally based in the environmentally most fertile lowlands and large river valleys in northern half of the country, and the intensity of regional archaeological fieldwork (*a*); map of sites and monuments in Bohemia documented by aerial photographs (project of the Institute of Archaeology, Czech Academy of Sciences). The dots represent both identified prehistoric to medieval sites (about 70%), and architectural monuments and urban units. The map shows that most aerial-surveyed sites have been recorded in the same territory as those identified previously on the ground. This is understandable as – in terms of environmental conditions – cropmarks territories generally corresponding to zones preferably settled by past populations. © *Institute of Archaeology, Czech Academy of Sciences.*

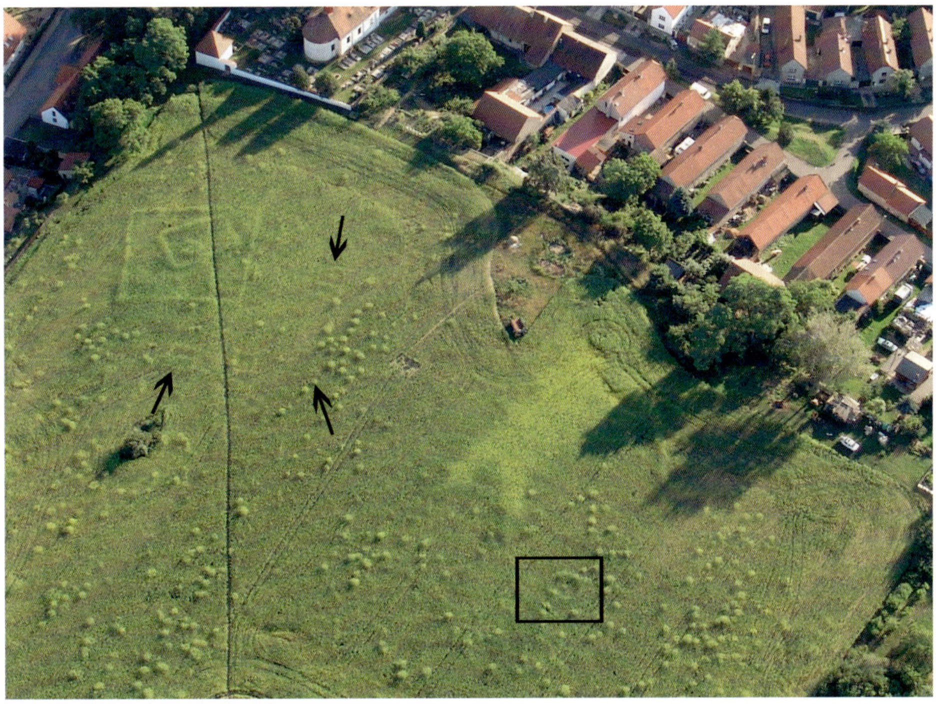

Figure 17.2: Ctiněves, northwest Bohemia. Part of a large late Bronze Age settlement displayed through hundreds of cropmarked pits. Clearly visible is also a large rectangular enclosure and a small one placed inside, and a central pit. Black arrows indicate a set of small enclosures, and in a black frame (bottom right) there is a ring ditch. Most prehistoric sites identified in Bohemia from the air so far include large numbers (10s to 100s) of sunken features such as pits and houses, and a few enclosures and linear ditches. © *Institute of Archaeology, Czech Academy of Sciences.*

All the data sources discussed above are the basis for identifying archaeological meaning and interpreting the Earth's surface recorded by various sensors. All such data, no matter whether performed from low altitude (aerial photographs, LiDAR) or from space (satellite) can be described under the comprehensive term of *remote sensing for archaeology*, including aerial archaeology (aerial survey, aerial reconnaissance and oblique photography), interpretation of vertical (orthorectified) photographs, panchromatic, multispectral and hyperspectral images and LiDAR image data (one can ask if the Aerial Archaeology Research Group will change its name accordingly, or whether will keep it forever, with respect to tradition).

Bohemia as a case study: retrospective overview

The vast majority of traces of past human activities in Central Europe are levelled by cultivation and in general terms a much smaller number of sites, mainly hillforts, have been preserved as earthworks; of the earthworks many are in woodland. Thus, since most archaeology is buried under the surface the only way to trace individual features, and especially large sites, without very expensive and time-consuming large-scale excavations or geophysical survey, is to identify and record them during observer-directed aerial survey or to detect them on existing aerial photographs deposited in archives, and available on the internet (orthophotographs) – simply because they show almost exclusively as cropmarks.

Since the beginning of the study of Bohemia reported on here, carried out by the Institute of Archaeology, Czech Academy of Sciences, Prague, since 1992 and since 2005 in cooperation with the University of West Bohemia, specialists have tended to include results of aerial survey into research on settlement development and dynamics in study areas and on the investigation of settlement structure in specific periods. The presence of specialists and a high standard equipment in the Institute allowed the development of an approach combining non-invasive methods (those operating both in large spatial units and on a site level) and sample excavations, which has turned out to be very effective. This approach has helped to create chronological frameworks for study areas, which is of primary significance for the study of complex settlement processes, underpinning better understanding of settlement strategies, the processes of stability and change in settlement history and preferences for site location and setting (for environmental and/or symbolic reasons). The programme also has been focused on developing the methodology of aerial archaeology. Some sites have been observed annually, and in individual phases of the year, informing an understanding of the role of climate, site geomorphology and plant types in the processes through which features are made visible as cropmarks and soilmarks (see also Czajlik *et al.* this volume). In addition, the effectiveness of aerial reconnaissance from quantitative and qualitative aspects (number and types of archaeological sites and features) has been compared to existing records on sites discovered in the same area over much longer periods by ground methods (Figure 17.1; for principal results of this approach see Gojda 2004a & 2004b).

Annual aerial survey campaigns over Bohemia have revealed about 1,000 sites of past settlement (and many hundreds of marks which were discarded as of either geological or recent origin during after post-reconnaissance aerial photo interpretation). Many of the archaeological remains are of otherwise unknown or rare types of sites and features. Most of them include non-linear features, such as pits and sunken houses which may be the only features detected on site or they are accompanied by linear ditches or enclosures (Figures 17.2 & 17.3).

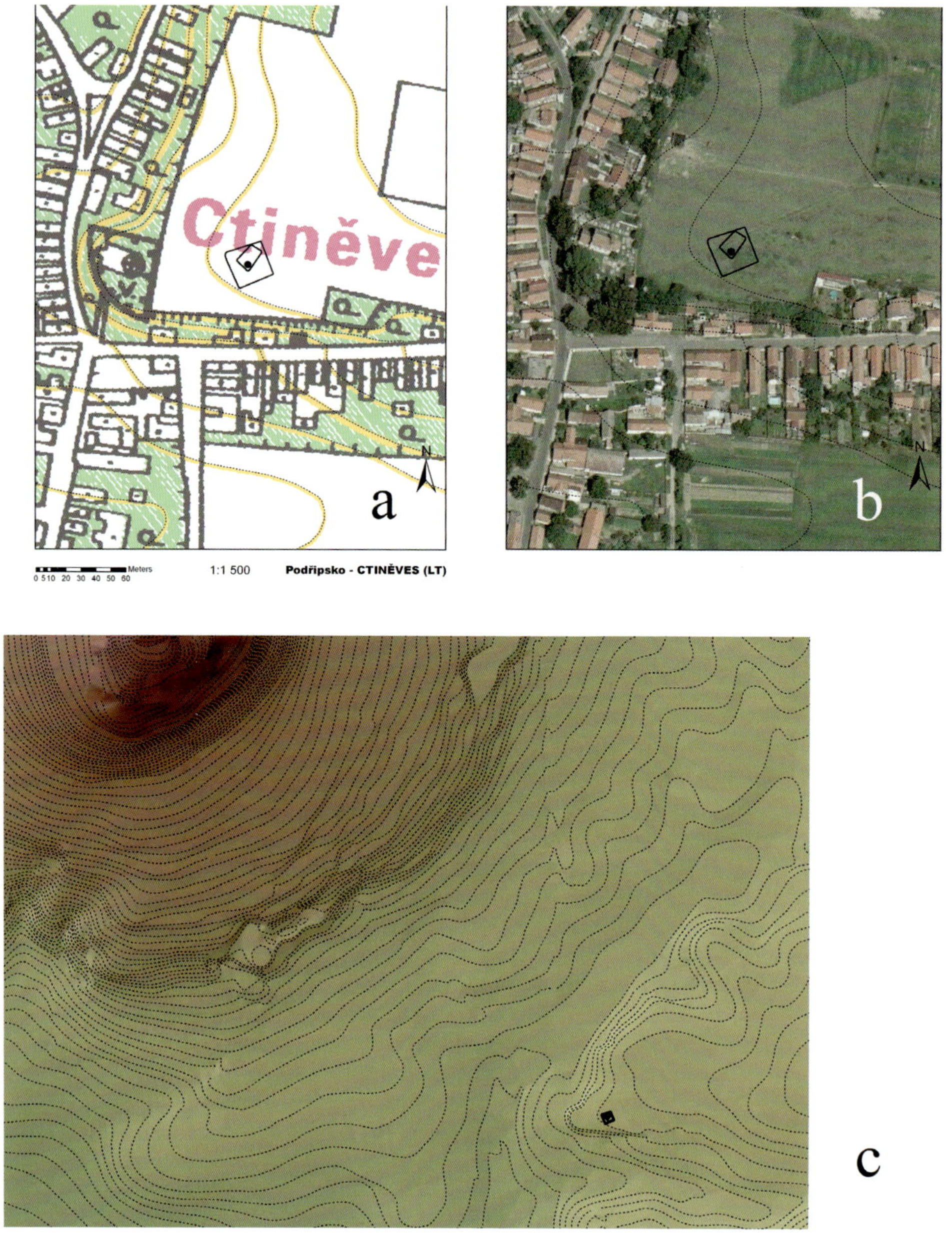

Figure 17.3: Ctiněves, northwest Bohemia. Large enclosure mapped from an orthophotomap (a, b - both files available at www.mapy.cz) and the setting of the enclosure shown in an elevation model (c) of the landscape around the Hill of Říp. This shows the site position on a local promontory facing a lowland area climbing slightly up the mountain's lower edge © *University of West Bohemia.*

In several cases large settlement areas, accumulations of residential and burial sites, spread over a few square kilometres have been identified through systematic annual reconnaissance over 10 to 15 years. These sites, which have been further investigated by extensive geophysical survey and small-scale sampling excavations, include some that have a range of remains indicative of a long settlement history. This has been confirmed by extensive analytical ploughed field-walking campaigns (surface artefact collections) in the 1990s, which demonstrated that in prehistory (i.e. from the Neolithic/Eneolithic to Roman periods) many of them were continuously settled (Kuna 2000). Evidence of multiphase prehistoric sites where use has varied between residential and funeral/ritual practice demonstrate the dynamics of settlement and illustrate the meaning of *genius loci* in the past. These areas occur both on terrace edges close to large river courses (Figures 17.4 & 17.5) and on plateaus several kilometres from the major rivers. In the past, however, these plateaus were crossed by minor watercourses which have since disappeared due to various factors, especially intensive agricultural practice. Systematic aerial survey of selected river basins and small stream valleys (few to few tens kilometres long) have lead to the discovery of dense linear concentrations of settlement areas situated a couple of kilometres apart. Some of them produced pollen data for environmental reconstruction of past landscapes (Figure 17.9).

Recently attention has also been focused on the analysis and interpretation of vertical aerial images (orthophotographs) and on the rectification and transformation of data identified on them into plans and maps. Progress has been made in 3-D analysis of verticals (Šmejda 2009), and also satellite images and their potential for Czech landscape and settlement study are now being evaluated (Figure 17.7b; Gojda & John 2009).

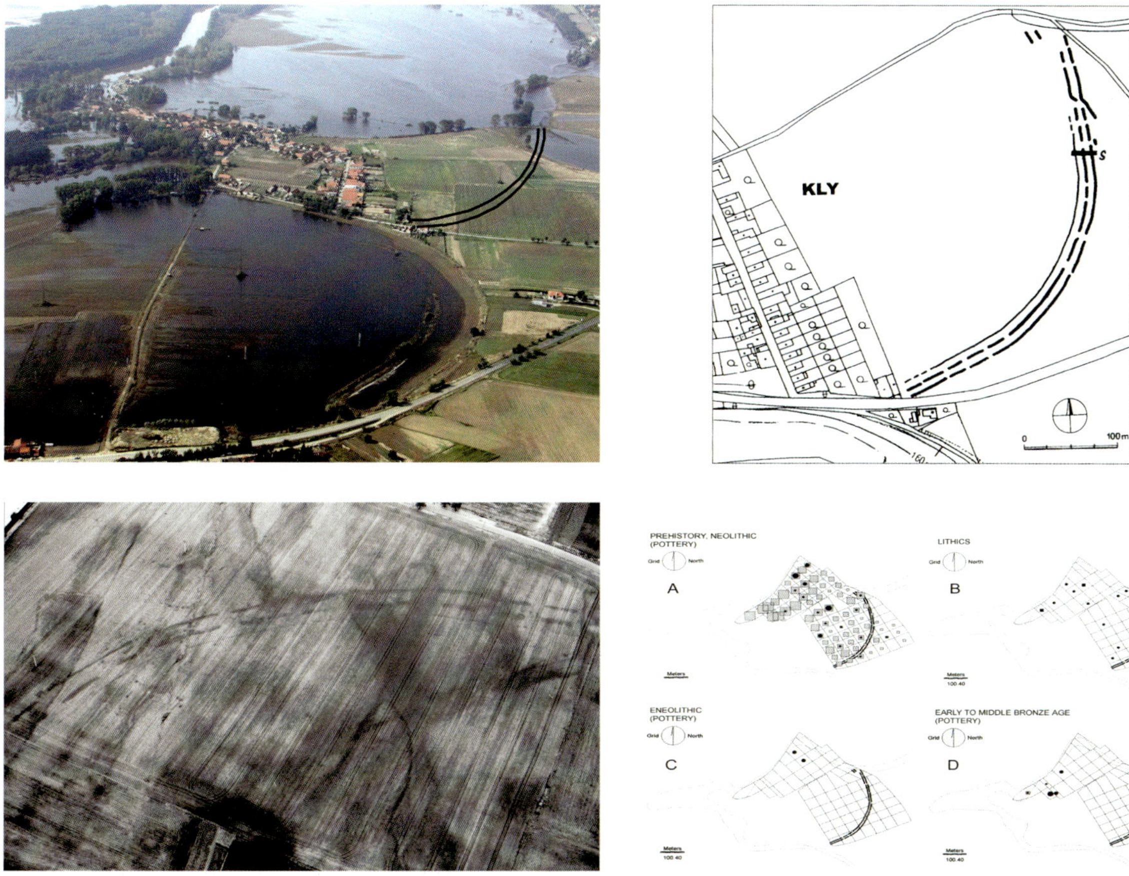

Figure 17.4: Kly, Central Bohemia. Many sites discovered from the air have become a target of fieldwork and research. A typical example of this approach is the site of Kly, an area enclosed by a double ditch and palisade trench placed on a low promontory raised above the alluvial zone of the River Labe (*Top left* – taken in August 2002 during flooding). The plan of the site (*Top right*) is based on interpreted aerial photographs and a magnetic survey, which has supplemented information on the northern end of the double ditch. A further part of the enclosing ditch system (almost 500m long) can be seen in a second aerial view (*Bottom left*). The site was also ploughed-walked, producing artefacts enabling its dating. Although the ditches date from to the early Eneolithic (Michelsberg culture, around 4000 BC) most pottery fragments (*Bottom right*) come from an earlier period (late Neolithic in Czech chronology scheme, second half of 5th millennium BC; see plan A). This very probably documents settlement continuity on the site from the Neolithic to the Eneolithic. A small excavation (section through the ditches – see letter S on Top right) produced in-situ artefacts dating the ditch system precisely. *© Institute of Archaeology, Czech Academy of Sciences.*

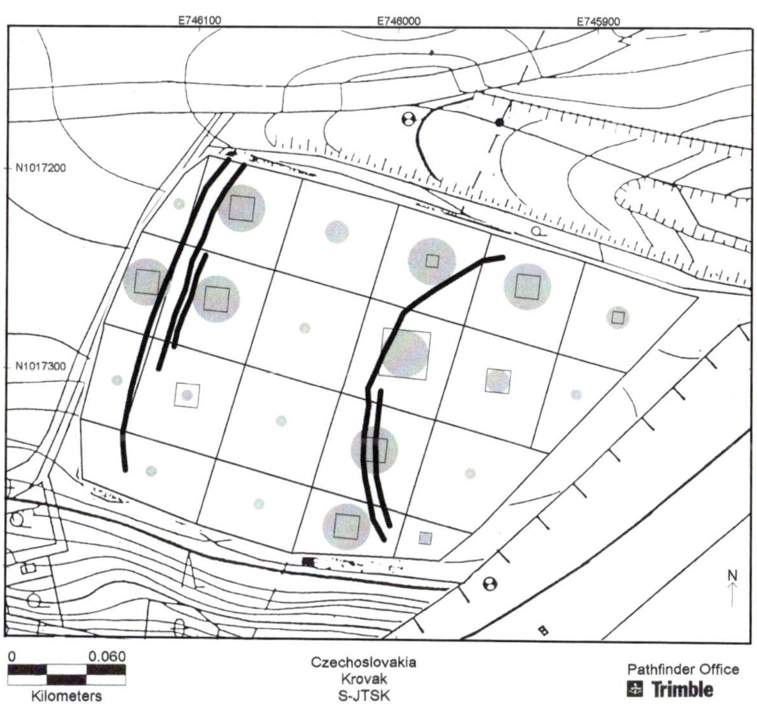

Figure 17.5: Vepřek, central Bohemia. Plan of a site placed on a strategic hilltop above the Labe. Three non-destructive methods have been applied to its study: aerial reconnaissance and geophysical survey identified two multiple ditched systems (black lines), and surface artefact collection (ploughed-field walking) produced data on the age of settlement activity (several prehistoric periods) and the distribution of artefacts on the surface. None of the field methods (including small-scale excavation) brought information about the exact age of the ditches. *© Institute of Archaeology, Czech Academy of Sciences.*

Figure 17.6: Nechanice, eastern Bohemia. Most of the late medieval to post-medieval earthworks in Bohemia lie in woodland, an environment more likely to allow the preservation of earthworks than open fields. The image shows a moated site enclosed by multiple ditch-and-bank system. The ditches are almost completely silted up and the banks largely levelled; unlike a few decades ago when the site was an earthwork and recordable as shadow marks it is now mainly visible as cropmarks. © *Institute of Archaeology, Czech Academy of Sciences.*

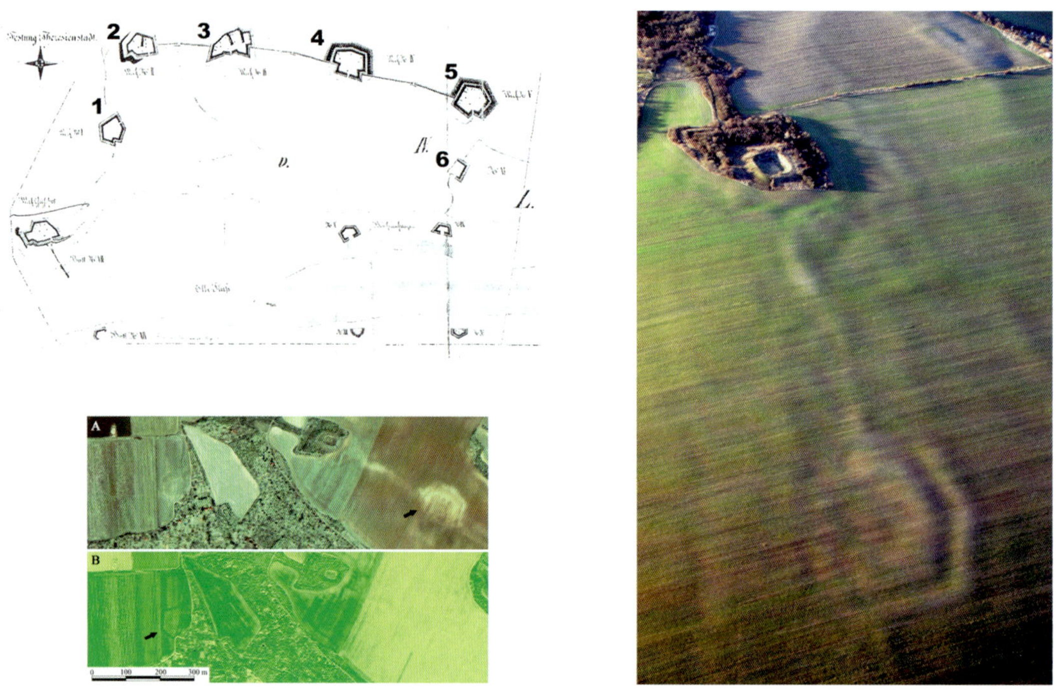

Figure 17.7: Třeboutice, northwest Bohemia. Recently orthophotographs and satellite images have been applied in Bohemia as an important remotely sensed data source for past landscape and settlement study and protection. The potential of satellite data has been tested at an early modern (mid-19th century) military installation near a large brick-walled 18th century fortress. The original plan of the fortification system (*Top left*) shows the layout, parts of which has also been recorded on QuickBird satellite imagery (*Bottom left*) of forts 1, 3 & 4: A = combination of multi-spectral images in the visible parts of the spectrum R+G+B, corrected by pan-sharpening; B = vegetation index NRVI, also corrected by pan-sharpening. ; Forts 3 & 4 have also been recorded on oblique views (*Right*). A late autumn aerial campaign in 2009 produced evidence, from slight shadows recorded in the late afternoon, that the site survives as a very low earthwork, in spite of lying in a regularly cultivated field. © *University of West Bohemia.*

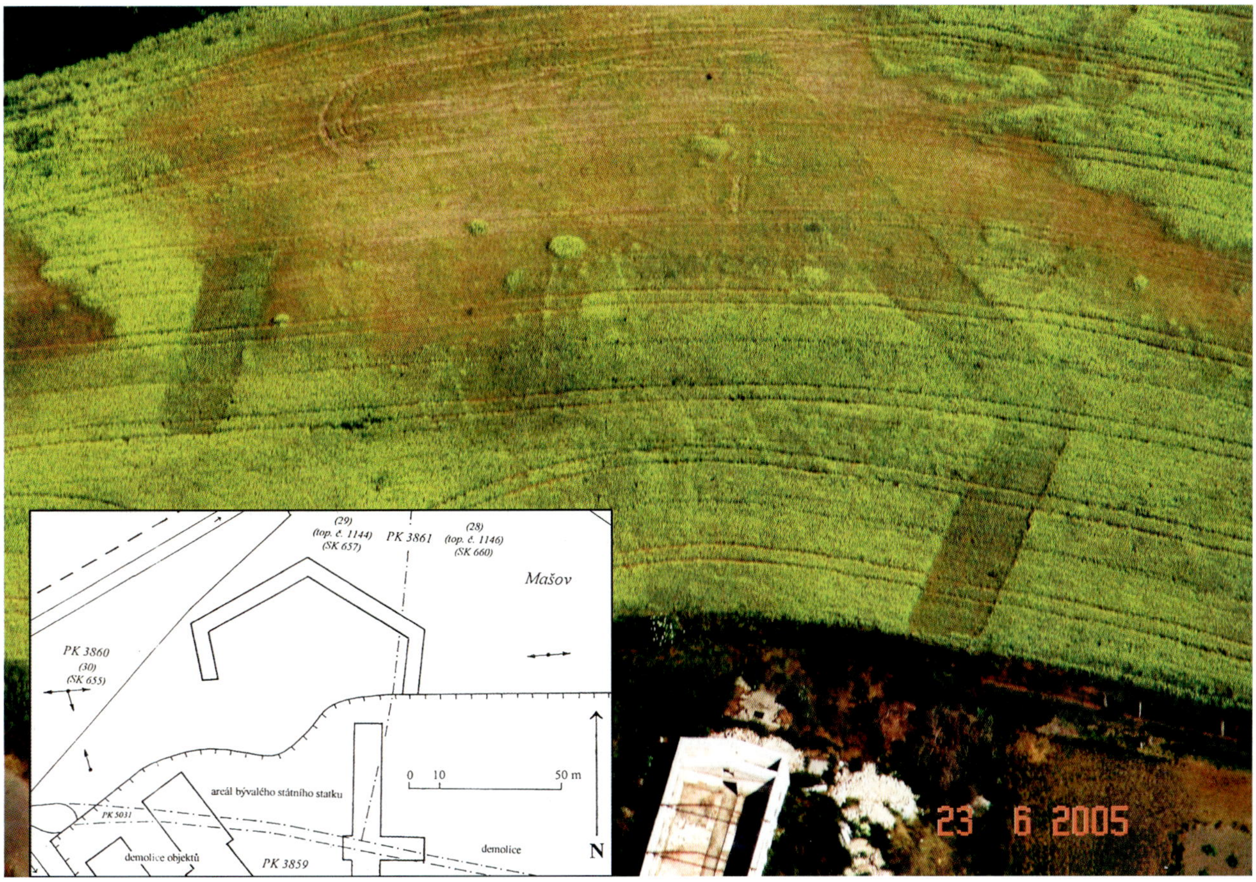

Figure 17.8: Louny, northwest Bohemia. Crop-marked small polygonal fort, a part of temporary field defences constructed in 1813 in response to the threat of military attack of Napoleonic troops from Saxony into Bohemia. Like many other prehistoric, ancient and later buried features and monuments this was identified during low altitude aerial survey, but has not been recorded on available satellite images and orthophotographs yet. The small point features are pits of prehistoric origin. © *Institute of Archaeology, Czech Academy of Sciences.*

Aerial monitoring and photography of cultural heritage: earthworks, architectural monuments and urban areas

Integrated into the aerial survey programme of the Institute since the beginning is also aerial photography of cultural heritage and monitoring changes and destructive processes, such as agriculture and construction. Systematic attention has been focused on sites and monuments in those parts of Bohemia over which annual aerial reconnaissance has been organized to identify past settlement through crop- and soilmarks. Consequently, documentation has been primarily in the most fertile lowland areas of central and northwest Bohemia (and which, in spite of continuous pressures from farming, industrial and

Figure 17.9: Mastířovice, northwest Bohemia. In some settlement areas recorded by remote sensing, environmental samples have been taken for pollen analysis where suitable soil layers exist, providing evidence for reconstruction of past land use. The photograph shows a hand auger being prepared for sampling sediments in a local stream basin situated close to a Roman period settlement which had been identified from the air through cropmarks. © *University of West Bohemia.*

construction activities, have a wide range of standing monuments and extremely well preserved medieval village cores and historic town centres). In these areas we have photographed various categories of cultural landscape, such as archaeological earthworks (e.g. Figures 17.6, 17.7 & 17.8), architectural monuments (e.g. Figure 17.10) and historic urban units.

Unfortunately, the repeated offer by the Institute to the institutions responsible for these aspects of cultural heritage for cooperation in systematic aerial photographic documentation of listed sites, for example, has not lead to regular collaboration. Only in the 1990s was a collection of aerial oblique photographs of listed archaeological sites deposited in the Sites and Monuments Record of the National Heritage Office. Recently, however, an agreement between the Institute and Prague City Council has allowed a transfer of data. A complete set of high resolution digital orthophotos of Greater Prague, consisting of periods of imagery taken between the late 1930s and the present, has been deposited in the Institute's Archive of Aerial Photographs (see below), while the Council obtained all oblique photographs taken since 1992 over both the city centre and the suburbs.

As the Institute is also one of the country's most active bodies in terms of large-scale rescue fieldwork, aerial photography of sites excavated in advance of developments, such as motorway constructions and aggregate extraction, is of great importance. It is a highly effective way of recording the work across the site as the fieldwork progresses (Figure 17.11).

The Institute's collection of aerial photographs of Czech historical landscapes and monuments – one of the largest in the country – is open to any kind of research and heritage management carried out on a professional basis. Its value will certainly be recognized in future as a source of information documenting the major changes that took place in the post-communist era in both rural and urban landscapes.

Data storage

All the available remotely sensed data has been deposited in the Archive of Aerial Photographs at the Institute. The traditional (analogue) collection includes negatives (6,500), slides (5,700) and printed enlargements filed by a town/village area and accompanied by maps and other relevant papers (850 files altogether). There are also 175 vertical images taken by Czech air forces between the late 1920s and the 1990s. The digital collection comprises photographs taken since 2002 (about 9,000 images), 15 hours of footage taken by semi-professional camcorder (12 hours in Bohemia, 3 hours abroad), and scans of the slide collection.

Perspectives for the future and conclusion

Undoubtedly the way forward in understanding past landscapes by remote sensing techniques will be through combinations of all methods mentioned above. Each of them can be used in specific conditions and consequently, their potential can be fully evaluated when they are integrated. While archaeological field methods will probably never be totally non-destructive, in the near future excavation will probably only be applied in rescue situations, and research projects on sites that are not threatened will focus completely on

Figure 17.10: Bezděz castle, northern Bohemia. Monitoring sites and documenting them with aerial photographs repeatedly since the early 1990s provides a good opportunity to record their state of preservation and to identify changes. This listed 13th century royal castle has been partly excavated during large-scale restoration, and the images show landscape setting and varying details of the monument. © *Institute of Archaeology, Czech Academy of Sciences.*

Figure 17.11: Documentation of large rescue projects in Bohemia is an important component of the aerial archaeology programme. Most rescue excavations take place during construction of new roads and motorways, such as that near the district town of Kolín, central Bohemia where a large single-ditched and part of a double-ditched enclosure are visible (*Top*), and during sand/gravel extraction, for example at Vliněves, central Bohemia where a Bronze Age burial site is being excavated (*Bottom*). *© Institute of Archaeology, Czech Academy of Sciences.*

non-invasive field methods. This will ensure that sites will be preserved relatively undamaged for future generations of archaeologists, whose methods and equipment will be much better than ours and will be able to reveal details we can hardly imagine.

For Bohemia the potential of ALS (LiDAR) for large-scale mapping of archaeological landscapes is of major importance. During 2010/11 a state-funded pilot project at the University of West Bohemia will scan a sample area, filter and classify the data, and finally evaluate the results against research objectives on the one hand, and of the requirements for sites and monuments record/management on the other. In the Czech Republic, and certainly also in other central European post-communist countries, a significant focus will be on the large-scale inclusion of LiDAR data into the mapping, study and management of archaeological (landscape) heritage. An ALS programme for the whole Czech Republic started in autumn 2009 and should be finished by 2012, although the primary unprocessed data from scanned regions will probably be available more or less immediately). It is the authors firm belief that a specialist centre for LiDAR data processing, mapping and subsequent ground-observation should be established, perhaps as a part of a university archaeological institute/department to fully exploit the potential of this technique (see Bofinger & Hesse this volume).

Finally, there are a number of theoretical and methodological issues to be addressed. Principle among them are the issues of the interpretation and classification of the vast amount of information on high resolution orthophotos and satellite data. These highlight the difficulties of interpreting the large number of features recorded as natural or cultural, and their integration into subsequent synthesis.

References

Bálek, M. 1995: Bisherige Ergebnisse der Luftprospektion in Südmähren und ihr Beitrag zur Erforschung der Feldmarschlagern aus der römischen Kaiserzeit. In Kunow, J. (ed.): *Luftbildarchäologie in Ost- und Mitteleuroopa – Aerial Archaeology in Eastern and Central Europe*. Forschungen zur Archäologie im Land Brandenburg 3, Potsdam, 241–6.

Bewley, R. 2005: Aerial Archaeology. The First Century. In Burgeois, J. & Meganck, M. (eds): *Air Photography and Archaeology 2003. A Century of Information*, Archaeological Reports Ghent University. Gent, 415–30.

Bewley, R.H. & Rączkowski, W. (eds) 2002: *Aerial Archaeology. Developing Future Practice*. IOS Press, Amsterdam.

Braasch, O. 1997: Bemerkungen zur archäologischen Flugprospektion in West und Ost. In Oexle, J. (ed.): *Aus der Luft. Bilder unserer Geschichte. Luftbildarchäologie in Zentraleuropa*. Landesamt fűr Archäologie, Dresden, 29–37.

Braasch, O. 2002: Goodbye Cold War! Goodbye Bureaucracy? Opening the Skies to Aerial Archaeology in Europe. In Bewley, R. & Rączkowski, W. (eds): *Aerial Archaeology. Developing Future Practice*. D.C: IOS Press, Amsterdam, 19–22.

Brophy, K. 2005: The hermeneutic spiral aerial archaeology and subjectivity. *AARGnews, Supplement* 1, 5–11.

Brophy, K. & Cowley, D. (eds) 2005: *From the Air: Understanding Aerial Archaeology*. Tempus, Stroud.

Burgeois, J. & Meganck, M. 2005: *Air Photography and Archaeology 2003. A Century of Information*. Archaeological Reports Ghent University 4. Gent.

Crawford, O.G.S. 1924: *Air Survey and Archaeology*. Ordnance Survey Professional Papers, New Series No. 7.

Crawford, O.G.S. & Keiller, A. 1928: *Wessex From The Air*. Oxford.

Deuel, L. 1969: *Flights into Yesterday. The Story of Aerial Archaeology*. London: Penguin Books.

Doneus, M. & Eder-Hinterleitner, A. & Neubauer, W. 2001: Archaeological Prospection in Austria. In Doneus, M., Eder-Hinterleitner, A. & Neubauer, W. (eds): *Archaeological Prospection. Fourth International Conference on Archaeological Prospection*, Austrian Academy of Sciences. Wien, 11–33.

Downey, R. 1980: A History of Archaeological Air Photography in Great Britain. *Orbit* 1, 1–20.

Fowler, M.J.F. 2010: Satellite imagery and archaeology. In Cowley, D.C., Standring, R.A. & Abicht, M.J. (eds): *Landscapes through the lens: aerial photographs and the historic environment*. Oxbow, Oxford. 99–110.

Gojda, M. 1993: Bohemia from the Air: Seven Decades after Crawford. *Antiquity* 67, 869–75.

Gojda, M. 1997: The Contribution of Aerial Archaeology to European Landscape Studies: Past Achievements, Recent Developments and Future Perspectives. *Journal of European Archaeology* 5(2), 91–104.

Gojda, M. 2004a: Prehistoric Bohemia: Landscape and Settlement in the Heart of Europe. *Landscapes* 5(1), 35–54.

Gojda, M. (ed.) 2004b: *Ancient Landscape, Settlement Dynamics and Non-Destructive Archaeology*. Academia, Prague.

Gojda, M. 2006: Large Prehistoric Enclosures in Bohemia: the Evidence from the Air. In Harding, A., Sievers, S. & Venclová, N. (eds): *Enclosing the Past: Inside and Outside in Prehistory*, Sheffield Archaeological Monographs 15, J.R. Collis Publications, Sheffield, 5–19.

Gojda, M. 2007: *Flights into the Past. Exhibition guide*. National Museum, Prague.

Gojda, M. & John, J. 2009: Dálkový archeologický průzkum starého sídelního území Čech – Konfrontace výsledků letecké prospekce a analýzy družicových dat – Archaeological Remote Sensing in Bohemia: Confrontation of Air Survey Results and Satellite Data Analysis. *Archeologické rozhledy* 61, 467–92.

Hanson, W.S. 2008: The Future of Aerial Archaeology (or Are Algorithms the Answer?). In Lasaponara, R. & Masini, N. (eds): *Remote Sensing for Archaeology and Cultural Heritage Management*, Rome, 47–50.

Hnojil, J. 2005: Od klu ke GISům. Jak historie formovala základy současných geoinformačních technologií. *GEOinformace* 3, 12–8.

Johnson, J.K. (ed.) 2006: *Remote Sensing in Archaeology. An Explicitly North American Perspective*. Alabama University Press, Tustaloosa.

Kennedy, D. & Riley, D. 1990: *Rome's Desert Frontier*. Batsford, London.

Kobyliński, Z. 2005: *Archeologia lotnicza w Polsce (Aerial Archaeology in Poland)*. Państwowe Muzeum Archeologiczne, Warszawa.

Kovárník, J. 1995: Luftbildarchäologie in Mähren (und in der ehemaligen Tschechoslowakei) 1983–1994. In Kunow, J. (ed.) 1995: *Luftbildarchäologie in Ost- und Mitteleuroopa – Aerial Archaeology in Eastern and Central Europe*. Forschungen zur Archäologie im Land Brandenburg 3, Potsdam, 247–50.

Krasnodębski, D. 2005: Polish Aerial Photos 1923–1929. In Bourgeois, J. & Meganck, M. (eds): *Air Photography and Archaeology 2003. A Century of Information*. Archaeological Reports Ghent University 4. Gent, 113–9.

Kuna, M. 2000: Surface artefact studies in the Czech Republic. In Bintliff, J., Kuna, M. & Venclová, N. (eds): *The Future of Surface Artefact Survey*. University Press, Sheffield, 29–44.

Kuzma, I., Hanzelyová, E., Rajtár, J. & Tirpák, J. 1996: New Results in Aerial Archaeology in Slovakia: Experience with Reconnaissance, Geophysical Measurement and Follow-up excavations. *Archaeological Prospection* 3, 71–9.

Lasaponara, R. & Masini, N. (eds) 2008: *Remote Sensing for Archaeology and Cultural Heritage Management*. Aracne, Rome.

Musson, C. 2005: Archeologia Aerea: Storia e Approcci Diversi. In Musson, C., Palmer, R. & Campana, S. (eds): *In Volo nel Passato. Aerofotografie e cartografia archeologica*. Siena, 15–33.

Nowakowski, J., Prinke, A. & Rączkowski, W. (eds.) 2005: *Biskupin...i co dalej? Zdjecia lotnicze w polskiej archeologii*. Ad rem. Poznań.

Palmer, R. 1989: Thoughts on Some Aspects of Air-Photo Archaeology. In Kennedy, D. (ed.): *Into the Sun*. J.R. Collis Publications, Sheffield, 53–60.

Palmer, R. 2005: 'If they used their own photographs they wouldn't take them like that'. In Brophy, K. & Cowley, D. (eds): *From the Air: Understanding Aerial Archaeology*. Tempus, Stroud. 94–116.

Parcak, S.H. 2009: *Satellite Remote Sensing for Archaeology*. Routledge, London – New York.

Poidebard, A. 1934: *La Trace de Rome dans le Désert de Syrie. Le Limes de Trajan á la Conquéte Arabe. Recherches Aériennes (1925–1932)*. Librarie Orientaliste Paul Geuthner, Paris.

Rączkowski, W. 2002: *Archeologia lotnicza – metoda wobec teorii (Aerial archaeology – method in the face of theory)*. Wydawnictwo naukowe UAM, Poznań.

Rączkowski, W. 2005: Theoretical dialogues – is there any theory in aerial archaeology?, *AARGnews, Supplement* 1, 12–22.

Schwarz, R. 2003: *Pilotstudien. Zwölf Jahre Luftbildarchäologie in Sachsen-Anhalt*. Landesmuseum für Vorgeschichte, Halle (Saale).

Šmejda, L. 2009: *Mapování archeologického potenciálu pomocí leteckých snímků*. Západočeská univerzita v Plzni, Plzeň.

Visy, Z. 1997: Stand und Entwicklung der archäologischen Prospektion in der DDR, der Tschechoslowakei und Ungarn in den Jahren 1945 bis 1990. In Oexle, J. (ed.): *Aus der Luft – Bilder unserer Geschichte*. Landesamt für Archäologie. Dresden, 23–27.

Zuk, L. 2005: Re-constructing 'bloody good mapping'. *AARGnews, Supplement* 1, 25–31.

18 | Airborne Laser Scanning for the management of archaeological sites in Lorraine (France)

Murielle Georges-Leroy

Abstract: This paper reports on two study areas in Lorraine recorded with Airborne Laser Scanning (ALS or LiDAR) for management of archaeological sites. The first, in the large Haye Forest near Nancy, recorded a very broad range of monuments of all types and periods, from which it was possible to propose measures to protect the sites during forest work. The second was undertaken within the framework of the construction of the second phase of the high-speed rail link between Paris and Strasbourg and illustrates the contribution of this method in development of preventive archaeology to all the phases of the process, but also its limits.

Introduction

Airborne Laser Scanning (ALS or LiDAR), a new-comer to the methods of remote sensing in archaeology, has been used in France since winter 2006-7. During this winter flights were carried out, mainly in the northern half of France (Alsace, Burgundy, Brittany, Lorraine) but also in Languedoc, within the framework of archaeological research programs. However, some of these results, and of later work, have been exploited for the management of archaeological sites. Among the LiDAR surveys either carried out specifically for management, or reused to this end, in Lorraine two examples will be briefly presented in this article. They relate to forest management, in the case of the large Haye forest beside Nancy, and to development or preventive archaeology, as with the case of the second phase of construction of the high speed Paris-Strasbourg railway line (Figure 18.1).

Methodology and study areas

The principles of ALS data collection and manipulation are described elsewhere (Joinville *et al.* 2003; Bofinger & Hesse, Doneus & Briese this volume), and for the purposes of this paper the important outcomes are a high-accuracy Digital Terrain Model (DTM) with a spatial accuracy of some 10s of cm. The ability to record micro-relief and to strip away forest cover is particularly valuable.

The forest environment lends itself particularly well to this method, because many remains are preserved there as slight variations in micro-relief. On the one hand, the forest areas provide a protection to archaeological remains as they are sheltered from erosion by agriculture or urbanization. However, because of the difficulties to their study (i.e. poor visibility due to vegetation and inaccessibility, imprecise dating because of the

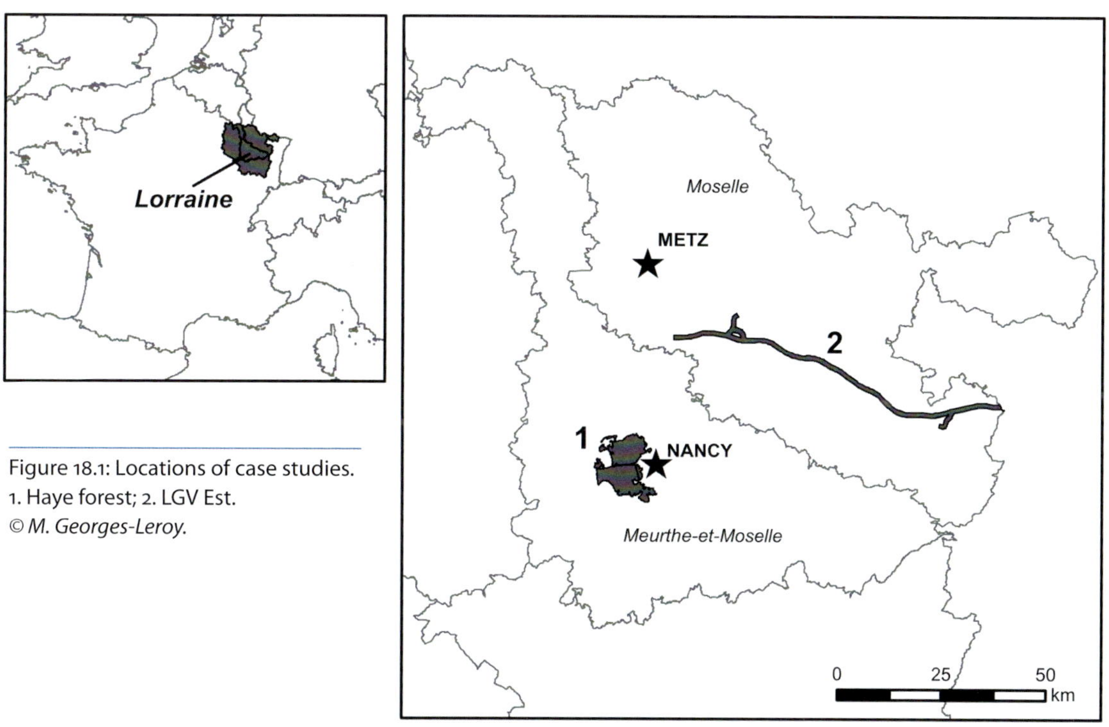

Figure 18.1: Locations of case studies.
1. Haye forest; 2. LGV Est.
© *M. Georges-Leroy.*

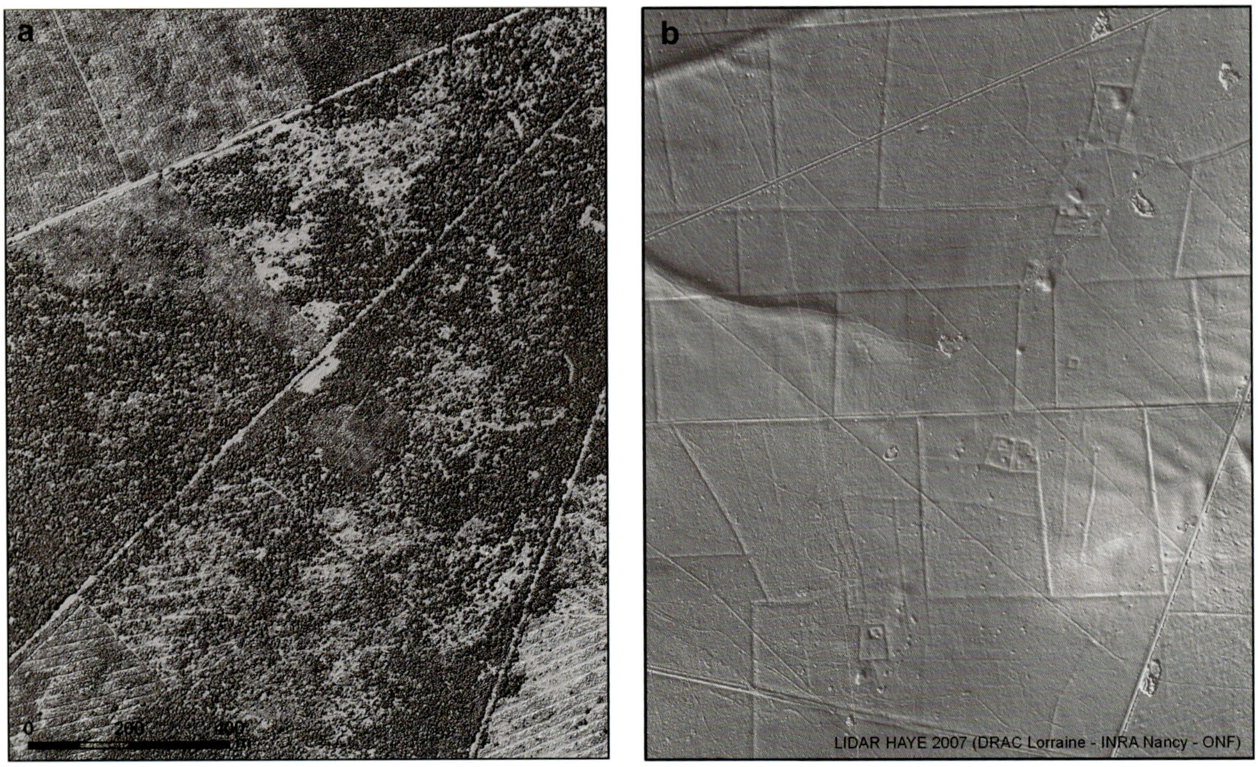

Figure 18.2: Comparative images of a sector in the south of the Haye forest – *a*) Model including the vegetation and *b*) DTM with vegetation stripped out. On the latter, in addition to the modern roads and forest tracks there are ancient track ways, marked by parallel stone banks, field boundaries and the settlements composed of enclosures and buildings. © *DRAC Lorraine - INRA Nancy - ONF – 2007.*

absence of artefacts, difficult topography etc.), the remains preserved in forests are not always studied by archaeologists. Hence, the absence of sites in large forests is often the result of a lack of archaeological prospection rather than a real absence (Georges-Leroy *et al.* 2010). The application of ALS overcomes some of these constraints and can contribute to the location and the mapping of archaeological sites. Applications in forests are thus numerous, in particular in Europe (Sittler 2004; Devereux *et al.* 2005; Doneus *et al.* 2008; Risbøl *et al.* 2006; Bofinger & Hesse this volume). The technique can also be of value in grassland and moorland where slight earthworks can be preserved, which can also be the case in ploughed fields (Romain & Burks 2008).

The two LiDAR surveys in Lorraine presented here were carried out in 2007 and 2008 and the results are still being worked on by the archaeologists. The first study area, in the Haye forest, covered an area of 116km² and is a collaboration between archaeologists of the Service Régional de l'Archéologie (SRA) de Lorraine (Direction régionale des Affaires Culturelles) and l'Institut National de Recherches Archéologiques Préventives (INRAP), as well as researchers in the Nancy office of the Institut National de Recherches Agronomiques (INRA) and l'Office National des Forêts (ONF). Started in 1998, this collaboration aims to study the fossilized Gallo-Roman landscapes preserved under forest on the limestone plateau between Pont-à-Mousson and Neufchâteau, including their impact on biodiversity and present day fertility (Dupouey *at al.* 2002; Georges-Leroy *et al.* 2007). The second LiDAR survey was financed by Réseau Ferré de France (RFF) within the framework of the second phase of construction of the high speed rail link between Paris and Strasbourg (LGV Est). This comprised a one-kilometer wide band along the entire distance of the link, including the areas of connection with rail lines. For Lorraine this covered an area of 102km².

These two flights were undertaken using multi-echo (from 2 to 4 points measured by emitted impulse) recording and were carried during April and March when vegetation cover was at its lowest. The densities of points emitted are important, at 5 and 8pts/m² respectively, providing a DTM with a resolution of 0.5m and 0.25m respectively. For the Haye forest this generated 2 billion points, of which 250 million recorded the ground surface while the remainder recorded vegetation (Georges-Leroy *et al.* 2008).

ALS and forest management: a case study from the Haye forest (Meurthe-et-Moselle)

While tree cover plays a protective part in the conservation of archaeological sites in forests, those remains are none-the-less fragile, in particular because of the significant development of forest machinery in the last 10 years (Dardignac & Dunoyer 2006). And before any protection plan can be developed, it is advisable to know what sites are present and their characteristics. The first protection measure thus consists of a detailed inventory and this is where an ALS survey is a powerful tool, as the survey in the Haye forest during 2007 demonstrates.

The forest primarily consists of deciduous trees. It was extensively damaged by the Lothar storm of

26 December 1999, when more than 50% of many plantings were destroyed making access on the ground difficult, or in places impossible (e.g. growth of brambles and dense regeneration). Ancient field systems in the forest were recorded archaeologically in 1979, with more systematic work from 1994. By the end of 2006, some 70 to 80km² had been mapped, comprising nearly 200km of stone banks and terraces, 45km of ancient track ways and about 50 settlements. This mapping was undertaken with GPS from 2000, but also includes data collected by tape measurements and prismatic compass. Thus, there is uneven metrical accuracy for the mapped data involved, varying from 1 m to several tens of meters according to the prospected zones. Moreover, the compilation of data and its integration in a geographical information system (GIS) is also characterized by a lack of consistency. The use of the GPS throughout would have improved mapping precision, with the advantage of easy assimilation into the GIS, but the human resources required for this task were not available, and anyway, certain areas remained inaccessible because of dense vegetation. This is why a very precise ALS survey was commissioned in order to accurately locate known archaeological remains, but also to explore the inaccessible zones.

In addition to creating the most complete possible mapping of archaeological remains, the aims of the ALS survey included the recording of the forest. Mapping the heights of trees, in particular, allowed the creation of an inventory of the woodland assets which could be used by foresters to manage the forest. In addition, the recognition that the heights of trees at a given age is also an indicator of the fertility of the ground allowed the integration of the forest data, archaeological information and ecological data and the analysis of the impact of past land use on the state and the operation of the current forest ecosystems (vegetation diversity, productivity of plantings).

For the archaeology, this survey revealed an extensive fossil landscape in the forest (Figure 18.2), adding 50% to what had been recorded using traditional methods of prospection on the ground. Moreover the mapping could be extended to the inaccessible zones on the ground (e.g. areas damaged by the 1999 storm, scrub vegetation). Remains appear remarkably well as the LiDAR allows the effective stripping away of the vegetation. Moreover, even very slight remains, almost undetectable on the ground, were visible and in these areas the project multiplied by four the mapped remains. Although linear features are particularly well represented (e.g. ancient field systems and track ways), the survey also showed its effectiveness on more specific remains and 26 additional ancient settlements were detected. This is 30% of the total, with the added benefit of an improved knowledge base.

Lastly, many other remains, previously known or unknown, which were not directly concerned with the research program could be studied thanks to this survey. Thus, in addition to the three protohistoric hillforts known in the study area (e.g. Figure 18.3), a new defended promontory site was discovered, never before recorded despite a strong tradition of research in this area since the 19th century. Other remains include extensive limestone quarries and lime kilns, and military structures dating to the late 19th century. While the most imposing military remains were previously known (e.g. forts, gun batteries, water tanks), many other less marked remains were located through the survey (e.g. trenches, huts), allowing a better understanding. However, not all types of site may be recorded. For example, while medieval and modern iron ore extraction sites are visible on the LiDAR surveys, the ore processing sites are not observable on these images. On the ground, these processing sites comprise little more than spreads of slag and charcoal in small valleys and illustrate the limits of this

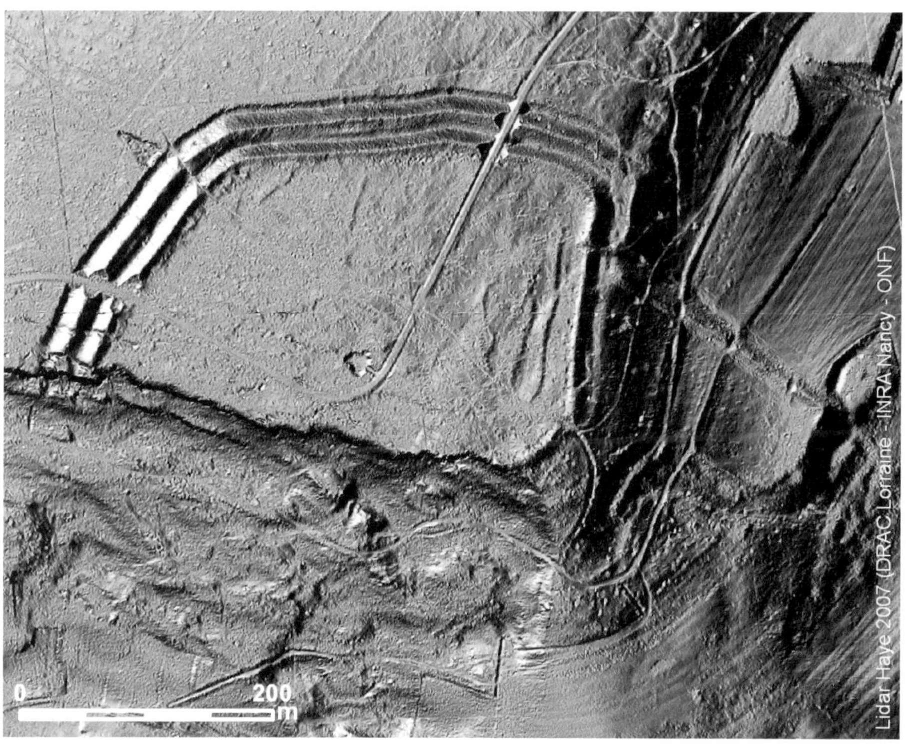

Figure 18.3: Camp d'Affrique in Messein, an Iron Age hillfort. © *DRAC Lorraine - INRA Nancy - ONF – 2007.*

kind of data. There are also probably other types of sites of which one suspects the existence of, but which also escape the LiDAR record, of which Gallo-Roman cemeteries provide another example.

The ALS survey, coupled with ground prospection, has allowed the creation of a very detailed inventory of the archaeological remains, from which it is possible to propose protection measures to be taken within the framework of forest work. To ensure that these measures are taken into account for the long term it is preferable that they are integrated in the forest management documents and plans (Dardignac & Dunoyer 2006) as they are progressively revised. Since archaeological remains are present throughout almost the whole of the forest (i.e. field boundaries and agricultural features cover approximately 80% of the forest floor) it is not possible to place constraints on the whole of these areas. The protection measures suggested by the SRA de Lorraine are thus graduated. In the zones where only agricultural remains and field boundaries are known, they are taken into account only during earth-moving projects such as the creation of roads. On the other hand, for certain types of sites (ancient settlements, medieval and modern metallurgical workshops, medieval mining) or areas with concentrations of sites greater constraints on forestry activity are asked for, such as the prohibition of heavy machines or of work that disturbs the ground surface (e.g. ploughing, planting).

In addition, a map of the archaeological sites and issues is also under development within the framework of a 'forest protection' plan for the whole area. This legal status defined by the Forest Code aims to protect this forest from the growth of nearby urban areas (in particular Nancy and Toul), both for the wellbeing of the population and the protection of land, but also for ecological reasons. The archaeological component is of relevance to this framework for several reasons. It contributes to ecological knowledge as the ancient land use plays a part in the operation of the current forest ecosystems. It can also contribute to the tourist development of the whole area because of presence of several remarkable archaeological sites (in particular hillforts – Figure 18.3).

In summary, even though it was not originally intended that the archaeological remains should be taken into account in forest management, the ALS survey has allowed it to be fully integrated. This process has been facilitated by the overall framework of a research program, which allowed the ground checks and analysis of remains.

ALS and preventive archaeology: a case study of the LGV Est in the Moselle

Experiments in the use of ALS survey in preventive archaeology are still rare, and this case study describes an example undertaken within the framework of the construction of the second phase of the Paris-Strasbourg LGV. This 106km long line connects Baudrecourt in the Moselle to Vendenheim, in the suburbs of Strasbourg. The focus here is in the 73km long 'section G', which was managed by the SRA de Lorraine (the other section (H) having been managed by the SRA d' Alsace).

In France, since 2002, the role of preventive archaeology is to *'assurer [...] dans les délais appropriés, la détection, la conservation ou la sauvegarde par l'étude scientifique des éléments du patrimoine archéologique affectés ou susceptibles d'être affectés par les travaux publics ou privés concourant à l'aménagement.'* (Art. L 521–1 du Code du Patrimoine: i.e. to ensure the detection, conservation or safeguard by scientific study of the elements of the archaeological inheritance affected or likely to be affected by public or private works). This is undertaken in two stages: the assessment and then excavation. The assessment phase aims, through desktop study, prospection or field work, to highlight and to characterize the elements of archaeological interest present on the site (Art. 14 du décret n° 2004–490 du 3 juin 2004). This is generally carried out as monitoring of topsoil stripping, although other methods are also used.

The work on the first phase of the rail link, just like other linear infrastructure projects, had shown the difficulties that forests present to archaeologists. Indeed, wooded areas are not available for mechanical topsoil stripping in the same way that cultivated areas are, because the trees are cut late in the programme of works. Moreover, on this new rail line tree clearance was delayed for as long as possible in the forests of the Parc Naturel Régional de Lorraine for ecological reasons. In addition, the removal of the trees can partially destroy sites preserved as slight earthworks in the forest.

The objectives of this LiDAR survey were twofold: firstly, to acquire a digital terrain model to assist the planning of the works, and secondly to obtain archaeological information in advance of the assessment phase. The densities of points for the flight (8pts/m^2) reflected the requirements of the archaeology. After delivery of the data in June 2008, the SRA de Lorraine addressed the archaeological questions. A rapid inspection, initially of the wooded areas, allowed a certain number of priority forest sectors to be identified, in particular in the sensitive forests of the Parc Naturel Régional de Lorraine. These data were then sent to INRAP, the operator in charge of the assessment of this section.

The assessment of the principal line began in April 2008 and was completed in June 2009. Approximately 470 hectares were examined identifying 22 major sites and 50 more minor site indicators dating from the Protohistoric period to modern times (Viller 2009, 2010). The excavations began in 2009 and still continue, alongside the assessment of the additional working areas (e.g. storage yards, site of buildings). While these works are still in progress and it is not possible to draw a final conclusion, the contribution of ALS survey to this operation, beyond helping to choose appropriate forest zones to cut down, is already clear on several points.

In all areas, one of the archaeological questions in the forest related to the presence of small closed wetland depressions or 'mardelles' (Figure 18.4). Their

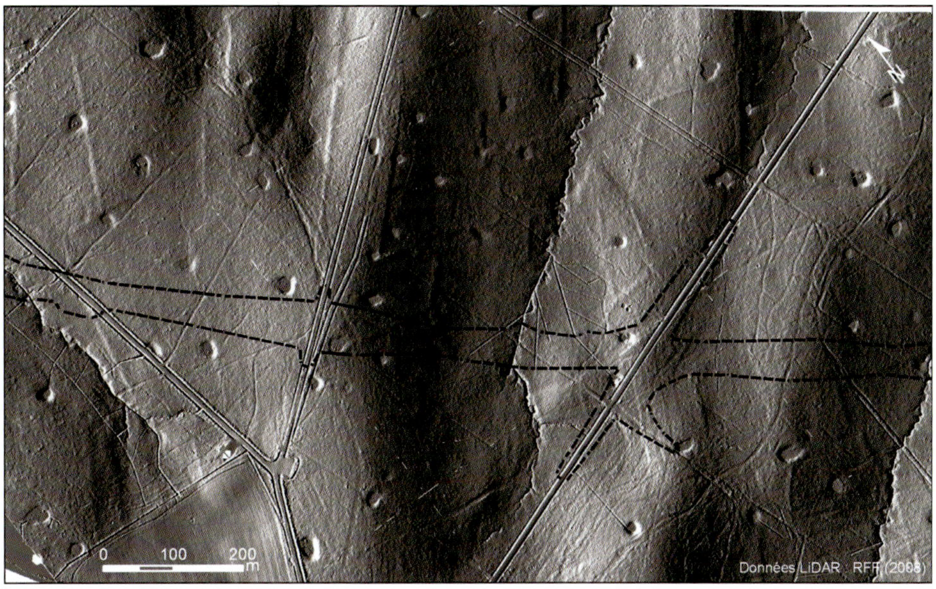

Figure 18.4: Mardelles and ancient field systems preserved in the national forest of Fénétrange. The footprint of the rail line works is outlined with a black pecked line. © *RFF – 2008.*

origin has been debated for more than a century by advocates of an anthropogenic origin and is now the subject of a research thesis (Etienne *et al.* 2010). The mardelles are particularly rich environments for vegetation and animals, some of which are protected. Their visualization on the LiDAR images created an inventory of 557 mardelles, of which two thirds were located in forest. Some occur in cultivated fields and having been levelled by the plough are almost invisible on the ground. This systematic inventory led to a programme of field investigation of the mardelles impacted by the rail line works before deforestation, including an inventory of vegetation and coring for pollen analysis. This then directed the assessment and monitoring of mechanical top soil stripping, and finally the excavation of four mardelles which will hopefully answer the question of their origins and functions (INRAP excavation in progress).

A second contribution of the ALS survey relates to the discovery of previously unknown ancient field boundaries in the national forests of Fénétrange and Albestroff (Figure 18.4). Nearly invisible on the ground before the deforestation, these were no more readable after the cutting of the trees, because of vehicle ruts and branches on the ground. Constructed only of soil, these features would not have been noted during mechanical topsoil stripping. LiDAR thus brings important information on this ancient cultivation system, whereas the surveys were regarded as 'negative'. They may be associated with the mardelles, and their mapping will add to the interpretation of these structures. In the same way, a probable old track was identified from the LiDAR in a wood in Conthil. It was not seen during mechanical topsoil stripping, which took place before the ALS survey. On the other hand, it should be noted that almost none the sites discovered during the mechanical top soil stripping appears on the LiDAR images – a rapid check of the 22 major sites showed that with one exception (a cluster of burial mounds), none would have been identified through the ALS survey. However, three of them have left marks in the landscape which are still visible on the LiDAR: slight depressions in ploughing for a track and a mardelle and a shapeless hill in grassland for a gallo-roman building.

A further contribution of the ALS survey is in providing context around archaeological sites. Thus, two important field terraces documented by earlier surveys could be attached to field systems visible in the LiDAR in the commune of Dolving (Viller 2010). Two ancient track ways observed in the ground surveys in the commune of Sarraltroff can be seen in the LiDAR leading to a large villa preserved in coppiced woodland (Viller 2009). Lastly, a new hillfort located outside the LGV line in the commune of Lesse was also identified on the LiDAR images and confirmed through a field visit.

Conclusion

We can state that ALS survey is a very powerful tool for detection of new sites, but also for the improvement of knowledge of already known sites (e.g. precision in location/morphology, contribution to the context). It can thus contribute to the development of protection measures for these sites, in particular during the preparation of forest management plans. It can also be successfully integrated at various stages in the process of preventive archaeology. In advance of works, it can be a tool for creating site inventories, but also as a decision-making aid. The example of the rail line showed its utility in the specific context of conservation and limited accessibility. Here it can be used to detect remains that would be difficult or impossible to discover during mechanical topsoil stripping. And finally, it makes it possible to better understand the topographic and archaeological context of sites. It is, however, necessary to be aware of its limits, because with some exceptions, this technique allows only the detection of micro-relief. Thus, in preventive archaeology in particular, it cannot replace the monitored mechanical topsoil stripping which makes it possible to locate remains that do not have any trace on the ground surface. In addition, a programme of field visits is necessary for the validation of observations made on the images and to obtain additional details (construction materials, dating artefacts etc.). It is also necessary to create a framework for interpretation, like what has been previously done in aerial prospection.

In France, with some exceptions, the services in charge of archaeological site management do not yet have the computer tools (software in particular) to process such data, nor the human competences for their analysis. But, with the increasing acquisition of data, this experience and competence will develop. For example further projects, such as the LiDAR specifically carried out by the SRA de Bretagne and the ONF for the national forest of Huelgoat in Finistère during 2009 (Dardignac 2009), allow the development of protection measures adapted to the specific site. Such ALS survey is also considered for the 'IIe charte forestière de territoire 2009–2014' (a working plan for forestry for 2009–14) of l'Arc Boisé in the Val-de-Marne with the aim of improving and developing the archaeological knowledge of this large area. Finally there is a project in progress undertaken by the SRA de Rhône-Alpes (coordinated by Benoit Helly) in the forest of Chambaran in Isère on the proposed site of a holiday centre on several hundreds of hectares.

Acknowledgements

The two case studies described here benefited from the contributions of many people whom I warmly thank: for the program on the Haye forest, Etienne Dambrine, Jean-Luc Dupouey (INRA Nancy), Jérôme Bock (ONF); for the project on the LGV Est, Eliane Mazerand (RFF), Sébastien Viller, Stéphanie Goepp (INRAP), David Etienne, Pascale Ruffaldi (UMR Chrono-Environnement Besançon), Etienne Dambrine (INRA Nancy), Stéphane Marion (SRA de Lorraine).

References

Dardignac, C. & Dunoyer, J.-L. 2006: Prise en compte de l'archéologie dans la gestion forestière. *Rendez-vous techniques* 14, 28–31.

Dardignac, C. 2009: *Rapport de prospection archéologique. Forêt domaniale de Huelgoat (Finistère, communes de Berrien, Huelgoat, Locmarria-Berrien)*, dact.

Devereux, B.J., Amable, G.S., Crow, P. & Cliff, A.D. 2005: The potential of airborne LiDAR for detection of archaeological features under woodland canopies. *Antiquity* 79, 648–60.

Dupouey, J.-L., Dambrine, E., Laffite, J.-D. & Moares, C. 2002: Irreversible impact of past land use on forest soils and biodiversity. *Ecology* 83(11), 2978–84.

Doneus, M., Briese, C., Fera, M. & Janner, M. 2008: Archaeological prospection of forested areas using full-waveform airborne laser scanning. *Journal of Archaeological Science* 35, 882–93.

Etienne, D., Ruffaldi, P., Goepp, S., Ritz, F., Georges-Leroy, M., Pollier, B. & Dambrine, E. 2010: The origin of closed depressions in Northeastern France: a new assessment. *Geomorphology*, doi: 10.1016/j.geomorph.2010.10.036.

Joinville, O. de, Saur, S. & Bretar, F. 2003: Le levé laser aéroporté : techniques, applications et recherche. *Bulletin d'information scientifique et technique de l'IGN* 74, 37–52.

Georges-Leroy, M., Heckenbenner, D., Laffite, J.-D. & Meyer, N. 2007: Les parcellaires anciens fossilisés dans les forêts lorraines. In Dupouey, J.-L., Dambrine, E., Dardignac, C. & Georges-Leroy, M. (eds): *La mémoire des forêts. Actes du colloque 'Forêt, Archéologie et Environnement' 14 – 16 décembre 2004*, ONF – INRA – DRAC Lorraine, 121–31.

Georges-Leroy, M., Bock, J., Dambrine, E. & Dupouey, J.-L. 2008: L'apport du laser scanneur aéroporté à l'étude des parcellaires gallo-romains du massif forestier de Haye (Meurthe-et-Moselle). *Bulletin de liaison AGER* 18, 8–11.

Georges-Leroy, M., Bock, J., Dambrine E. & Dupouey, J.-L. 2010: Le massif forestier, objet pertinent pour la recherche archéologique. L'exemple du massif forestier de Haye (Meurthe-et-Moselle). *Revue géographique de l'Est* [En ligne], Le massif forestier, objet géographique, mis en ligne le 07 Octobre 2010, consulté le 14 octobre 2010. URL : http://rge.revues.org/1931.

Risbøl, O., Gjertsen, A.K. & Skare, K. 2006: Airborne laser scanning of cultural remains in forests: some preliminary results from a Norwegian project. In Campana, S. & Forte, M (eds): *From Space to Place. 2nd International Conference on Remote Sensing in Archaeology. Proceedings of the 2nd International Workshop*, CNR, Rome, Italy, December 4–7, (BAR International Series 1568), 107–12.

Romain, W.F. & Burks, J. 2008: LiDAR Analyses of Prehistoric Earthworks in Ross County, Ohio. *Current Research in Ohio Archaeology 2008*. http://www.ohioarchaeology.org/joomla/index.php, accessed 19 November 2009.

Sittler, B. 2004: Revealing historical landscapes by using airborne laser scanning. A 3-D model of ridge and furrow in forests near Rastatt (Germany). In *Proceedings of the ISPRS working group VIII/2 'Laser-scanners for forest and landscape assessment' Freiburg 03–06 October 2004*, 258–61.

Viller, S. (dir.) 2009: *Rapport de diagnostic archéologique. Indices de sites de la Seconde Phase de la L.G.V. Est. De Dannes-et-Quatre-Vents à Marthille*, dact.

Viller, S. (dir.) 2010: *Rapport de diagnostic archéologique. Indices de sites de la Seconde Phase de la L.G.V. Est. De Mathille à Morville-sur-Nied. De Rorbach-lès-Dieuze à Dolving*, dact.

19 | Aerial archaeological survey of a buried landscape: The Tóköz project

Zoltán Czajlik, László Rupnik, Máté Losonczi and *Lőrinc Timár*

Abstract: The GIS Research Laboratory and Aerial Archaeological Archive of the Archaelogical Institute of the Eötvös Loránd University, created in 1993 in order to help the rescue excavations along the M3 motorway project, and the French-Hungarian archaeological co-operation, has a collection of more than 30,000 aerial photographs covering more than 1,500 archaeological sites. Field data is also collected alongside full-scale archaeological interpretation of aerial photographs in small regions and on compilation of photo-maps. The research area reported on in this paper lies in northwest Hungary in the micro-region of Tóköz. Using aerial survey we can identify large sites even on surfaces covered by very late Holocene alluviation.

Introduction

The main task of the GIS Laboratory, created in 1993 as a part of the Archaeological Institute of the Eötvös Loránd University (Budapest), was to process the GIS background of the rescue excavations on the M3 motorway in Hajdú-Bihar County (Czajlik & Holl 1996; Czajlik *et al.* 1997; Raczky *et al.* 1998, 2002). The creation of the laboratory made it possible to start complex scientific projects, for example in managing the background research for the Hungarian-French aerial archaeological project run between 1993 and 2000. That research was initiated by Prof. Miklós Szabó (on the Hungarian side) and René Goguey (on the French side) and the main sponsors were the Ministry for Foreign Affairs of the French Republic, the Regional Conseil of Burgundy (France), the Archaeological Institute of the Eötvös Loránd University of Budapest and the Balaton Programme. The project staff also includes Anne Violot-Richeton documentarist and Zoltán Czajlik, archaeologist, who have participated in the project from its very beginning. René Goguey, a pilot-archaeologist from Burgundy, has conducted eight campaigns until 2000, spending 30–40 hours in the air in Hungary every year, identifying more than 400 new aerial archaeological sites (*cf.* Goguey & Szabó 1995; Goguey *et al.* 2003). Since 2001, Zoltán Czajlik has continued this project with aerial photography of hillforts, tumuli, tell-settlements and aerial archaeological prospection. By 2010, the Aerial Archaeological Archive of the GIS Laboratory has amassed more than 30,000 photos, covering more than 1,500 aerial archaeological sites (Figure 19.1).

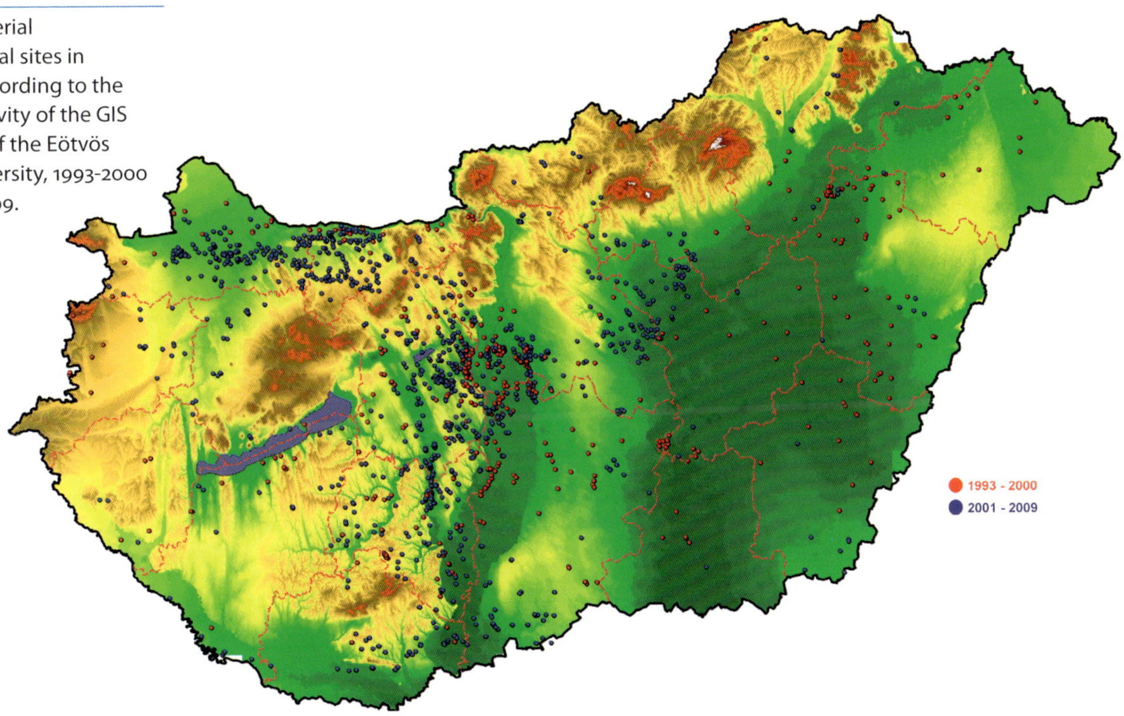

Figure 19.1 Aerial archaeological sites in Hungary, according to the research activity of the GIS Laboratory of the Eötvös Loránd University, 1993-2000 and 2001-2009.

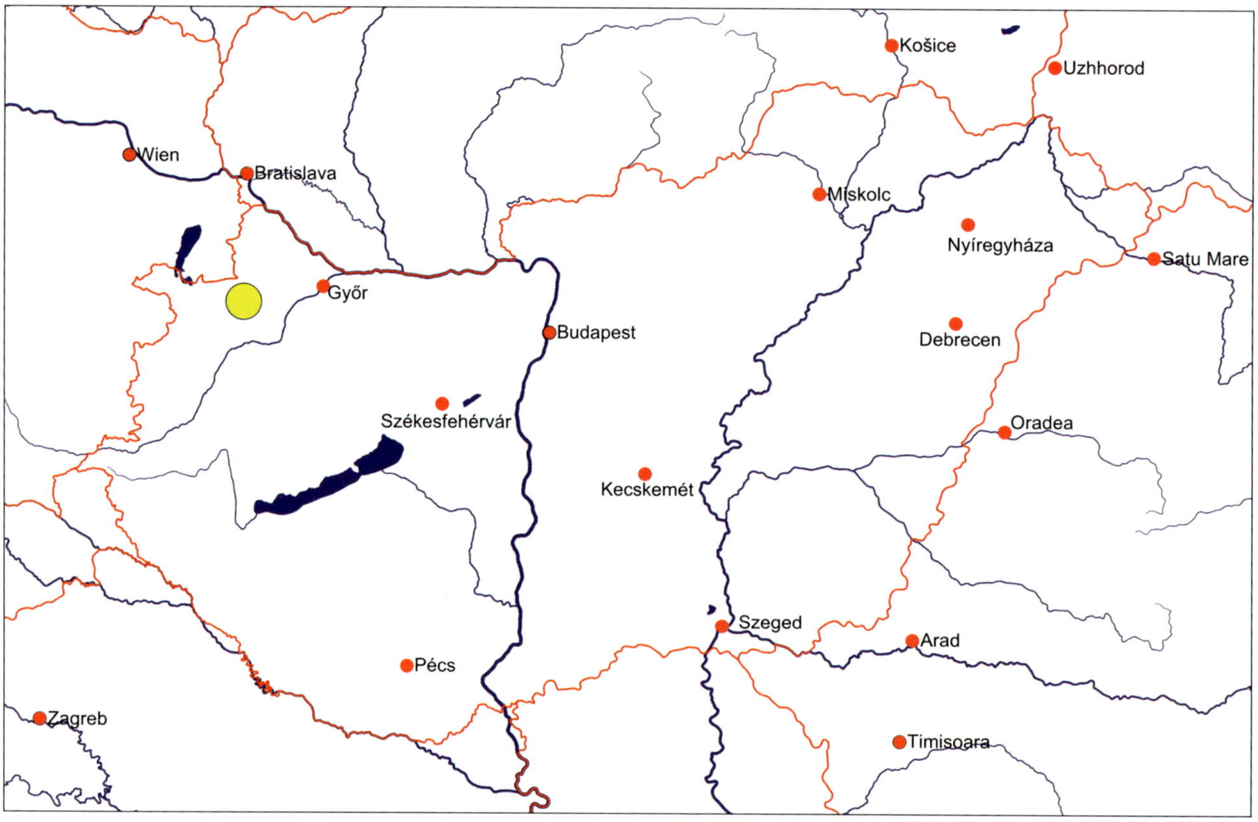

Figure 19.2 The geographical situtation of the Tóköz.

The experience of the last 15 years shows that in Hungary conditions are extremely good for this traditional method of remote sensing, especially in areas of alluvial gravel cones which can produce excellent cropmarking (for example: to the south from Budapest, see: Goguey *et al.* 2003, fig. 6). There are similar experiences from Burgundy (Goguey *et al.* 2003, 81), from Switzerland (Lekebusch & Nagy 1991, 6–7) and from Bavaria (Christlein & Braasch 1990, 33–7, Abb. 18, 20–1).

The Tóköz project

The Tóköz area is located in northwest Hungary to the southwest of Lake Fertő (Figure 19.2). The area of approximately 20 by 20 km lies at the border of two larger geographical units: the marshlands of Hanság and the plain of Kapuvár which consists of alluvial gravel deposits. Smaller and larger sand and gravel mounds are typical for both of them although they are of different origin. In the Hanság the development of dry surfaces suitable for human settlement began in the Pleistocene period, while on the plain of Kapuvár the activity of rivers created terrace-islands in the Holocene period.

This region became of interest in 1998 when it produced some cropmarked sites for the first time. In 2003, there was a pronounced summer drought and many crop marks were detected showing very fine details – and that is why we started a large-scale survey (Czajlik 2004, 112–5). According to our 15 years of research experience in this region there were no significant aerial archaeological results to be expected in 2005 and 2006 when the weather was not so dry. The continuation of the research was possible due to the droughts in 2007 and 2009. During these campaigns we identified many new aerial archaeological sites as well as observing sites discovered in 2003 and recording new information on them. Despite the fact that the cropmarks in 2007 were of different intensity (usually weaker) compared to 2003, good additional information was obtained about the sites.

One of the most important observations is that areas covered by ripening vegetation of yellowish colour are identical with the extents of the Pleistocene gravel bed, which rises in some cases to the surface, such as in areas south of Budapest. This has been confirmed during our field surveys. The thick gravel layer has developed through the activity of the ancient Danube and Rába rivers, and has been covered by younger sediments (sand and alluvial soils) brought there by the Rába river. The thickness of the younger sedimentary layers varies between 30 and 90cm. The interesting issue in the case of the Tóköz is the strong sensitivity to yearly precipitation (Figure 19.3) and the significant relationship between the locations of the gravel layers and the disposition of sites.

The evaluation of oblique aerial photographs – a comparison with field survey data

After the evaluation of all the aerial photographs and field surveys undertaken in the spring of 2008 we had to face the fact that we had significantly more knowledge about the researched area from the aerial

Figure 19.3 Spring flood in the Tóköz.

photos than that from field survey data. An important result of our field surveys is that the phenomena identified on photographs are often not accompanied by finds collected from the ploughed field surfaces. As a result, it is not possible to suggest dates for sites, even the larger ones. We have also seen that the information from aerial photographs can fill the gaps between the zones of artefact scatters in ploughed fields.

One of the best examples for this phenomenon is the result of the field survey of Veszkény – Keleti csapásra-dűlő (Figure 19.4). This area is flat and covered by alluvial soil. Although the aerial photographs show different types of structures (buildings with foundation trenches, ring-ditches) belonging to different archaeological periods, we did not make any finds during field-walking. A number of parallels for the buildings with foundation trenches recorded on the photograph were discovered in the neighbouring area during aerial survey. The photo map which has been made from the interpretation and transformation of the photographs taken at the site Szárföld-Átaljáró in 2003 and 2007 shows a good example for this (Figure 19.5). The settlement consists of buildings with foundation trenches that are surrounded by a double ditch and a palisade. It is notable that the orientation of the

Figure 19.4 Veszkény – Keleti csapásra-dűlő on 22. June 2003 Aerial archaeological site and its interpretation viewed from the south.

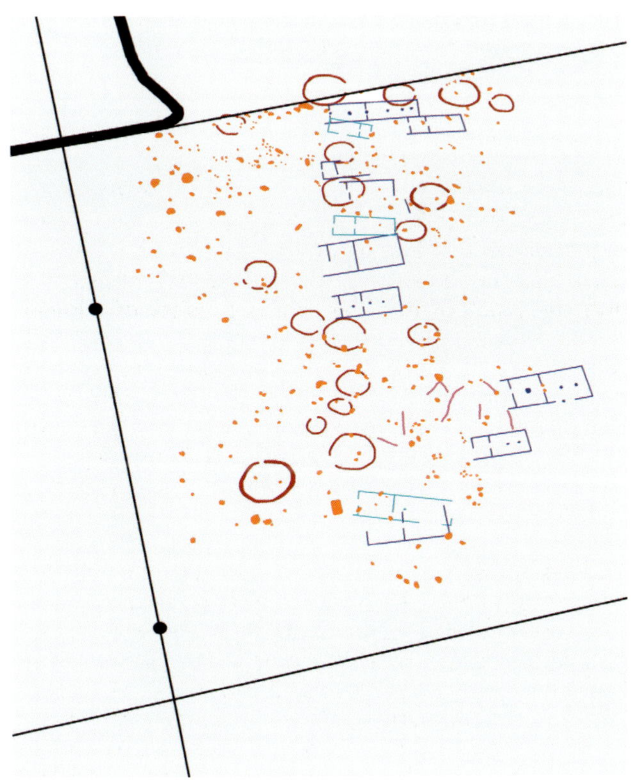

buildings is east-west and it is the same or similar in the entire zone no matter if they come in groups or stand alone (Figure 19.6). We can assume that the settlement units were uniform and possibly contemporary, even though we lack the surface finds to corroborate this.

In the northwest Transdanubian region there are a number of excavated buildings that show a bipartite structure of foundation trenches similar to the ones we know from aerial photographs. At Győr – Szabadrét-domb (Virág & Figler 2007, fig. 2.3–11) and Lébény – Bille-domb (Németh 1994, fig. 10–12) there are foundation trenches of groups of buildings as well as at Szárföld – Átaljáró and Osli. Bipartite buildings with foundation trenches are known also from Mosonszentmiklós – Egyéni-földek (Egry 2003). All the three excavated structures mentioned above belonged to the Balaton-Lasinja Culture. A number of buildings from Börcs – Paphomlok-dűlő (Figler 1994) and Győr-Marcalváros, Bevásárlóközpont (Egry 2001, fig. 59) have the same dating but they are known from preliminary reports only.

Another building type is known from Szárföld-Átaljáró – a nearly rectangular building consisting of one room. Its close parallels come from the Copper Age sites of Zalavár – Basasziget (Virág 2007) and Zamárdi – Kútvölgyi-dűlő (Kiss & Réti 2007).

We also know some well documented settlements lying at some distance from Szárföld but showing similar features. The settlements of Branč (Vladár & Lichardus 1968) and Nyitra Jelšovce near Nitra (Pavúk & Bátora 1995) are dated to the Copper Age Ludanice Culture according to their excavators. There is a fortification similar to the one at Szárföld – Átaljáró published by Pál Patay at Tiszalúc-Sarkad (Patay 2005). The disposition and the forms of the buildings inside the fortified area are similar to the ones at Szárföld, but instead of the foundation trenches there are rows of stakes.

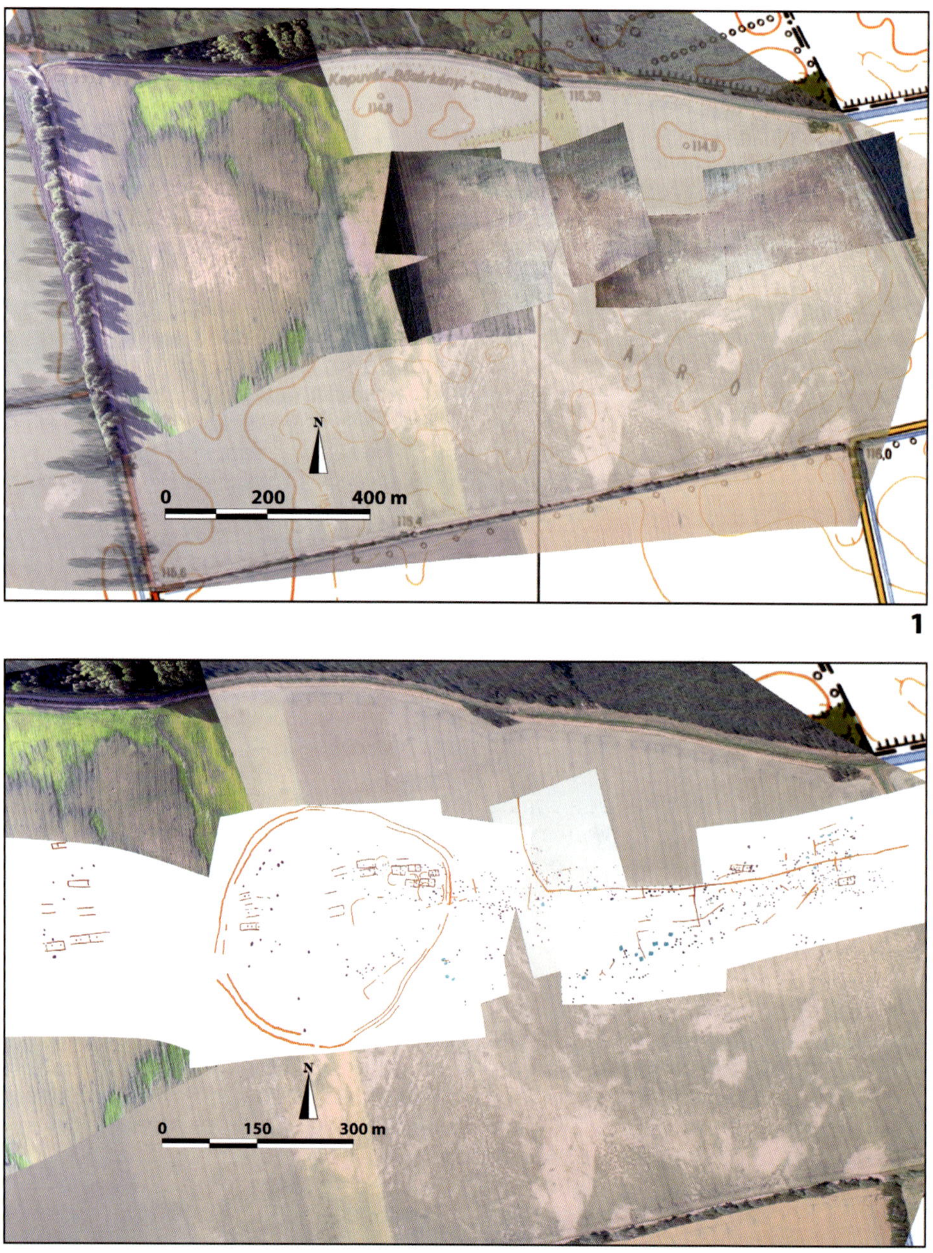

Figure 19.5 Szárföld – Átaljáró on 22 June 2003 and 10 June 2007. interpreted and transformed aerial photos and photo map.

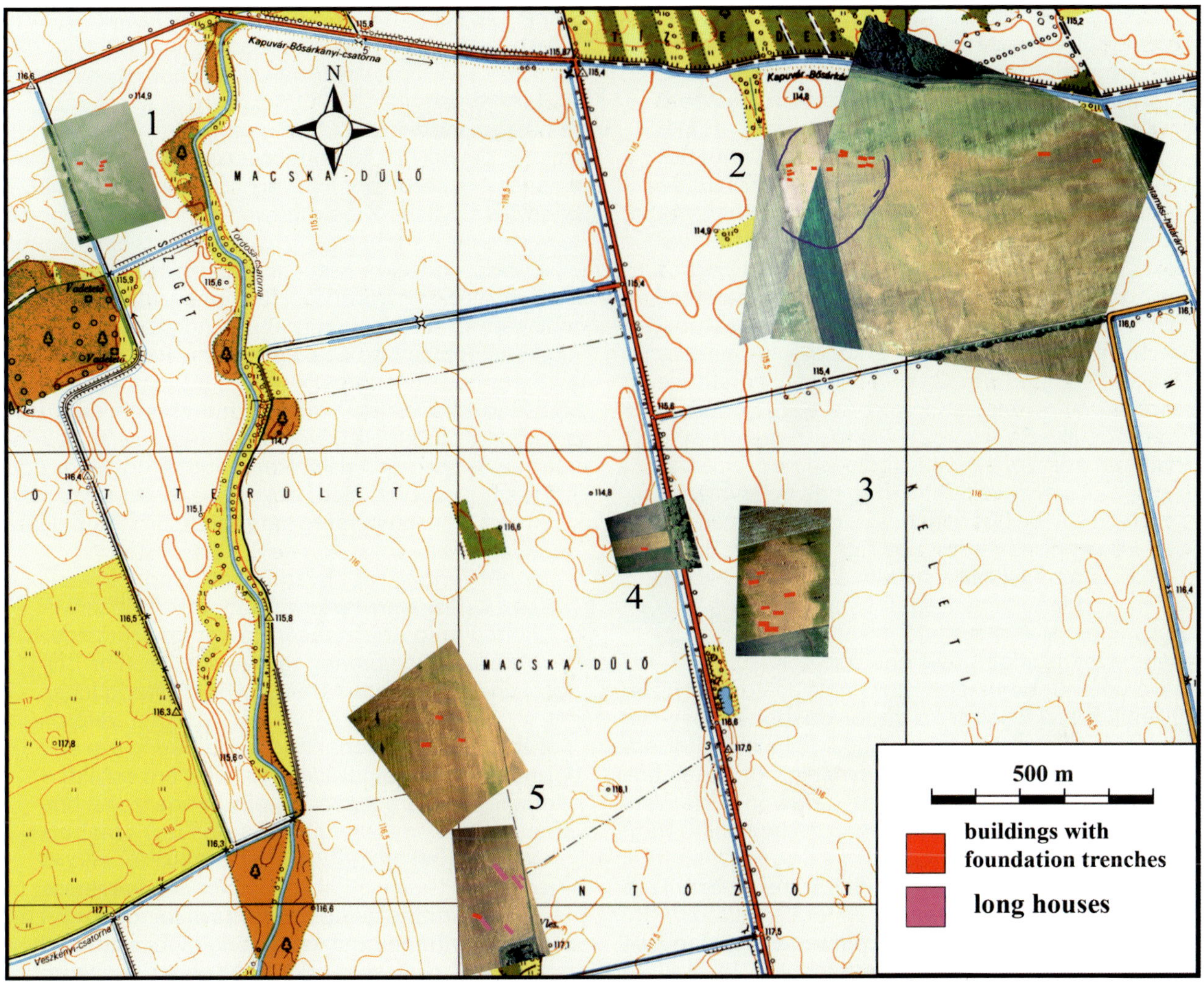

Figure 19.6 A photo-map showing Neolithic Age long-houses and Copper Age buildings with foundation trenches identified by aerial archaeology, north of Szárföld.
The sites on the rectified map: Osli – Tordosa-sziget (1), Szárföld – Átaljáró (2), Veszkény – Keleti csapásra-dűlő (3), Szárföld – Macska-dűlő (4) és Szárföld – Felső-tag (5). (Photos taken by Zoltán Czajlik on 22.06.2003 and 15.06.2007, rectified by László Rupnik).

North of Szárföld we have identified some structures near to each other. They form a scattered network of buildings, building groups and a fortification of a uniform east-west orientation. We can assume they belong to the Copper Age too, although it is necessary to carry out excavations in order to justify our hypothesis.

There are at least two archaeological periods present at Veszkény – Keleti csapás as we have mentioned above. The cropmarks can refer to geomorphological processes too; the yellow -brown zone is showing the top of a smaller mound of gravel that is covered by alluvial layers deposited in historic periods.

This archaeological research has explored the unusual properties of the alluvial zones of the Danube-Rába river system (Figler *et al.* 1997, 224). For the first time Eszter Szőnyi and Péter Tomka have identified an alluvial layer between the prehistoric and medieval layers at Csorna – Gázfogadó. Later they have established a more precise date between the early and late Roman Age for the same layer at Jobaháza – Borsody-dűlő. A similar alluvial layer has been identified by Károly Takács at Kapuvár and by Máté Losonczi at Öttevény. According to the excavations of Károly Takács at Mosonmagyaróvár – Vámház tér and Péter Tomka at Himod-Káposztáskertek it can be dated to the period of the Turkish occupation. This means that we have at least two periods (2nd and 3rd centuries AD and the period of the Turkish occupation in the 16–17th century AD) when a mosaic of land was flooded. The floods have covered with their deposits the earlier archaeological layers, and that is why the field surveys have partially no results there (Figure 19.7).

Summary

The problem of the sites inundated by silt is not unknown in archaeological research, for example in some regions of the Mediterranean, the Rhône and other parts of continental Europe (Howard *et al.* 2003). The methods used for the exploration of those zones are rather conventional: soil sampling by boring (Price *et al.* 1964) or the combined use of archaeological, geomorphological and GIS research. The latter method is used in the research of the alluvial zones of the Oise

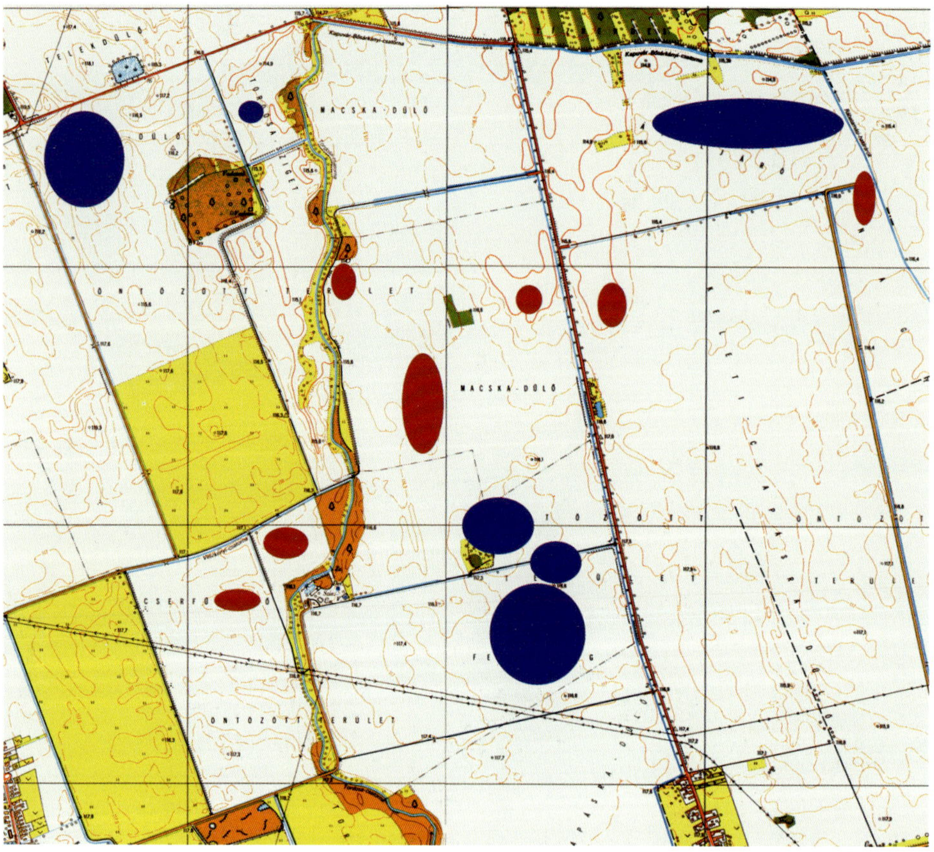

Figure 19.7 Overview of the researched area. Sites identified by field surveys and aerial photos (with blue colour). Sites identified on aerial photos only (with red colour).

(Rodriguez & Foucault 2005), the Loire (Cubizolle & Georges 2001) and the Rhône (Landuré *et al.* 2004). The use of the aerial photography is not widespread and we know of only a few attempts (Petit 1999).

The Tóköz project underlines the importance of using aerial photographic mapping in archaeological topographical research. It is very important for future archaeological and geographical research as it gives us more precise knowledge on the correspondence between the alluvial gravel cones and the cropmark phenomena.

The examples mentioned above show that the region of the Tóköz originally had an alluvial surface with gravel mounds appropriate for human habitation which silted up in historic periods. During this progress there was a significant change not just in the hydrological but in the geomorphological features of the region. The only way to obtain information on these developments is aerial reconnaissance and analysis of its results. The great number of structures which are only revealed by aerial reconnaissance has also been illustrated, demonstrating that the aerial photographic evidence does not just reinforce the field survey data but can complement it, giving more precise and detailed information on the types and extents of the sites than can be gathered from artefact scatters alone.

Acknowledgements

The research has been funded by the Hungarian Scientific Research Fund (proposals T043762 and NK68824) and supported by the Hungarian National Office for Cultural Heritage – Regional Department of the Western-Transdanubia. We have to thank Eszter Szőnyi, Péter Tomka and Károly Takács for sharing their observations. (Data collection of Xántus János Múzeum in Győr: RA.9–99., RA. 8–99., RA. 13–99., RA. 93–2003., RA. 52–2005., RA 36–200., and the report of the National Office for Cultural Heritage 410/1061/2007).

References

Christlein, R. & Braasch, O. 1990: *Das unterirdische Bayern: 7000 Jahre Geschichte und Archäologie im Luftbild*, Stuttgart, 1990.

Cubizolle, H. & Georges, V. 2001: L'évolution holocène de la plaine alluviale de la Loire dans le bassin du Forez (Massif Central, France). Mise en évidence du potentiel géoarchéologique. *Quaternaire* 12, 2001, 53–67.

Czajlik, Z. 2004: Légirégészeti kutatások Magyarországon 2003-ban (Rövid beszámoló az ELTE Régészettudományi Intézetének Térinformatikai Kutatólaboratóriumában folyó munkáról.) – Aerial archaeological investigations in Hungary in 2003 (A short report on the work done in the GIS Research Laboratory of the Institute of Archaeological Sciences, Eötvös Loránd University). *Régészeti kutatások Magyarországon – Archaeological Investigations in Hungary, 2003.*, KÖH – Magyar Nemzeti Múzeum, Budapest, 111–25.

Czajlik, Z. & Holl, B. 1996: The Archaeological Chronology and the Excavation Data Bases. In Marton, E. (ed.): *The Future of Our Past '93–'95*, MNM, Budapest, 144–50.

Czajlik, Z., Marton, Á. & Holl, B. 1997: Az M3-as autópálya régészeti leletmentéseinek térinformatikai feldolgozása Hajdú-Bihar megyében – The GIS Processing of the Rescue Excavations Associated with the M3 Motorway in Hajdú-Bihar County. In Kovács, T. Anders, A. & Raczky, P. (eds): *Utak a Múltba – Paths into the Past – Az M3-as autópálya régészeti leletmentései – Rescue Excavations on the M3 Motorways*, Budapest, 153–5.

Egry, I.M. 2001: Beszámoló a Győr-Marcalváros – bevásárlóközpont területén végzett megelőző régészeti feltárásról – Report on the Preliminary Exploration Carried out on the Area of Győr-Marcalváros – Shopping Centre. *Arrabona* 39, 57–78.

Egry, I.M. 2003: Rézkori településrészlet Mosonszentmiklós – Egyéni-földek lelőhelyen – Das Detail einer kupferzeitlichen Siedlung auf dem Fundort Mosonszentmiklós – Egyéni-földek. *Móra Ferenc Múzeum Évkönyve – Studia Archaeologica* 9, 95–100.

Figler, A. 1994: Börcs – Paphomlok-dűlő (M1 Börcsi pihenő). *Régészeti Füzetek* Ser. 1 No 46. 1994, 7.

Figler, A. Bartosiewicz, L. Füleky, Gy. & Hertelendi, E. 1997: Copper Age Settlement and the Danube Water System: a case study from North Western Hungary. In Chapman, J. & Dolukhanov, P. (eds): *Landscapes in Flux. Central and Eastern Europe in Antiquity*. 209–30.

Goguey, R. & Szabó, M. 1995: *L'histoire vue du ciel. Photographie aérienne et archéologie en France et en Hongrie – Légi fényképezés és régészet Franciaországban és Magyarországon*, Budapest, 1995.

Goguey, R., Czajlik, Z. & Bödőcs, A. 2003: Magyar-francia légi régészeti kutatások Magyarországon – Recherches franco-hongroises d'archéologie aérienne en Hongrie en 2000. *Régészeti kutatások Magyarországon – Archaeological Investigations in Hungary, 2000*, KÖH – Magyar Nemzeti Múzeum, Budapest, 2003, 75–86.

Howard, A.J., Macklin, M.G. & Passmore, D.G. 2003: *Alluvial archaeology in Europe: proceedings of the Alluvial Archaeology of North-West Europe and Mediterranean, 18–19 December 2000, Leeds, UK*, Lisse, 2003.

Kiss, V. & Réti, Zs. 2007: Középső rézkori alapárkos ház Zamárdiból (A háromdimenziós számítógépes grafika alkalmazásának lehetőségei régészeti rekonstrukciók készítésénél) (The Middle Copper Age house at Zamárdi. The potentials of 3D computer graphics in archaeological reconstructions. *Ősrégészeti Levelek* 7(2005), 73–90.

Landuré, C., Pasqualini, M. & Guilcher, A. (dir.) 2004: Delta du Rhône. Camargue antique et médiévale. *Supplément au Bulletin Archéologique de Provence*, 2, Aix-en-Provence, 2004.

Lekebusch, J. & Nagy, P. 1991: *Prospektionsmethoden in der Archäologie*, Zürich, 1991.

T. Németh, G. 1994: Vorbericht über spätneolitische und frühkupferzeitliche Siedlungsspuren bei Lébény (Westungarn) – Késő neolit és rézkori településnyomok Lébény határában. *Jósa András Múzeum Évkönyve* 36, 241–61.

Patay, P. 2005: *Kupferzeitliche Siedlung von Tiszalúc*. Inventaria Praehistorica Hungariae XI., Budapest, 2005.

Pavúk, J. & Bátora, J. 1995: *Siedlung und Gräber der Ludanice-Gruppe in Jelšovce*. Nitra, 1995.

Petit, Ch. (dir.) 1999: *Occupation, gestion des plaines alluviales dans le Nord de la France de l'âge du fer à l'époque gallo-romaine. Table ronde de Molesme, 17 et 18 sept. 1999*. Besançon, 2005.

Price, J.C., Hunter, R.G. & McMichael, E.W. 1964: Core Drilling in an Archaeological Site. *American Antiquity* 30/2, 1964, 219–22.

Raczky, P., Czajlik, Z., Holl, B. & Marton, Á. 1998: GIS and the evaluation of rescue excavations along the M3 Motorway in Hungary. *Porocilo* 24, 1998, 157–70.

Raczky, P., Puszta, S., Czajlik, Z., Holl, B. & Marton, Á. 2002: Investigation of archaeological sites with the integrated application of aerial photography, magnetometer surveys and archaeological excavation. In Jerem, E.T., Bíró, K. & Rudner, E. (eds): *Archaeometry 98. Proceedings of the 31st Symposium Budapest, April 26 – May 3 1998, Volume I*, BAR International Series 1043(I), 135–43.

Rodriguez, P. & Foucault, M. 2005: La plaine alluviale de l'Oise : milieu et système fluvial; stratigraphie des formations alluviales et archéologie. *Revue Archéologique du Centre de la France* 44, 2005, 97–107.

Virág, Zs. 2007: Középső rézkori épületek rekonstrukciós lehetőségeiről (Reconstruction of a Middle Copper Age house). *Ősrégészeti Levelek* 7, 2005, 60–72.

Virág, Zs. & Figler, A. 2007: Data on the settlement history of the Late Lengyel Period of Transdanubia on the basis of two sites from the Kisalföld (Small Hungarian Plain). In Kozlowski, J. & Raczky, P. (eds): *The Lengyel, Polgár and related cultures in the middle/late Neolithic in Central Europe*. Kraków, 345–64.

Vladár, J. & Lichardus, J. 1968: Erforschung der frühäneolithischen Siedlungen in Branč. *Slovenska Archaeologica* 16, 263–352.

20 | The archaeological landscape of northeast Iceland: a ghost of a Viking Age society

Árni Einarsson and *Oscar Aldred*

Abstract: A current mapping project of medieval archaeology in an area of over 3,000km^2 in northeastern Iceland makes use of aerial photography on a large scale, including over 2,000 low altitude oblique photographs. Aided by ground surveys and other fieldwork, the project has revealed a whole system of well-preserved remains dating to the medieval period. An extensive network of turf walls is most prominent, but farmsteads, churches, assembly sites, pagan grave fields, peat cutting, charcoal pits and horse tracks are also clearly visible. This medieval archaeological landscape belongs largely to the Viking period and is a relic of the settlement pattern of the first few generations in Iceland. Its high conservation value and future preservation is discussed in the paper.

Introduction

Iceland was discovered and colonised in the late 9th century, following the westward expansion of the Norse seafaring Viking culture. The settlement period and the subsequent three centuries have been immortalised in the rich medieval saga literature of the Icelanders. This period was remarkable for the creation of a new society with a hierarchical system of assemblies with legislative and judicial functions, and the distribution of power among a number of chieftains instead of one central authority. This political system was to last more than three centuries, until Iceland became part of the Norwegian and later, Danish kingdom.

Much of the early archaeological work in Iceland was concentrated on the Viking period (Vésteinsson 2004a; Friðriksson 1994) because it represents a 'glorious past' and also because the identity of the Icelandic nation is rooted in events and processes in this period. The Viking period is still a major focus in the archaeology of Iceland but, in line with global trends in science, the emphasis is now on the dynamic interaction of humans and their environment. This entails a greater focus on off-site archaeology, primarily associated with the use of resources. This wider scope calls for a landscape approach with large scale dating and registration of monuments, preferably aided by remote sensing.

This paper provides the context for and the description of the findings from a research project that has used aerial photography on an unprecedented scale for Iceland. Initial research began as a study of an extensive system of medieval turf walls in northeastern Iceland, using vertical aerial photographs, revealing a complex division of the landscape (Einarsson *et al.* 2002). The project then continued, involving wide ranging oblique low altitude aerial photography supported by archaeological excavations and ground surveys. Augmented by related projects, as well as existing knowledge (McGovern *et al.* 2007), this effort gradually revealed a whole network of walls and other features (Figure 20.1) that were roughly contemporaneous and contained important elements of the Viking period society. Apart from turf walls, these features included assembly sites, burial sites, churches/chapels and farm sites of various characters, and also features that have not been properly dated yet, like routes, peat cuttings and charcoal pits. Many of the oldest features are so well preserved, widespread and prominent that we have been tempted to describe this archaeological landscape as a relic of the Viking period.

Figure 20.1: The great turf wall at Fljótsheiði. Note how the wall traces the edge of the bog to the right. The wall continues in the upper left corner of the photograph. Some of the main horse tracks can be seen as winding linear features and there are peat pits to the right.

In this paper we first give an overview of the use of aerial photography in Iceland for cartography and archaeology. We then focus on our study area in the northeast of Iceland, listing some of the most prominent types of ancient remains we see from the air, Viking period or later, and finally discussing issues relating to heritage management and the conservation value and preservation of the archaeological landscape.

The use of aerial observations in Icelandic archaeology

The history of aerial photography in Iceland can be traced back to oblique photographs taken of Reykjavík in 1919. However, aerial photography was not systematically carried out until 1937-8, when it was used to compliment the ground surveys for the first detailed maps of Iceland at a published scale of 1:100,000 (Bragason & Guðmundsson 1988). World War II marked an expansion in the areas covered by aerial photographs, again for cartography. However, these series of vertical aerial photographs taken by the German, British and American military, as well as later by the National Land Survey of Iceland (Landmælingar Íslands), are an invaluable resource in two related but different ways. The first is tracing landscape change over the last 60 years. This is a period that has seen radical changes in the landscape, especially around farms and urban centres. Secondly these aerial sources are invaluable in the mapping of archaeological sites. The retrospective mapping of archaeology from the earliest available sources often provides a much better understanding of landscape formation and organisation and the status of archaeological features than using conventional map sources (Aldred *et al.* 2010).

In the late 1970s and early 1980s Sveinbjörn Rafnsson used oblique aerial photography as an integral part of his archaeological survey of deserted valleys in the east of Iceland (Rafnsson 1990). This was the first practical use of aerial survey specifically for archaeology. Sveinbjörn used both conventional and infra-red images. However, the flights were largely directed at previously known sites in order to obtain a basic record, and no detailed transcription mapping was involved. In the 1980s, Guðrún Sveinbjarnardóttir conducted regional studies of farm abandonment in Iceland for her doctoral thesis (Sveinbjarnardóttir 1992). She made occasional use of aerial photographs, although this was hampered by the high altitude of the photography and the difficulty in recognising distinct features. In the 1990s, the Institute of Archaeology, Iceland (Fornleifastofnun Íslands), amongst other professional archaeologists, began extensive field surveys. At first, aerial reconnaissance was incorporated into the survey, taking oblique shots in both conventional and infra-red format. This programme, however, was short-lived.

The first publication that presented the explicit use of aerial photographs in archaeology in Iceland was in 1995 (Ísaksson & Helgason 1995). The article discussed the use of the technique in relation to features on farms and land around Reykjavík and particularly highlighted the advantages of photography under light snow cover. Photographs had been taken of the same site under different conditions, using both conventional and infra-red film. The article also attempted to discuss the specific nature of aerial survey in Iceland as well as promote its more systematic use in archaeology.

Although archaeological survey has a long tradition in Iceland, beginning in the 19th century (Friðriksson 1994), it is only in the last quarter of the 20th century that the first systematic archaeological surveys of the country commenced. In the first decade of the 21st century tremendous progress has been made and the number of sites registered increased dramatically. In the same period, usage of aerial photographs in archaeological survey multiplied and currently most archaeological surveys use aerial photographs at some stage. In most cases, these are vertical aerial photographs, often from

Figure 20.2: A medieval turf wall.

a fairly high altitude of 18,000ft or 5,486m, with a focal length of 152mm (a scale of 1:36,000).

Recent survey, driven by new legislation in 1989, has aimed to achieve total coverage of districts and regions (Ólafsson 1991; Friðriksson & Vésteinsson 1998). The underlying premise of contemporary practice is to systematically register all surviving remains to a set standard. The registration is informed by textual and oral histories. Although these are invaluable sources in discerning the presence of archaeology, textual sources can be fragmentary in coverage, and the use of oral sources, derived from place-name surveys, is based on an assumption of continuity in the social memory from one generation to the next. Aerial photography provides valuable additional records of sites which may have been identified by other means. It reveals certain types of sites (e.g. boundaries and charcoal pits) that tend to be under-represented in the other types of evidence and which are difficult to find or comprehend through conventional field walking (Figure 20.2).

While there has been an increase in the use of aerial photographs and other remote sensing in recent years, these techniques are still not routinely used in archaeology in Iceland. Aerial sources have been used *ad hoc* for both research and heritage management, and there are several projects conducted in the last few years that have been laying the foundation for more routine use. These have developed best practices and demonstrate the benefits for archaeology and landscape studies. A few of these projects are discussed below.

A project on the history of the human habitation of the Skagafjörður region in northern Iceland used low altitude oblique aerial photography to register the deserted farmsites in the most remote valleys (Pálsson 1999–2010). Another project on the early arable cultivation in Iceland made extensive use of conventional and infra-red aerial photographs, as well as other sources, to demonstrate the presence or absence of ancient sites with cultivation remains, and to study the discontinuity of arable cultivation in relation to the deteriorating climate from the 14th century onwards (Guðmundsson *et al.* 2004). In addition, another project, conducted in 2007 by an undergraduate student at the University of Iceland, assessed the utility of high altitude vertical aerial photographs for archaeological survey. The photographs were viewed at a magnification of 10x and while new features were mapped, the results showed some limitations. For example, mainly linear and larger features such as areas of peat cutting and boundaries were added to the records, but the ability to recognise smaller and more discrete sites was limited by the photograph scale and resolution (Sveinbjarnarson 2007). In another project, four landscape areas covering about 200km^2 were mapped using DigitalGlobe, pan-sharpened Natural colour and DRA contrast enhanced 0.6m pixel resolution satellite imagery, captured between 2002 and 2006 (Lárusdóttir & Aldred 2008). These projects demonstrated the relative ease of both creating and using aerial and satellite sources to expand and enhance our existing knowledge of archaeology, which can be further illustrated in the main case study of this paper – the ongoing work in northeast Iceland.

The aerial archaeology of northeast Iceland

The main project that encapsulates the foundation work and best practices for aerial survey is focused on northeast Iceland. It was carried out by the Institute of Archaeology and the Mývatn Research Station and funded by the Icelandic Centre for Research (RANNÍS) and Þjóðhátíðarsjóður (a fund that commemorates the 9th century settlement of Iceland) to map the extent and preservation of the ancient wall systems in northeast Iceland. The aim was to determine their date, spatial extent, structure and possible function. The project systematically employed aerial photographs, both verticals and obliques, as well as satellite imagery, to map the archaeology of a large area. At last count over 3,000km^2 have been covered and about 400km of walls have been mapped.

The study area

Our study area covers 3,164km^2 in the Counties of Suður- and Norður Þingeyjarsýsla (Figure 20.3). The area can conveniently be divided into four main geographical units. Two of them (Kelduhverfi and Mývatnssveit) are within the so-called neo-volcanic zone characterised by extensive postglacial lava fields and flat landscape interrupted by Pleistocene ridges of hyaloclastite, indicative of subglacial eruptions. The Mývatn area (Mývatnssveit) is dominated by Lake Mývatn, a 37km^2 shallow eutrophic lake about 50km from the coast, surrounded by dry lava fields on one side and extensive wetlands on the other. Modern day farms are situated around the lakeshore and around fertile wetlands on the River Kráká delta to the south. Kelduhverfi, down by the coast, is dominated by flat lava fields originating in the shield volcano of Þeistareykjabunga to the south and on the north side by sandur plains deposited by the glacial River Jökulsá á Fjöllum. The modern day farms lie on the border between the lava and the sandur plain, but a large number of ancient deserted farms lie on the lava field to the south.

A third landscape area is the valley and moorland between Lake Mývatn and the coast. This is outside the zone of present day volcanic activity although big prehistoric lava streams did flow down two of the valleys. The bedrock is moraine-covered interglacial basalt carved by ice and with a rather thin layer of organic soil. The shallow valleys are orientated north-south and provide the focus for most of the present day farms. They are separated by ridges of moorland that extend like fingers from the highland plateau to the south. The fourth landscape area is the peninsula of Tjörnes that lies between Kelduhverfi and Húsavík. This area has a low rocky coastline with small streams at regular intervals, and at present only the western half is inhabited. The bedrock is the same as described before but with thick banks of raised marine sediments on the coast.

The whole habitable area of northeast Iceland was apparently covered with birch (*Betula pubescens*) scrubland at the time of first settlement. Today it is almost devoid of woodland. The vegetation on the moorland and lava areas, and in some of the uninhabited parts of the valleys, is mostly heath-like (Nielsen 1995), dominated by dwarf birch (*B. nana*) and

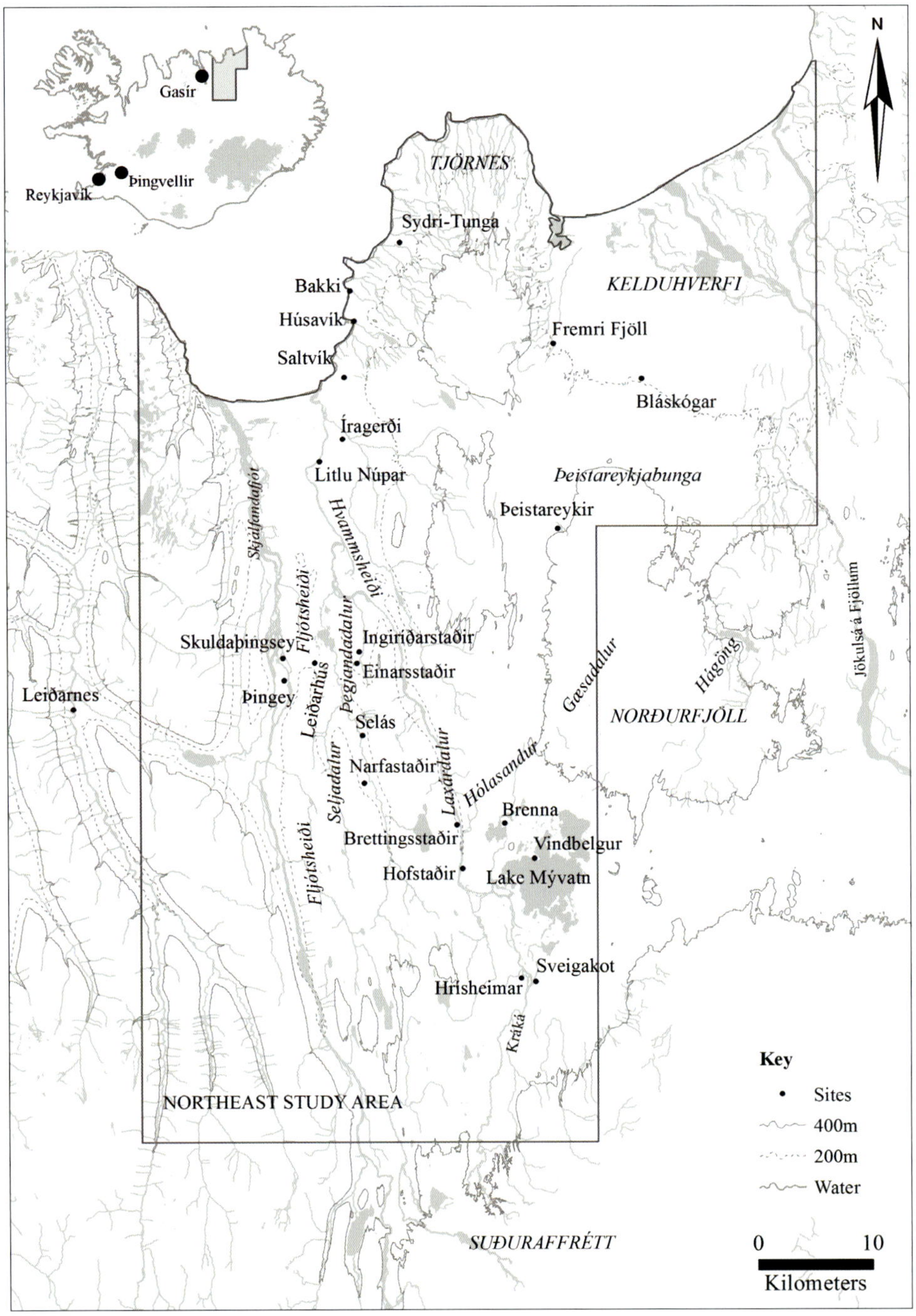

Figure 20.3: The study area in northeast Iceland.

crowberry (*Empetrum nigrum*). The inhabited parts of the study area have much more grassland. Hay for winter fodder is the only substantial crop produced on the hayfields around the farms. Bogs and other wetlands are scattered throughout the area except in the largest lava fields. Extensive blanket and string bogs are a characteristic of the southernmost moorlands. The soil is minerogenic and rather thin (commonly about 1m). The soil cover is mostly continuous but is locally eroded in many exposed and steep places. A massive erosion front migrating from the highlands reached the area south and east of Lake Mývatn in the 17th – 18th century and another erosion front is active in the area northwest of Lake Mývatn (Hólasandur).

The study area has been inhabited from the very beginning of settlement and like elsewhere in Iceland was based on dispersed single-household farms, often accommodating tenant farms within their territory. Wildlife resources include fish such as salmon, trout and Arctic charr in the lakes and rivers and abundant sea fish, mostly cod. Rock ptarmigan, Arctic fox and gyrfalcon occur in the dry upland and coastal areas, and seabirds (and their eggs) on the coastal islands. Duck eggs are utilised extensively in Lake Mývatn and there is a big eider colony on the coast. All these resources were utilised in the Viking period, and the coastal harvest was brought inland (McGovern *et al.* 2006).

Categories of archaeological features

Although the main task of the project was to map the extensive wall system, the aerial observations have also registered a large number of other archaeological features. Below are some of the main characteristics of the archaeology derived from the aerial surveys. Some of the listed features below date from the Viking period.

Walls

The ancient turf walls are the most prominent feature of the archaeological landscape (Figure 20.4). They run long distances, criss-crossing the moorlands and heaths (a total of about 400km at the last count; Einarsson *et al.* 2002; Aldred 2008). They seem roughly contemporaneous, though there are indications of multiple phases of construction in some areas (e.g. Figure 20.5). For the most part, a basic pattern can be discerned (Figure 20.6), modified by only minor repairs, rebuilds and additions, and it appears that the walls went out of use a few generations after they were built. Dating of some, using tephra (volcanic ash) from 27 trenches to determine the latest and earliest times a wall may have been constructed or fallen out of repair, places the majority in the 10th to 11th centuries. Their maintenance was discontinued sometime before the 13th century and all the dated walls, except one, had collapsed long before a characteristic tephra layer from AD 1477 was deposited. The focus of the study has been on the outfield walls, less on infield ones encircling the hayfields which may have a more complex history due to wall building activities associated with the 18th – 19th century agricultural reformation.

The walls form a basic pattern of home range enclosures (Figure 20.4), each one corresponding to an individual farm but subdivided for more local management of grazing, stock manipulation and protection of growing crops. In the valley landscape the walls run uphill on the probable boundaries between neighbouring farms. A horizontal wall on the hillside divides the rangeland above the farm into a near and far section. It is uncertain if the horizontal wall marks a limit of ownership or is just conveniently placed for stock management. The walls,

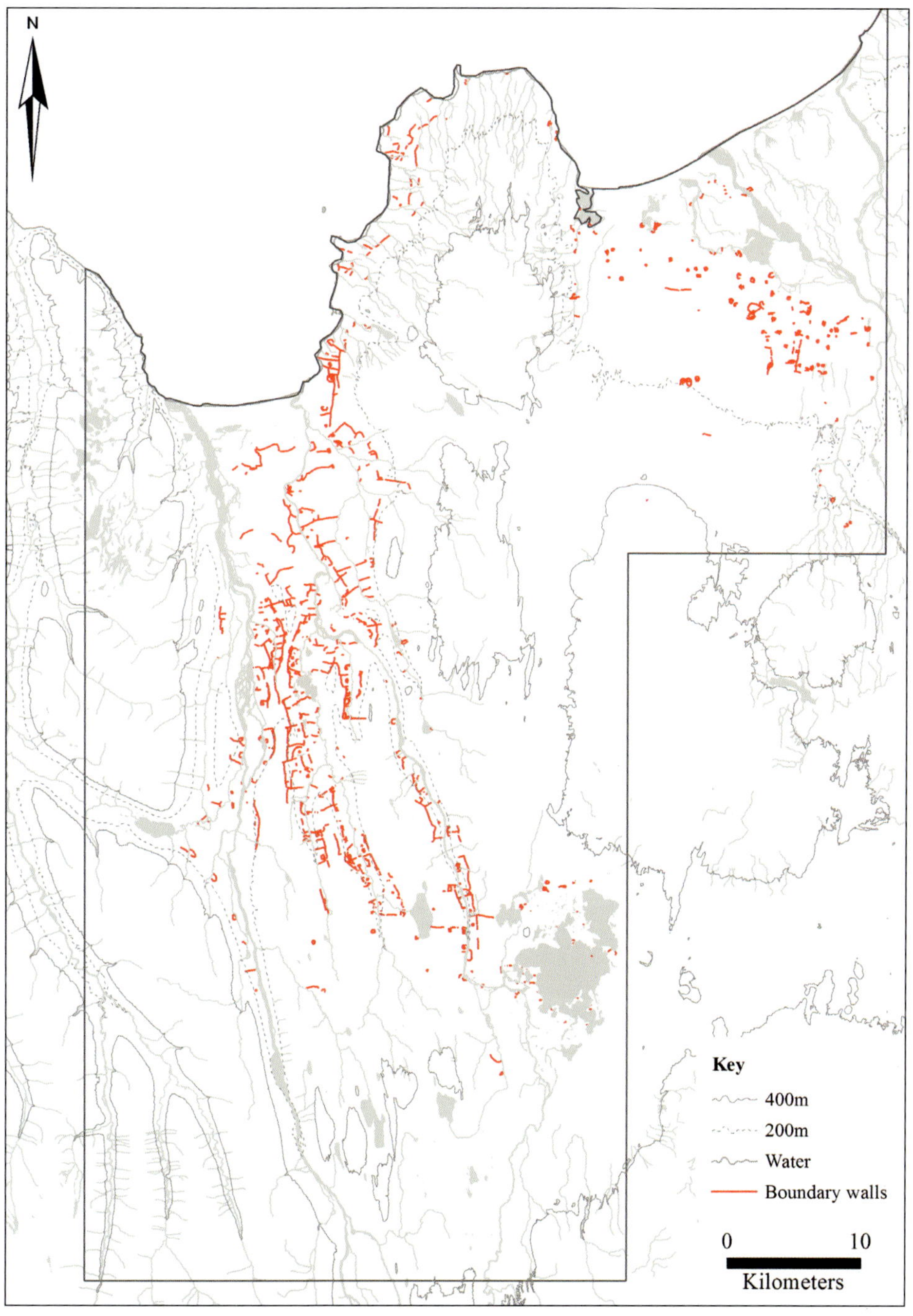

Figure 20.4: A map of the study area in northeast Iceland showing the extent of the medieval wall system.

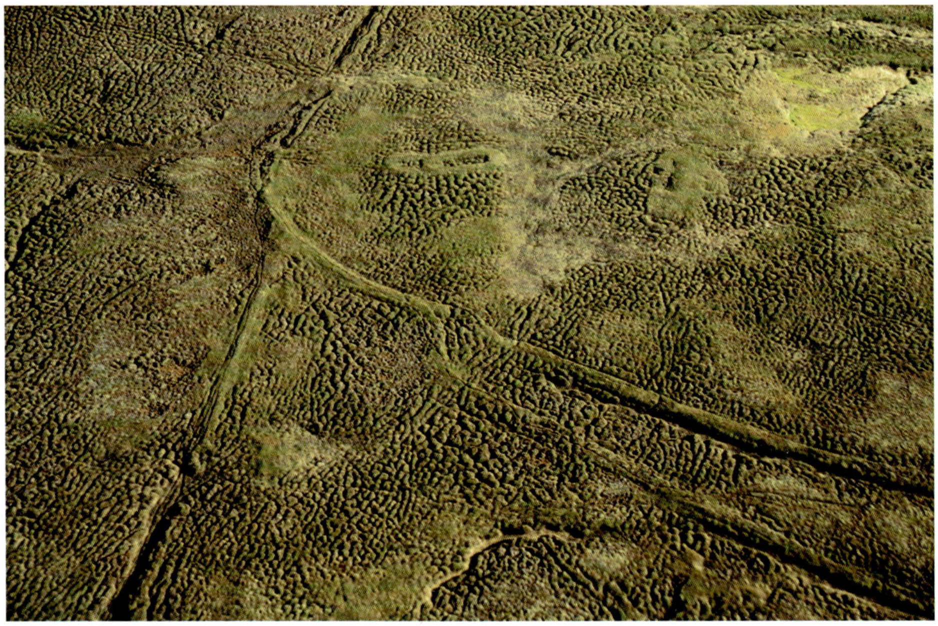

Figure 20.5: An ancient farm site with walls at Höskuldsstaðir at Fjótsheiði.

enclosing the farm on three sides (the river typically closes the boundary on the downhill side), look like they have been built as a single entity. The horizontal wall usually joins similar ones at the neighbouring farms, creating a continuous structure that fences entire valleys from the surrounding hills.

Most of the walls have collapsed and can only been seen as low earthworks in the landscape. What makes them prominent, especially from the air, is that the collapsed wall is very broad (commonly 4–6m across) and that the ditches on each side, from where the turf had been dug, are still quite deep (often expanded by erosion) and their vegetation differs from that on top of the collapsed wall. The result is that from the air many of the walls look like huge wheel tracks (Figure 20.7).

The wall system is very extensive, and clearly the most extensive archaeological phenomenon in Iceland. The challenge to get it mapped was a major impetus for the aerial surveys presented in this paper. After trying several methods for effective mapping we decided that none of them was working successfully on its own. We ended up with a combination of methods involving both oblique and vertical photographs (the latter being the standard source for geodetic purposes) and field walking. The oblique photos turned out to be essential for the detailed interpretation of the walls in the landscape and often also to ascertain their absence.

Small enclosures

A variety of small enclosures can be seen from the air, but two particular types deserve attention because of their well defined geometry. One type is square on plan, measuring about 20m across and usually located on the infield side of the wall system, sometimes isolated, sometimes combined with the walls and

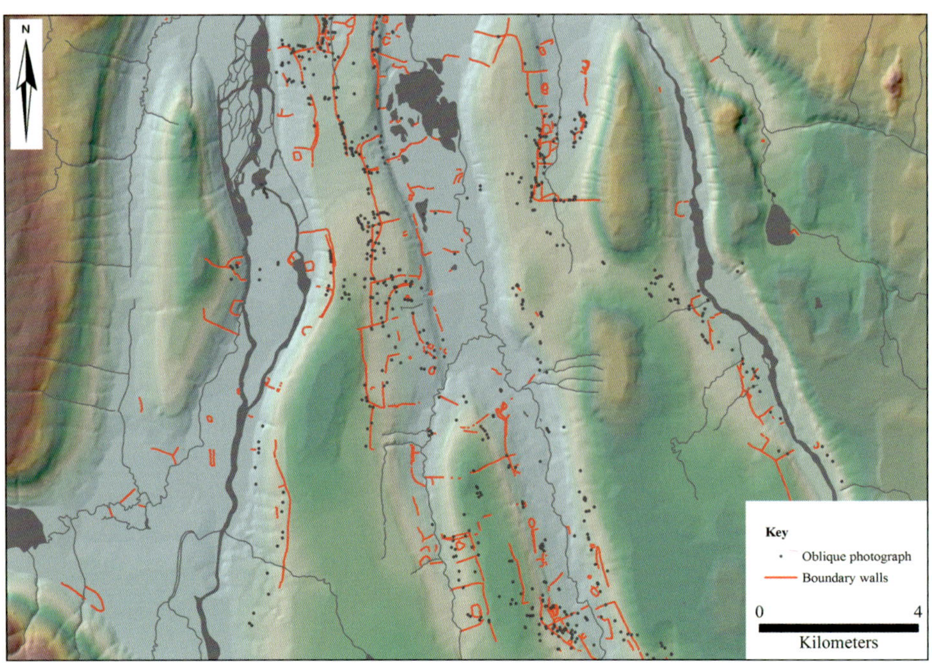

Figure 20.6: A close-up of the medieval turf wall system at Fljótsheiði.

Figure 20.7: Walls in Laxárdalsheiði by the river Laxá.

can then be assumed to be contemporaneous. These square enclosures can for example be seen at the farm sites of Narfastaðir, Ingiríðarstaðir, Einarsstaðir and Brettingsstaðir (see Eldjárn (1981) for similar structures in Skíðadalur, North Iceland). An archaeological excavation underway at Ingiríðarstaðir has discovered ard-marks inside one of the enclosures (Howell Roberts pers. com.). The other type is represented by perfectly circular enclosures, 6–12m in diameter. Further studies are needed to determine if these two types of enclosures represent functional groups or not.

Charcoal pits

The natural climax vegetation in most of the study area is birch woodland. Most of this disappeared rather quickly after settlement (landnám), or in less than a century in the south of Iceland but more gradually in the west and north, including our study area (Hallsdóttir 1987; Lawson *et al.* 2007; Lawson 2010; Gathorne-Hardy *et al.* 2009). Our aerial surveys have disclosed hundreds of charcoal pits scattered between Lake Mývatn and the coast. They tend to occur in clusters on low ridges or flat slopes not far from major horse tracks. Individual pits within a cluster are spaced some 50–100m apart. Sometimes, the pits occur in pairs. Most are about 2m across on the surface (max. 4m) and about 50cm deep and look square-shaped from the air.

Excavations reveal an original circular outline and the squarish shape must be a secondary feature. On the ground a raised rim of soil upcast is clearly visible, but this is less apparent from the air (Figure 20.8). Nine pits in one cluster have been dated, and all fall within the period AD 1000–1200, except one that is slightly younger, but still dates before 1300 (Church *et al.* 2007). Charcoal was produced throughout the history of Iceland, up to the middle of the 20th century. The product was used locally for smithy work (including scythe sharpening), and in medieval times also for the smelting of bog iron.

Figure 20.8: Charcoal pits on Laxárdalsheiði.

Horse tracks

Overland transport in Iceland has taken place either on foot or on horses, and wheeled transport did not exist until the late 19th century when the first roads were made for wheeled carriages. Most road construction was in the form of 'bridges', i.e. short dykes of sod or stone across rock clefts or marshes. Travellers on horseback kept to traditional routes that tended to follow the contours of the landscape. Aerial surveys easily identify horse tracks. The hooves of the horse cut quickly through the topsoil and soon a deep track was formed and then parallel ones when the earlier tracks became too deep to use. The most travelled routes thus developed multiple tracks, with tens or even a hundred furrows in a braided spread across a wide area (Figure 20.9). There are two types of tracks inscribed into the landscape. Inter-region tracks allowed for the movement of goods and people between a resource area and a distribution centre, while intra-region tracks connect individual settlements, as well as activity areas within farmland. The intensity, as well as duration of use, is represented by how deeply incised the track is, as well as its breadth. Fainter traces of tracks, perhaps less used rather than any later in date, are evident in many areas. Occasionally tracks run along the line of walls or on top of them.

The horse tracks are obviously an accumulation of over 1,000 years of transport history and some may retain their medieval period locations and some may not. Many of the tracks seem to postdate the walls as they appear to cut across them. Interestingly, some tracks can still be seen leading to farms that were deserted centuries ago. The most impressive horse tracks in the study area are south of the trading harbour of Húsavík and branches of this massive track can be traced into all the main valleys and all the way to the Lake Mývatn region. Some of the tracks may relate to transport of sulphur from the areas east and southeast of Lake Mývatn to Húsavík and perhaps Gásir by Eyjafjörður for export to continental Europe. Sulphur was mined in this region over seven to eight centuries. In one large area, Hólasandur between Mývatn and Húsavík, the topsoil has been blown away and the underlying sandy subsoil has been exposed. Here the main track expands into a 6–9m broad road where rocks have been cleared to the side to create a soft substrate for riding. The age of this unique construction is unknown but it is probably post-medieval, judging from preliminary dating of the erosion (unpublished data).

Hay stores

Early 20th century cultivated hayfields only provided about half of the total hay production in the study area (Hólmgeirsson 1978). Most of the other hay came from wetlands. According to 19th and 20th century ethnographic sources, the hay was stacked on raised ground on the edge of the marsh or, if the marsh was large, on specially built platforms (in Icelandic = *heystæði*), until it could be transported by sledge in winter. These hay platforms, square or oval in shape, were used extensively until the mid 20th century and are clearly visible from the air (for example, south of Lake Mývatn, or on the islands within the Lake Mývatn). None of them has been dated but it is not unlikely that their use goes back to medieval times. The hay was covered by turf strips for protection. The remains of the turf tended to accumulate at the edge of the platform, forming a rim that can easily be mistaken for a house ruin.

The medieval law books frequently mention enclosures for storage of hay (Icelandic = *heygarður*). No structures have been identified from the air that fit their description, but it seems likely that such hay enclosures were small infield features and attached to other wall structures.

Herding structures

Some of the walls incorporate variously shaped enclosures that may be related to stock management. Ancient documents, including Grágás, the medieval law code of Iceland, refer to pens for herding (in Icelandic = *rétt*) but their design is not described and they most likely varied with both purpose and landscape. None of the observed structures resembles the modern sheep folds used for managing sheep driven from the upland commons in the autumn. The date of the early medieval phase of the herding structures has yet to be demonstrated archaeologically, but it can be intimated that they were similar to those that have been excavated; dating to the early-17th century (Aldred 2010).

In the Mývatn area, the earliest sheep folds seem to be located away from the settlements in the grazing areas, both to the south in Suðurafrétt, and in Norðurfjöll (Gæsadalur and east of Hágöng), but are also evident in other places across Iceland (Aldred & Madson 2009; Aldred 2010). The earliest structures are simple in their form (c. 500m^2), or utilised topography, rather than the multi-compartment folds of 19th century date and later.

Churches or chapels

In the beginning of Christianity in Iceland churches were private property and most were built close to farmhouses. The church or chapel was oriented east-west and enclosed within a circular wall that was both symbolic and practical (to keep animals away). Most churches were small and the diameter of the circular wall was only 20–30m. Square churchyards became the norm in the 19th century (Jónasson 1961, 347). Only four ancient (but undated) churches/chapels have been

Figure 20.9: Horse tracks by Lake Mývatn.

Figure 20.10: Brettingsstaðir, an ancient farm site with a church or chapel (inside the larger circle) and a small circle with unknown function. Remains of a 20th century farmhouse above and an apparently Viking Age longhouse below the circular walls.

seen from the air in our study area: at Ingiríðarstaðir and Einarsstaðir in Þegjandadalur valley, Brettingsstaðir in Laxárdalur valley (Figure 20.10) and Saltvík by Húsavík. A church at Hofstaðir was located through a ground resistance and magnetometer survey (Horsley & Dockrill 2002) and a few other medieval churches are known from historical sources.

Assembly sites

Assemblies (Icelandic = *þing*) were an important part of the social organisation from the very early settlement in the late 9th century and were formalised on a nationwide scale in AD 930 when the general assembly at Þingvellir was established. It is normally assumed that the whole assembly structure collapsed along with the judicial system in the 13th century, to be replaced by a more centralised judicial system and royal executive power in the wake of Iceland's union with Norway in 1262.

Although the assemblies were the stage for many epic events in the Saga literature, surprisingly little is known about their spatial organisation, and there is no clear typology to support aerial observations (cf. Vésteinsson *et al.* 2004; see also Friðriksson 1994). A cluster of small and evenly spaced house ruins in protected locations close to water may qualify as possible assembly sites. Some sites with these characteristics correlate with locations given in the ancient literature and are sometimes supported by place name evidence. The visible layout of the only documented *þing*-site in our study area, Þingey, differs in having a series of house ruins that seem connected gable to gable to form two parallel rows. Extra complexity is added to the archaeology of the Þingey site because of 19th century farming activities.

The nearby site of Skuldaþingsey, however, has a dispersed cluster of 30 ruins (Figure 20.11) that can only

Figure 20.11: An assembly site at Skuldaþingsey where 26 of the 30 booths can be seen.

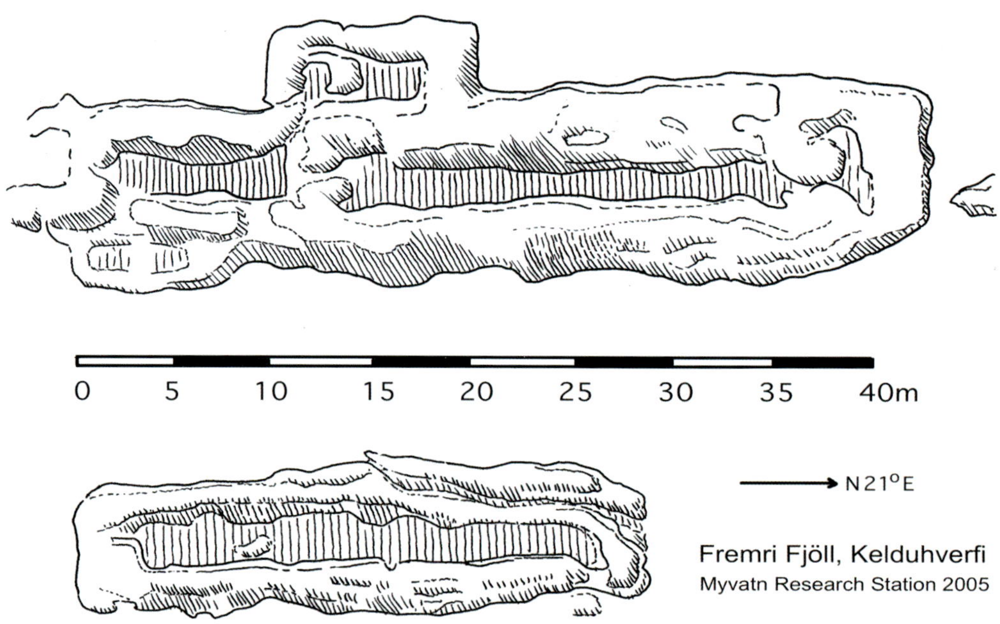

Figure 20.12: Fremri Fjöll, a major farm site in Kelduhverfi. *Top*: The large house from the air. *Bottom*: A line drawing based on detailed measurements of the same house and another longhouse close by.

be interpreted as booths for some kind of assembly (an interpretation also assisted by the *þing* element in the name). The location of this large assembly site so close to the traditional major site in Þingey is somewhat puzzling. Both assembly sites date to the medieval period and an excavation in Skuldaþingsey shows evidence of prolonged but seasonal use (Vésteinsson *et al.* 2004).

The lack of clear typology exemplified by the striking dissimilarity of the two major and adjacent sites Þingey and Skuldaþingsey makes it difficult to assign a function to two other suspected assembly sites, Leiðarhóll by Helgastaðir and Leiðarnes by the River Fnjóská. The names suggest minor assemblies (Icelandic = *leið* refers to a local assembly for the announcements of decisions made at regional assemblies). The Leiðarhóll site is also complicated by later farming activities – in fact there is nothing assembly-like about it other than the name. The Leiðarnes site, however, certainly qualifies as an assembly site, judging by the name and by the typology discussed above. It has a group of evenly spaced ruins of similar size, located on a peninsula created by a meander in the river. Both sites are awaiting aerial recording.

Farm sites

The set of ancient farmsteads observed from the air is certainly a biased sample, as the best preserved sites are the marginal ones that were abandoned early. The 12–13th century saw large-scale abandonment of the marginal, mostly upland and interior settlements (Thorarinsson 1976; Rafnsson 1990; Sveinbjarnardóttir 1992). Although farm abandonment may have taken place in other periods, most of the farms that survived the medieval period were still in use in the early 20th century and the ruins of the original settlement have been superseded by modern development. Also, the marginal upland farm sites will tend to have lower soil accumulation rates, and therefore be more visible, than those of the richer lowland.

There are two notable exceptions where major farm sites have survived. One is the well known archaeological site of Hofstaðir by Mývatn, a large 10th century farm with an oversized Viking type longhouse, showing evidence of ritual feasting (Lucas 2010). The wall system associated with the farm is also well preserved. The other is Fremri Fjöll in the Kelduhverfi district. It has not been dated but the ground plans of the houses suggest 9th–11th century (Christian Keller pers. com.). The site is beautifully preserved and the constructions are still prominent above ground with two exceptionally large halls (Figure 20.12), a set of at least seven smaller houses and a complex system of walls.

Two areas deserve special mention because of good preservation. One is the valley of Þegjandadalur that was deserted early (probably long before the 1477 tephra) and has well-preserved farmhouse complexes, walls, a couple of churches and a Viking period grave field. The other site, Brettingsstaðir, is a well-preserved farm site in the valley of Laxárdalur and has what seems to be a double Viking period hall (Figure 20.10), a church or chapel ruin, some outhouses and a system of walls. The ridge above Brettingsstaðir has a large cluster of charcoal pits.

The farm sites in the marginal and usually better preserved settlements are of various types but most share the feature of one or more quasi-circular or sub-rectangular walled enclosures (Figure 20.13). If there are multiple rings of enclosure they are often concentric, and frequently connected by transverse walls. The house complexes are usually in the innermost circle. The district of Kelduhverfi has an impressive concentration of this type of settlement and some of those are truly marginal in every sense, like those at Bláskógar with three large-diameter quasi-circular fences adjacent to each other in the mid-slope lava fields of the shield volcano of Þeistareykjabunga.

The origin of multiple concentric quasi-circular fenced enclosures is not understood, but it has been suggested that they represent stages in an expanding hayfield from the time of first settlement (Líndal 1951). Alternatively they may reflect a contracting hayfield in response to deteriorating environmental conditions, multiple periods of settlement and abandonment, and finally, and perhaps most commonly, they may be a way of subdividing the home range for management purposes.

Many of the well preserved abandoned farm sites show signs, sometimes backed up by historical documentation, of much later use as grazing stations. If this later activity is within the last one hundred years or so the fertilising effect of the activity is easily recognizable from the air by a rich growth of grass. The vegetation on older ruins conforms with the natural vegetation, usually dwarf shrubs. It is possible that a site has oscillated between permanent and seasonal use in response to changing population size

Figure 20.13: A probably medieval deserted farm site with a triple enclosure by the River Reykjakvísl. There is a riding path to the right. The yellow colour inside the innermost enclosure is indicative of farming activity in modern times.

Figure 20.14: A peat pit in Fljótsheiði.

or environmental conditions. Much work remains to correlate the surface appearance of the various houses to their original structure and function.

Single houses

Quite a few ruins of single houses have been located from the air. Some are very small (4 by 4m) and occur along the main riding paths. These may be shelters for travellers (Icelandic = *sæluhús*) or shepherds (Icelandic = *smalakofi*), shielings (summer grazing stations) or temporary stores of bog iron or peat, just to mention a few possibilities.

Pre-Christian graves

Christianity was made the official religion in AD 1000 and in the approximately 130 years before that non-Christian Viking Age burial customs were prevalent. We have been able to see a number of looted pre-Christian graves from the air. Looted graves have a telltale scar and signs of upcast soil, even if the looting dates to the Middle Ages (Vésteinsson 2004b). Untouched graves have no known external features that allow their location either at ground level or from the air.

Peat pits

Peat was used as fuel along with wood from an early date (Simpson *et al.* 2003). The peat was cut in certain peat pits (locally known as '*svarðargrafir*') in the wetlands and dried on nearby hillocks before transport. The pits are easily recognisable from the air by their squareness, but they vary in size and shape. Clusters of small (less than 5m in diameter) pits are frequent. Larger pits (over 10m across) have a characteristic L-shaped internal profile (Figure 20.14). If the wetland is situated on a slope there is usually a narrow cut out of the pit for draining. None of the pits has been dated, but some were used till the mid-20th century.

Discussion

Cultural value of the monuments and their protection

Clearly, the spatial extent of the northeast landscape, its good preservation, as well as the antiquity and importance of its archaeology for Iceland, if not Europe, calls for a policy of protection and other management. There are several challenges that need to be met in order for the archaeological landscape to be protected or otherwise managed to ensure its cultural value for future generations. The northeast landscape, like many areas where archaeology is prevalent, is both a relict and a living landscape. The first challenge is to find ways in which the demands of the relict and the living can be managed to complement each other. The extensive character of the archaeology, especially the wall systems, also presents another challenge of how best to protect or manage a resource that is so widely distributed. Clearly, there is a need to be well informed, and to continue to monitor the resource for any potential or active threats, such as by development associated with infrastructure projects, agriculture, afforestation, as well as natural threats such as volcanic activity or erosion. Providing the opportunity for protection and management that is connected with both natural *and* cultural factors, and archaeology's relationship to both, is an important challenge to meet.

A clear road-map for protection and management lies in two complementary bodies of legislation – firstly Icelandic heritage law (e.g. Law number 107/2001 Þjóðminjalög) and second, European frameworks, such as the European Landscape Convention (ELC, Council of Europe 2000; ESF Science Policy Briefing 41 2010). Both provide a strategic policy to examine how a landscape, such as the one described in this paper, can be protected and managed in a way that accounts for both the historical landscape and the dynamic living one occupied by people depending on it for their livelihood.

Knowledge is a prerequisite for efficient management of the historic landscape. Gaps in our knowledge need to be identified and filled, and a monitoring programme of the status of the landscape should be established, such as Historic Landscape Characterisation (Aldred & Fairclough 2003; Fairclough & Macinnes 2003; Fairclough 2007). These monitoring programmes should be conducted with the aid of different stakeholders, including landowners.

While direct protection is perhaps suited to exceptional landscapes, such as the ones listed below, the majority of the landscape has no such protection and needs to be effectively managed. Protection alone, therefore, leaves some ambiguity over areas not afforded special protection, and one could view this as a rather passive form of heritage management that is subject to contemporary economic and socio-political conditions (Aldred & Friðriksson 2008). A management approach, however, recognises the value of *all* landscape, as indicated through the ELC framework, and a (pro)active approach to land management. Direct protection should then be seen as an extreme and localised case of more wide-ranging management.

The mapping of archaeology from aerial sources, combined with field based archaeological survey, and small-scale excavation designed to date and characterise specific parts of the resource, give the north-eastern landscape a good basis for the development of an exemplar for best practice of landscape management and protection in Iceland.

Much work is needed to classify the archaeological landscape in terms of preservation or management value. Nevertheless, our aerial surveys already allow us to pinpoint ten outstanding areas in this respect. They have a high density of well-preserved and clearly visible remains, often interconnected by a system of walls and horse tracks, in an attractive landscape that is not interrupted by large-scale modern activities.

Those ten areas are:

1. Þegjandadalur, a deserted valley with a multitude of medieval farms with a complex wall system, two churches and a large pagan grave field (Hreiðarsdóttir & Roberts 2009);
2. The upper part of Laxárdalur, including the area between Lake Mývatn and Laxárdalur, with highlights including Hofstaðir, Brettingsstaðir and an unnamed farm 2km north thereof, all with well preserved wall systems, the west side of Lake Mývatn with three well preserved medieval farm sites (Brenna, Selás and Vindbelgur), shielings (summer grazing stations) and a pagan grave field;
3. Fljótsheiði with an extremely long wall and side walls and associated remains of medieval buildings;
4. Hvammsheiði moorland with the farm sites of Litlu Núpar and Íragerði and a few massive walls crossing the moorland, also a boat burial;
5. The concentration of deserted encircled farms above the main row of present day farms in Kelduhverfi, including the extraordinarily well preserved site of Fremri Fjöll;
6. The well preserved system of walls immediately south of Húsavík, including a church, many house ruins and a pagan grave field (cf. Lárusdóttir 2007);
7. The archaeological hotspot of Bakki just north of Húsavík with a dense complex of walls, irrigation structures and peat pits;
8. Seljadalur, a deserted valley with a row of well-preserved farm ruins with an impressive system of walls;
9. Þingey and Skuldaþingsey, the well preserved dual assembly site;
10. The sulphur mining area and farm site of Þeistareykir.

The list above includes areas where pre-modern landscapes remain more intact than elsewhere and covers some of the hotspots of the Viking period society. Of course there is a multitude of more recent archaeological monuments that deserve attention and should influence any plans for regional scale management. These include the extensive (undated) walls built of lava rock by Lake Mývatn and along the River Laxá, and the road on Hólasandur. The various structures associated with sheep farming, folds and pens for milking sheep and feeding lambs many of which are integrated in the lava landscape in an interesting way, should also be included. Unique to the Mývatn area is the extensive use of lava caves as retreats for sheep from the bloodsucking blackfly (*Simulium*).

General remarks

It is rare to have a large scale, one thousand year old archaeological landscape that is so well preserved, as well as one that was moulded within such a short period and still shows the pattern of landholdings, resources and fine scale management. The potential for using the wall system to analyse the settlement pattern in relation to landscape characteristics is enormous. We not only see the pattern of fences but also the communication routes and the distribution of some resources like peat and woodland. Also, religious centres and other assembly sites can be pinpointed. This information is augmented by detailed archaeological and palaeoecological studies of individual sites yielding a good record of diet, fuel and trade (Lucas 2010).

The geography of any human society is the result of dynamic processes that depend on the productivity, dispersion and seasonality of resources and also on population pressures and a range of cultural responses. The Icelandic farming society has always been a dispersed territorial system based largely on an exclusive home range for summer grazing of sheep and cattle and haymaking for winter fodder (see e.g. Vésteinsson 1998, 2000, 2005). Most farms also had outstations for summer or winter grazing and access to upland commons. Territoriality and animal husbandry stimulates the erection of fences, given the capacity to invest in such structures. Fences are needed to keep your animals within reach and those of your neighbours at bay. The home range may be subdivided for more local management of species, sexes and age groups and also to protect crops or to manure the ground. The configuration of the walls and the large investment involved must therefore tell us a great deal about the geography, structure and resources of the Icelandic society of the Viking period (Aldred 2008). The wall configuration shows evidence of being influenced mainly by landscape characteristics and population density. Tightly packed home ranges tend to have straight boundaries and polygonal shapes. Linear landscapes, like narrow valleys or coastlines, tend to have a single row of square enclosures, while home ranges on a flat terrain tend towards circular or polygonal shapes.

The archaeological remains in the study area may hold the key to understanding the dramatic changes that clearly occurred all over Iceland in the 12th–13th century and involved large scale contraction of the settlements, perhaps preceded by the disuse of the wall system. Hypotheses as to the causal factors include landscape degradation, climatic cooling and an economic shift (see e.g. Dugmore *et al.* 2000, 2005). None of these hypotheses are mutually exclusive, but they involve either migration of people due to shifting resources or the decimation of the population by high mortality rates. Either way, they concern the interaction between people and environment, and the dynamics of the change may be important not only for our perception of the history of Iceland but also that of other settlements in the North Atlantic.

Acknowledgements

The Medieval Wall Project, of which the aerial survey was part, is a joint undertaking of a group of people from the Institute of Archaeology, Iceland, the Myvatn Research Station and the University of Oslo. Apart from the present authors they are: Birna Lárusdóttir, Christian Keller, Elín Hreiðarsdóttir, Oddgeir Hansson, Orri Vésteinsson and Stefán Ólafsson. We would like to thank our pilots Leifur Hallgrímsson and Birgir Steingrímsson, and also our many field assistants for hard work during ground surveys and excavations. The project was supported by the Icelandic Centre for Research (RANNÍS), Vegagerðin (the Icelandic Road Administration) and Þjóðhátíðarsjóður (a fund that commemorates the 9th century settlement in Iceland). Finally we thank Orri Vésteinsson for critical reading of the manuscript. All photos are by the first author.

References

Aldred, O. 2008: Unfamiliar landscapes: infields, outfields, boundaries and landscapes in Iceland. In Chadwick, A.M. (ed.): *Recent Approaches to the Archaeology of Land Allotment.* BAR International Series 1875, Archaeopress, Oxford. 299–321.

Aldred, O. 2010: *Réttir in the landscape. Archaeological investigations of sheep-folds in Skútustaðahreppur and other neighbouring districts.* Unpublished report, International Polar Year.

Aldred, O. & Fairclough, G. 2003: *Historic landscape characterisation: Taking stock of the method – The National HLC Method Review 2002.* English Heritage, London.

Aldred, O. & Friðriksson, A. 2008: *Iceland.* In Fairclough, G. & Møller, P.G. (eds): *Landscape as Heritage. The management and protection of landscape in Europe, a summary by the COST A27 project 'LANDMARKS'.* COST, University of Berne, Institute of Geography (G79), Bern. 145–55.

Aldred, O., Hreiðarsdóttir, E.Ó. & Sveinbjarnarsson, Ó.G. 2010: On the precipice: aerial archaeology in Iceland. *Archaeologia Islandica* 8, 111–21.

Aldred, O. & Madson, C. 2009: *Réttir in the landscape. A study on the interactions between humans and animals through sheep-fold monuments.* International Polar Year unpublished report.

Bragason, Þ. & Guðmundsson, M. 1988: *Fifty Years of Change and Development: Aerial Photographs from Iceland.* Hörpuútgáfan, Landmælingar Íslands.

Church, M.J., Dugmore, A.J., Mairs, K.A., Millard, A.R., Cook, G.T., Sveinbjarnardóttir, G., Ascough, P.A. & Roucoux, K.H. 2007: Charcoal production during the Norse and early medieval periods in Eyjafjallahreppur, southern Iceland. *Radiocarbon* 49, 659–72.

Council of Europe, 2000: *European Landscape Convention.* Council of Europe (Treaty series No. 176), Florence.

Dugmore, A.J., Newton, A.J., Larsen, G. & Cook, G.T. 2000: Tephrochronology, environmental change and the Norse colonization of Iceland. *Environmental Archaeology* 5, 21–34.

Dugmore, A.J., Church, M.J., Buckland, P.C., Edwards, K.J., Lawson, I.T., McGovern, T.H., Panagiotakopulu, E., Simpson, I.A., Skidmore, P. & Sveinbjarnardóttir, G. 2005: The Norse landnám on the North Atlantic islands: an environmental impact assessment. *Polar Record* 4, 21–37.

Einarsson, Á., Hansson, O. & Vésteinsson, O. 2002: An extensive system of medieval earthworks in NE-Iceland. *Archaeologia Islandica* 2, 61–73.

Eldjárn, K. 1981: Uslaréttir. *Árbók Hins íslenska fornleifafélags 1980.* Reykjavík. 101–10.

Fairclough, G. 2007: The contemporary and future landscape: change & creation in the later 20th century. In McAtackney, L., Palus, M. & Piccini, A. (eds): *Contemporary and Historical Archaeology in Theory, Paper for the 2003 and 2004 CHAT conferences*

(Studies in Contemporary and Historical Archaeology 4, BAR International Series 1677). Archaeopress, Oxford. 83–8.

Fairclough, G. & Macinnes, L. 2003: *Understanding Historic Landscape Character, Topic paper 5, Landscape Character Assessment Guidance for England and Scotland*. Countryside Agency, Scottish Natural Heritage, Historic Scotland and English Heritage.

Friðriksson, A. 1994: *Sagas and Popular Antiquarianism in Icelandic Archaeology*. Aldershot, Avebury.

Friðriksson, A. & Vésteinsson, O. 1998: Fornleifaskráning – Brot úr íslenskri vísindasögu. *Archaeologia Islandica* 1, 14–44.

Gathorne-Hardy, F.J., Erlendsson, E., Langdon, P.G. & Edwards, K.J. 2009: Lake sediment evidence for late Holocene climate change and landscape erosion in western Iceland. *Journal of Paleolimnology* 42, 413–26.

Guðmundsson, G., Snæsdóttir, M., Simpson, I., Hallsdóttir, M., Sigurgeirsson, M.A. & Árnason, K. 2004: Fornir akrar á Íslandi. *Árbók hins íslenzka fornleifafélags* (2002-3), 79–106.

Hallsdóttir, M. 1987: *Pollen Analytical Studies of Human Influence on Vegetation in Relation to the Landnám Tephra Layer in Southwest Iceland*. Lund University, Department of Quaternary Geology (Lundqua Thesis 18), Lund.

Hólmgeirsson, G. 1978: Fóðurbirgðaskoðun í Reykdælahreppi 8.–22. mars 1919. *Árbók Þingeyinga* (1978), 159–66.

Horsley, T.J. & Dockrill, S.J. 2002: A preliminary assessment of the use of routine geophysical techniques for the location, characterisation and interpretation of buried archaeology in Iceland. *Archaeologia Islandica* 2, 10–33.

Hreiðarsdóttir, E.Ó. & Roberts, H. 2009: Þögnin rofin. Fyrstu niðurstöður fornleifarannsókna á eyðibyggð á Þegjandadal. *Árbók Þingeyinga* 2008, 5–24.

Ísaksson S.P. & Helgason, Þ.S. 1995: Vetrarmyndir frá Nesi við Seltjörn og Laugarnesi. *Árbók hins íslenzka fornleifafélags* (1994).

Jónasson, J. 1961: *Íslenzkir þjóðhættir* (The Ethnography of Iceland). 3rd edition. Ísafoldarprentsmiðja, Reykjavík.

Lárusdóttir, B. 2007: Settlement organization and farm abandonment: The curious landscape of Reykjahverfi, North-East Iceland. In Davies, W., Halsall, G. & Reynolds, A. (eds): *People and Space in the Middle Ages, 300–1300* (Studies in the Early Middle Ages 15). Brepols, Turnhout. 45–63.

Lárusdóttir, B. & Aldred, O. 2008: *Kortlagning fornleifa af gervihnattamyndum*. Fornleifastofnun Íslands (FS389–08181), Reykjavík.

Lawson, I. 2010: The palaeoenvironment of Mývatnssveit during the Viking age and Early Medieval period. In Lucas, G. (ed.) *Hofstaðir. Excavations of a Viking Age feasting hall in North-east Iceland*. Institute of Archaeology Monograph Series (1), Reykjavík, 26–54.

Lawson, I.T., Gathorne-Hardy, F.J., Church, M.J., Newton, A.J., Edwards, K.J., Dugmore, A.J. & Einarsson, Á. 2007: Environmental impacts of the Norse settlement: palaeoenvironmental data from Mývatnssveit, northern Iceland. *Boreas* 36, 1–19.

Líndal, J.H. 1951: Um forn mannvirki og örnefni á Lækjamóti í Víðidal. *Árbók hins íslenzka fornleifafélags* 1949–50, 78–101.

Lucas, G. (ed.) 2010: *Hofstaðir. Excavations of a Viking Age feasting hall in North-east Iceland*. Institute of Archaeology Monograph Series (1), Reykjavík.

McGovern, T.H., Perdikaris, S., Einarsson, Á. & Sidell, J. 2006: Coastal connections, local fishing, and sustainable egg harvesting: patterns of Viking Age inland wild resource use in Mývatn district, Northern Iceland. *Environmental Archaeology* 11, 187–205.

McGovern T.H., Vésteinsson, O., Fridriksson, A., Church, M., Lawson, I., Simpson, I.A., Einarsson, Á., Dugmore, A., Cook, G., Perdikaris, S., Edwards, K.J., Thomson, A.M., Adderley, W.P., Newton, A., Lucas, G., Edvardsson, R., Aldred, O. & Dunbar, E. 2007: Landscapes of settlement in northern Iceland: Historical ecology of human impact and climate fluctuation on the millennial scale. *American Anthropologist* 109, 27–51.

Nielsen, Ó.K. 1995: Karrar og gróðurfar. (English summary: Rock ptarmigan density and heathland vegetation in Suður-Þingeyjarsýsla, NE-Iceland). *Náttúrufræðingurinn* 65, 81–102.

Ólafsson, G. 1991: Fornleifaskráning og fornleifavernd. *Sveitarstjórnarmál* II, 80–2.

Pálsson, H. 1999–2010: *Byggðasaga Skagafjarðar*. Vols. 1–5. Sögufélag Skagafjarðar, Sauðárkrókur.

Rafnsson, S. 1990: *Eyðibyggð í Hrafnkelsdal og á Brúardölum: Brot úr byggðarsögu Íslands*. Rit hins íslenzka fornleifafélags, Reykjavík.

Science Policy Briefing 41 2010 *Landscape in a changing world*. European Science Foundation & ESF COST (Accessed 10/12/2010: http://www.esf.org/publications/ science-policy-briefings.html).

Simpson, I.A., Vésteinsson, O., Adderley, W.P. & McGovern T.H. 2003: Fuel resources in landscapes of settlement. *Journal of Archaeological Science* 30, 1401–20.

Sveinbjarnardóttir, G. 1992: *Farm Abandonment in Medieval and Post-Medieval Iceland: an Interdisciplinary Study*. Oxbow (Monograph 17), Oxford.

Sveinbjarnarson, Ó.G. 2007: *Aerial Survey: Archaeological survey through high altitude aerial photographs*. Unpublished BA thesis. University of Iceland, Reykjavík.

Thorarinsson, S. 1976: Gjóskulög og gamlar rústir. Brot úr íslenskri bygðasögu. *Árbók hins íslenzka fornleifafélags* 1976, 5–38.

Vésteinsson, O. 1998: Patterns of settlement in Iceland. A study in prehistory. *Saga-Book of the Viking Society* XXV, 1–29.

Vésteinsson, O. 2000: The archaeology of landnám. Early settlement in Iceland. In Fitzhugh, W.W. & Ward, E. (eds): *Vikings. The North Atlantic Saga*. Washington DC, Smithsonian Inst. Press. 164–74.

Vésteinsson, O. 2004a: Icelandic farmhouse excavations: field methods and site choices. *Archaeologia Islandica* 3, 71–100.

Vésteinsson, O. (ed.) 2004b: *Archaeological studies in Saltvík 2003*. Unpublished report, Fornleifastofnun Íslands (Institute of Archaeology, Iceland), Reykjavík.

Vésteinsson, O. 2005: Archaeology of economy and society. In McTurk, R. (ed.): *A Companion to Old Norse – Icelandic Literature and Culture*. Blackwell, Oxford, 7–26.

Vésteinsson, O., Einarsson, Á. & Sigurgeirsson, M.Á. 2004: A new assembly site in Skuldaþingsey in NE-Iceland. In Guðmundsson, G. (ed.): *Current issues in Nordic Archaeology. Proceedings of the 21st Conference of Nordic Archaeologists, 6–9 September 2001, Akureyri, Iceland*. Society of Icelandic Archaeologists, Reykjavík, 171–9.

21 | Reserved optimism: preventive archaeology and management of cultural heritage in Slovenia

Gašper Rutar and *Matija Črešnar*

Abstract: Following legislative changes in 2008, considerable modifications have marked the last two years in the management of cultural heritage in Slovenia. Some of the more profound changes occurred in the broad field of preventive archaeology. Firstly a new, mainly research oriented, operational body, the Centre for Preventive Archaeology (CPA), was founded within the Institute for the Protection of Cultural Heritage of Slovenia (IPCHS). Its duties are largely connected with the execution of archaeological research preceding the National Spatial Plans (NSP) and subsidised research, the last also a novelty introduced by new legislation. The work already carried out in the first 15 months of CPA's operation are encouraging, but also show possibilities for improvement.

Introduction

There have been considerable changes over the last two years in the management of the cultural heritage in Slovenia, especially in the status and development of preventive archaeology. This is due firstly to changes in Slovenian legislation regulating cultural heritage in 2008 (i.e. the Cultural Heritage Protection Act: ZVKDS-1; UL RS 16/2008, 123/2008) and its implementation in practice. The Act has been in use for over two years and this paper reflects on its impact, reviewing both the favourable results achieved and the problems encountered.

The legislative foundations

For a clear understanding of the recent developments, we shall focus on only the most striking changes. The first is the definition of terms of reference such as 'archaeological finds', 'archaeological remains' and 'archaeological sites'. These are applied to all finds and remains recovered from the field which are at least 100 years in age, but also to all military equipment and other remains of (military) conflicts recovered from the field which are at least 50 years old. Archaeological sites are identified as all the original places of deposition and recovery of archaeological remains and after professional identification and entry in the register of cultural heritage the site receives the status of heritage, as outlined elsewhere (Djurić *et al.* 2009).

In addition, the Institute for the Protection of Cultural Heritage of Slovenia (IPCHS) responsible for protection of the unmovable heritage has undergone major change. Besides regional offices, it also houses the Centre for Conservation, within which the Centre for Preventive Archaeology has been founded. This Centre is financed from the state budget to undertake archaeological research in advance of development. Firstly, this is research connected to the National Spatial Plans (NSP). Secondly, and most importantly for the broader public, there are subsidised investigations. The ZVKD-1 legislation allows for subsidised archaeological research within registered heritage sites, where the investor is an individual and is building private housing for personal use on a plot inside a residential area. The construction of not-for-profit housing and buildings for public use are also included in this subsidised approach.

However, outside of NSP and subsidised research, the IPCHS is no longer the sole legally competent institution for conducting preventive and rescue archaeological research and private archaeological companies, freelance archaeologists, other archaeological research institutions[1] are equally eligible to undertake archaeological research in this framework. Alongside the state subsidised research described above, the financing of such works is predominantly based on the principle of the 'polluter pays'.

The first large scale NSP project, the building of motorways, has been a valuable learning experience for Slovenian archaeology. Here, archaeologists confronted with the already planned courses of roads developed a new methodology, which continues to be used today, with some slight modifications. This approach produced remarkable results, with the identification of 100 previously unknown archaeological sites in the 250km long and 50m wide motorway corridor (=12.5km²). A surprising number of these sites were registered in environments that traditionally were not targeted by systematic archaeological research. These results, demonstrate that the number of known archaeological sites (2,938[2]) registered by the Heritage Information and Documentation Centre (INDOC) are just a fraction of the potential sites, of which an estimated 100,000 may exist across the whole country (Djurić 2007).

1 The law foresees a register of all companies and individuals which are capable of executing archaeological excavation.

2 Information acquired in November 2010.

The motorway work has produced valuable results, but it is clear that with research in the earliest phases of planning in advance of the works many of the sites could have been spared and their excavation avoided. As foreseen by Djurić (2007, 186) there was an imperative to move towards preventive archaeology, where works could be diverted away from sites through planning, rather than rescued through excavation. In this sense, the new Act has taken a natural step forward and implements archaeology in the earliest phases of spatial planning in order to reduce the amount of salvage and rescue excavations and to minimize the overall impact of development on the cultural heritage.

The Centre for Preventive Archaeology

Immediately after its establishment in March 2009, the Centre for Preventive Archaeology had to cope with its first big challenge, the preparation of guidelines for standardised research methodology and techniques, which also have to enable us to manage large-scale projects, which can cover over 100km^2 (Figures 21.1 & 21.2). In their preparation we have, on the one hand, followed the predominantly field-oriented experiences already established in Slovenian archaeological practice, especially systematic field survey (extensive or off-site field walking) covering all areas of large-scale projects (e.g. motorways). Their results shifted attention to areas previously ignored by archaeological investigation, such as broad river plains for example. On the other hand, however, we also took into consideration up-to-date methods being used in the broader European field (i.e. remote sensing techniques).

In 2009, a series of meetings were organised with all major (state) developers, firstly to present our new research guidelines and secondly to build a timetable for all National Spatial Planning activities. The developers were also asked to provide all data that they generated during the first stages of planning (e.g. LiDAR, aerial photographs, thermal images etc.) that could also be used in archaeological research.

However, it was apparent that the available archaeological data were insufficient to assess the archaeological potential of areas under development in the first phases of research. The Heritage register (RKD) was never conceived as a tool in the archaeological evaluation process (see Djurić *et al.* 2009) but rather as a guideline for planners, in the sense that registered units are subject to certain restrictions.

In order to ensure good collaboration with planners and builders, we must provide data which are simple and easily read by all parties in the form of widely accessible plans developed in GIS. However this has proved to be the main obstacle for the Centre. In recent decades a huge amount of research was conducted, the results of which exist as unpublished briefs and reports gathered by INDOC in paper format, but have never been published. All of these, and other sites which were published but never registered, have to be digitized and geo-referenced to ensure quick access to map data in combination with development plans in order to provide first assessments of archaeological potential in these areas. The number of past investigations in our database is currently approaching 4,000 for just half of Slovenia, with the other part still in process. Investigations vary from simple archaeological topographic survey conducted throughout most of the 20th century, producing notes, observation and simple test trenching, to a variety of other investigations, expanding especially in the last 20 years, that include remote sensing (aerial surveys, LiDAR), systematic extensive (off-site) and intensive (intra-site) surveys, geophysical prospection, manual test trenching and mechanical trenching, excavation and so on. They should all result in full reports with detailed plans showing the areas covered. As the next step, all the reports are incorporated into a cumulative plan of investigations, and will be integrated with the existing GIS database (i.e. RKD). This combined data on all known registered and unregistered sites, will be a powerful tool for all parties involved in heritage protection and spatial planning.

There are two different areas of its use. Firstly, there are residential areas near or within registered sites, where subsidised investigations may be applied. These areas show an exponential rise of investigations in the last 20 years. When a plot is intended for building the whole parcel in investigated, and either registered or released for building as a whole. Since the trend in spatial planning favours continuing development within residential areas, there is also a considerable amount of data from previous investigations on adjacent or nearby areas, which facilitates the planning process for further research. It should be stressed that subsidised investigations within registered sites do not mean that the core of the site is destroyed to enable building. We are dealing with marginal areas of registered sites, where conservation guidelines in the first place can allow or deny such investigations and further building, so in a way these investigations validate the extent of the existing entries in RKD. The other areas are large open spaces between residential areas, on the one hand, and/or registered sites on the other. These have tended not to be located through site-oriented archaeological topographic survey, through which most of the known and registered sites were detected and which can be investigated only by systematic field walking, remote sensing or even better with a combination of the two. Furthermore, integrating the results of any systematic research and interpretation with environmental data is crucial to enable the assessment of archaeological potential for these areas.

Self-reflection

In the 15 months since the CPA has been fully operational, work with the NSP is in the first phase of recognition of archaeological potential (Figure 21.1). The total area on NSP documents for our current year programme (2010) covers 38,546 hectares. The study area for the first phase of historical analysis and collation of known data and their integration into GIS was, however, considerably larger, covering 450,000 hectares (Figure 21.3). In the course of investigation 1,712 archaeological sites and potential archaeological sites were identified in this area, 43% of which were

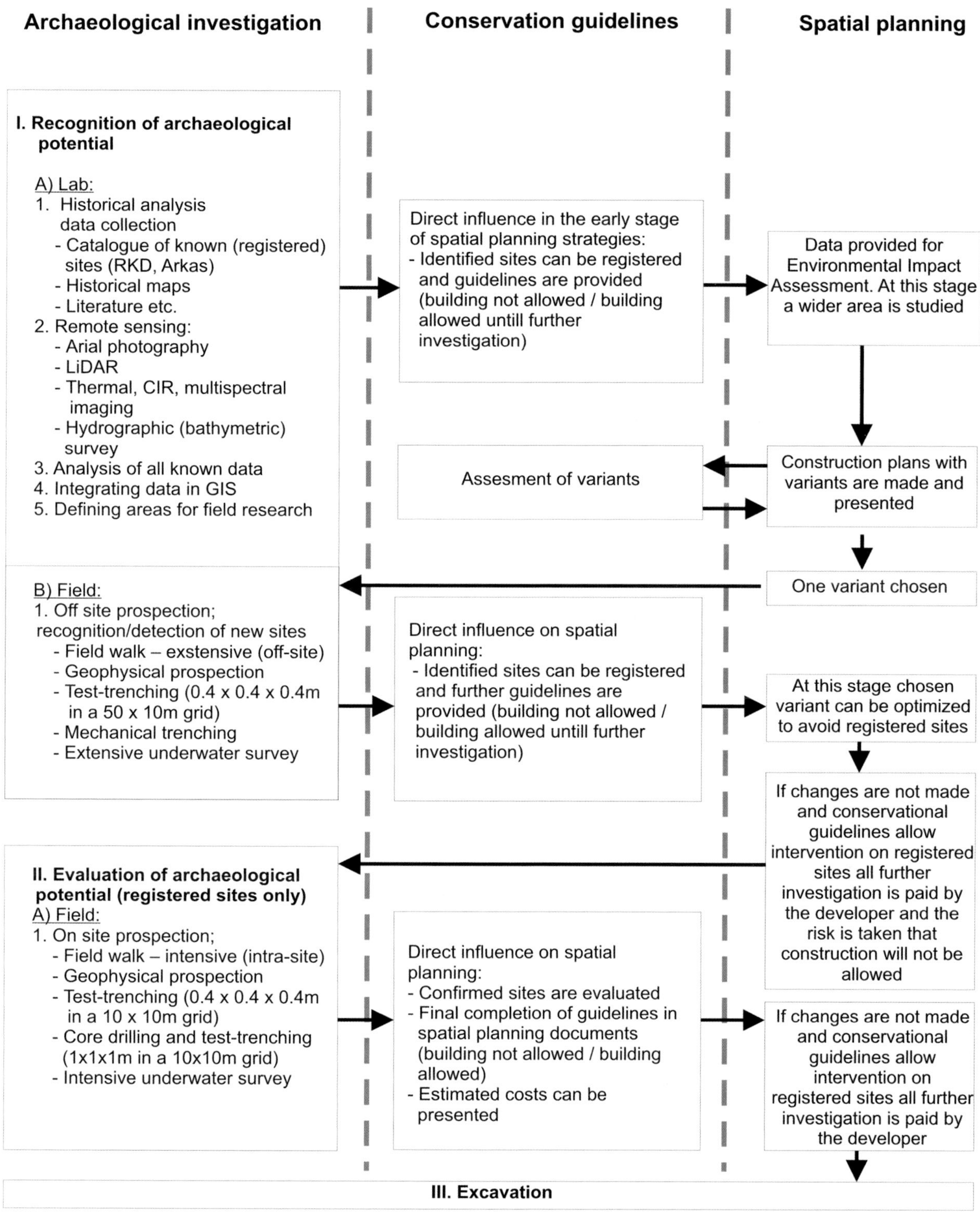

Figure 21.1: Simplified scheme for National Spatial Planning procedure.

already registered archaeological sites. A further 21% lay within sites registered as non-archaeological sites (e.g. pertaining to the history of art, history and ethnography) and are now pointed out, whether because of the change in the legislation or because of new data having been discovered. Finally 36% are newly discovered sites or potential archaeological sites. Most of these were identified through systematic study of the available data, including unpublished research, literature and historical maps. Following this, some of these sites need to be registered, although not all have this potential; however they do provide an important clue for understanding and assessing the potential of certain areas. Furthermore, almost 7% of the newly recognised sites were identified during the analysis of LiDAR data, although only 1% of the study area was covered by LiDAR (Figure 21.4). Most of the anomalies identified require ground observation to clarify their nature, but LiDAR is undoubtedly the most effective airborne remote sensing technique in Slovenia, since

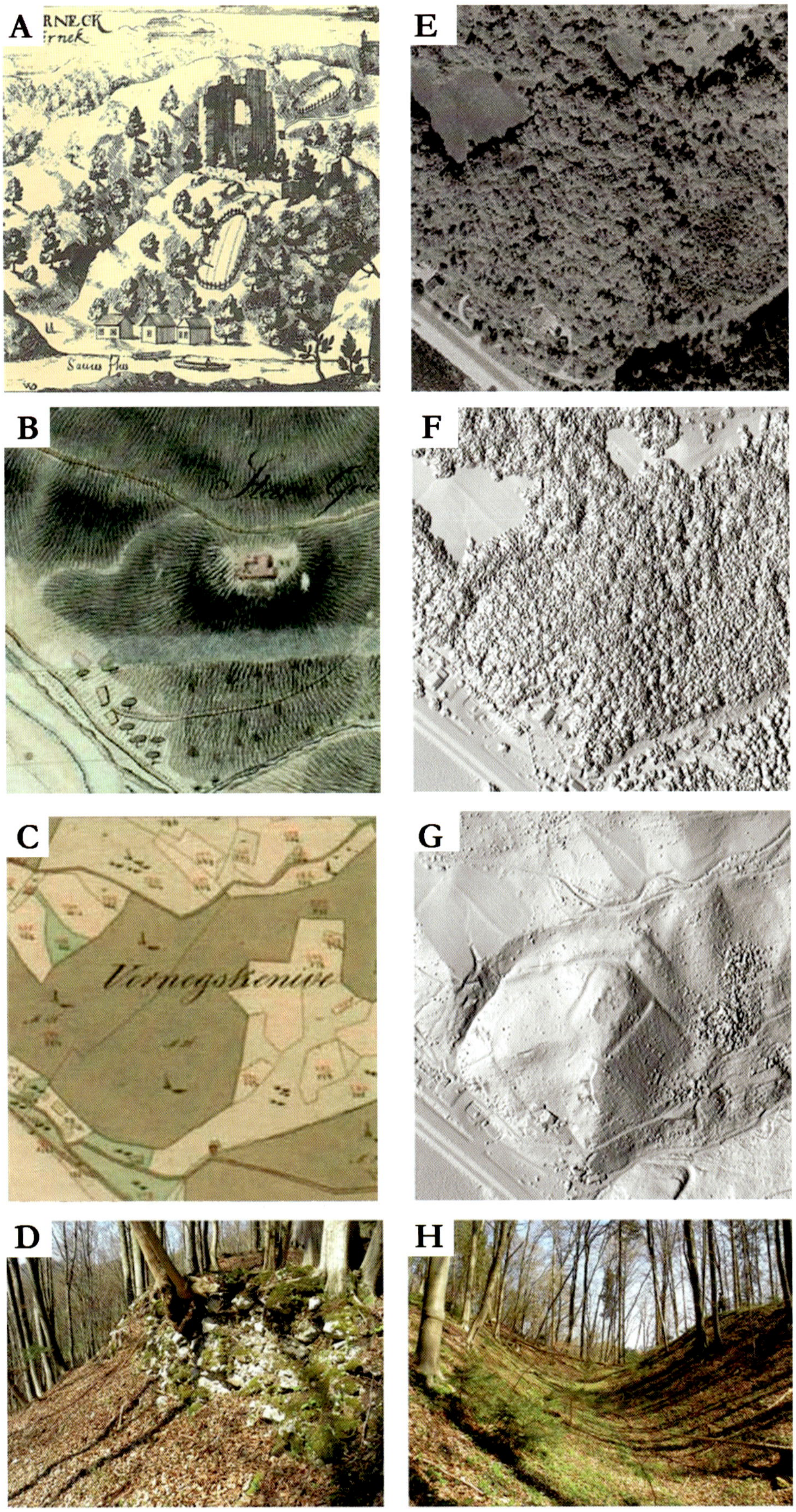

Figure 21.2: The castle of Vernek was newly registered during the Hydroelectricity Plant project on the River Sava, and investigated using multiple methods. A) Werneck/Vernek castle depicted by J.V. Valvasor in 1679; B) 1807 map labelled as ˝Stari grad˝(Old castle) (AS 1068-2); C) Castle does not appear on the Franciscan map from 1826 (AS 176, L077); D) Photo of the remains - walls; E) Same area on 1999 orthophoto (GURS); F) LiDAR – digital surface model; G) LiDAR – digital terrain model; H) Photo of the remains – defensive ditch.

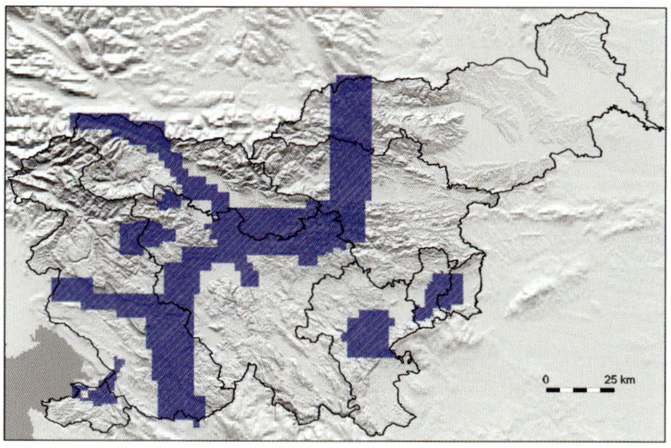

Figure 21.3: Areas covered by historical analysis and analysis of known/existing data.

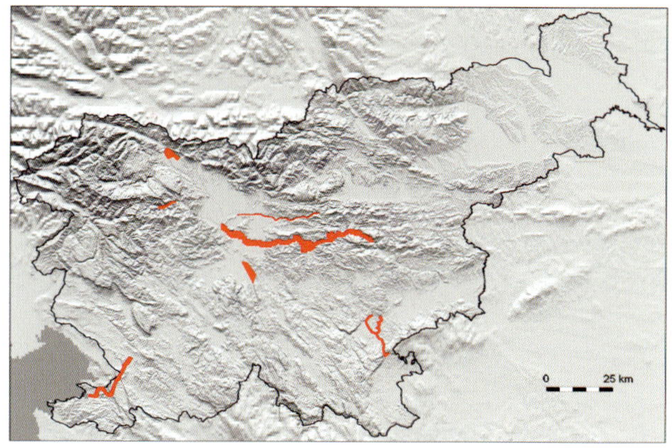

Figure 21.4: Areas of LiDAR coverage.

60% of the country is covered by forest and the rest of the countryside is divided into small parcels of land, making it less favourable for other airborne prospecting such as aerial reconnaissance.

However, there are some drawbacks when putting the presented NSP scheme to action. LiDAR scanning is not compulsory for developers and, where it is undertaken, it is commissioned in later stages of planning, when archaeological investigation moves into the field. Another problem when conducting archaeological research (i.e. extensive field surveys) in the early stages of planning can arise when landowners, and possibly the public in general, can refuse access for investigations since they are not informed about these plans, or they do not agree with them.

Research relating to subsidised investigations for private building covered almost six hectares for 57 individual buildings in the last eight months of 2010 alone. These investigations covered all levels of work up to the final excavation that cleared the way for building works to begin. Although we are dealing with smaller areas, the timeline from design to construction is short, especially for private construction, even more so because many individuals have held building plans for years, but had been unwilling to pay the extra price for archaeological investigations, and so have now eagerly taken the opportunity for subsidised works. Indeed, very quickly the number of applications for subsidised investigations has surpassed the budget or capabilities of the Service. Waiting times can now be up to six months, depending on the season. This has had the effect that people can quickly forget the benefit of fully subsidised research and some may see the IPCHS and cultural heritage above all as an obstacle.

Conclusion

We have made huge steps forward in unifying and standardising archaeological prospection and in building a central database, which will incorporate all the available archaeological data previously dispersed or even unused. With these steps unknown sites are systematically listed and the register of sites updated, both by adding new sites and by updating the precision of registered information. These procedures are working well within the NSP, which includes the environmental impact assessment stage (EIA). The results and guidelines gathered in EIA, including those dealing with cultural heritage, must to be taken into account before further steps and construction can begin (Figure 21.1). However, problems are still encountered in implementing developments, and at the local level, where smaller scale plans are developed by local authorities (municipalities) and environmental impact assessments are not undertaken, the protection of cultural heritage and listing of new sites has proved more difficult. The IPCHS does not have sufficient resources to conduct investigations for hundreds of local development plans, since local lobbies and interests can easily prevail. However the new, more centralised basis for operation has yielded its first results and gives hope that the next stage, a more holistic and integrated heritage conservation underpinned by a range of sources, is obtainable.

References

Some general information about NSP procedure can be found at: http://www.coe.int/t/dg4/cultureheritage/heritage/cemat/compendium/CompendiumSlo_en.pdf.

Djurić, B. 2007: Preventive Archaeology and Archaeological Service in Slovenia. In Bozóki-Ernyey, K. (ed.): *European preventive archaeology: papers of the EPAC Meeting, Vilnius 2004*. National Office of Cultural Heritage, Hungary, Council of Europe.

Djurić, B., Mason, P., Mlakar, B., Kovačec Naglić, K. & Petek, B. 2009: Listing archaeological sites – integrating heritage: case of Slovenia. In Schut, P.A.C. (ed.): *Listing Archaeological Sites, Protecting the Historical Landscape*. EAC Occasional Paper No. 3, 87–93.

ZVKDS-1, Zakon o varstvu kulturne dediščine (Cultural Heritage Protection Act). *Uradni list RS (Official Gazette of the RS) 16/2008 and 123/2008.*

22 | World War I Heritage in Belgium: combining historical aerial photography and EMI

Birger Stichelbaut, Timothy Saey, Fun Meeuws, Jean Bourgeois* and *Marc Van Meirvenne

Abstract: Millions of aerial photographs were taken by all fighting nations during World War I. These photographs are a remarkable and previously underexploited source of information for the study of this conflict heritage. This paper describes interdisciplinary research combining archival research, the interpretation and detailed mapping of historic aerial photographs and the application of electromagnetic induction. These techniques are combined in two case studies (Geluveld & Oostduinkerke) to evaluate the archaeological heritage of World War I.

Introduction

During World War I, for the first time, aerial photography rapidly developed as an intelligence tool that saw large scale application by all fighting nations (Grand Quartier Général des Armées 1916; Kommandierender General der Luftstreitkräfte 1917). Huge numbers of these photographs have survived in archives all over Europe, the United States and Australia. These are a remarkable primary record of the progress of World War I, but are also a unique record of the landscape at the beginning of the 20th century and a valuable source of data for archaeologists, landscape historians and cultural resource managers (Chielens *et al.* 2006; Decoodt 2007). The approaching centennial of the start of World War I in 2014 has triggered a renewed interest in the history and memory of this important violent episode in European history. The archaeological research of this conflict is slowly becoming a part of mainstream archaeology in Europe, whereas in the last twenty years it was the domain of amateur archaeologists.

This paper describes the possibilities and outcomes of combining the interpretation and GIS mapping of historical aerial photographs and geophysical prospection techniques. The intention of this paper is to further contribute to the rapidly developing field of 20th century conflict archaeology with an innovative approach based on a combined interrogation of the 1914–18 aerial photographs with an electromagnetic induction soil sensor as a ground based method to evaluate whether material remains of World War I are still preserved beneath the surface.

This paper will first of all describe the framework of World War I conflict archaeology (as opposed to traditional battlefield archaeology) and its evolution in Belgium. Secondly the aerial photograph research is discussed followed by the application of electromagnetic induction (EMI) in Flanders. The outcome of the two methods is brought together in two case studies.

Conflict archaeology

The aerial photographs taken during World War I contain a wealth of information relevant to the study of the earlier landscapes on the Western Front, and enable previously unknown archaeological sites of different time periods to be detected. Yet the photographs are also one of the unrivalled sources for the study of a new and rapidly developing scientific archaeological field of research: the archaeology of World War I. This *'first global industrialised conflict'* (Saunders 2002) scarred the landscape, and some features, mine craters and bunkers for example, are still visible in the current landscape, although the majority of features now lie beneath the surface.

Archaeological research into World War I remains is often designated by outsiders as 'battlefield archaeology'. Other terms which are frequently used are 'military archaeology', 'World War I or Great War archaeology', and 'conflict archaeology'. In this research, it has been decided to use the latter term, mainly because the phrase 'battlefield archaeology' has a pejorative connotation created by John Laffin's *Battlefield Archaeology* (Laffin 1987), which resembles a looter's manual, and excavations by amateur organisations digging for militaria. In addition to this, battlefield archaeology can be described as 'digging battlefields' and is often a continuation of traditional military history (Saunders 2009). The term 'battlefield archaeology' is also very much associated with the research of medieval and post-medieval battlefields by means of metal-detecting and analysing arrowhead or musket balls patterns; a discipline which is completely different from the research goals in this paper.

Conflict archaeology on the other hand is not solely focused on the actual battlegrounds and strives for a multidisciplinary approach using archaeology, anthropology, cultural geography and other disciplines (Saunders 2009). Some examples of conflict archaeological approaches are *The Plugstreet Project* in Belgium (Osgood & Brown 2007) and *The Great Arab Revolt Project* in Jordan (Saunders 2007). This approach involves researching not only the material remains of the conflict through archaeological field work, but also the relationship between the conflict, the landscape and the memory of the conflict. This paper is embedded within the rapidly expanding and interdisciplinary investigation of conflict zones as defined by Saunders (2009). Thus, it is not only the areas in which the actual fighting took place (see for instance case study Geluveld), but also areas towards the rear that had different functions (case study Oostduinkerke), that are subject to investigation with non-destructive techniques.

Over the last decade, there has been a shift in perspectives as at least part of the professional archaeological world has begun to deal with military archaeology and conflict archaeology. In the UK, this has resulted in an acceptance in academic circles of this field of study, following the example of the USA where 'battlefield archaeology' had become generally acceptable by the end of the 20th century (Pollard & Banks 2006). In Belgium, this acceptance arrived somewhat later. In 2002, professional archaeologists of the VIOE (*Vlaams Instituut voor Onroerend Erfgoed* – Flemish Heritage Institute) in Belgium were required by Belgian Minister Paul Van Grembergen to investigate the material remains of World War I along the proposed route of the A19 motorway near Ieper (Dewilde *et al.* 2007). Around the same time, the potential of World War I archaeology for academic research was explored, and resulted in a small number of Master's dissertations (Stichelbaut 2004; Wackenier 2004; Verdegem 2007). Both developments resulted in a more general acceptance of World War I military heritage in the Belgian archaeological community.

The professional and scientific archaeological research of the material remains of World War I can add whole new layers of otherwise hidden information, such as: unrecorded behaviour and activities, dealing with the dead (Desfossés *et al.* 2003); detailed typology of trenches and other features; daily life at the front; analysing the multi-layered conflict landscape and much more. The archaeology of World War I, and 20th century conflict archaeology in general, is still in its infancy but is certainly a rapidly evolving field of research, encouraged by the approaching centennial of the start of the War. We aim to develop the discipline of analysis of historical aerial photographs combined with geophysical prospection techniques as an essential part of the multidisciplinary approach of conflict archaeology.

The current developments towards the protection and management of World War I military heritage makes it necessary to investigate what this heritage comprises. Therefore, it is important to know exactly what can be expected to have survived, in archaeological terms, beneath the surface. This is supported by the recent development of various inventories of archaeological and military heritage in Flanders (de Meyer 2005; de Meyer & Demeyere 2004; Dewilde 2006; Decoodt 2007). However, these inventories do not make systematic use of large numbers of historical aerial photographs and often take the current surface of the landscape as a starting point without looking at what might be preserved beneath the surface.

A multidisciplinary aerial photographic study using modern techniques can provide more robust answers than any other source. The broad perspective provided by aerial photographs enable a redirection of focus from a site-directed approach to research on a landscape scale. By combining the aerial photographic interpretation data with geophysical prospection techniques such as EMI we are able to detect and reveal conflict archaeological heritage where it has survived beneath the surface and provide a rapid and rigorous interpretation and evaluation of the geophysical data.

Historical aerial photographs (1914–18)

Shortly after the Battle of the Marne at the end of 1914, the war came to a standstill in the trenches. The traditional eyes of the army – cavalry and espionage – failed to provide necessary intelligence and thus created an opening for aerial photography (Carlier 1921). Aerial photographs recorded far more information than the aerial observers could see, providing images that were indisputable documents able to confirm the accounts of the aviators. What was once a hobby for a few enthusiasts during the first months of the war soon developed into a completely new science that was used on a large scale by all combatants. As progress was made in the technical aspects of aerial photography, so the art of reading, or interpreting, aerial photographs also advanced (Wrigley 1932).

These extraordinary sources document a pan-European landscape of horror that stretches from the North Sea in Belgium to the French-Swiss border, from the Black Sea in the south to the Baltic in the north, parts of Italy, the Balkans and even more distant areas of the world. Aerial photographs were taken all over these theatres of war, documenting a cultural landscape from which the relicts often remain visible as scars on the landscape, but are frequently concealed from the untrained eye. These photographs, used for the first time now as primary sources, hold a wealth of useful data for archaeologists, (war) historians and landscape specialists.

Archives

Many archaeological papers have acknowledged the importance of World War I for the development of aerial photography, but without going into detail. At the very least, this is rather curious since the father of archaeological aerial photography, O. G. S. Crawford, was engaged in photographic reconnaissance missions along the Western Front in France and Belgium (Crawford 1955). Furthermore, the historiography of the war in the air during World War I has always been overshadowed by – in the eyes of the general public – the heroic battles of fighter pilots such as Ball, Guynemer or von Richthofen (Streckfuss 2009). Aerial photography was believed to be a passive and less colourful occupation than that of the fighter pilots and has therefore received little attention in historical research. However, the astronomical number of photographs taken shows the real importance of this information. Until recently (see, for instance, Finnegan 2006) World

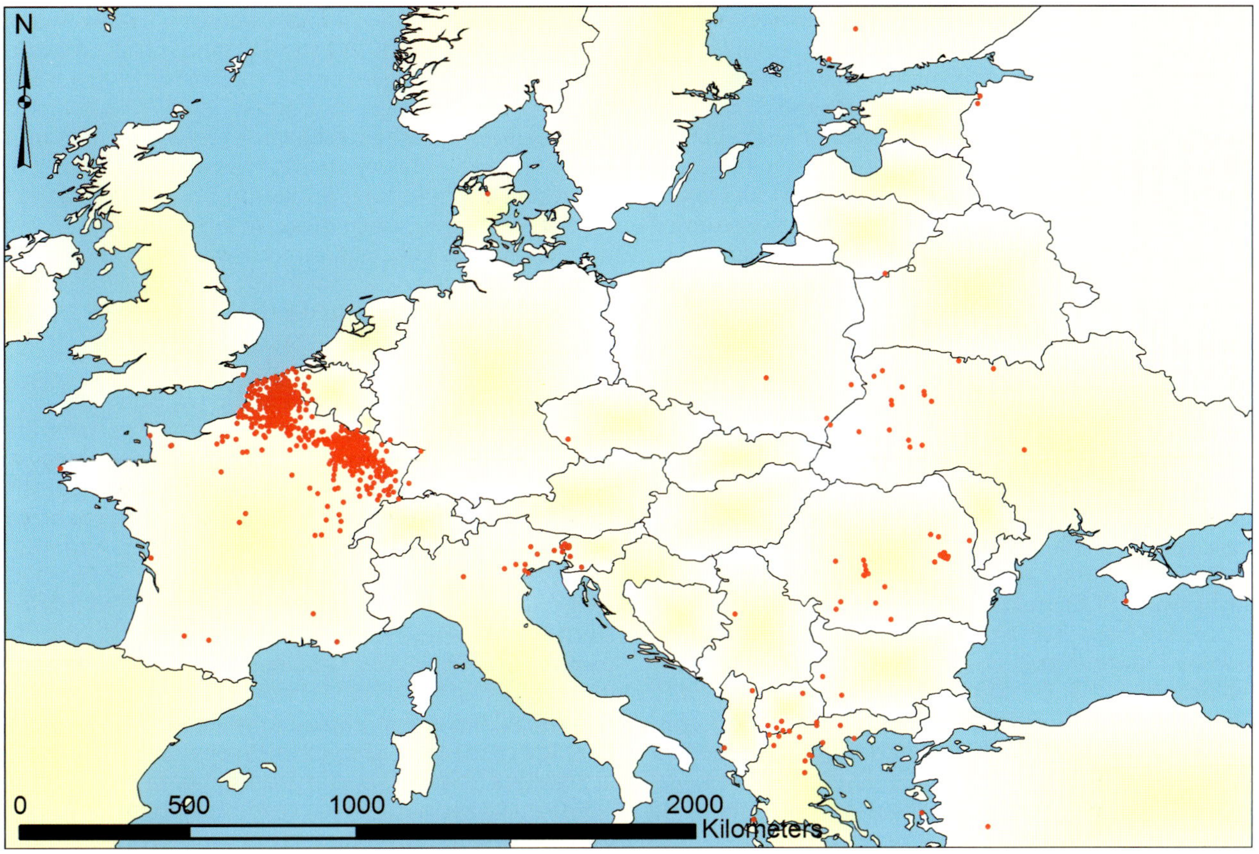

Figure 22.1: Distribution of German Aerial photographs in Europe (red dots represent entries in the BS-*Aufklärung*).

War I aerial reconnaissance has rarely been studied as a whole. In addition, nobody has ever even considered looking at the overall picture in which World War I aerial photographs have been preserved. Making the content of the collections accessible will benefit a wide range of researchers, including archaeologists, historians and geographers. Consequently, aerial photographic archives in general contain a hidden potential. Or, as Bewley and Rączkowski put it: '*Unlocking these archives is perhaps the single most important development which would dramatically improve our understanding and knowledge of Europe's historical environment.*' (Bewley & Rączkowski 2002).

Aerial reconnaissance on the Western Front was conducted by a variety of nationalities. The photographs were produced in an almost industrial process, brought together over four years, and survive in large quantities in collections spread out across Europe, the United States and even Australia. The numbers of photographs, their quality but most of all their accessibility varies enormously from archive to archive. The following National/military archives have significant sources available:

- *Bayerisches Hauptstaatsarchiv* (Germany)
- Belgian military archives (Belgium)
- National Archives and Record Administration (United States)
- Russian State Military History archive (Russia)
- *Service Historique de l'Armée* (France)
- Army museums and memorials:
- Australian War Memorial (Australia)
- *L'Historial de la Grande Guerre* (France)
- Imperial War Museum (United Kingdom – see Stichelbaut *et al.* 2010)
- In Flanders Fields Museum (Belgium)
- Royal Museum the Armed Forces and Military History (Belgium)

Photographic coverage of Europe numbers more than half a million aerial images, the majority of which are focused on the Western Front in France and Belgium (Stichelbaut 2009; Stichelbaut & Bourgeois 2009), though large numbers are also to be found in other parts of Europe. For instance, the aerial photographic collection of the Bavarian war archives (*Bildsammlung-Aufklärung* at the *Bayerisches Kriesarchiv*) in Munich hold records for large parts of Europe, covering at least 22 different countries (Figure 22.1). Most of the coverage is for the Western Front in Belgium and France, followed by Romania, Ukraine, Greece, Slovenia, Italy and Macedonia.

Mapping the Western Front in Flanders (Belgium)

The use of a detailed inventory of war features, based on historical remote sensing data, and analysis of the conflict landscape, can support and anticipate current needs in heritage management and inform those heritage agencies. Several initiatives are currently being explored, including a strong desire to protect the battlefields of World War I as UNESCO World Heritage sites.

In Belgium approximately 14,000 historical aerial photographs have already been collected from various archives. To date, approximately 11,000 have been georectified (Figure 22.2) in the framework of a joint research project of the Department of Archaeology (Ghent University), Province of West-Flanders and the In Flanders Fields Museum (Ypres). The aim of this project, amongst others, is to map all war features constructed during World War I in the western part of Belgium.

The aerial archaeological research of the Western Front using contemporary aerial photographs provides a temporally accurate insight into the density, distribution and diversity of war features. This approach enables an analysis of how the conflict landscape was organised and where certain types of features were situated, providing a detailed level of information which cannot be found in any other historical source.

An example of such a detailed inventory of war features is a ridge near Bellewaarde (Figure 22.3), which was heavily fought over in the first years of the war, mainly in between the Second and Third battle of Ypres. This small area is covered by 127 aerial photographs ranging in date from early 1915 to the end of the war in the autumn of 1918. The GIS mapping of the aerial photographs resulted in an accurate inventory of all war features which were constructed during the entire conflict by all combating parties. The density of features such as trenches is extremely high and clearly illustrates the highly contested nature of these grounds. The position of the 1915–17 frontline is marked by 36 craters from mines which were detonated beneath both the German and British frontline.

The strengths of using contemporary aerial photographs to study the conflict landscapes of World War I are to:

- accurately locate possible archaeological remains
- provide a framework to date remains as a result of the high temporal resolution (some areas have frequent coverage in different years and months)
- show what really happened on the ground with a bird's eye view
- provide an insight into the density, distribution and diversity of war features at a landscape scale.

Although these kinds of inventory are quite detailed they only indicate where possible material remains are located on the basis of what was present during 1914-18. So, to gain insights into what material remains are still preserved beneath the surface we have to look at other prospection techniques.

Geophysical survey with electromagnetic induction of World War I heritage

The possibilities for exploring the buried potential heritage of the war using other prospection techniques are somewhat limited by the specific nature of World War I, which leaves few options. Field walking, for instance, is in this case a useless method because the area of the frontlines is littered with battlefield debris, while aerial reconnaissance has proved to be unsuccessful in the area around Ypres. This is due to the heavy, poorly drained soils in this area of Flanders, which limit the potential for cropmark formation revealing buried trenches, for example. The area has certainly been frequently

Figure 22.2: Density of 11,000 georectified aerial photographs (1914-18) in Belgium.

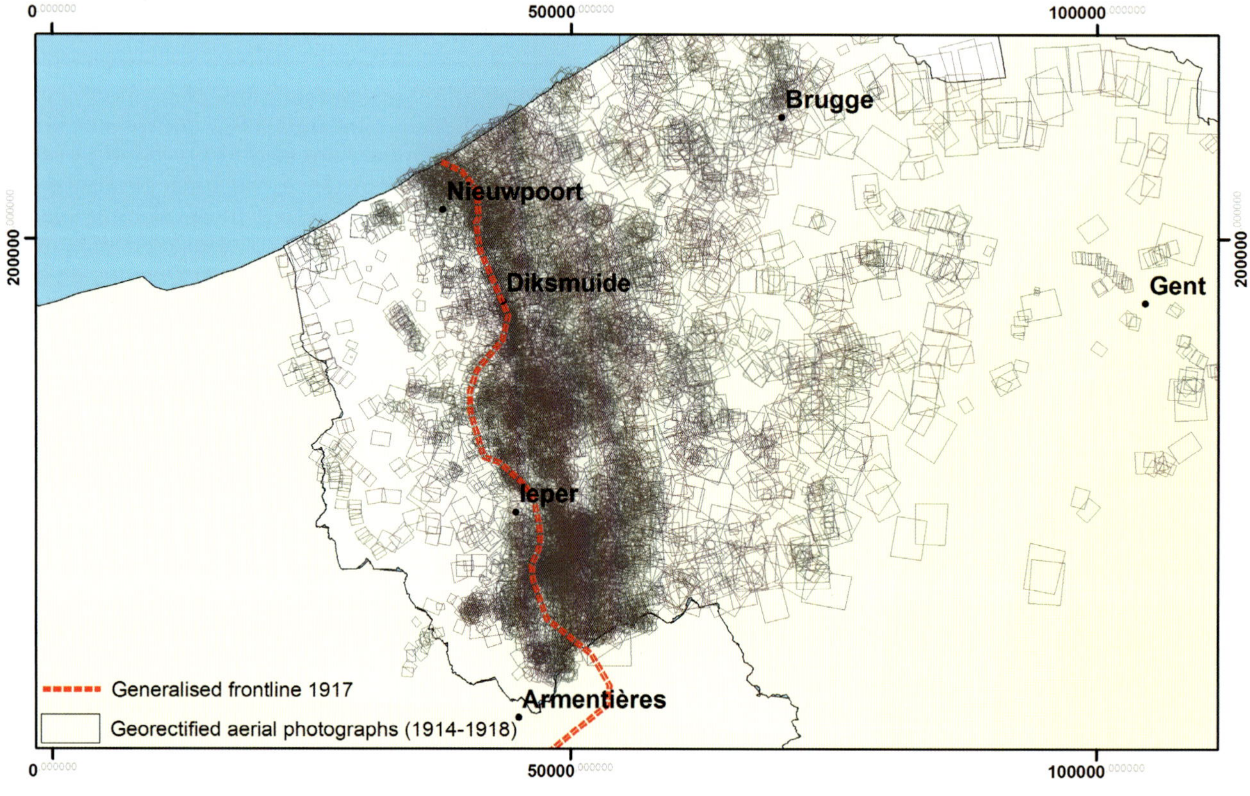

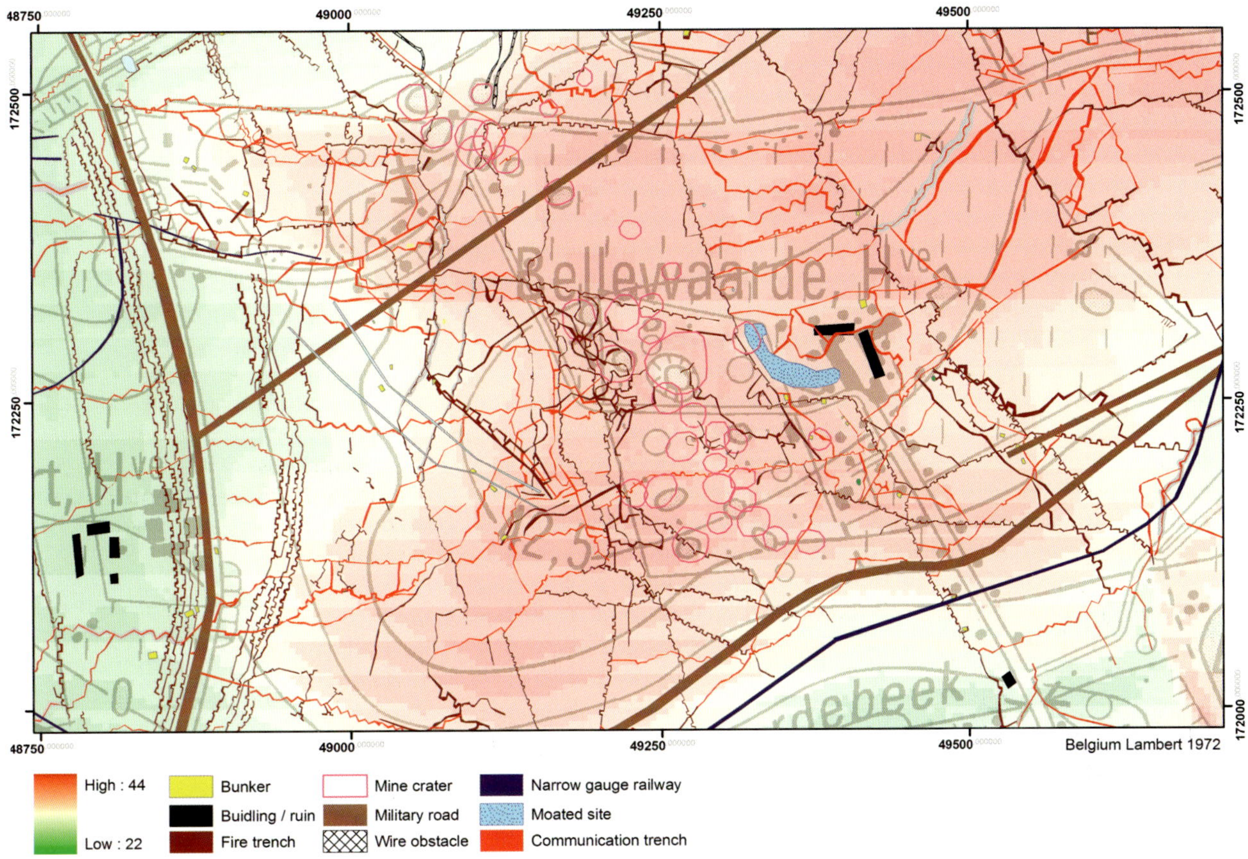

Figure 22.3: GIS mapping of war features on the infamous Bellewaarde Ridge, 3km to the east of Ypres. This presents a diachronic overview of all war features, but also takes the temporal data into account, making it possible to represent certain discrete time periods.

prospected from the air, but only very few war features have been discovered in this way, in contrast to the huge number of archaeological remnants of the war revealed by archaeological excavations around Ypres (de Meyer & Pype 2004; Dewilde 2006; Dewilde *et al.* 2007; Osgood & Brown 2007).

Geophysical prospection techniques are in our opinion the only non-destructive techniques to successfully locate and evaluate subsurface World War I features, a view supported by the good results obtained with electromagnetic induction (EMI) in two cases studies. In EMI sensing, an alternating current is passed through a transmitter coil, which produces a primary magnetic field, which induces eddy currents in the soil. These currents create a secondary magnetic field proportional to the strength of the currents. This secondary magnetic field induces again alternating currents in the receiver coil. In the quadrature-phase response, the ratio of the secondary over the primary field is proportional to the apparent (depth-weighted) electrical conductivity (ECa) of the soil; in the in-phase response, this ratio is proportional to the apparent magnetic susceptibility (MSa) (McNeill 1980). The great advantage of using EMI is the low cost of sampling data at sufficient intensities to provide accurate mapped information. It is a rapid, non-invasive method for collecting soil ECa and MSa information (Saey *et al.* 2009).

EMI frequently functions as a guide to the placement of excavation trenches, thus reducing the costs of exploration stages (Venter *et al.* 2006). Generally, the ECa measurements are strongly related to different physical soil factors of which clay content is the main parameter influencing the conductivity measurements. Other important parameters are soil moisture content, organic matter content and soil compaction. Therefore, all disturbances of any great volume and deviating ECa (e.g. by brick walls, cesspits, infilled trenches) influence the measurements to a certain extent. MSa is considered useful for archaeology because it successfully images traces of human occupation, such as foundations, ditches, pits and stakes. However, the success of magnetic mapping depends on the contrasts in magnetic soil properties (Lück *et al.* 2003). Both ECa and MSa are strongly influenced by metallic objects in the subsoil, and extreme values can therefore be produced by buried metallic objects, making it impossible to detect subtle variations in ECa and MSa.

Little research has been done to exploit the possibilities of EMI soil sensors for locating World War I remains, or for metal detection and characterisation.

Geluveld case study

ECa measurements were undertaken on a battle field at Geluveld, in the southwest of the Province of West-Flanders, Belgium. It was bombed intensively as a result of which a large amount of shell fragments and unexploded ammunition was expected in the subsoil of this arable field. Indeed, the farmer frequently unearths unexploded shells during tillage, confirming this hypothesis. EMI measurements were performed with the DUALEM-21S (DUALEM Inc., Milton, ON, Canada) instrument. In its mobile configuration, the DUALEM-21S sensor is mounted on a sled pulled by an all terrain vehicle (ATV), which drove at a speed of approximately 5km h^{-1}. Every eighth fraction of a second, ECa and MSa measurements were simultaneously recorded by a field computer (Simpson *et al.* 2009). A Trimble AgGPS332, with Omnistar correction, was used to georeference the measurements with a pass-to-pass accuracy of ± 0.10m. Navigation was supported by a Trimble Lightbar Guidance System. The large number of measurements, collected in a relatively short time, provides a comprehensive coverage of the sites (10,000 measurements per hectare; 10 hectares per day).

To remove the influence of natural soil variability on the measurements and focus on the local anomalies in the data, a filtering procedure was followed. The gradual natural trend was subtracted from the original ECa measurements to highlight the local anomalies. As an outcome, positive unique anomalies for metal objects down to 1.0m in depth were found. Metal is expected to be present as positive peaks, while negative values suggest an absence of metal objects in the soil profile.

Figure 22.4a shows the filtered ECa measurements. The brown speckles are probably all due to bomb wefts, shrapnel and unexploded ammunition remaining in the soil. Comparing this DUALEM-21S conductivity map with the war-time aerial photograph demonstrates that the fire trench visible on the aerial photograph is also represented in the geophysical measurements. To both sides of the trench, a 'metal-empty' zone with low ECa was observed, contrasting with the high ECa of the trench, a contrast due to battlefield clean-up operations after the war, when, most metal debris around the trench was collected into it.

The comparison of the aerial photographs and EMI measurements not only confirms the interpretation of a trench but also allows us to refine the image of ECa survey and provides more detail. The aerial photographs reveal, for instance, the presence of a zigzag communication trench leading towards the south which has not been recorded on the ECa survey. Clearly, this smaller trench was not filled with metal debris and the zone around it had not been cleaned-up. Small ECa differences originating from the infilling of the trench with topsoil material are masked by the large and extensive deviations from the numerous buried metal objects.

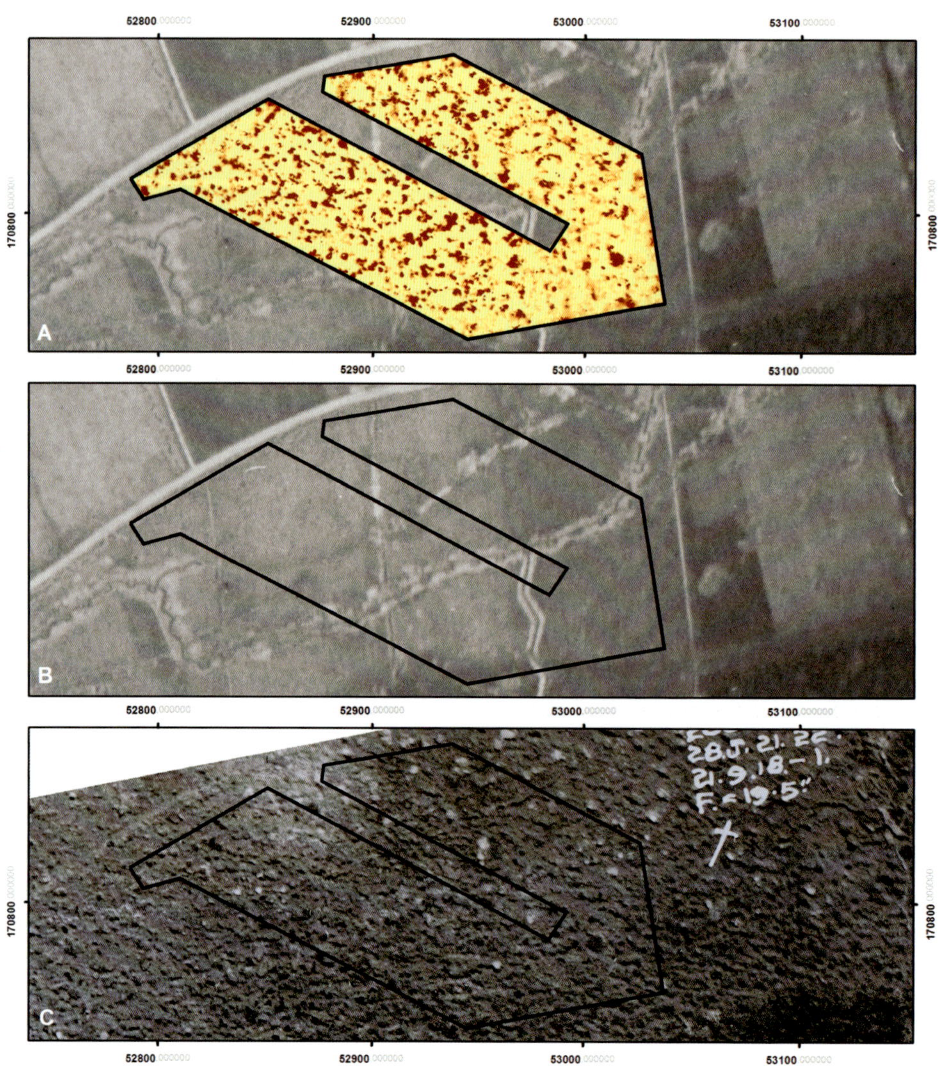

Figure 22.4: Geophysical survey (ECa) superimposed on aerial photograph (A) aerial photograph 15-9-1916 (B), aerial photograph 21-9-1918 (C). *Aerial photograph source: Australian War Memorial.*

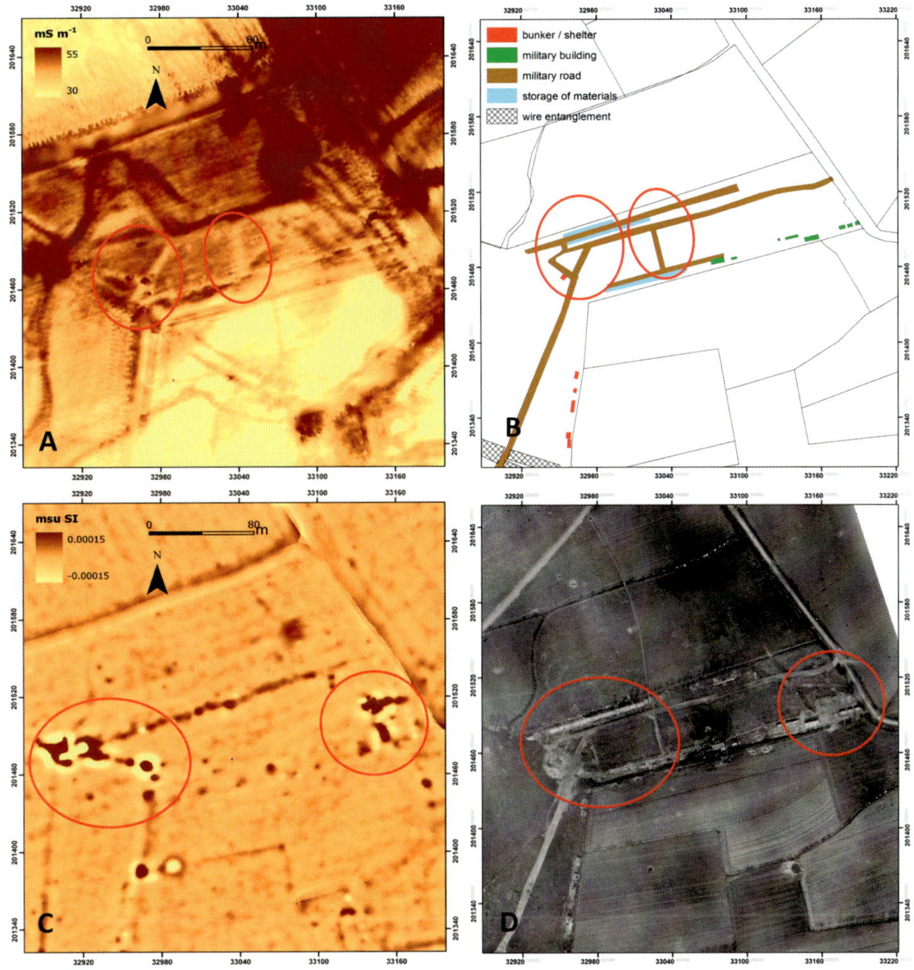

Figure 22.5: MSa (A) and ECa (C) survey near Oostduinkerke combined with aerial photograph (D) and GIS mapping and interpretation (B). *Source: Australian War Memorial and VIOE).*

Aerial photograph research indicates that this German trench (indicated on British trench maps as Jackson trench[1]) was constructed around September 1915 (Figure 22.4b). The aerial photographs provide a better understanding of what happened there from the time of construction until November 1918 and place this single trench into its wider surroundings, contextualising this single trench as a small part of a dynamic wider conflict landscape.

In the period between November 1914 and the second phase of the Third Battle of Ypres (September 1917) the study area was part of a defensive position approximately 4km behind the German lines. During British preparations for the Passchendaele offensive the entire area was intensively shelled. By the end of November 1917 the offensive had stalled and the frontline fossilised at the survey area where it remained until the start of the Lys offensive when the Germans pushed the Allies back towards the west. An aerial photograph of September 1918 (Figure 22.4c) shows the area as completely devastated and peppered with uncountable shell holes. This highlights the importance of the 'time-lapsed' view of this area offered by the aerial photographs, since if we only looked at this single aerial 1918 photograph, there would be no possibility to detect the trench at all.

1 British 1:10,000 trench map of sheet 28NE3, edition 6b, 12-09-1917 – Gheluvelt (*sic*).

Oostduinkerke case study

This study site is located near Oostduinkerke near the coast and far behind the Allied lines. The ECa and MSa survey (Figure 22.5a & 22.5c) revealed a significant number of geophysical anomalies. Comparison of the data with Belgian aerial photographs indicates that these features were constructed during World War I for storage, with access to the materials facilitated by a small network of specially built roads. A number of the anomalies in the geophysical data can be identified on the photograph (Figure 22.5d), including Belgian military roads and a number of small bunkers which were presumably used for the storage of ammunition.

The ECa measurements show both metal objects and the military road structure. The visibility of the roads can be attributed to the greater compaction of the soil, lowering the ECa values. Bordering the roads, bands with higher ECa were detected as infilled ditches. This phenomena occurs when the topsoil material (clay rich = high conductivity) contrasts with the subsoil material (sandy = low conductivity) and when the infilled ditches extend through the topsoil and into subsoil.

The MSa measurements are strongly influenced by buried metal in a similar way to the ECa measurements. Apart from this, some linear features were found, originating from old foundations, asphalt and other magnetic debris in the subsoil. This case study is important because it shows that not only can trenches be recorded on a former battlefield, but also features with potentially less impact on the soil such as roads and bunkers which were mainly constructed on the surface, rather than dug into it. This case study also illustrates not only the infrastructure of war far behind the frontline, but also demonstrates that they can be identified with geophysical prospection techniques. The potential of combining the two prospection techniques in these back areas behind the battle zones certainly needs to be emphasised again with cultural heritage managers dealing with this kind of fragile heritage.

Future research?

The use of EMI sensors in these archaeological contexts still requires research, since the limitations and optimal operating circumstances have not yet been fully determined. Using the aerial photographs to pinpoint areas of archaeological interest, several parameters such as influences of different soil types on the detection of these remains can be examined.

Combining these approaches with other geophysical survey techniques such as Ground Penetrating Radar (GPR) certainly would give added value to this kind of research. A GPR operates by emitting broadband electromagnetic pulses, which reflect from subsurface structures. As the EMI sensors mainly reveal electromagnetic susceptibility and conductivity of the material, GPR is complementary by giving a good understanding of depth and electromagnetic permittivity. The combination of both techniques can thus provide measurements of the three principal physical parameters in archaeological prospection. Another complementary aspect of this combination is measurement of depth as GPR can improve the understanding of complex structures by providing depth information, which enlarges the possibilities for combining the conductivity responses of the EMI sensors to accentuate those buried remains. On the other hand, the conductivity measured with EMI sensors will indicate the effective penetration depth of the GPR waves, which is limited by the presence of highly conductive material such as clay.

Conclusion

The combination of two different non-destructive prospection techniques described above is, in our opinion, an ideal means of documenting, accurately locating and validating the conflict archaeology of the former battlefields of World War I in support of management policy. The combined data is an ideal starting point for developing a clear vision of what is worth protecting for future generations.

For World War I the combination of aerial photographs and geophysical prospection techniques provides an accurate view of what was originally constructed during the war and what has survived in the present day landscape beneath the surface. This technique can also be used in other theatres of war, and can help in dealing with one of the main issues of World War I conflict archaeology – that is the sheer quantity of data and archaeological remnants. This makes evaluation processes to determine what is important to preserve *in situ* or by record extremely important. The combination of techniques outlined in this paper provides the basis to address the need for firm heritage management based on robust documentation and understanding.

References

Bewley, R. & Rączkowski W. 2002: Future Development of Aerial archaeology. In Bewley, R. & Rączkowski, W. (eds): *Aerial Archaeology. Developing Future Practice. NATO Science Series vol. 337.* Amsterdam, IOS Press: 1–8.

Carlier, A. 1921: *La Photographie Aérienne Pendant la Guerre.* Paris, Librairie Delagrave.

Chielens, P. Dendooven D. & Decoodt, H. (eds) 2006: *De Laatste Getuige. Het Landschap van Wereldoorlog I in Vlaanderen.* Tielt, Lannoo.

Crawford, O.G.S. 1955: *Said and Done: the Autobiography of an Archaeologist.* London, Wiedenfeld.

de Meyer, M. 2005: Houthulst and the A19-Project: Inventory of World War I Heritage Based on Wartime Aerial Photography and Trench Maps. In Bourgeois, J. & Meganck, M. (eds): *Aerial Photography & Archaeology 2003. A Century of Information.* Ghent, Academia Press, 87–99.

de Meyer, M. & Demeyere, F. 2004: *De Inventarisatie van de Gemeente Houthulst (prov. West-Vlaanderen). De Opbouw van een Archeologisch Beleidsinstrument.* CAI1 (IAP-Rapporten) Brussel, VIOE, 43–74.

de Meyer, M. & Pype, P. 2004: *The A19 Project. Archaeological Research at Cross Roads.* Zarren, A.W.A. Publications.

Decoodt, H. 2007: De Sporen van 'Den Grooten Oorlog. Monument, *Landschappen en Archeologie* 26(1), 4–36.

Desfossés, Y., Jacques, A. & Prilaux, G. 2003: Arras 'Actiparc', les Oubliés du Point du Jour. *Sucellus* 54, 84–100.

Dewilde, M. 2006: De Eerste Wereldoorlog en Archeologie. In Chielens, P., Dendooven, D. & Decoodt, H. (eds): *De Laatste Getuige. Het Landschap van Wereldoorlog 1 in Vlaanderen.* Tielt, Lannoo, 137–42.

Dewilde, M., de Meyer, M. & Saunders, N. 2007: Archeologie van de 'Groote Oorlog'. De Vlaamse Situatie. *Monumenten, Landschappen en Archeologie* 26(1), 37–54.

Finnegan, T. 2006: *Shooting the Front. Allied Aerial Reconnaissance and Photographic Interpretation on the Western Front – World War I.* Washington DC, National Defence Intelligence College Press.

Grand Quartier Général des Armées 1916: *Notes sur l'Interprétation des Photographies Aériennes.* S.l., s.n.

Kommandierender General der Luftstreitkräfte 1917: *Bildmeldung der Luftschiffer.* Charleville, s.n.

Laffin, J. 1987: *Battlefield Archaeology*. Hersham, Ian Allen.

Lück, E., Callmer, J. & Skånberg, T. 2003: The house of Bailiff of Sövestad, Sweden – a multi-method geophysical case-study. *Archaeological Prospection* 10, 143–51.

McNeill, J.D. 1980: *Electromagnetic terrain conductivity measurement at low induction numbers. Technical Note TN-6*. Missisauga, Geonics Limited.

Osgood, R. & Brown, M. 2007: Au Champ d'Honneur… Recherche Archéologique à Comines-Warneton, Eté 2007. *Les Cahiers de l'Urbanisme* 65 (Septembre 2007), 85–8.

Pollard, T. & Banks, I. 2006: Preface. *Journal of Conflict Archaeology* 1, vii–xi.

Saey, T., Van Meirvenne, M., Vermeersch, H., Ameloot, N. & Cockx, L. 2009: A pedotransfer function to evaluate the soil profile textural heterogeneity using proximally sensed apparent electrical conductivity. *Geoderma* 150, 389–95.

Saunders, N. 2002: Excavating Memories: Archaeology and the Great War, 1914–2001. *Antiquity* 76, 101–8.

Saunders, N. 2007: *Killing time. Archaeology and the First World War*. Gloucestershire, Sutton.

Saunders, N. 2009: *Conflict Archaeology*. Retrieved 30–1–2009, from http://www.jordan1914-18archaeology.org/conflictarch.htm.

Simpson, D., Van Meirvenne, M., Saey, T., Vermeersch, H., Bourgeois, J., Lehouck, A., Cockx, L. & Vitharana, U. 2009: Evaluating the multiple coil configurations of the EM38DD and DUALEM-21S sensors to detect archaeological anomalies. *Archaeological Prospection* 16, 91–102.

Stichelbaut, B. 2004: *Belgische Militaire Luchtfotografie tijdens de Eerste Wereldoorlog, Historische Studie, Luchtfoto-Interpretatie en Inventarisatie van Zichtbare Sporen. Een Bijdrage tot de Vlaamse 'Battlefield Archaeology'* (Unpublished dissertation). Gent, Universiteit Gent.

Stichelbaut, B. 2009: The Interpretation of Great War Air Photographs for Conflict Archaeology & Overview of the Belgian Royal Army Museum's Collection. In Stichelbaut, B., Bourgeois, J., Saunders, N. & Chielens, P. (eds): *Images of Conflict: Military Aerial Photography and Archaeology*. Newcastle-upon-Tyne, Cambridge Scholars Publishing, 185–202.

Stichelbaut, B. & Bourgeois, J. 2009: The Overlooked Aerial imagery of World War One: a Unique Source for Conflict and Landscape Archaeology. *Photogrammetrie – Fernerkundung – Geoinformation* (3), 231–40.

Stichelbaut, B., Gheyle, W. & Bourgeois, J. 2010: Great War aerial photographs: The Imperial War Museum's Box Collection. In Cowley, D.C., Standring, R.A. & Abicht, M.J. (eds): *Landscapes through the Lens. Aerial Photographs and Historic Environment*. Oxbow, Oxford. 225–36.

Streckfuss, J. 2009: Why the Air War Really Mattered: a Guide to Understanding the Significance of Aviation in World War I. In B. Stichelbaut, Bourgeois, J., Saunders, N. & Chielens, P. (eds): *Images of Conflict: Military Aerial Photography and Archaeology*. Newcastle-upon-Tyne, Cambridge Scholars Publishing. 41–54.

Verdegem, S. 2007: *Caesar's Nose, 1914–1918. Archeologisch Onderzoek van Oorlogspatrimonium langs de Ieperboog* (Unpublished dissertation). Gent, Universiteit Gent.

Venter, M., Thompson, V., Reynolds, M. & Waggoner, J. 2006: Integrating shallow geophysical survey: archaeological investigations at Totógal in the Sierra de los Tuxtlas, Veracruz, México. *Journal of Archaeological Science* 33, 767–77.

Wackenier, W. 2004: *Archeologie van Staal en Beton. Bunkers en minierte Unterständte uit de Eerste Wereldoorlog in Staden, Westrozebeke en Oostnieuwkerke. Vakgroep Archeologie* (Unpublished dissertation). Gent, Universiteit Gent.

Wrigley, H. 1932: *The Battle Below. Being the History of No. 3 Squadron*. Sydney, Errol Know.

23 | An aerial view of the past – aerial archaeology in Denmark

Lis Helles Olesen

Abstract: Holstebro Museum in western Denmark has secured funding for a four-year aerial archaeology research project. The project is set to run until 2013, comprises nine sub-projects and has a budget of € 650,000. This article deals with the preliminary results of aerial survey and monitoring of scheduled ancient monuments from the air. In 2008 and 2009 the project totalled about 130 flying hours and located some 560 sites. This paper presents an overview of the first 100 fully recorded localities: among other results house sites were found at 33 of the 100 sites and pit-houses provide a good opportunity to obtain an overview of Late Iron Age and Viking Age settlements (20% of the total number of sites date from these periods). Results of the first aerial monitoring of 50 scheduled monuments are also presented. Even these early results allow us to say that there is great potential inherent in both sub-projects.

Introduction

Holstebro Museum in western Denmark has been successful in securing funding for a four-year research project on aerial archaeology entitled *An aerial view of the past*. The aim of this project is to draw attention to, and to analyse, the potential inherent in aerial archaeology for research, communication and education and in the planning process. The ultimate aim is to obtain a fuller understanding and better protection of the archaeological heritage. Target groups are archaeologists, planners and the general public.

The budget is € 650,000, and the project is funded by the Danish Ministry of Culture, The Heritage Agency of Denmark, Holstebro Municipality, Central Denmark Region, Holstebro Museum, two foundations and, to a lesser degree, the six museums responsible for the areas we survey systematically from the air. The project is, in 2009 and 2010, part of a regional cultural agreement, and it has been designated by the Danish Ministry of Culture as a national assignment. The project is set to run until the middle of 2013.

An aerial view of the past: brief project description

The project comprises nine sub-projects, as outlined briefly below:

1. Basic recording
An overview will be produced of existing historical aerial photograph series covering Denmark, mostly verticals taken for mapping purposes. A description will be produced of where these photos are kept, their accessibility, their state of preservation and an evaluation of their archaeological potential.

2. Basic investigations: aerial photographs and aerial reconnaissance
This sub-project is the most important. Nine areas, each of at least 155km², will be analysed systematically through the historical photo series. The same areas will be systematically surveyed from the air together with other areas (e.g. Figure 23.1). The most informative images will be rectified (from oblique to vertical) and the sites will be mapped digitally and entered into the national *Sites and Monuments* database. An overview of recorded ancient monuments for the whole of Denmark is freely available on www.dkconline.dk.

3. Archaeological excavations
Minimal, targeted excavations will be carried out at selected sites or structures where an interpretation on the basis of aerial photographs alone proves difficult. Radiocarbon dating may be undertaken. The aim is to clarify the nature of the sites or structures that have been detected.

Figure 23.1: Cropmarks of a characteristic long house with curved side walls of Viking date together with other house sites and pit-houses. Lydum 2 July 2008. © *Photo: Lis Helles Olesen.*

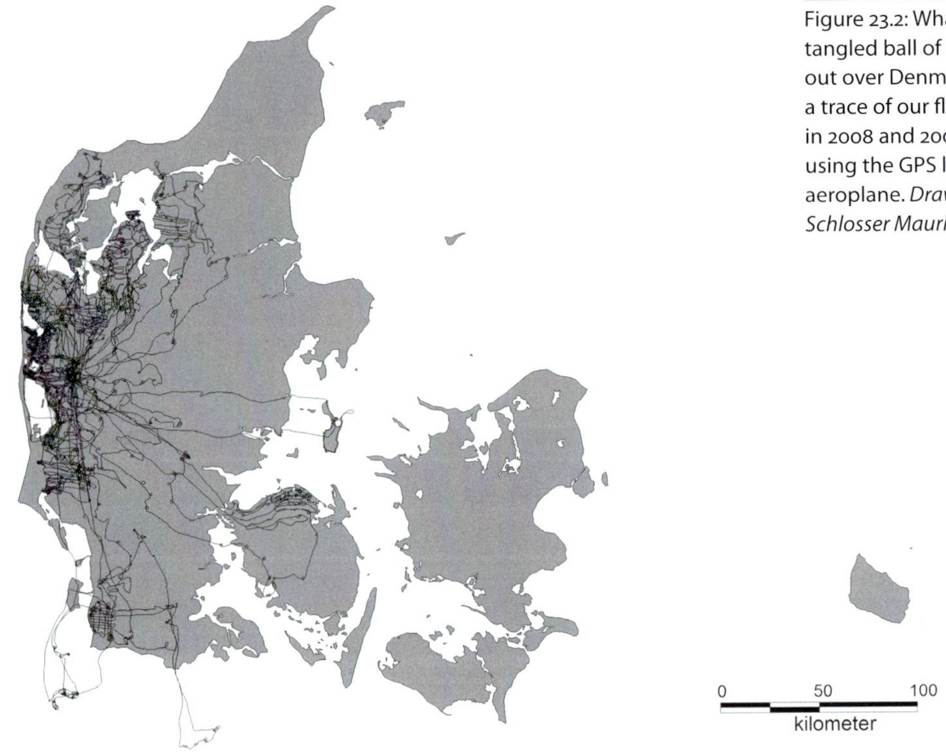

Figure 23.2: What looks like a tangled ball of wool spread out over Denmark is, in fact, a trace of our flight paths in 2008 and 2009 drawn using the GPS logger in the aeroplane. *Drawing: Esben Schlosser Mauritsen.*

4. Scheduled ancient monuments
A large number of scheduled ancient monuments in the rural landscape will be experimentally surveyed from the air. The responsibility for the monitoring of scheduled ancient monuments has been shared out between ten museums across the whole country, of which Holstebro Museum is one. We are obliged to visit all the ancient monuments on the ground during the first five-year period (2007–11). The present project will attempt to establish whether it is possible, from the air, to carry out some future monitoring and re-inspection where there has been damage to the monuments.

5. National monuments
Well-known scheduled ancient monuments, such as dolmens and passage graves, Bronze Age barrow groups, tells, war booty sites, ring forts, churches, monasteries, medieval strongholds and castle mounds, as well as non-scheduled special structures such as Iron Age boundaries and Neolithic causewayed enclosures, will be surveyed from the air in order to obtain further information on and understanding of these sites.

6. Underwater archaeology from the air
To a lesser degree, areas of coastal waters will be surveyed from the air in search of shipwrecks, defensive barrages, submerged settlements and so on. The aerial archaeological potential of Danish coastal waters is very poorly understood in contrast to the southern and eastern Baltic Sea where many previously unknown underwater structures have been detected by aerial surveys.

7. Modern technology
The possibilities offered by LiDAR (Light Detection And Ranging, see Doneus & Briese this volume) will be tested on five known archaeological sites. The potential inherent in the national digital terrain model (DK-DEM) will also be investigated in the nine selected areas.

8. National and international collaboration
Measures will be taken to develop national and international collaboration in aerial archaeology. These will include the organisation of meetings, conferences and an international aerial archaeology course in Denmark, and participation in international collaboration and aerial archaeology conferences.

9. Communication and education
An important part of the project is communication of the results with the aim of promoting greater knowledge about the protection and understanding of our hidden cultural heritage. Aerial archaeology provides us with a unique opportunity to display prehistoric sites and monuments which otherwise no-one can see, and so communicate the importance of these otherwise rather unknown remains. We will make active use of the media and the results of some of the sub-projects will be published in articles and pamphlets. At the end of the project a book will be produced. A project website has already been set up: www.aerial-archaeology.dk.

Preliminary results: the first 100 sites

Aerial reconnaissance has proved very effective (Figures 23.2 & 23.3). In 2008 we undertook about 60 flying hours and found about 280 prehistoric and historic sites. In 2009, 70 hours were flown, recording the same number of sites as in 2008. In 2009, we also undertook spring and autumn flights with the aim of investigating traces revealed by soil colour differences in uncultivated fields as well as submerged features, and to test the potential for monitoring scheduled ancient monuments. On these flights we did not find as many new sites.

Figure 23.3: The red dots show all the sites discovered in 2008 and 2009. *Drawing: Esben Schlosser Mauritsen.*

Find circumstances

All 560 sites have been entered into our database. In order to give an impression of what we have found, an overview is presented here of the first 100 fully recorded sites. A total of 79 of the 100 localities were new discoveries by the project. Of the 21 previously known sites, four were found by St Joseph of Cambridge University in the late1960s/early 1970s (Eriksen & Olesen 2002, 14). At that time, he surveyed Jutland from the air several times for Moesgård Museum. Nine localities had previously been found by the local museums from the air, while eight localities are long-known sites.

The majority of remains (i.e. 89%) have been recorded as positive cropmarks, revealing the presence of ditches and pits dug in to the ground, providing better growing conditions for the cereal or grass than in the rest of the field. A further 6% are negative cropmarks, which may be produced by buried stone foundations or poorer growing conditions over sub-surface negative features due to them being filled, for example, with shifting sand. Only 1% of the sites have been recorded as variations in soil colour in ploughed fields. This very low percentage is due to us not having flown very much in the spring and autumn when the fields are not covered by crops and may have been ploughed. Furthermore, we have not included in this figure ploughed-down barrows that are already known. The remaining 4% of sites are the remains of slight earthworks surviving on the surface and recorded in raking oblique sunlight as shadows and highlights.

Different types of structure

Below is an overview of the first 100 fully recorded sites (Table 23.1). It is apparent from this how many times the various monument types such as house sites, pits and fences occur at the various localities. Account has also been taken of how many examples have been found of each monument type. The fact that the total number of sites does not add up to 100 is due to the potential presence of several different monument types at one individual locality.

Type of structure	Number of sites	Number of structures
House site	33	111
Pit-house (SFB)	19	about 320
Pit	55	about 600
Fence/ditch	18	about 60
Ring ditch	6	about 15
Postholes (many)	7	many
Farm site (of recent date)	3	3
Castle mound (medieval)	3	3
Ancient sunken road	3	8
Ridge and furrow systems	7	many
Ploughed-down barrow	4	6
Animal pen	3	6
World War II structure	1	1
Shipwreck	1	1
Unknown structure	2	5

Table 23.1: Overview of the first 100 fully recorded sites.

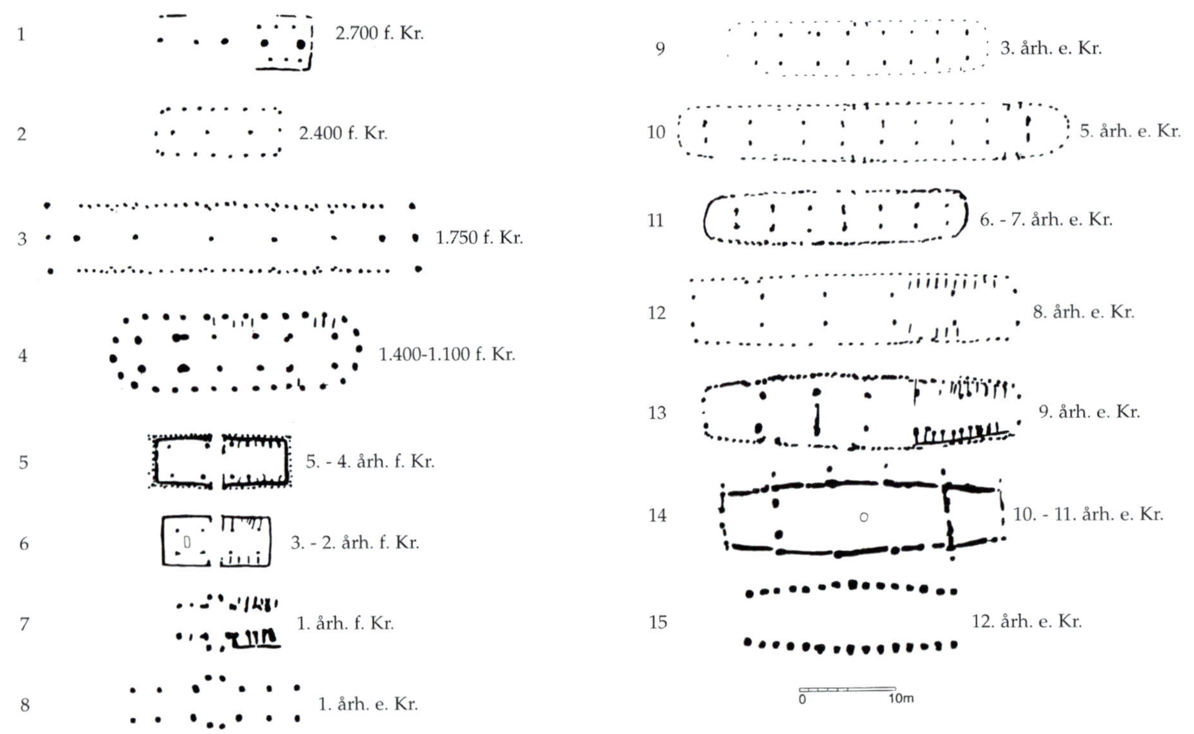

Figure 23.4: The development of the long-house from the Neolithic to the early Middle Ages. *After Hvass & Storgaard 1993.*

Pits

The most common type of structure are pits found at 55 localities with around 600 pits in total. We have attempted, as far as is possible, to exclude modern diggings in the form of, for example, sand-extraction pits. These are, as a rule, quadrangular and often lie in dense clusters on ridges. A few pits in the vicinity of an existing farm are similarly not usually included, illustrating the many choices that must constantly be made in an evaluation of the located traces. Single pits are not deemed as important as house sites, and while the pits cannot be dated without excavation, they may be all that can be seen from the air of a hidden ancient settlement. They are therefore still important.

House sites

It was an exciting discovery that houses could be recorded in so many places, at 33 out of 100 sites, and in such numbers, 111 houses located. In 15 cases, traces of only one house site were found, in nine cases there were between two and four, in seven cases there were five to eight house sites and in two further cases there were 10 and 15 house sites, respectively.

As the development of house types through prehistory is very well understood, a large proportion of the house sites can be dated (Figure 23.4). Not surprisingly, the majority (62) are from the Early Iron Age (500 BC–AD 200) spread across 13 locations. From around 100 BC–200 AD the houses had very large earth-set posts and often a sunken byre at one end as a result of mucking out the animals, factors that make it easier to locate them.

Periode	Underperiode		År f./e. Kr.
Nyere tid *Recent times*			
Middelalder *Middle Ages*			1536
Vikingetid *Viking Age*			1100
Jernalder *Iron Age*	Yngre germansk	*Germanic*	700
	Ældre germansk		550
	Yngre romersk	*Roman*	400
	Ældre romersk		200
	Førromersk	*Pre-Roman*	0
Bronzealder *Bronze Age*	Yngre bronzealder		- 500
	Ældre bronzealder		- 1100
Yngre stenalder *Neolithic*	Sen neolitisk tid *Late Neolithic*		- 1700
	Enkeltgravskultur *Battle-Axe culture*		- 2400
	Tragtbægerkultur *Funnel-necked beaker culture*		- 2800
Ældre stenalder *Mesolithic*	Ertebøllekultur		- 4000

Figure 23.5: Chronological table. *After Eriksen et al. 2009.*

Period	Number of dates as % of total
General prehistory	11 %
Neolithic/Bronze Age	4 %
General Iron Age	4 %
Early Iron Age	14 %
Late Roman/Germanic Iron Age	4 %
Late Iron Age/Viking Age/ Early Middle Ages	20 %
Middle Ages	2 %
Middle Ages/recent	3 %
Post-Reformation/recent	10 %
Undated	28 %
Total	100 %

Table 23.2: Dating of the 100 sites.

A total of 16 house sites, distributed across five locations, date from the Late Iron Age in general and seven house sites from six different places from the Late Germanic Iron Age or Viking period. One location with around 15 houses dates from the Middle Ages. From the 16th/17th century there is a single house site which was known already. As it lies within the moat of a known farmstead, we can date it to this time. Five house sites at two localities cannot be dated more closely than to the Early Iron Age, and five other house sites, occurring singly, are only dated as prehistoric since the traces are not very clear (Figure 23.5).

Pit-houses or Sunken Floored Buildings

Pit-houses are small workshop huts some 2–3m in diameter. From the 100 locations, traces of pit-houses have been found on at least 19. Their interiors are dug some way down into the subsoil and when they were abandoned they were filled up with refuse and topsoil. Due to their size and the often-deep pit associated with them, they are seen more often as crop marks than house sites (Figures 23.1 & 23.6), and more than 300 pit-houses have turned up on these 19 sites. The aerial perspective clearly has great potential to provide a good overview of the location of Late Iron Age and Viking settlement in the landscape as well as the size of these settlements, specifically on account of these small workshops. For example, after several years of intensive searching in archives, stores/collections and excavation records, we had records of 39 Viking Age settlements in the former Ringkøbing County (4,600km^2) by 2007/2008 (Eriksen *et al.* 2009). Of these settlements, nine were recorded from the air. Subsequently, 11 new localities were found by this project in 2008 and 2009, and a further five sites were found by excavation. So we now know 55 Viking Age settlements in the former Ringkøbing County, of which 20 were discovered from the air (i.e. 36%). To these can be added several possible localities where it could not be ascertained securely whether there were pit-houses or just pits present.

Dating

The broad dating of house sites is dealt with above. Here, an account will be given of the dating of the 100 sites in general. It is clear that the Iron Age and Viking periods are the best represented, accounting for around 44% of the total number of localities (Table 23.2). It is surprising that around half of these can be assigned to the Late Iron Age and Viking times, when it is taken into consideration that so many of the house sites are from the Early Iron Age. The many dates in the Late Iron Age and Viking times are due primarily to sites with pit-houses.

As expected, there are few monuments from the Neolithic and Bronze Age. We could easily increase this total by including all the ploughed-down barrows we see, but for these classes of sites, an examination of earlier national series of aerial photographs and the new national digital terrain model (DK-DEM) is more

Figure 23.6: Cropmarks of many pit-houses and a few long-houses from the Late Iron Age or Viking times. Rødding 30 June 2009. © *Photo: Lis Helles Olesen.*

Figure 23.7: Cropmarks of a large square structure of unknown function. Jersore 3rd July 2009. © *Photo: Lis Helles Olesen.*

profitable. This source gives us a truer picture of known and unknown ploughed-down barrows. Traces of house sites from the Stone Age have so far not been detected from the air in Denmark. These house sites have only a single row of post-holes running down through the middle of the house, and, especially when damaged by ploughing they will be difficult to detect. A few house sites have, on the other hand, turned up from the Bronze Age.

The Middle Ages and later centuries are also poorly represented. This is due to the fact that we only include structures that lie within the scope of the project. Thus sand- and clay-extraction pits, recent roads, drainage, cable and pipe trenches, filled-in ponds, grubbed-out shelter belts and water courses and the like are not recorded.

A relatively large number of structures are un-dated. This category covers, among others, sunken roads and many localities with pits, which even though they lie close to a known Iron Age fortification cannot be specifically dated. Many interesting structures that we so far have not been able to classify further are also concealed within this category, for example a large square ditch, measuring about 45 × 45m, at Jersore on Northern Funen (Figure 23.7).

The results from the first 100 sites are not comprehensive as, for example, the Bronze Age house sites have not been included in this overview. Even so, these results give a good impression of the direction that the aerial surveys will take with respect to date, type of structure, find context and so on. Our minor excavations at selected sites and the valuable assistance from metal detectorists supplement our knowledge at a number of the localities (Figure 23.8). These provide an identification of the structure type and a good indication of date so we can minimise the number of unidentified and un-dated structures. While, however, there are discoveries which we cannot immediately identify and date, the majority can be assigned to categories and it will be interesting to see what the development in the number of discoveries in the coming years will be, relative to the first 100 localities dealt with here.

Preliminary results: scheduled ancient monuments

Scheduled ancient monuments reflect special character and value in the landscape, and are also a treasure chest of information about our history. In Denmark, there are about 33,000 scheduled ancient monuments, of which by far the greatest number are barrows. The majority are in arable land and they are often damaged by ploughing. In addition it has always been time consuming and expensive to monitor these ancient monuments adequately. A new law lead, in 2007, to the responsibility for monitoring of scheduled ancient monuments being transferred to ten museums around the country – in the first instance for a five-year period. In our project we thought it was obvious to investigate

Figure 23.8: We enjoy good working relations with volunteer metal detectorists. They often find artefacts which give a good indication of the date of sites such as here at Naur Church where they found two small fibulas of late Viking date. © *Photo: Peder Laursen.*

the possibilities of being able to do some of this work from the air, and that new methods are required in order to keep costs down so that monitoring could be carried out at intervals of a maximum of five years.

It should be mentioned, in explanation, that there are two zones around the Danish scheduled ancient monuments that relate to their scheduling. Around the actual monument itself there is a 2m zone within which ploughing, manuring or sowing/planting are not permitted (Figure 23.9). Beyond this is a 100m zone within which no changes may be made in the existing state of the area. For example, it is not permitted to erect fences, plant trees, park caravans and the like.

We have initiated the first trials involving aerial survey of scheduled ancient monuments. The very preliminary results from this trial are included here, comprising one survey in which we surveyed 50 scheduled barrows in less than 1½ hours. We chose an area where we had monitored the state of the monuments from the ground 1½ years ago.

There were two aims of the aerial survey:

- to investigate whether it was possible to record monument condition with the criteria required for traditional monitoring on the ground,
- to investigate whether there have been changes in the state of the monuments since the monitoring carried out on the ground in 2008.

Of the 50 sites, 45 could be seen from the air, while four others lay in woodland/garden, and one barrow in arable land was so heavily overgrown as not to be visible. Thus for the 45 barrows condition and potential damage could be evaluated both within the 100m cordon and within 2m (actually visible in 46 cases).

The evaluation of the aerial photographs of the barrows demonstrated that although there were no infringements of the monuments within a 0.5–1m cordon, there were 11 infringements within 1–1.5m and 17 infringements within 1.5–2m. These three infringement categories originate in the recording required when monitoring at ground level. The reason that no damage was recorded within 0.5–1m could be that it is difficult to judge these distances precisely from the air. For example, in judging distance from the aerial photographs we can make use of tractor wheel tracks, which are 2m apart, and the distance between rows of maize stubble, which is 70–80cm, and possibly other features that occur in the vicinity of the monument, so it can clearly not be completely accurate. In some instances it is also difficult to judge the exact position of the barrow edge, but this may also be the case ground during a ground visit.

The need for management, for example felling of trees, could be established in 42 cases, and in seven the tree-cover was too dense to even estimate how many trees or shrubs should be removed. On one of the barrows, the trees were so slender and without leaves that they were difficult to see, but use of a longer telephoto lens would solve this problem. At five mounds the evaluation from the aerial photographs was judged unsatisfactory.

Assessment of changes in monument condition

It proved possible to compare change in monument condition between the recent survey and that in 2008 for 47 barrows (three lay in dense woodland and were therefore not visible in detail from the air). For 31 cases the situation appeared to be unchanged, although five barrows lack data from the earlier survey. However, for 11 barrows (23%) of the total the situation had changed.

Figure 23.9: Many farmers do not respect the 2m cultivation-free zone around barrows. © *Photo: Lis Helles Olesen.*

At five barrows the infringements had become worse, although in four cases there had been an improvement. In two cases, the barrows lay fallow in February 2008, at one there was now an infringement and at the other the situation was unchanged. The result obtained, with changes in the state of 23% of the ancient monuments over a period of less than two years, shows that it is important to follow up the survey quickly by making contact with the owner when it has been established that there has been an infringement. The damage was reported to The Heritage Agency of Denmark, which, unfortunately, has not followed this up with contact to the landowners.

Time expenditure

The time expended in this first 'monitoring survey' of 50 ancient monuments was 41 hours split between two people broken down into eight hours preparation, two hours travel, three hours flying and 28 hours examining photographs. Time expenditure will be much reduced in the next monitoring surveys as we streamlined procedures and agreed on which map formats and database to use. Thus, it should be realistic to survey a greater number of monuments in one operation – at least 100.

In general, on the basis of this first trial, there appears to be great potential for monitoring scheduled ancient monuments in rural arable areas from the air, as in about 80% of cases it was possible to obtain the relevant information. We will continue these flights in the coming years.

Conclusion

After a little more than a year several parts of the project are now well under way. Two relatively dry summers have ensured good results from aerial survey. The examination of vertical photos from 1954 and the digital terrain model (DK-DEM) in all nine study areas have also provided a large number of new monuments and a greater understanding of past landscapes and settlement patterns. We believe we are already in a position to say with certainty that there is great potential for aerial archaeology in Denmark.

References

Eriksen, P., Egeberg T. & Helles Olesen, L. & Rostholm, H. 2009: *Vikinger i vest. Vikingetiden i Vestjylland.* Højbjerg, Denmark.

Eriksen, P. & Helles Olesen, L. 2002: *Fortiden set fra himlen*. s. 14. Holstebro, Denmark.

Hvass, S. & Storgaard, B. (eds) 1993: *Da klinger i muld ... 25 års arkæologi i Danmark*. Århus, Denmark.

24 | Knowledge-based aerial image interpretation

Rog Palmer

Abstract: Knowledge-based image interpretation is a subjective process that converts the images recorded on aerial photographs, satellite images and other remotely-sensed sources to information extracted during question-orientated examination. This contribution examines the ways that we look at photographs and how our reasons for looking will influence what we see. An example demonstrates the value of a photo reading exercise as it progresses through a series of questions that show how much information can be extracted from a single aerial image. Ways of working with photographs are noted and different levels of interpretation and mapping – and appropriate uses for these – are outlined. A final section links the results of image interpretation with other sources and shows how these can assist our understanding of the past.

Introduction: why interpretation?

Without interpretation, aerial photographs, satellite images, airborne laser scans and products of other remotely-sensed sources remain little more than illustrations – although this is one of their occasional uses. Many manuals identify several stages in the process of photo examination of which 'interpretation' is the highest level. In most cases they note that interpretation must end with a map as this is a necessary part of interpretation. Some twenty years ago I outlined three stages of photo examination for archaeology, culminating with interpretation (Palmer 1989, 54-6). Since that date, the natural processes of development and change to the theory and practices of our interpretation of aerial images make it worth expanding and re-emphasising those aspects that take us towards archaeological understanding. This form of self reflection needs to be done periodically (Palmer & Cowley 2010).

Figure 24.1 is an attempt to sketch out and link some of the processes of data collection and its analysis in the search towards archaeological understanding. As is common knowledge, our data – aerial and satellite photographs/images[1] – come in two types. Some are *biased* – those, for example, taken by an airborne observer to record what has been recognised and thought to be of interest. Others are *unbiased* and will usually record a large extent of the ground as completely as possible at the time of capture. The unbiased group includes most of the data collected by earth observation satellites. Both types of image can be examined stereoscopically although our flying observers often do not take such suitable photographs. Image interpreters have to work with a mix information and, if they are able, to link this with knowledge gained by other means as noted in the lower part of Figure 24.1.

We interpret to clarify for ourselves what we have seen on those images and to show this clearly to others. Knowledge-based interpretation is, and will remain, a subjective process and thus interpretations will vary with the experience of the interpreter and for myriad other reasons (e.g. see also Horne this volume). Our interpretations must also be expected to differ in response to the questions being asked of the images. 'Is there a burial mound in this field?' will produce different results to asking what the tonal variations or colours mean in terms of what may be under, or on, the ground that is causing them. Our interest in the UK and temperate Europe ought not to be to identify phenomena such as 'crop marks' but to understand the reasons that are causing crops to grow differently. In such crops we should be able to distinguish those variations in growth that are due to buried archaeological features, those due to natural disturbances and others that have a more recent origin. Similarly, with upstanding features, 'what caused this lump or hollow to look as it does?' is an interpreter's basic question. We need to see the whole picture and to understand certain of the post-depositional events and their outcomes. From there we progress, by elimination, to those parts of a picture that interest us.

Learning to see

This process of elimination can be illustrated by a photo-reading exercise of the kind that has been a standard part of all courses run under the aegis of Culture 2000 (Culture 2000). These begin to open students' minds to the amount of information that can be seen and how the non-archaeological content can help them understand any archaeological features. The example that follows is in Romania and was published there in a chapter that aimed to introduce aerial photographs and aerial photography to archaeologists in that country (Palmer 2009). A written example misses the essential guidance that occurs in class when a teacher's feedback and questions can help direct a student's

1 From this point, unless specifically stated, the term 'aerial images' will be used to include pictures taken from aircraft and satellites.

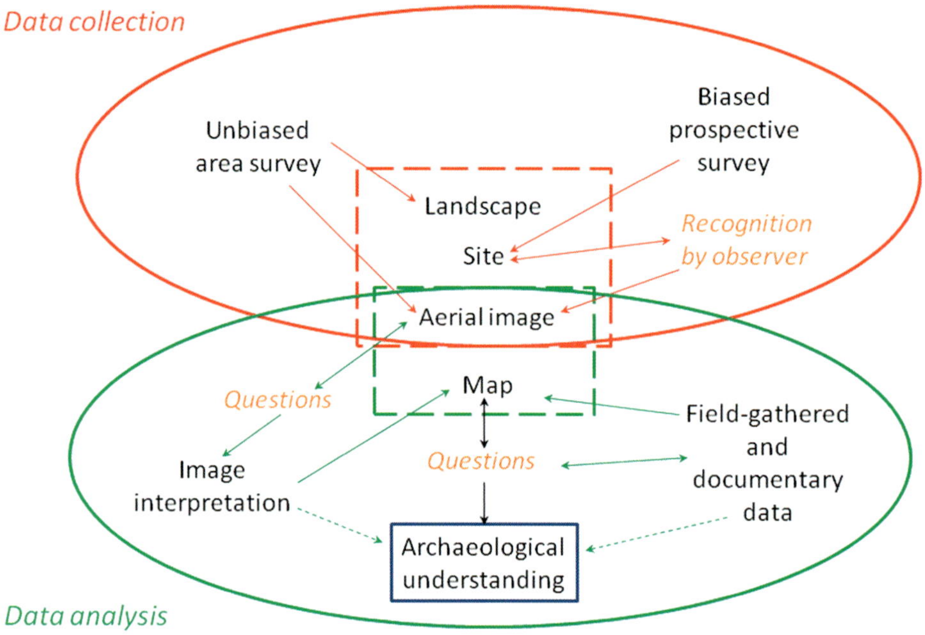

Figure 24.1: The diagram outlines the processes of data collection (aerial and satellite photography) and its analysis (photo interpretation) and draws in other sources of information to seek greater archaeological understanding. *Source: author.*

thinking. But the description that follows will attempt to recreate the classroom feel in the hope that a reader may be encouraged to learn to see more.

Describe the main components of the picture (Figure 24.2). The photograph shows part of a village (A) with a mixture of cultivated land (B) and pasture (C). A small river (D) meanders through the pasture fields.

What time of year do you think the photograph was taken? The trees in the village have full canopies, grass is green and beginning to burn dry in places and some of the cultivated crops are ripe (yellow or brown coloured). Some of the grass has been cut (E) and at least one of the strips of cultivated crop have been harvested (F). So a date when this could happen in southern Romania (where the photo was taken) is mid to late June depending on the weather that year.

And what was the time of day?
Shadows from trees are short, so the photograph was probably taken between 10am and 2pm. We could make a closer estimate if we had a map and could identify north on the photograph and so estimate the position of the sun from the direction of shadows.

Can you see any shape to the ground? Are there high or low places?
The river has to flow along the lowest ground and its meandering course, tributaries (G) and small pockets of wet ground (H) tell us that there is quite a broad floodplain and that much of the grassland is likely to be fairly level ground as is the cultivated ground closest to the river (F-B). We can see that the river is not deep because the track that crosses it (at J) has no bridge. So it is likely that tractors and animals can cross without problem, although there is a small narrow bridge next to this crossing so that pedestrians do not need to get their feet wet.

There is an impression, helped by the curves in the farm tracks, that from B the ground rises to the road. This change to higher ground can also be seen in the strip of land just outside the village which appears to rise up to the village from the pasture fields (K-K). There is also a D-shaped island of higher ground in the centre of the photograph (L). In summary, the photo shows a floodplain that has higher ground below the modern village and the road at the top of the picture.

What can you tell me about the island?
This could be an archaeological site – or a significant local place. The island may have started life as a small area of higher ground, perhaps seasonally isolated by the flooding river, and as such it could have been settled by earlier communities. In fact, the shape of it and its height suggest that this is a tell and so it is likely to indicate a site of Neolithic occupation. Much later, it became a cemetery for the village with its own church and this use would have damaged the upper layers of Neolithic features.

This short series of questions and answers has not been just an intellectual exercise as we have created a mental picture of a tell being formed by long-term use of a small island in an area of what was probably seasonal wetland. By so doing we have put the occupation site and its inhabitants in an environmental context that may help explain how they lived. The alternative description made by going directly to the archaeological site on the photo and describing it as a tell by a river would have given us less information and a lower level of understanding of what may have happened in the past.

With a responsive class and a skilled teacher, this series of questions and answers can go on almost indefinitely and can probe the depth of students' knowledge. Specialist knowledge of non-archaeological subjects can also take the responses in new directions. Imagine, for example, the different answers to the above questions that would come from a group of farmers and a group of bank managers.

For those who really want to know, the photograph was taken on 15 June 2009 at 12.12pm. The road next to A is aligned north-south with north at the top of the photograph. The archaeological site is Tell-ul Mgura at Teleorman (approx. 44.040° N, 25.389° E).

Working with aerial images

Alison Deegan recently (2010, 57) outlined the ideal process of knowledge-based aerial photo interpretation as the following sequence of actions, which are here slightly modified and extended. An interpreter will examine all available prints and digital images together, select those required for mapping, scan the prints and then transform the scans and digital images using appropriate software. Next, the transformed images are imported into CAD or GIS where the features of interest will be digitised from them while making reference back to the original images which, ideally, will be viewed as stereo pairs. The final step is to collate all the information from those chosen images into a map and then check any that were not transformed to see if they hold extra information that can be added to the map.

In the 40 or so years that I have been working with aerial images the methods of creating an accurate map have changed from hand drawing using the Mobius network and its variations, through hand digitising of traced interpretations, to drawing on screen using a mouse to position polygons and lines over digital copies of transformed images. Use of computers and the development of software have considerably eased the technical part of the mapping process, enabling us to get features in the right place with a little thought and a few clicks of the mouse. Work that took several weeks in 1970 can now be accomplished in a few minutes and we cannot imagine the changes that the next 40 years might bring. However, the process of knowledge-based interpretation has not changed and still requires a person to examine images and make decisions about what is relevant to the questions being asked. This is also the case when automated processes have been used in the hope of detecting archaeological features. Human intervention is required at some stage to verify what has and has not been identified and to judge whether or not the results are of archaeological character.

In the early 21st century most archaeological image interpretation is done on screen as part of the process of making a map that accurately depicts what has been identified and locates it with a precision that matches that of the base map. There are good points to working this way – for example, a viewer can zoom in and out to look, to see, to understand and to draw with the accuracy required – and there are bad points. In particular, the ability to zoom in may mean that an interpretation is made to show detail that is excessive for the intended scale of the end product. One loss to the digital age may be in the lack of paper prints that can be viewed stereoscopically. Use of a stereoscope is crucial to the production of high-quality interpretations from aerial images. Not only does this allow the viewer to recreate the dimension of height, but stereoscopic examination also adds confidence to an interpretation and produces a more reliable end product.

Figure 24.2: An oblique aerial photograph used to demonstrate the 'talk through' process of photo reading. Tell-ul Mgura, Teleorman, Romania. *Photograph © Rog Palmer: 20090615-026.*

A grave disadvantage to those of us working on screen with vertical sources in the northern hemisphere is that we are likely to perceive a reversal of topography. It is a well known fact to those who study visual perception, and frequently illustrated in their publications, that the brain understands a view better if the light is coming towards the viewer. We have acclimatised through millions of years of evolution in a world where the sun has shone from above and to suddenly reverse that norm can cause confusion to the eye-brain network (Snowden *et al.* 2006, 217-8). This is the situation we are presented with on screen if north remains in its traditional position to the top (Figure 24.3 Top). So to aid interpretation – which can be difficult at the best of times – images should be appropriately rotated when they are being examined and interpreted (Figure 24.3 Bottom). As a guide to whether this has been done successfully I have added a Sun Point to these figures which, based on the concept known from children's drawing, is smiling when it is high in the sky. Adding a Sun Point to show the direction of the sun will, thus, indicate by a frown or smile whether an image has been suitably orientated. North-to-the-top is a fairly recent *cartographic* convention and there is no reason why it needs to be adhered to when vertical photographs and satellite images are being displayed.

Figure 24.3: *Top* – vertical photograph of part of Iceland (northern hemisphere) aligned with north to the top. The eye-and-brain tends to read this with reverse topography, hence the miserable-looking Sun Point that has been placed to show the direction of the sunlight. *Bottom* – Exactly the same image as above but here rotated so that the sunlight falls towards the viewer. With this lighting, topography should be correctly read by a viewer as is indicated by a happier Sun Point that is now in its usual place in the sky. *Source: National Land Survey of Iceland: http://atlas.lmi.is/myndaskjar/.*

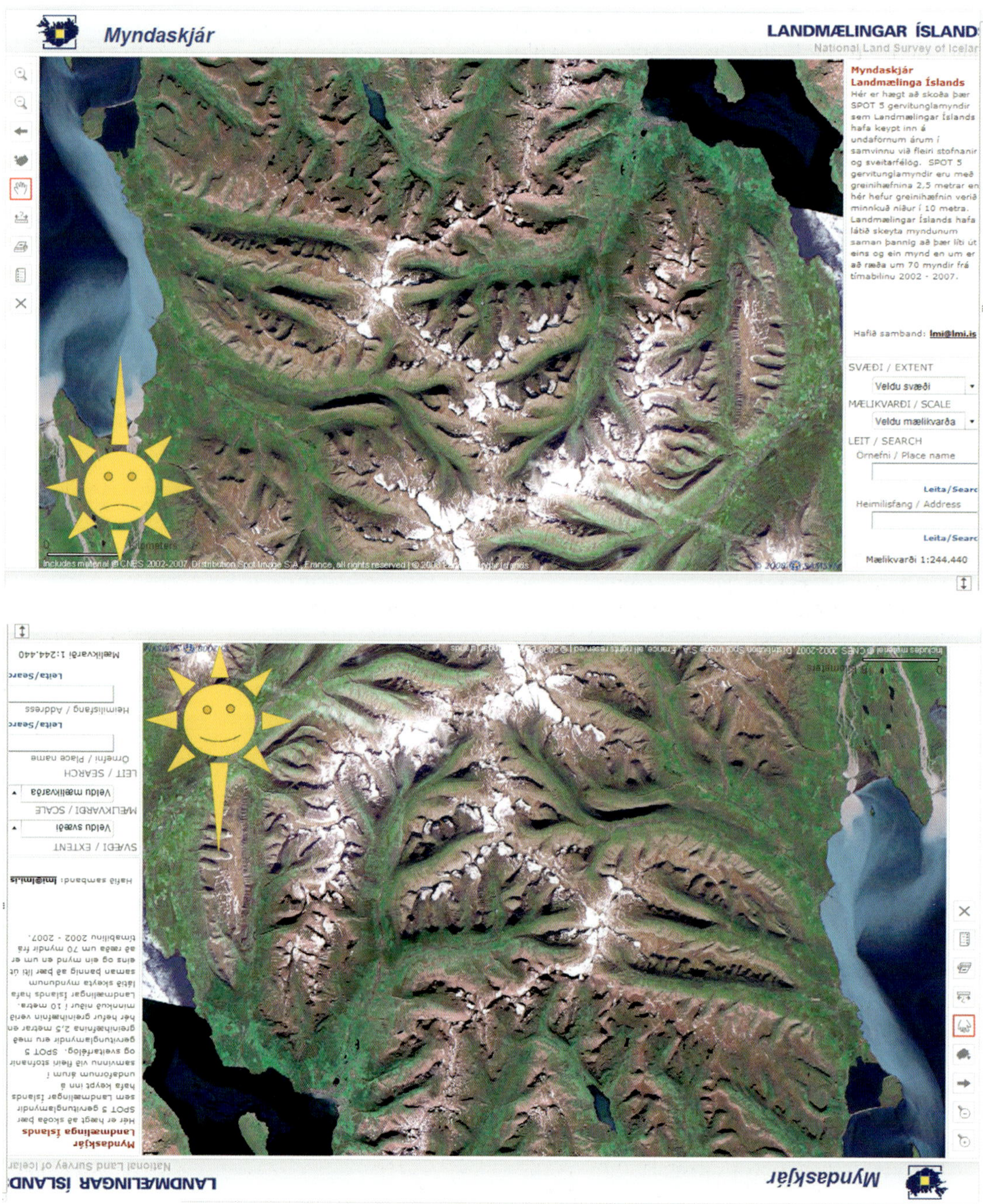

Levels of interpretation, scales and maps

Knowledge-based image interpretation is a subjective process and thus there can be no such thing as a definitive interpretation of an aerial image or of a set of aerial images. As well as the experience of an interpreter leading to different outcomes, it is a basic premise that image interpretation is carried out with a specific aim in mind. Features which are relevant to one aim are likely to be filtered out by eye and brain when interpreting the same images for a different purpose. An example relevant to archaeological work could be the decision whether or not to depict recently-removed field boundaries that are visible on aerial images. If the intent of the work was to produce an interpretative plan that showed the relationships between archaeological features in one area, the addition of non-archaeological information may cause confusion. But, if a plan is being made to aid and guide field work, these additional details may help inform work and decisions on the ground. In this example, two different results can come from examination of one set of images and both can be equally valid although one may contain a wider spectrum of information than the other.

There has to be a reason for making any interpretation and an aim that will be achieved by its completion. A request to 'interpret this image' will result in as many different results as there are people and these results will vary with the backgrounds of different interpreters. An extreme case would be the farmers and bank managers we met earlier, but even within a group of archaeologists we must expect what each person sees and understands will be guided and limited by their personal knowledge and experience.

In addition, it is a basic assumption that the scale at which a map is to be compiled and used has been decided before starting to make an interpretation. This affects not only accuracy and the level of detail that can be portrayed but is also another basis for making decisions about what types of information are to be sought and mapped. Some seventeen years ago a Technical Paper was written for the UK Institute for Archaeologists (Palmer & Cox 1993). This is now technologically antiquated, but its comments about the philosophy that directs an interpreter's decisions remain valid. For example, the Paper outlined the differences in what may be included and expected in two maps made of the same area: one at a scale of 1:10,000, the other at 1:2,500. These were the medium- and large-scale paper maps then available in Britain that were commonly used as base maps on which to prepare and show the image interpretations.

Different scales of maps, regardless of whether these are paper or digital, are surveyed and compiled to different degrees of absolute and relative accuracy. Details and figures need not concern us here (but should be available on a mapping agency's web site) and it is sufficient to note that the larger the scale, the greater is the accuracy to which a map or plan is surveyed and drawn. Consequently, measurements made from those maps are subject to similar degrees of precision. It follows that work prepared at 1:10,000 will give a false degree of reliability if it is enlarged. Conversely, details shown on maps made at 1:2,500 scale may fuse together and become unreadable if they are reduced.

In the age of digital maps, the scale of base maps and their accuracy is just as relevant as it was in the days of paper maps, so continued use of the Representative Fractions (RFs) 1:10,000 and 1:2,500 provides us with two norms, or levels, to which interpretative maps still are prepared in the UK. These RFs can be used as a guide to what information we may expect to be shown on each and to elaborate the differences between them. I use the term 'levels' to explain the different ways we think as we interpret images rather than to propose a set of standards for our work.

Interpreting at different levels will affect the range of content that is extracted from aerial images and will also affect the amount of detail that it is necessary to depict. Use of the two RFs as names for these levels helps to indicate what information may be redundant at the smaller scale because reproduction at that scale (or at a reduced scale) will render it meaningless. As a guide, the thinnest line width that I used when drawing in ink, and which I still use in computer graphics, is 0.18mm. This is the equivalent of 1.8m on the ground at 1:10,000 scale or 45cm at 1:2,500. In my opinion, this is perfectly adequate for our subject matter which, at its clearest, will be a hard-edged feature such as a wall. In most cases we are using aerial images as source material from which we are interpreting irregular and undulating earthworks or bands of differently-coloured soil or differently-growing crops. When we interpret earthworks it is necessary to decide where to mark the break of slope and sometimes we need to decide whether we are going to mark 'up' or 'down' – although this description may not make much sense until you try it. These decisions can be difficult enough on the ground with a full-scale earthwork. On a stereo pair of 1:10,000 verticals, even when these are viewed at 4x magnification, the accurate positioning of a line along a change of slope is more challenging and relationships between two adjoining features, that may be decipherable on the ground, are likely to remain unseen or indistinct to the image interpreter. Marks in soil or crops can also be quite diffuse. There are limits, therefore, to the accuracy to which we should interpret 'edges' of soil and crop changes especially as we know that these may appear slightly differently on images taken on another date. Furthermore, in these situations, the interpreter should be visualising the underlying archaeological features and not drawing differences in (for example) the tops of cereal plants that are up to 1m above the ground.

This is not to suggest that we don't show small differences that may relate to sub-surface features, but that we impose a degree of editing on what is, in any case, a subjective interpretation. An example for discussion is shown in Figure 24.4 which illustrates crop responses above a fairly typical palimpsest of levelled features on a gravel deposit in Cambridgeshire, UK. Dark green crops indicate the presence of deeper soils resulting from a range of natural and anthropogenic

Figure 24.4: This oblique aerial photograph is used to illustrate some of the thought processes and reasoning that is required to interpret the ring ditch and end up with a circle, or arc, on a map. Photograph © *Rog Palmer: 94.149/28.*

activities in what is now a small modern cultivated field. I have used this photograph as a teaching example for many years and am fairly certain that the only pre-medieval archaeological feature on it is the ring ditch which here can be used to pose some of the questions that an interpreter needs to consider when that feature is being converted from pixels on a digital copy of the photograph to a vector drawing that will show the results of interpretation. There are two immediate problems relating to how this ring ditch will be drawn. Firstly, the crop response varies around its circumference and is weak, or poorly defined, in the arc closest to the camera; and secondly, can the ring ditch really be seen where it crosses (or is crossed by) the linear features or does the eye imagine a continuation?

To fully discuss the reasoning necessary to answer these two problems would require several pages of text, a few explanatory drawings and, probably, a higher resolution picture than can be reproduced in a book – so it will not be done here. Suffice to say that, with reference to other images taken on the same day and one dated 6 March 2006 – which is almost certainly an incorrect date – in Google Earth™ (at 52.559° N, 0.089° W), I would draw the ditch circuit complete, and at a fairly consistent width, but for the part where it is masked by the periglacial feature on its lower right side. There I cannot be definite about the continuity of the ditch so it is omitted from my map and a note in my report will highlight this omission.

For any scale of drawing or reproduction, there is a sensible limit at which to stop messing around with trivial details and trying to answer questions which may not be resolved by the existing images, and produce something that is reasonably accurate but edited. An extreme example of what not to do was when I caught a student who had zoomed into a picture to such a degree that she was tracing around individual pixels. By zooming in that far her work had changed from interpretation to tracing as there was minimal awareness of any archaeological features or their context. After zooming out it was possible to see the archaeological feature and draw a smoother outline along groups of pixels that provided a much better representation of the edge of a tonal difference on an image.

With the above comments and generalities in mind, we can look at how these two levels may provide a guide to what is interpreted from images and what, therefore, is shown on a map. I should add that my own philosophy of interpretation is that natural and non-archaeological features can, in places, be as important as the archaeological ones.

Interpretations made at 1:10,000 level are likely to be undertaken to build up a map of an area that will show different types of archaeological features in their topographical contexts and indicate, if possible, the relationships between them. The mental attitude of the image interpreter when preparing material for medium scale (1:10,000) mapping, will be concerned with overviews and observing interactions between 'sites' rather that trying to identify every pit or minor length of ditch within a settlement complex. This scale is ideal for showing, for example, field systems and settlements – the topographical location of these and the spatial relationships between them. At this scale of drawing or reproduction there may be small breaks in ditch circuits, for example, that cannot be shown unless the width of causeway is exaggerated. It is also unwarranted to draw every pit to scale although it may be possible to show the difference between those of large and small size. This is not to say that these cannot be drawn – because the interpreter may do this by zooming in to see a large-scale view on screen – but that it is unnecessary to draw them because it exceeds the detail that is expected, and can be usefully portrayed, at 1:10,000 scale. It is expected that, at any scale, finished maps will include, for example, a layer to show 'possible archaeological ditch' to indicate features of which the interpreter was uncertain but which may be of archaeological.

Depending on the reasons for making the interpretations and compiling maps at this scale, it may be advantageous to show relevant natural features such as palaeochannels as these can indicate minor topographical differences and may also highlight where settlement sites may or may not be expected.

Work at 1:2,500 level will be expected to interpret and map considerably more information. Archaeological features are likely to be drawn to show as much detail as can be seen on the images and to include, for example, changes in ditch width, accurate sizes of breaks in ditches, all pits and postholes, buttresses on walls…and so on.

There are two main reasons for working at this level – to prepare an accurate and detailed plan of individual sites within a landscape study area that has been mapped at 1:10,000 level; or to produce a map that can be used as the basis for further investigation which may include field walking, geophysical survey and targeted evaluation trenches, or which may be necessary for management purposes.

At this level, non-archaeological and uncertain features are expected to be interpreted and mapped with equal accuracy. The content may vary according to the needs of the map but at its most detailed (e.g. in advance of field investigation) the range of features shown is likely to include any that will indicate damage to archaeological contexts (e.g. old field boundaries, field drains, backfilled quarries and pipelines) and natural features that may help understanding (e.g. palaeochannels and areas of locally-high ground) or that may be confused with archaeological deposits after topsoil has been removed (e.g. geological disturbances). All of these features, archaeological or not, may be visible on aerial images in exactly the same way – through different growth of crops above them – so mapping areas of deeper soil will indicate where archaeological features may not have been visible and thus help explain blank areas or other absences of information. Reference back to Figure 24.4 will show that it includes periglacial fissures and other, probably related, geological noise, hand-dug quarry pits and, at its lower edge, a band of deeper soil that marks a former fen edge. All these are likely to be included on a map that is derived from interpretation at 1:2,500 level.

A final difference between these two levels returns us to the accuracy of the base maps on which the interpretative maps are based. The British Ordnance Survey publish figures for relative and absolute accuracy of their maps which show that work undertaken using 1:10,000 maps (or the more accurate but now outdated 1:10,560 maps) cannot be of sufficient accuracy to enable precise positioning of evaluation trenches, and possibly are not good enough to mark the boundaries of a managed site.

Moving towards understanding?

The route towards archaeological understanding is enhanced if we can add information from other sources to that interpreted from aerial images as there are limits to what can reliably be said on the basis of aerial evidence alone.

Image interpreters need to know the limitations of the data they use and that these criteria may vary from site to site and the most important missing element is provision of a reliable dating framework. Other limitations derive from the visibility of features on aerial images and knowing, and indicating, where these are not visible, or not likely to be visible, on an image is as important as identifying and mapping that which is visible. Similar words have been written by others and some of us try to indicate 'blank' areas on our maps.

The diagram in Figure 24.1 identified, in general terms, two main sources of supplementary information: that gathered in the field and that coming from documentary sources. The two, of course, may be complementary and may also duplicate information. The main sources of information that help us to understand information from aerial images come from field walking surveys and excavation.

The value of surface collection of artefacts is highly dependent on depth of topsoil above buried archaeological deposits. But, in appropriate areas surface collection may provide a new level of information – dates – if a sufficient quantity of identifiable objects has been collected and if there is fair certainty that these may be derived from layers of features that have been mapped from aerial images or using geophysics. A survey of the East Anglian Fenlands of England gave the opportunity to merge surface collection data with that derived from aerial photographs although this was not an original concept of that project. A series of maps was prepared that showed selected areas of the Cambridgeshire Fens and the results, against a relevant environmental background, of integrating these surveys (Figure 24.5). If this integration had been conceived at the outset of the project there would have been opportunities to resurvey areas where only one method had provided information.

Analytical field survey can be an effective tool to add detail in areas of upstanding archaeology to that mapped from aerial photographs. This was been demonstrated in England on Bodmin Moor (Johnson & Rose 1994) and parts of Bokerley Dyke (Bowen 1990), for example. Interpretation and mapping from aerial images has to be done at high level to make these projects effective and economical.

Aerial and geophysical surveys have an uneasy relationship which often appears to be competitive. They provide very similar data from non-intrusive

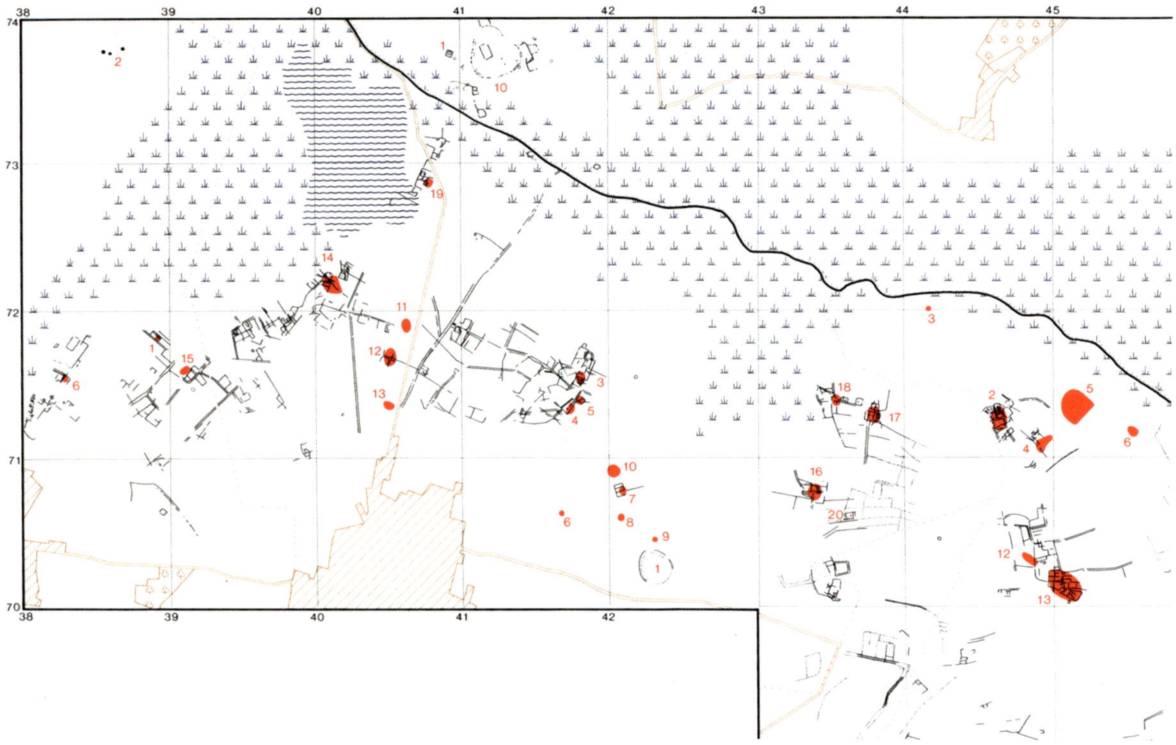

Figure 24.5: Part of a map that integrates results of field walking survey (red blobs) with features interpreted and mapped at 1:10,560 from aerial photographs. This admixture provides crude dates for the photographed evidence and provides context for the surface collections. Questions should arise for further work when there are not coincidental locations. The map also shows the extent of peat deposits and the 'negative' areas caused by modern villages and woodland. Grid lines are at 1km intervals. *Source: Hall & Palmer 1996, Figure 67*.

methods and it sometimes seems unnecessary to use both in the same place. This theme will not be expanded here, but an example that compares the two sources of data comes from a recent survey of the Traprain Law environs in which the results, but not the interpretations, of aerial photography and geophysical survey were compared at 30 discrete sites in the area (Hale & Cowley 2009; Cowley *et al.* 2009, table 2.1, figure 2.6). However, in some parts of the UK, comparisons of results from the two sources have shown that areas, or soil deposits, that are blank on aerial photographs are often equally uninformative to the range of geophysical techniques. So we are getting confirmation, rather than clarification, of these negative zones. To maximise use of our resources we should aim to achieve complementary results rather than duplication.

Documentary sources, or printed records, have been used by medieval scholars for many years and provide detail that otherwise would be unobtainable – such as identifying ownership of specific furlongs within a strip field system (Beresford & St Joseph 1979, 21–48). More recently, the so-called 'grey literature' – reports with very limited circulation that result from evaluations and excavations made by commercial archaeological units in Britain and parts of Europe – has the potential to add a vast amount of new information to that available through traditional published sources. This is very useful on a local scale and a recent attempt has been made to synthesise these 'grey' results on a national level (Bradley 2007). The relevance to this paper is that many of these sites in rural areas have commissioned specialist aerial image interpretation prior to undertaking field work and thus dating information, as well as other useful 'facts' can firmly be attached to those maps derived from aerial images.

Conclusions

There is more to image interpretation than drawing lines on a map because we are able to identify the nature of that drawn feature (for example, as a ditch, pit, wall or periglacial fissure). However, in some cases this is the limit of our understanding which can only be extended with reference to information from other complementary sources. This is vital for any research projects that use aerial images as a major source and it is necessary when we are considering or designating sites or landscapes as 'important' for purposes of preservation or management. In the past it seemed that sites recorded on aerial photographs may be scheduled if they appeared to be very complex or if they provided yet another example of a well-known type (Roman camps come to mind). Yet knowledge-based aerial image interpretation has the power to identify and record much more of past landscapes than the complex or the simply obvious and is at its strongest when integrated with other sources. And it is only by understanding the breadth of what has been identified that we should be able to design preservation and management strategies that will encompass a complete and representative sample of past settlement and landuse.

References

Beresford, M.W. & St Joseph, J.K.S. 1979: *Medieval England: an aerial survey.* Cambridge.

Bowen, H.C. 1990: *The Archaeology of Bokerley Dyke.* RCHME, London.

Bradley, R. 2007: *The Prehistory of Britain and Ireland.* University Press, Cambridge.

Cowley, D.C., Hale, D., Hunter, F. & Macleod, K.H.J. 2009: Survey in the Traprain Law Environs Project area. In Haselgrove, C. (ed.): *The Traprain Law Environs Project: fieldwork and excavations, 2000–2004.* Society of Antiquaries of Scotland. 11–21.

Culture 2000: http://muzarp.poznan.pl/EuLandscapes/ (accessed 26 April 2010).

Deegan, A. 2010: Old photos and new imagery – archaeological interpretation and mapping. In Cowley, D.C., Standring, R.A. & Abicht, M.J. (eds): *Landscapes through the lens: aerial photographs and the historic environment.* Oxbow, Oxford. 55–64.

Hale, D. & Cowley, D.C. 2009: Appendix 1. Cropmark evidence and geophysical survey: a comparison of results from sites investigated by the TLEP. In Haselgrove, C. (ed.): *The Traprain Law Environs Project: fieldwork and excavations, 2000–2004.* Society of Antiquaries of Scotland. 239–58.

Hall, D.N. & Palmer, R. 1996: Cambridgeshire Fenland Survey: the silt fens. In Hall, D. (ed.): *Fenland Landscapes, Cambridgeshire, Volume 3. East Anglian Archaeology* 79.

Johnson, N. & Rose, P. 1994: *Bodmin Moor: an archaeological survey. Volume 1: the human landscape to c.1800.* English Heritage Archaeol Rep 24/RCHME Supplementary Series 11.

Palmer, R. 1989: Thoughts on some aspects of air photo-archaeology. In Kennedy D. (ed.): *Into the Sun: essays in air photography in archaeology in honour of Derrick Riley*: Sheffield. 53–60.

Palmer, R. 2009: Implicații ale fotografiei aeriene pentru arheologia din Romnâia. In Palmer R., Oberländer-Târnoveanu, I. & Bem, C. (coord.): *Arheologie Aerianâ în România şi în Europa.* CIMEC: Buchureti. 8–61.

Palmer, R. & Cowley, D. 2010: Interpreting Aerial Imagery – developing best practice. In Campana, S., Forte, M. & Liuzza, C. (eds): *Space, Time, Place. Third International Conference on Remote Sensing in Archaeology, 17th–21st August 2009, Tiruchirappalli, Tamil Nadu, India.* BAR -S2118, 129–35.

Palmer, R. & Cox, C. 1993: *Uses of aerial photography in archaeological evaluations.* IFA Technical Paper 12.

Snowden, R., Thompson, P. & Troscianko, T. 2006: *Basic Vision: an introduction to visual perception.* University Press, Oxford.

25 | Training for the future: cooperation across Europe

Chris Musson

Abstract: From the 1990s onwards attention has frequently been drawn to the limited opportunities across Europe for education and practical training in aerial archaeology and other forms of remote sensing, and hence of their application and potential contribution to discovery, interpretation and conservation within the fields of heritage and landscape studies. Some of the steps taken in the last 15 years to improve this situation are outlined in the first part of this contribution. The second part then turns to future prospects within a recently commenced 5-year scheme of international cooperation within the Culture Programme of the European Union.

Introductory note

Although addressed principally at aerial survey and photography most of the patterns described in this contribution have been, or could be, applied equally effectively to training and education in other forms of remote sensing, from geophysical prospection to LiDAR and satellite recording.

Training initiatives in the 1990s and early years of the 21st century

Ever since archaeologists and aerial photographers first began to cooperate with one another on a European scale in the early 1990s there has been a continuing concern with education and training for specialist work in aerial archaeology, geophysics and other forms of remote sensing (Cowley & Palmer 2009, *passim*). In the present decade there has also been a progressive move towards a 'landscape' approach to the use of these techniques in heritage exploration, conservation and public appreciation.

The concern with education and training first of all focussed on the near-total absence of practical training across Europe in the use or acquisition of aerial photographs for archaeological purposes, whether in the university sector or in professional practice. True, in Britain there were a few universities which provided elementary teaching about the archaeological uses of these techniques – among many others. But coverage on the Continent was virtually absent, especially so in central and eastern Europe, then emerging from a period in which private flying was almost impossible and maps and aerial photographs were treated as military and/or bureaucratic secrets. Indeed, it was the emergence of the countries of the former Soviet zone that prompted the seminal gathering of archaeologists and aerial photographers at Kleinmachnow, near Potsdam, in 1994 (Kunow 1995).

From that meeting, and from the initiatives that it prompted in individuals like Bob Bewley, Otto Braasch and other members of the Aerial Archaeology Research Group (AARG, then UK-based but now truly international), there emerged a series of initiatives which have grown and diversified over the following 15 years, in many cases with generous financial help from the Culture Programme of the European Union (see, for instance, www.muzarp.poznan.pl/EuLandscapes).

Conferences, training schools and workshops

Leaving aside a series of conferences which helped to advance cooperation and mutual understanding across Europe, the most obvious educational initiative during those years lay in a number of training schools in aerial archaeology for up to 20 or more students at a time (Figures 25.1 & 25.2). Run for the first time in Hungary and Poland in 1996 and 1998, they then moved between 2001 and 2007 to Italy (four times), Germany (three times, on a slightly smaller scale) and most recently the UK (for the 1996 training school see Bewley *et al.* 1996; notes on many of the other schools, workshops and related meetings have appeared in issues of *AARGnews*, now freely available through the AARG website at www.univie.ac.at/aarg/).

Figure 25.1. Excited students returning from one of the first flights at an aerial archaeology training school at Foggia, southern Italy, in 2003.

Figure 25.2. Pre-flight briefing before one of the training at the Cirencester aerial archaeology school in the UK in 2006.

Over time there has been a progressive elaboration of the 'standard' teaching programme at these events but one basic principle has remained unchanged – the more or less equal attention given to ground-based and to in-air experience for the students over their week or so at the school. Inevitably the balance tips slightly in favour of the in-air element, since some of the ground-based instruction relates directly to the flying activities, leaving a little less than half of the time for instruction and practice in the photo-interpretation, mapping and *uses* of aerial photographs. In the early days, between 1996 and 2001, this was understandable since one of the key objectives was to initiate the application of 'aerial archaeology' in parts of central/eastern Europe and the Mediterranean which up till then had made little if any use of the technique, whether in the office or through exploratory survey and oblique aerial photography undertaken by archaeologists themselves.

In later schools efforts have been made to achieve a more equal attention to in-air and ground-based aspects, though the predictable excitement of the participants' first and subsequent flights have still tended to dominate their engagement with the course. Not for all, however, and at the UK school in 2006 the great majority declared that their prime reason for attending the school was to learn how to *use* aerial photographs in their everyday academic or professional work. It remains true, however, that even a brief taste of 'life in the air' can provide an important background for those whose daily work will involve them in the study and exploitation of aerial evidence rather than in its initial acquisition.

Individual initiatives

A considerable impact has also been made in the past few years through individual personal initiatives. For instance, Otto Braasch, one of Europe's finest aerial photographers, has continued to provide in-air experience – in Germany and many other parts of Europe – for any who have asked or responded to his invitations. There is no better way to learn than seeing a master at work. In the UK several students have completed placements with Rog Palmer at Air Photo Services in Cambridge. These have involved students from Poland, Italy, Romania, Latvia and Armenia, usually one at a time over periods of a few weeks to several months, during which they have gained sufficient experience in photo interpretation and mapping to work on money-earning commissions or to contribute to research projects with collaborating organisations.

There have been other initiatives, too. Following an introductory workshop in Romania in 2005, Rog Palmer has returned three times for periods of 1–3 weeks to instruct and work alongside staff at the Institute for Cultural Memory (CIMEC) in Bucharest. During these visits training has extended from photo interpretation and the acquisition of existing vertical photographs from military sources, to the incorporation of mapped information into the Institute's GIS and the structuring of documentation to relate interpretations of photographs to known sites and locations. One of the staff members who has become the principal aerial specialist at the Institute has also spent time at Cambridge to increase her experience in photo interpretation and mapping. In 2008 Palmer's visit to Bucharest was timed to coincide with a period of airborne survey, during which he was able to help with procedures in the air and to ensure that photographs were securely documented after the flights. Flying from the same airfield, in the same aircraft but going in a different direction, the present author has cooperated as visiting expert in another aerial survey project, photographing sites and landscapes in an archaeologically rich valley south-west of Bucharest.

This kind of cross-country sharing of skills can be relatively easily organised and financed. It will undoubtedly play a significant role in the pattern of practical training over the coming years.

New perspectives: *ArchaeoLandscapes Europe*, 2010–2015

Exciting longer-term perspectives have now been opened up by the approval of a further EU-assisted project, *ArchaeoLandscapes Europe* which will facilitate a five-year programme of cooperative work in the fields of aerial archaeology, remote sensing and landscape studies. Two of the key themes will be education and training for specialists and the promotion of a better appreciation of the contributions that these techniques can make in all aspects of heritage investigation, conservation and public appreciation.

Running for five years from September 2010 and involving a total expenditure of €5 million, shared equally between EU grant and the participants' own resources, the new project initially involves 34 educational and heritage organisations from almost all parts of Europe (Figure 25.3; see Appendix and www.archaeolandscapes.eu). Its central aim is *to increase public appreciation, understanding and conservation of the landscape and archaeological heritage of Europe through the application and international sharing of skills and experience in airborne and other forms of remote sensing*. It will seek to achieve this through eight key objectives or 'Actions', in which the project partners will participate according to their needs and capacities at the time. The Actions are as follows:

- The creation of an ultimately self-supporting *ArchaeoLandscapes Europe Network*, with a small secretariat, to provide leadership, coordination and advice in the use for heritage purposes of aerial photography, remote sensing and landscape studies.
- The use of traditional and innovative methods to publicize the value of aerial survey, remote sensing and landscape studies amongst the general public, students, teachers and all those who explore, enjoy or care for cultural landscapes and heritage sites across Europe.
- Promotion of the pan-European exchange of people, skills and understanding through meetings, workshops, exchange visits, placements and opportunities for specialist training and employment.
- Enhancing the teaching of remote sensing and landscape studies through courses for students and teachers, and in the longer term through the foundation of a European Masters degree in remote sensing and heritage management.
- Securing the better exploitation of existing air-photo archives across Europe by researching, assessing and publicizing their potential for heritage interpretation and landscape conservation.
- Supporting aerial survey, remote sensing and landscape exploration in countries relatively new to their use, especially in northern, eastern and southern Europe.
- Exploring the uses of laser, satellite and other forms of remote sensing and web-based geographical systems in archaeological and landscape research, conservation and public education.
- Providing technical guidance and advice on best practice in aerial survey, remote sensing and landscape studies, with a particular emphasis on conservation and heritage management.

The breadth of scope and geographical coverage in this new project, along with its assured funding over a five-year period, obviously create new possibilities for the intensification and diversification of educational and training initiatives on a pan-European scale. Inevitably, however, these possibilities arise in part from the initiatives undertaken and experience gained in the decade and a half since the inspirational days of the Kleinmachnow meeting in 1994.

The new project will promote several more training schools of the kind already held in various parts of Europe and the Mediterranean, again with an emphasis on parity between ground-based and in-air

Figure 25.3. Representatives of 34 institutions from 27 European countries at the launch of the *ArchaeoLandscapes* project in Bucharest, Romania, in September 2010.

experience and with providing an impetus to countries or organisation where the uses of aerial survey and photographs have yet to become every-day parts of academic and professional practice in the field of landscape and heritage studies.

But the last few years have seen an increasing use of ground-based workshops and intensive training courses – in Finland, Romania, Slovenia, Poland and Denmark. These have concentrated far more intensively on the study, interpretation and mapping of information derived from aerial photographs, whether newly obtained from exploratory survey or from the vast archives of mainly-vertical aerial photographs that exist across Europe. At the last of the 'traditional' training schools, in Italy in 2007, there was also a parallel course for a small group of more advanced students, concentrating on intensive interpretation and mapping, with only limited but carefully targeted aerial sorties as part of the overall experience.

So, over time, a more varied pattern of purpose and instruction has begun to emerge. Undoubtedly this trend will continue within the new project, if only because an all-embracing ground-and-aerial school for (say) 24 students and 10–12 instructors will cost a serious amount of money to organise, even with the help of European Union funding. It may be that entirely ground-based workshops will assume a more dominant role in providing a kick-start to the use of aerial survey and existing aerial-photograph collections by countries, organisations or individuals taking part in the new project.

That said, the future is likely to see a continuing pattern of two- or three-day conferences and one-day specialist meetings of the kind that have featured in the last two decades, supplemented in the coming years by smaller training schools and workshops of one kind or another and by a new round of initiatives involving short-term staff or student exchanges and on-the-job training. As originally planned, and without the tremors caused by financial problems and volcanic eruptions, the EAC meeting that has prompted this volume would have been allied to a student training school involving 2–3 days of ground-based discussion and instruction, followed by a single day of 'taster' flights to round off the experience without distracting too much attention from the rest of the course.

In Denmark in 2008 the funding for a successful ground-based meeting on aerial archaeology was organized by students from earlier training schools (Helles Olesen 2009). The pattern set by the two Polish workshops in 2007 may continue, in Poland and hopefully elsewhere across Europe, with a concentration on the landscape and archaeological interpretation of existing material from the huge archives of oblique and vertical photographs accumulated over the decades. Meetings and workshops of this kind are infinitely less expensive to organize than those involving in-air experience. There may be other patterns that could be built upon, too, such as the 19 days of intensive photo-interpretation and mapping undertaken through a variety of funding by an international group of 10 staff and 33 students, several of them 'graduates' of earlier flying schools, in Slovenia in 2006.

A different approach, however, either through the *ArchaeoLandscapes* project or through exchange agreements or fund-raising of other kinds, might involve work-placements for students or young professionals to learn by working alongside more experienced colleagues in organizations that are already engaged in aerial survey, photo interpretation and mapping. These might operate either within individual countries or internationally. Some heritage organizations might wish to improve the 'aerial consciousness' of their staff by bringing in the skill of experienced external practitioners who would work in their offices and instruct their staff for periods of secondment – anything from a few days to a couple of months. There may be other patterns, too – English Heritage, for instance, is currently providing funding for flying programmes in four county councils in England where the training of new 'flyers' by experienced practitioners forms a key part of the exercise.

Looking further ahead, distance-learning, already an established pattern in certain disciplines, might form a useful part of an overall strategy for raising skill levels in the wider archaeological profession. The *ArchaeoLandscapes* project, for instance, will aim to put together an effective Internet-based introduction to aerial archaeology, in all its aspects, and then to support this through web-based specialist courses and discussion groups in which the participants could receive feedback or seek advice from experienced practitioners, maybe even with real-time visual contact through skype or other internet conferencing methods.

Coda: international cooperation and innovation

A consistent message throughout this contribution has been the urgent need to improve the teaching of landscape studies and remote sensing in many parts of Europe, and to increase the practical opportunities for students, professionals and cultural resource managers to acquire specialist skills. The *ArchaeoLandscapes* project will address this need by creating contacts, facilitating exchanges and prompting discussion between teachers and professionals working in these fields or wishing to apply these techniques in their research or conservation work. The improvement and broadening of course-content will be a priority, with the shared and compared experience of existing teachers and professionals acting as key contributory factors. There will be a particular concentration on establishing intensive courses in existing institutions throughout Europe and (as noted above) in creating web-based material for distance learning. A cherished but inevitably long-term objective will be the creation of a year-long European Masters degree involving students in work and study in at least three countries of Europe, broadening their perspectives and opening up possibilities of future employment outside their own native countries.

The particular role of the *ArchaeoLandscapes* project, through a series of carefully targeted Working Groups, will be to bring initiatives of these and other kinds to reality, both in individual countries and through international cooperation across Europe. In all probability the pattern of undergraduate teaching in aerial archaeology, remote sensing and heritage studies is likely to remain heavily influenced by institutional patterns within individual countries. But the sharing of attitudes and experience, and the creation of new forms of teaching and training on an international basis, can only benefit from the pooling of ideas and skills that lie at the core of the *ArchaeoLandscapes* project.

Similar initiatives in related fields such as geophysical prospection or specialist aspects of satellite imagery or airborne scanning might add significantly to these developments in the acquisition and uses of aerial photographs. Bilateral or multilateral international cooperation will surely lie at the root of successful approaches to such projects as the exploration and conservation of 'Doggerland' in the sea between the UK and Continental Europe or the initiation of new approaches to preventive and rescue archaeology in Italy (Fitch *et al.* and Campana this volume). These in turn may lead to educational initiatives that will enhance the range of skills and understanding across Europe in the fields of archaeological and landscape exploration, conservation and public presentation.

References

Bewley, R., Braasch, O. & Palmer, R. 1996: An aerial archaeology training week, 15–22 June 1996, held near Siófok, Lake Balaton, Hungary. *Antiquity* 70, 745–50.

Cowley, D. & Palmer, R. (eds) 2009: *Education in Aerial Remote Sensing for Archaeology*. AARG Occasional Publication No 1.

Helles Olesen, L. 2009: An aerial archaeology school in Denmark 2008. In Cowley, D. & Palmer, R. (eds) 2009: *Education in Aerial Remote Sensing for Archaeology*. AARG Occasional Publication No 1. 48–51.

Kunow, J. (ed.) 1995: *Luftbildarchäologie in Ost- und Mitteleuropa*. Aerial Archaeology in Eastern and Central Europe, Potsdam.

Musson, C. & Palmer, R. 2009: Training schools and workshops in the past decade – and their future in the next?. In Cowley, D. & Palmer, R. (eds) 2009: *Education in Aerial Remote Sensing for Archaeology*. AARG Occasional Publication No 1. 52–6.

Appendix: INITIAL membership of the ArchaeoLandscapes project

The Project, which forms part of the European Union's Culture 2007–2013 Programme, will bring together the following 34 organisations from 24 different European countries. It is hoped that other organisations will join as Associated Partners during the lifetime of the project.

Co-ordinator/Project Leader:

The Roman-Germanic Commission, German Archaeological Institute, Germany.

Co-organisers:

Belgium: The In Flanders Fields Museum.

Cyprus: The Cyprus Research and Education Foundation (STARC).

Denmark: The Holstebro Museum.

Germany: The Landesamt für Denkmalpflege, Baden-Württemberg.

Greece: The Institute for Mediterranean Studies (FORTH).

Hungary: The Baranya County Museum Authority.

Iceland: Fornleifastofnun Íslands – Institute of Archaeology.

Ireland: The Discovery Programme; and University College Dublin.

Italy: The Universities of Foggia, Salento (Lecce) and Siena.

Lithuania: The University of Klaipeda.

Netherlands: The University of Leiden.

Norway: The Norwegian Institute for Cultural Heritage Research (NIKU).

Poland: The Adam Mickiewicz University (Institute of Prehistory), Poznań.

Romania: The Institute for Cultural Memory (CIMEC).

Serbia: The Institute of Archaeology, Belgrade.

Slovakia: The Archaeological Institute of the Slovak Academy of Sciences.

Slovenia: The Slovenian Academy of Sciences and Arts; and the University of Ljubljana.

Spain: The Heritage Laboratory (LaPa), Instituto de Estudos Galeos Padre Sarmiento.

United Kingdom: English Heritage; the University of Exeter; the University of Glasgow; and the Royal Commission on the Ancient and Historical Monuments of Scotland.

Associated Partners

Austria: The Institute of Pre- and Proto-History, University of Vienna.

Czech Republic: The University of West Bohemia, Pilsen.

Estonia: The Estonian Heritage Society.

Finland: The Helsinki University of Technology.

International: The Aerial Archaeology Research Group (AARG).

Latvia: The Latvian Academy of Sciences.

Spain: The University of Granada.

Contributors

Dr Oscar ALDRED
University of Iceland
Department of Archaeology
Sæmundargata 2
101 Reykjavík, Iceland
oscar@hi.is

Dr Anthony BECK
School of Computing
University of Leeds
Leeds, LS2 9JT, United Kingdom
arb@comp.leeds.ac.uk

Rebecca BENNETT
School of Applied Sciences
Bournemouth University
Talbot Campus, Fern Barrow
Poole, BH12 5BB, United Kingdom
rbennett@bournemouth.ac.uk

Dr Jörg BOFINGER
State Office for Cultural Heritage Baden-Württemberg
Berliner Strasse 12
73728 Esslingen am Neckar, Germany
joerg.bofinger@rps.bwl.de

Professor Jean BOURGEOIS
Department of Archaeology
Ghent University
Sint-Pietersnieuwstraat 35
9000 Gent, Belgium
Jean.Bourgeois@UGent.be

Dr Christian BRIESE
Ludwig Boltzmann Institute for
Archaeological Prospection and Virtual Archaeology/
Christian Doppler Laboratory "Spatial Data from Laser Scanning and Remote Sensing"
Institute for Photogrammetry and Remote Sensing
Vienna University of Technology
Gußhausstraße 27-29/E122
A-1040 Vienna, Austria
christian.briese@archpro.lbg.ac.at

Professor Stefano CAMPANA
Landscape Archaeology and
Remote Sensing Laboratory
Department of Archaeology and the History of Arts
University of Siena
via Roma 56 53110, Siena, Italy
campana@unisi.it

Anthony CORNS
The Discovery Programme
63 Merrion Square
Dublin, D2, Ireland
Anthony@discoveryprogramme.ie

David C COWLEY
RCAHMS
16 Bernard Terrace
Edinburgh, EH8 9NX, United Kingdom
dave.cowley@rcahms.gov.uk

Dr Matija ČREŠNAR
Centre for Preventive Archaeology
Institute for the Protection of
Cultural Heritage of Slovenia
Department of Archaeology, Faculty of Arts
University of Ljubljana, Slovenia
matija.cresnar@ff.uni-lj.si

Dr Zoltán CZAJLIK
GIS-Laboratory, Aerial Archaeological Archive
Institute of Archaeological Sciences
Eötvös Loránd University
Budapest, Hungary
czajlik@ludens.elte.hu

Professor Michael DONEUS
Ludwig Boltzmann Institute for
Archaeological Prospection and Virtual Archaeology
c/o Department for Prehistoric and
Medieval Archaeology
University of Vienna
Franz-Kleingasse 1
A-1190 Vienna, Austria
michael.doneus@archpro.lbg.ac.at

Dr Árni EINARSSON
Myvatn Research Station
IS-660 Myvatn, Iceland
arnie@hi.is

Lesley FERGUSON
RCAHMS
16 Bernard Terrace
Edinburgh, EH8 9NX, United Kingdom
lesley.ferguson@rcahms.gov.uk

Dr Antony FIRTH
Coastal and Marine
Wessex Archaeology
Portway House
Old Sarum Park
Salisbury, SP4 6EB, United Kingdom
a.firth@wessexarch.co.uk

Dr Simon FITCH
IBM Visual and Spatial Technology Centre
Institute for Archaeology and Antiquity
University of Birmingham Edgbaston
Birmingham, United Kingdom

Dr Andrew FORD
School of Applied Sciences
Bournemouth University
Talbot Campus, Fern Barrow
Poole, BH12 5BB, United Kingdom

Dr Chris GAFFNEY
Division of Archaeological,
Geographical and Environmental Sciences
University of Bradford
Richmond Road
Bradford, BD7 1DP, United Kingdom
c.gaffney@bradford.ac.uk

Professor Vincent GAFFNEY
IBM Visual and Spatial Technology Centre
Institute for Archaeology and Antiquity
University of Birmingham Edgbaston
Birmingham, B15 2TT, United Kingdom
v.l.gaffney@bham.ac.uk

Dr Benjamin GEAREY
Birmingham ArchaeoEnvironmental
Institute for Archaeology and Antiquity
University of Birmingham
Birmingham, United Kingdom

Murielle GEORGES-LEROY
DRAC de Lorraine
Service régional de l'Archéologie
6, place de Chambre
57045 Metz, France
murielle.leroy@culture.gouv.fr

Professor Martin GOJDA
Department of Archaeology
University of West Bohemia,
Sedláčkova 15
306 14 Plzeň, Czech Republic
martin.gojda@seznam.cz

Lis HELLES OLESEN
Holstebro Museum
Museumsvej 2
7500 Holstebro, Denmark
lho@holstebro-museum.dk

Dr Ralf HESSE
State Office for Cultural Heritage Baden-Württemberg
Berliner Strasse 12
73728 Esslingen am Neckar, Germany
Ralf.Hesse@rps.bwl.de

Dr Ross A HILL
School of Applied Sciences
Bournemouth University
Talbot Campus, Fern Barrow
Poole, BH12 5BB, United Kingdom

Pete HORNE
Aerial Survey and Investigation
Heritage Protection Department
English Heritage
37 Tanner Row
York, YO1 6WP, United Kingdom
pete.horne@english-heritage.org.uk

Dr Kristín HULD SIGURÐARDÓTTIR
Forstöðumaður
Suðurgötu 39
101 Reykjavík, Iceland
kristinhuld@fornleifavernd.is

Máté LOSONCZI
National Office for Cultural Heritage
Regional Department of the Western-Transdanubia
Győr, Hungary

Fun MEEUWS
Department of Soil Management
Ghent University
Coupure Links 653
9000 Gent, Belgium
Fun.Meeuws@UGent.be

Chris MUSSON
Tanyffordd
Pisgah
Aberystwyth, SY23 4NR, United Kingdom
abermusson@btinternet.com.

Rog PALMER
Air Photo Services
21 Gunhild Way
Cambridge, CB1 8QZ, United Kingdom
rog.palmer@ntlworld.com

Professor Dominic POWLESLAND
The Landscape Research Centre
The Old Bridge Barn
Yedingham
Malton
North Yorkshire, YO17 8RY, United Kingdom
d.powlesland@btinternet.com

Professor Włodek RĄCZKOWSKI
Institute of Prehistory
Adam Mickiewicz University
Poznań, Poland
wlodekra@amu.edu.pl

Dr Eleanor RAMSEY
IBM Visual and Spatial Technology Centre
Institute for Archaeology and Antiquity
University of Birmingham Edgbaston,
Birmingham, United Kingdom

Dr Fabio REMONDINO
3D Optical Metrology (3DOM)
Bruno Kessler Foundation (FBK)
Trento, Italy
remondino@fbk.eu

László RUPNIK
GIS-Laboratory, Aerial Archaeological Archive
Institute of Archaeological Sciences
Eötvös Loránd University
Budapest, Hungary
rupnik.laci@gmail.com

Gašper RUTAR
Centre for Preventive Archaeology
Institute for the Protection of
Cultural Heritage of Slovenia, Slovenia
gasper.rutar@cpa-rs.si

Timothy SAEY
aDeDe bvba
Antwerpsesteenweg 56
9000 Gent, Belgium
t.saey@adede.com

Robert SHAW
The Discovery Programme
63 Merrion Square
Dublin, D2, Ireland
Robert@discoveryprogramme.ie

Dr Birger STICHELBAUT
Department of Archaeology
Ghent University
Sint-Pietersnieuwstraat 35
9000 Gent, Belgium
birger.stichelbaut@ugent.be

Lőrinc TIMÁR
GIS-Laboratory, Aerial Archaeological Archive
Institute of Archaeological Sciences
Eötvös Loránd University
Budapest, Hungary

Professor Marc VAN MEIRVENNE
Department of Soil Management
Ghent University
Coupure Links 653
9000 Gent, Belgium
marc.vanmeirvenne@ugent.be

Kate WELHAM
School of Applied Sciences
Bournemouth University
Talbot Campus, Fern Barrow
Poole, BH12 5BB, United Kingdom

Résumés

Catherine Fruchart

1 | Introduction: la télédétection pour la gestion du patrimoine archéologique en Europe

David C Cowley et ***Kristín Huld Sigurðardóttir***

Cette introduction définit quels sujets relatifs à la télédétection sont abordés dans cet ouvrage et présente les thèmes fondamentaux associés: paysage, gestion, intégration, communication. On y détaille également le plan de l'ouvrage.

2 | Identifier l'inimaginable – gérer l'ingérable

Dominic Powlesland

Cet article identifie les rôles de la télédétection dans la découverte et l'enregistrement des sites et des contextes archéologiques et, en établissant des liens entre eux, des paysages. En se référant au travail réalisé depuis trente ans par le Centre de Recherches sur le Paysage du Nord de l'Angleterre, l'article examine comment on devrait intégrer les données de fouilles à grande échelle, les différentes techniques de télédétection et de prospection pour mieux comprendre les indices archéologiques, en particulier pour prendre en compte le fait que la visibilité change d'une zone à l'autre. En utilisant des techniques et des méthodes différentes on enregistre une variété d'informations, qui confirment ce que d'autres sources ont détecté ou qui apportent une information unique; l'ensemble contribue à l'établissement d'une grande carte du paysage qui contient de multiples niveaux d'information. L'enjeu est de comprendre ce paysage dans son étendue et sa complexité avant de pouvoir planifier une stratégie de gestion efficace. Il est vital d'évaluer, dans le paysage qui a fait l'objet d'opérations de télédétection, quelle étendue a été préservée d'une destruction due à l'agriculture moderne ou à toute autre raison pour mettre en place des stratégies visant à préserver ou à limiter les destructions.

Les ressources du paysage identifiées par le programme Vale of Pickering, qui utilise de multiples facettes de la télédétection sur le long terme, met en évidence que des dépôts archéologiques étendus et complexes, où les rapports entre visibilité, information reçue et valeur potentielle ne sont pas simples, défient les procédures de désignation des statuts actuels.

3 | "L'Archéologie totale" pour réduire le recours à l'Archéologie de Sauvetage: le projet BREBEMI (Italie)

Stefano Campana

En Italie le débat va grandissant à propos des méthodes qui peuvent ou qui devraient être employées lorsque l'archéologie préventive ou de sauvetage se retrouve face à des projets d'infrastructures de grande envergure. La pression s'est accrue à cause de l'impact potentiel (mais pas encore totalement réalisé) de la nouvelle législation du droit privé relative aux procédures à adopter pour l'évaluation des implications des potentiels archéologiques de projets de développement. Cette contribution décrit les méthodes mises en oeuvre préalablement au développement d'une autoroute de premier ordre dans le Nord de l'Italie, et parallèlement les progrès qu'on entrevoit par l'usage d'approches nouvelles, à la fois pour extraire l'information qui provient des sources existantes et pour le déploiement d'un large éventail de méthodes de prospections et de détection sur le terrain. Il est cependant clair que, dans le cadre de tels développements, les conflits demeurent entre l'approche "traditionnelle" et le potentiel offert par la perspective nouvelle d'un gain maximal accompagné de pertes minimales.

4 | La télédétection pour gérer le patrimoine archéologique - découverte, interprétation et enregistrement de sites

David C Cowley

Il est vital, pour la mise en place de connaissances sur lesquelles on peut compter pour une gestion efficace, qu'on comprenne ce qu'on connaît et l'impact de facteurs tels que les usages passés et présents des terres ou des centres d'intérêts des archéologues. Cet article aborde ces questions à partir d'exemples choisis en Ecosse afin d'identifier les défis posés pour la compréhension de paysages particuliers ; ils mettent également en évidence l'importance de définir des caractères « dans les grandes lignes » pour fournir un cadre permettant d'adapter les méthodes de prospection aux contextes locaux.

5 | Relevé par laser aéroporté sous couvert forestier – potentiel et limites d'une technique de prospection archéologique

Michael Doneus et ***Christian Briese***

On voit de plus en plus de projets archéologiques utilisant le relevé par laser aéroporté (ALS). Comme la production de modèles numériques de terrain

(MNT) réalisés à partir de relevés ALS coûtent très cher, la plupart des projets utilisent généralement des ALS basiques, en général moins coûteux et parfois même gratuits pour un usage dans le cadre de travaux scientifiques. Le problème principal que pose l'utilisation de ce type de données est le manque de métadonnées. Souvent, l'archéologue ne dispose pas des informations relatives à la densité de points, à la durée de vol, aux instruments de mesures, aux types de vol, filtres, procédures employées pour la génération de MNT etc. En conséquence l'ALS devient une sorte de "boîte noire" et le MNT est employé sans bien connaître la technologie mise en oeuvre, les algorithmes et les métadonnées. Il en découle que les données utilisées risquent de ne pas être adaptées à un usage archéologique.

En s'appuyant sur l'expérience acquise par un projet qui s'est déroulé sur deux ans "Prospections archéologiques en forêt avec l'aide du LiDAR", l'article fait une synthèse sur l'usage de l'ALS à des fins archéologiques; il explique les procédures de base, montre son potentiel pour l'archéologie du paysage et tout particulièrement pour les zones fortement boisées, et attire l'attention sur certains paramètres capitaux concernant l'ALS, que les utilisateurs devraient connaître. Enfin, il parle d'autres problèmes, qui seront à résoudre dans le futur proche.

6 | LiDAR à haute résolution pour un usage spécifiquement archéologique: utilise-t-on pleinement cette précieuse ressource ?

Robert Shaw* et *Anthony Corns

Le Programme Découverte a été pionnier en Irlande dans l'utilisation des données LiDAR à haute résolution pour cartographier et modéliser les paysages archéologiques, en intégrant ces données aux travaux de recherche sur la modélisation 3D d'un certain nombre de sites. Globalement, les résultats ont reçu un accueil favorable, suscitant un enthousiasme immédiat pour les MNT, MNE et autres modèles ombrés qu'on leur associe. La conséquence en a été que les organismes gouvernementaux nous ont suivis et ont missionné d'autres projets de relevés LiDAR à vocation archéologique, et ont parfois "re-découvert" des sets de données considérés comme insignifiants faute de traitements experts.

Cet article montre, sur la base d'exemples pris sur le sol irlandais, l'exceptionnelle qualité des modèles obtenus à partir de données LiDAR à haute résolution; on les y compare aux modèles issus d'un LiDAR "normal" et on montre ce qu'ils apportent à notre connaissance des paysages. Mais cet article se penche aussi sur les problèmes de gestion de l'information et d'accès aux données LiDAR, qui freinent sérieusement les capacités des organismes à tirer pleinement les fruits de leur investissement. Ceci pourrait à l'avenir entraîner des réticences à financer des projets. L'article examine l'éventuelle importance, pour résoudre en grande partie ces problèmes, de développer des infrastructures pour l'information spatiale (SDI), et d'offrir à ces données LiDAR de qualité une chance d'être un produit utilisable par une communauté aussi large que possible.

7 | Utilisation en archéologie des données multi et hyper spectrales: enjeux et possibilités

Anthony Beck

Les capteurs multi et hyper spectraux offrent un potentiel immense dans le domaine de la prospection archéologique. Leur principe est de capter le rayonnement émis ou réfléchi suivant différentes caractéristiques (longueurs d'ondes) du spectre électromagnétique. Leurs deux atouts majeurs sont leur potentiel à détecter des sites archéologiques et des monuments (réduits à l'état de traces archéologiques) indétectables dans les longueurs d'onde visibles, et une ouverture des champs d'investigation pour la détection. Par exemple, les variations locales de la santé des cultures (marquées par un stress ou une vigueur particuliers), prémices aux anomalies de croissance des végétaux, ressortent parfois mieux dans le proche infra-rouge que dans le visible. De plus, des données multi ou hyper spectrales recueillies par des moyens techniques différents (aériens ou satellites) et suivant des contextes qui varient peuvent être utilisées pour générer des couches d'information auxiliaires qui aident à l'interprétation (couches qui renseignent sur les sols, la géologie, l'aménagement du territoire, par exemple) et qui sont importantes pour "l'archéologie totale". Toutefois, les relevés multi et hyper spectraux coûtent assez cher et pour obtenir un résultat, il faut que les prospections soient réalisées de façon systématique et dans de "bonnes conditions". Ce dernier point est essentiel: comme on comprend mal comment les dynamiques spatiales, environnementales et saisonnières génèrent ces contrastes qui créent des "conditions adéquates", on ignore si des entités intéressantes pour l'archéologie pourront être détectées.

8 | Tirer le meilleur parti des techniques de télédétection aéroportée pour la prospection et l'interprétation en archéologie

Rebecca Bennett, Kate Welham, Ross A Hill* et *Andrew Ford

Au cours des dix dernières années, l'usage de la télédétection aéroportée a connu une popularité croissante dans le domaine de l'environnement historique. Nombre de projets sur les paysages à travers l'Europe prennent en compte les archives de prospections aériennes existantes, et de plus en plus, on missionne des prospections spécifiques aux projets. Ceci est particulièrement vrai pour le relevé laser aéroporté (ALS); mais malgré quelques opérations prometteuses, l'acquisition de données spectrales a été nettement moins utilisée. Notre compréhension de tout le potentiel porté par ce type de données en est à ses tout débuts. Cet article apporte un résumé des applications actuelles et met en évidence les besoins pour la recherche dans trois domaines clés pour améliorer l'utilisation de la télédétection aéroportée pour les recherches en environnement historique. En conclusion, il présente un projet porté par l'Université de Bournemouth qui vise à développer une approche des environnements liminaux par télédétection aéroportée avec capteurs multiples.

9 | Relevés 3D à l'usage du patrimoine culturel

Fabio Remondino

On comprend bien, à un niveau international, qu'il est important d'être documenté sur les patrimoines culturel et naturel, et on reconnaît de plus en plus la valeur des relevés numériques en trois dimensions (3D). En permanence, de nouveaux capteurs, de nouvelles méthodes d'acquisition des données et de représentation en 3D à différents niveaux de résolution sont mis au point, et les méthodes existantes sont améliorées. La recherche dans ces domaines progresse et peut contribuer de manière significative à la gestion, la conservation et la représentation numérisée du patrimoine. Cet article passe en revue quelques points importants relatifs aux spécificités et nécessités liées à la documentation; il envisage également les possibilités et les limites des relevés 3D ainsi que celles des techniques et des méthodologies de modélisation. Une attention particulière est accordée aux techniques de prospection fondées sur la réalité et destinées aux objets et structures. En exemple, on présente le cas de sites et d'objets pris dans le patrimoine existant, qui ont bénéficié d'un relevé 3D.

10 | A travers un filtre imparfait: techniques géophysiques et gestion du patrimoine archéologique

Chris Gaffney et *Vincent Gaffney*

Cela fait plus d'un demi-siècle que les techniques géophysiques font partie de la trousse à outils de l'archéologue. Au cours de cette période on est passé de méthodes de prospection à faible échelle faisant appel à une seule technique à toute une batterie de stratégies d'investigations en subsurface. Cette nouvelle dimension a permis à l'archéologue-géophysicien de passer de la production d'un "contexte" propre aux sites ou aux structures à un travail de prospection, d'analyse et d'interprétation des sites qui produit du sens archéologique. Le potentiel pour la gestion, et par conséquent pour la protection du patrimoine archéologique enfoui est évident, en particulier pour les zones agricoles nivelées par les labours ou pour les zones "sensibles" qui ne peuvent être fouillées.

Suite aux récents développements techniques on peut maintenant collecter en un seul passage avec un véhicule motorisé ou tracté à la force des bras des sets de données multiples. En général, celles-ci sont géoréférencées grâce à un GPS embarqué, et de plus certaines procédures d'acquisition s'affranchissent d'un carroyage, ce qui accélère les vitesses d'enregistrement tout en diminuant les coûts. La mise en oeuvre de nouvelles stratégies géophysiques a permis des relevés qui couvrent souvent des surfaces plus étendues tout en augmentant la densité de données. On a franchi un cap: il y a encore cinq ans, il fallait choisir entre large échelle ou haute résolution.

Cet article passe en revue quelques développements techniques récents et examine comment les résultats géophysiques peuvent être utilisés pour évaluer un potentiel archéologique et apporter des informations sur les problèmes de gestion de ces ressources. La discussion porte sur la question de "vitesse" opposée à "impact minimum" et sur l'évaluation de l'importance de l'environnement des données numériques dans lequel les relevés géophysiques sont faits. Pour illustrer la direction prise actuellement par les recherches, des sites sont donnés en exemple, avec des problématiques de gestion pertinentes.

11 | Géophysique en milieu marin: approches intégrées pour relever les fonds marins

Antony Firth

Cet article s'intéresse à l'utilisation de la géophysique et de la télédétection en archéologie sous-marine, en mettant l'accent sur leur emploi dans le cadre d'évaluations sur le terrain. Les écho-sondeurs simples et multifréquences, les radars, les sonars et les magnétomètres sont classés dans les méthodes géophysiques; le carottage, le prélèvement d'échantillons de sols marins, les caméras vidéos et de prises de vues, les véhicules téléguidés et autres sont considérés comme des formes de télédétection. La discussion porte sur divers motifs, thèmes, échelles ou types d'opérations, et également sur les implications en terme d'innovations méthodologiques dans trois domaines: le choix des positions, l'enregistrement basé sur un événement et l'enregistrement guidé par un choix.

12 | Le Programme National pour la Cartographie du Patrimoine Anglais

Pete Horne

Le Programme National pour la Cartographie du Patrimoine Anglais a déjà réalisé la photo interprétation à des fins archéologiques des données de prospections aériennes pour plus de 40% de l'Angleterre. On estime que ce projet a permis l'enregistrement de plus de 100 000 sites archéologiques, dont la moitié n'était pas recensée auparavant. Les méthodologies continuent à se développer avec les changements de technologies et de sources disponibles, mais le principe de base reste identique: maintenir en place des équipes spécialisées dans l'interprétation d'images aériennes et dans l'analyse des données déjà disponibles. C'est un programme sur le long terme, et c'est un défi d'assurer une continuité aux financements, de maintenir l'accès à un fonds toujours croissant d'archives photographiques et autres sources de données, tout en augmentant le nombre de professionnels de l'interprétation. Mais, en ayant soin de fixer les objectifs sur ces zones dont on tirera le plus de profit, des résultats bien visibles assureront une position clé du Programme National dans la gestion de l'environnement historique en Angleterre.

13 | Intégration des données de prospection: le recensement polonais AZP et au-delà

Włodek Rączkowski

Le recensement des sites archéologiques polonais AZP a déjà été présenté maintes fois. Il est considéré comme une base de données simple et objective des sites archéologiques, et comme ce projet est achevé pour certaines régions, on en vient à s'interroger: que faire après? Un nouveau concept appelé projet AZP_2 est en cours de discussion. Les méthodes non invasives ont déjà été acceptées comme partie intégrante du projet. Jusqu'à présent quatre catégories différentes d'informations qui découlent des méthodes appliquées sont collectées dans quelques projets pilotes: données des prospections au sol (comme dans AZP), photographies aériennes, données géophysiques et sondages. Il est nécessaire d'intégrer toute l'information dans une base de données relationnelle. Les divergences entre les sources de données donnent lieu à une réflexion sur l'origine des différences. En concevant AZP_2, on génère au moins deux autres problèmes. Il n'est pas suffisant d'intégrer l'information et les méthodes. Les agents de gestion du patrimoine tout comme les archéologues de terrain doivent comprendre le nouveau concept et être en mesure d'interpréter toute la gamme des données intégrées. Les pratiques et les mentalités doivent se développer, et tous les groupes devraient se former aux nouvelles technologies (y compris les SIG) et aux questions complexes liées à l'intégration des données.

14 | Aux confins de ce que peut atteindre le laser... analyse laminaire des structures détectées par LiDAR comme un outil de fort potentiel pour la gestion du patrimoine archéologique du Baden-Würtemberg, Allemagne

Jörg Bofinger et *Ralf Hesse*

Un nouveau projet a été initié par l'Office National de la Gestion du Patrimoine du Baden-Würtemberg pour l'analyse laminaire des données LiDAR sur le Baden-Würtemberg. Ce projet fait partie d'un ensemble de prospections archéologiques en Baden-Würtemberg. De nouvelles techniques pour la détection de structures archéologiques utilisant des méthodes particulières de génération des données à partir de l'information LiDAR ont été développées. Le projet a pour but de déterminer dans l'imagerie LiDAR de l'ensemble de l'état la totalité des structures visibles qu'on suppose être des sites archéologiques ou au moins des éléments du paysage historique. Le projet vise, sur la base d'un set de données LiDAR intégralement à haute résolution pour toute la surface du Baden-Würtemberg, à dresser en trois ans une cartographie complète de tous les sites potentiels de cette zone de 35 751km^2.

15 | Entre les lignes – améliorer les méthodologies pour l'exploration de vastes paléopaysages inondés

Simon Fitch, Vincent Gaffney, Benjamin Gearey et *Eleanor Ramsey*

Depuis quelques années, le potentiel archéologique des reliefs côtiers européens est mieux pris en considération. Cet intérêt ne découle pas simplement des progrès de l'histoire maritime traditionnelle, mais résulte plutôt d'une ouverture de l'archéologie en milieux marins à la recherche de phases de peuplement dans l'immensité du paysage préhistorique qui gît sous nos côtes. Encore inaccessibles il y a peu, ces paysages inhospitaliers sont désormais à notre portée grâce au développement de toute une panoplie de techniques de télédétection. L'information apportée par leur utilisation est en train de changer complètement notre perception de l'archéologie du Paléolithique et du Mésolithique. Cet article se penche sur l'utilisation des données témoignant des catastrophes naturelles du passé pour cartographier les paléopaysages, et montre les différences de valeur entre la 2D et la 3D pour arriver à cette fin. Les résultats de nouvelles recherches en Mer du Nord et au large des côtes ouest du Royaume-Uni sont présentés et utilisés pour mettre à l'épreuve un large échantillon de stratégies permettant l'exploration des zones sous-marines non cartographiées avec des techniques 3D. A notre sens, il faudrait que les prospections de ces surfaces soient faites en employant un éventail de stratégies qui devraient être adaptées pour être utilisées de façon optimale afin de répondre à des aspects spécifiques de la recherche. Au vu du développement du domaine maritime et des incertitudes économiques actuelles, il faut que les conservateurs du patrimoine et les chercheurs profitent pleinement d'une information qui fournit à la gestion du patrimoine et à la recherche une opportunité sans pareille et dont la constitution a coûté des milliards.

16 | Les archives aériennes pour la gestion du patrimoine archéologique: Les Archives de Reconnaissance Aérienne – une ressource européenne partagée

Lesley Ferguson

Aujourd'hui hébergée par la Commission Royale des Monuments Historiques et Antiques en Ecosse (RCAHMS) à Edimbourg, les images de reconnaissances aériennes militaires issues des Archives de Reconnaissance Aérienne (TARA) constituent un enregistrement exceptionnel des paysages et des activités humaines passées saisies entre les années 1940 et 1990 à travers le continent européen. Cet article fait un rapide survol du contenu archéologique de TARA et présente quelques-unes des découvertes et réalisations faites depuis son transfert à la RCAHMS en 2008. Cette contribution envisage aussi dans une perspective d'avenir les collaborations possibles, essentielles à un futur développement; elle se penche également sur les obstacles à surmonter pour donner libre accès à l'exploration et à la découverte aux communautés archéologiques et même au-delà.

17 | La télédétection pour l'étude intégrée et la gestion de sites et de monuments – perspectives en Europe Centrale et cas de la République Tchèque

Martin Gojda

Les récentes innovations dans les techniques de télédétection ont transformé les possibilités de ce qu'on appelle communément l'archéologie aérienne. On étudie et on ressent ces changements aussi bien en Europe de l'Ouest qu'en Europe Centrale, où la plupart des pays ont eu un accès tardif à l'archéologie aérienne dans les années 1990 après la chute des régimes communistes européens. Les obstacles légaux et administratifs opposés pour un motif ou un autre aux prospections aériennes, à la prise de photographies et à leur publication, ont dans la plupart des cas été la source du retard à lancer, dans cette partie de l'Europe, des programmes de survols continus. Malgré cela, les activités de télédétection ont, au cours des deux dernières décennies, considérablement enrichi les méthodologies et la compréhension archéologiques dans l'étude des peuplements et des paysages du passé; elles ont aussi favorisé une ouverture de débats théoriques. Cette contribution illustre l'influence d'un programme d'archéologie aérienne en cours en Bohème sur les conceptions des dynamiques et des formes de peuplement du passé, et montre l'importance de la photographie aérienne pour le suivi et la documentation du patrimoine culturel tchèque. La Bohème considère comme une priorité à venir de tester le potentiel de l'ALS pour une cartographie extensive des paysages archéologiques (projet pilote financé par l'état en 2010-2011) et ensuite l'acquisition d'un LiDAR pour l'ensemble de la République Tchèque.

18 | Bref aperçu de l'utilisation du laser scanneur aéroporté dans la gestion des sites archéologiques en Lorraine (France)

Murielle Georges-Leroy

Deux vols de télédétection par laser scanneur aéroporté (ou LiDAR) ont été exploités à des fins de gestion des sites archéologiques en Lorraine. Le premier a concerné le massif forestier de Haye à côté de Nancy. Il a permis un inventaire très large de vestiges de tous types et de toutes périodes, à partir duquel il a été possible de proposer une hiérarchisation des mesures de protection à prendre dans le cadre d'aménagements et de travaux forestiers. Le deuxième a été réalisé dans le cadre de la construction de la seconde phase de la ligne ferroviaire à grande vitesse Paris-Strasbourg. Il illustre les apports de cette méthode pour l'archéologie préventive, à toutes les phases du processus, mais également ses limites.

19 | Prospection archéologique aérienne d'un paysage enfoui: le projet Tóköz

Zoltán Czajlik, László Rupnik, Máté Losonczi et *Lőrinc Timár*

Le Laboratoire de Recherches en SIG et les Archives Archéologiques Aériennes de l'Institut Archéologique de l'Université Eötvös Loránd, créé en 1993 pour venir en aide aux fouilles de sauvetage le long du projet autoroutier M3, et la coopération archéologique franco-hongroise disposent d'une collection de plus de 30 000 photographies aériennes couvrant plus de 1500 sites archéologiques. Des données de terrain sont également collectées par le biais d'une interprétation archéologique totale des photographies aériennes sur de petites régions et grâce à la création de cartes à partir d'une compilation de photographies. La zone d'étude décrite ici se situe dans le nord-ouest de la Hongrie, dans la micro-région de Tóköz. Grâce à la prospection aérienne, nous pouvons identifier des sites étendus même sur des surfaces recouvertes par des alluvions Holocènes très tardives.

20 | Paysage archéologique du nord-est de l'Islande: un fantôme des sociétés Vikings

Árni Einarsson et *Oscar Aldred*

Un projet en cours de cartographie des vestiges archéologiques médiévaux sur une surface de plus de 3000km² au nord-est de l'Islande utilise la photographie aérienne à large échelle, en incluant plus de 2000 photographies obliques prises à basse altitude. Avec l'aide de prospections au sol et autres travaux sur le terrain, le projet a révélé les restes bien préservés d'un ensemble daté de la période médiévale. Le plus remarquable est un réseau étendu de murs en mottes de terre, mais des emplacements de fermes, des églises, des espaces de réunion, des cimetières païens, des zones de prélèvement de tourbe, des fosses à charbon et des chemins empruntés par des chevaux sont également bien visibles. Ce paysage archéologique médiéval appartient en grande partie à la période Viking; il est un vestige du mode de peuplement des premières générations de populations islandaises. L'article montre la valeur de ce paysage, due à son excellente conservation et parle de la question de sa préservation.

21 | Un optimisme modéré: archéologie préventive et gestion du patrimoine en Slovénie

Gašper Rutar et *Matija Črešnar*

A la suite de changements législatifs en 2008, des modifications très importantes dans la gestion du patrimoine ont marqué ces deux dernières années en Slovénie. Quelques-uns des changements les plus profonds sont apparus dans le domaine très général de l'archéologie préventive. D'abord, une nouvelle cellule opérationnelle, le Centre de Prévention en Archéologie (CPA), principalement orientée vers la recherche a été fondée au sein de l'Institut pour la Protection du Patrimoine de Slovénie (IPCHS). Les tâches qui lui incombent sont largement liées à la réalisation de recherches archéologiques préalables au Plan sur l'Espace National (NSP) et à la recherche subventionnée, cette dernière étant aussi une nouveauté issue de la nouvelle législation. Les travaux faits par le CPA sur les 15 premiers mois de son existence sont encourageants, mais montrent également que des améliorations sont à apporter.

22 | Le patrimoine de la Première Guerre Mondiale en Belgique: en associant photographie aérienne et EMI

Birger Stichelbaut, Timothy Saey, Fun Meeuws, Jean Bourgeois et *Marc Van Meirvenne*

Des millions de photographies aériennes ont été prises par les nations qui ont combattu pendant la Première Guerre Mondiale. Ces photographies sont une source remarquable d'information auparavant peu exploitée pour l'étude du patrimoine issu de ce conflit. Cet article décrit des recherches croisant diverses disciplines, combinant les recherches archivistiques, l'interprétation et la cartographie détaillée des photographies aériennes historiques et l'utilisation des prospections par induction électromagnétique.

Ces techniques sont associées pour deux zones d'étude (Geluveld & Oostduinkerke) afin d'évaluer leur patrimoine archéologique de la Première Guerre Mondiale.

23 | Une vue aérienne du passé – l'archéologie aérienne au Danemark

Lis Helles Olesen

Le Musée Holstebro à l'ouest du Danemark a assuré le financement d'un projet de recherches archéologiques sur quatre ans. Prévu jusqu'en 2013, il comprend neuf sous-projets bénéficiant d'un budget de 650 000 €. Cet article fait un point sur les premiers résultats des prospections aériennes et du contrôle aérien des monuments anciens attendus. En 2008 et 2009 quelques 560 sites ont été localisés en 130 heures de vol environ. L'article fait un tour d'horizon des 100 premiers lieux entièrement relevés: entre autres résultats, on a découvert des maisons sur 33 des 100 sites, et ces habitats semi-enterrés offrent une bonne chance de nous donner une vue d'ensemble du peuplement pour les dernières périodes de l'Âge du Fer et pour la période Viking (20% du nombre total des sites datent de ces périodes). On présente également les résultats du contôle des 50 premiers monuments attendus. Même s'ils sont précoces, ces résultats nous permettent de souligner tout le potentiel qui réside dans les sous-projets présentés dans cet article.

24 | L'interprétation d'images aériennes basée sur les connaissances

Rog Palmer

L'interprétation d'images aériennes basée sur les connaissances est un processus subjectif qui convertit les images enregistrées sur des photographies aériennes, des images satellites et autres sources de télédétection en information extraites au cours d'un examen fait en fonction de question posées. Cette contribution examine de quelle manière nous regardons les photographies et comment les raisons pour lesquelles nous observons influencent notre regard. A l'aide d'un exemple, on démontre la valeur d'un exercice de lecture de photos au fur et à mesure de sa progression dans une série de questions qui montrent à quel point on peut extraire de l'information à partir d'une seule image aérienne. On note les façons de travailler avec les photographies et on souligne les différents niveaux d'interprétation et de cartographie – et à quels usages ils sont destinés. Une dernière partie fait un lien entre les résultats des interprétations d'images et d'autres sources; on montre comment celles-ci peuvent nous aider à comprendre le passé.

25 | Se former pour l'avenir: coopération à travers l'Europe

Chris Musson

Depuis les années 1990 on se rend de mieux en mieux compte qu'on a en Europe des possibilités limitées d'éducation et de formation pratique en archéologie aérienne et autres formes de télédétection, et donc de leur potentiel à contribuer à la découverte, l'interprétation et la préservation dans les domaines du patrimoine et des études sur le paysage. Dans un premier temps, cette contribution souligne quelques-unes des étapes franchies au cours des 15 dernières années pour améliorer cette situation. Une deuxième partie se tourne vers l'avenir avec un projet sur 5 ans commencé récemment dans le cadre d'une coopération internationale au sein du Programme Culturel de l'Union Européenne.

Zusammenfassungen

Johanna Dreßler

1 | Einführung: Fernerkundung für archäologisches Kulturerbemanagement

David C Cowley und *Kristín Huld Sigurðardóttir*

Der einleitende Beitrag beschreibt das Anwendungsgebiet der Fernerkundung und führt an die Schlüsselthemen „Landschaft", „Verwaltung/Schutz", „Eingliederung" und „Kommunikation" heran. Es wird der Inhalt des gesamten vorliegenden Bandes mit seinem breit gefächerten Inhalt umrissen.

2 | Das Unvorstellbare erkennen und den Überblick über das Unüberschaubare bekommen

Dominic Powlesland

Der vorliegende Beitrag beschäftigt sich mit der Rolle, welche die Fernerkundung beim Entdecken und Erfassen von archäologischen Fundplätzen und der Landschaft, in der sie liegen, spielt. Basierend auf der mittlerweile 30-jährigen Arbeit des Landscapes Research Centre in Nordengland wird gezeigt, wie großflächige Ausgrabungen und eine Vielzahl von Fernerkundungs- und Surveymethoden integriert und so das Verständnis für archäologische Hinterlassenschaften verbessert werden können. Vor allem muss man sich bewusst werden, dass es Bereiche unterschiedlicher Sichtbarkeit gibt, die unterschiedlicher Herangehensweisen bedürfen. Unterschiedliche Ansätze und Messgeräte erfassen unterschiedliche Informationen und liefern unterschiedliche Daten. Zusammen tragen sie jedoch einen großen Teil dazu bei, um ein vielschichtiges und umfassendes Bild einer Landschaft zu kreieren. Diese komplexe Landschaft muss verstanden werden, bevor effektive Managementstrategien entwickelt werden können. Von grundlegender Bedeutung ist, dass man feststellen muss, welcher Anteil einer durch Fernerkundungsmethoden erfassten archäologischen Landschaft noch nicht durch moderne Landwirtschaft oder andere moderne Nutzung beeinträchtigt oder zerstört ist. Nur so kann eine effektive Vorgehensweise in Hinblick auf präventive oder zumindest abmildernde Maßnahmen erreicht werden.

Die Landschaft im Vale of Pickering ist durch ein vielfältiges und langfristiges Fernerkundungsprogramm bereits erforscht. Die Beziehungen zwischen Sichtbarkeit, Informationsgewinn und zukünftigem Nutzen der ausgedehnten und komplexen archäologischen Ablagerungen sind oft nicht überschaubar. Hieraus ergeben sich Herausforderungen für die Anwendung der derzeitigen denkmalrechtlichen Einstufungen.

3 | "Allumfassende Archäologie" zum Reduzieren der Notwendigkeit von Rettungsgrabungen: The BREBEMI Projekt (Italien)

Stefano Campana

In Italien gibt es eine immer größer werdende Debatte, welche archäologischen Methoden im Vorfeld großer Bauvorhaben eingesetzt werden sollen, um betroffene archäologische Fundstellen durch präventive, planmäßige Maßnahmen schützen zu können. Dies gewinnt immer mehr an Bedeutung, da eine neue Gesetzgebung die Möglichkeiten bezüglich präventiver Maßnahmen regeln soll (die bislang jedoch noch nicht realisiert wurde).

Anhand des Beispiels eines Autobahnbaus im Norden Italiens soll gezeigt werden, um welche Methoden es sich dabei handelt. Darüber hinaus werden neue Erkenntnisse, welche während der dortigen archäologischen Tätigkeiten gewonnen wurden, vorgestellt. Hier wurden Informationen nicht nur von den bereits bekannten Quellen, sondern auch von einer großen Breite an Survey- und Erforschungsmethoden gewonnen. Der Konflikt zwischen „traditionellen" Herangehensweisen und neuen Sichtweisen, die nach maximalem Erkenntnisgewinn und minimalem Informationsverlust streben, muss am Ende jedoch ungelöst bleiben.

4 | Fernerkundung zum Schutz und zur Verwaltung des Kulturerbes – Entdeckung, Erfassung und Interpretation

David C Cowley

Um wirklich zu verstehen, was man sieht und weiß, muss man anfangen, über die Probleme nachzudenken, die auf einen beim Versuch, das Wissen zu erweitern, zukommen. Es ist von entscheidender Bedeutung, die Faktoren zu kennen, welche den Daten, mit denen man arbeitet, zu Grunde liegen. Denn es gibt viel mehr, als nur die, die darauf zurück zu führen sind, was früher, zum Entstehungszeitpunkt der archäologischen Hinterlassenschaft, war. Dabei handelt es sich in erster Linie um die alte sowie die rezente Landnutzung und das Interesse des jeweiligen Archäologen, aber auch um die moderne politische Linie, um nur einige zu nennen. Diese Problematik wird während des Vortrags anhand von Beispielen aus Schottland umrissen. Man muss sich bewusst werden, dass es immer wieder aufs Neue auf den jeweiligen Kontext ankommt: man muss sich im Vorfeld über den Typ der zu untersuchenden Landschaft klar sein, um sich die richtige, passende Untersuchungsstrategie aussuchen zu können.

5 | Airborne Laser Scanning in bewaldeten Gebieten – Potential und Grenzen einer archäologischen Prospektionsmethode

Michael Doneus* und *Christian Briese

Airborne Laser Scanning (ALS) wird in der Archäologie immer häufiger angewendet. Da die Anfertigung von ALS-abgeleiteten digitalen Geländemodellen (DGM) eine beträchtliche Menge an Geld benötigt, werden häufig ALS Daten der allgemeinen Vermessung genutzt. Diese sind in der Regel billiger, mitunter werden sie für wissenschaftliche Zwecke sogar umsonst zur Verfügung gestellt. Eines der Hauptprobleme, welches mit dieser Art von Daten einhergeht, ist das Fehlen von Metadaten. Information zur Punktdichte, Flugzeit, den genutzten Instrumenten, Flugzeugtyp, Filter sowie das Verfahren, mit dem das DGM berechnet wurde, liegen oftmals nicht vor. Somit wird ALS zu einer Art „black box". Die DGM werden ohne weiteres Verständnis für die zugrunde liegende Technik, die Algorithmen und die Metadaten angewandt. Damit kann das Risiko einhergehen, dass die genutzten Daten nicht passend für die jeweilige archäologische Anwendung sind.

Anhand der Erfahrungen und Ergebnisse eines zweijährigen Projekts „LiDAR-gestützte archäologische Prospektion im Wald", wird im Vortrag ein Abriss über archäologisches ALS gegeben, der zugrunde liegende Prozess erklärt und das daraus resultierende Potential für landschaftsarchäologische Untersuchungen vor allem in dicht bewaldeten Gebieten dargestellt. Ebenso werden einige kritische Parameter des ALS, welche dem Nutzer bekannt sein sollten, aufgezeigt. Am Ende werden weitere Fragestellungen, die in naher Zukunft einer Lösung bedürfen, diskutiert werden.

6 | High Resolution LiDAR in der Archäologie: schöpfen wir diese nützliche Quelle voll aus?

Robert Shaw* und *Anthony Corns

Die Arbeit an den Grundlagen der 3D-Modellierung im Rahmen des irischen *Discovery Programmes* anhand mehrerer Fundplätze, hat den Weg bereitet, hochauflösende LiDAR Daten zum Erfassen und Modellieren von archäologischen Landschaften zu nutzen. Bislang sind die Ergebnisse, d.h. die daraus resultierenden DTMs, DSMs und die entsprechenden Hillshade-Modelle, mit einem sofortigen „Wow" sehr gut angekommen. Daraufhin sind mehrere staatliche Dienststellen dem Beispiel gefolgt und haben weitere archäologierelevante LiDAR Projekte in Auftrag gegeben, zum Teil wurden auch Daten wieder aufgegriffen, die aufgrund von fehlenden Bearbeitungsmöglichkeiten bisher nicht berücksichtigt werden konnten.

Die außergewöhnliche Qualität der aus den hochauflösenden LiDAR Daten gewonnenen Modelle wird im Vortrag an einigen Beispielen aus Irland aufgezeigt werden, betont durch den Vergleich mit „normalen" LiDAR Modellen. Dadurch wird der daraus resultierende Wert im Verstehen von archäologischen Landschaften hervorgehoben. Trotzdem gibt es Probleme in der Verwaltung und dem Zugang der Daten, was die Möglichkeiten im Einsatz mitunter beschränkt. Dies kann dazu führen, zukünftige Projekte nicht zu bewilligen. Der Vortrag möchte zu bedenken geben, ob nicht das Entwickeln von Geodateninfrastrukturen einen Teil dazu beitragen könnte, dieses Problem zu lösen. Darüber hinaus muss dafür Sorge getragen werden, dass die Möglichkeiten und der Nutzen solcher LiDAR Daten einer größeren Gemeinschaft näher gebracht werden.

7 | Anwendung von Multi/Hyper-Spektraldaten in der Archäologie – Herausforderung und Potential

Anthony Beck

Als archäologisches Prospektionsinstrument bieten Multi/Hyper-Spektral Sensoren ein immenses Potential. Die Sensoren sind sensibel genug, um Strahlungen unterschiedlichster Wellenlängen des elektromagnetischen Spektrums zu empfangen oder zu reflektieren. Dabei gibt es zwei maßgebliche Vorteile: die Fähigkeit, archäologische Fundstellen zu entdecken, welche im visuellen Bereich nicht sichtbar sind, und die Gelegenheit die Entdeckungsmöglichkeiten neuer Fundstellen zu verbessern. So ist zum Beispiel der durch Veränderungen in Wuchs und Vitalität hervorgerufene Stress von Pflanzen, welcher wiederum zur Ausbildung von Bewuchsmerkmalen führt, oftmals im Bereich des Nahen Infrarot deutlicher oder überhaupt erst sichtbar. Darüber hinaus können Multi/Hyperspektraldaten unterschiedlicher Art (Satelliten- und Luftbilder), die unter unterschiedlichen Bedingungen aufgenommen wurden, von Nutzen sein. Zusätzlich können weitere berücksichtigte Faktoren (z.B. Boden, Geologie, Landnutzung) bei der Interpretation helfen und sind somit wichtig für eine „Gesamte Archäologie". Leider sind Multi/Hyper-Spektralsensoren sehr teuer und bedürfen, um erfolgreich zu sein, einer systematischen Datenerhebung unter entsprechenden Bedingungen. Vor allem der letzte Punkt ist kritisch: es gibt oft nur ein geringes Verständnis der räumlichen, ökologischen und saisonalen Prozesse, die entsprechende Bedingungen hervorrufen und somit, ob Merkmale von archäologischem Interesse ausgebildet und entdeckt werden können.

8 | Der größte Nutzen der flugzeuggetragenen Fernerkundung für archäologische Prospektion und Interpretation

Rebecca Bennett*, *Kate Welham*, *Ross A Hill* und *Andrew Ford

Während des letzten Jahrzehntes erfreute sich die flugzeuggetragene Fernerkundung immer größerer Beliebtheit. Viele Landschaftsprojekte in ganz Europa beziehen die Anwendung flugzeuggetragener Untersuchungen mit ein, und immer mehr geben systematische Surveys in Auftrag. Dies gilt vor allem für ALS. Digitale Spektraluntersuchungen jedoch werden trotz einiger viel versprechender Ausnahmen zu wenig genutzt. Das Verständnis der gesamten Bandbreite dieser reichen Informationsquellen steckt immer noch in den Kinderschuhen. Im Vortrag werden alle derzeitigen Anwendungsgebiete zusammenfassend vorgestellt. Die Notwendigkeit, die Hauptfunktionsgebiete der Luftbildfernerkundung weiterzuentwickeln, wird besonders hervorgehoben. Er schließt mit der Vorstellung eines Projekts der Bournemouth Universität zur Entwicklung einer Multisensortechnologie zum Erfassen kaum wahrnehmbarer Unterschiede in den flugzeuggetragenen Surveydaten.

9 | 3D-Aufnahmen zur Erfassung von kulturellen Hinterlassenschaften

Fabio Remondino

Der Bedeutung der Dokumentation von Kultur- und Naturerbe ist man sich international bewusst, und digitale, dreidimensionale Aufnahmen der Denkmäler gewinnen immer mehr an Wert und Ansehen. Die Entwicklung und Erforschung neuer Sensortechniken, neuer Datenerhebungsmethoden und von hochauflösenden 3D-Darstellungen sowie die Verbesserung bereits existierender Techniken wird immer weiter vorangetrieben. Dies leistet einen nicht geringen Beitrag für die Verwaltung, die Instandhaltung und die digitale Präsentation von Kulturstätten. Im Vortrag sollen Potential als auch Grenzen von 3D Modellierungs- und Erfassungstechniken sowie wichtige Voraussetzungen der Dokumentationsmethoden und deren Anwendungen diskutiert werden. Besondere Bedeutung wird hierbei einer realitätsnahen Darstellung und der Erfassung zukommen. Veranschaulicht wird der Vortrag anhand von digital aufgenommenen Beispielen kultureller Stätten und Objekte.

10 | Unvollständig gefiltert: geophysikalische Messmethoden und archäologisches Kulturerbemanagement

Chris Gaffney und *Vincent Gaffney*

Geophysikalische Techniken sind nun seit über einem halben Jahrhundert Bestandteil der archäologischen Forschung. Im Laufe dieser Zeit hat sich die Anwendung von kleinflächigen Untersuchungen, bei denen eine einzige Technik angewandt wurde zu kleinmaßstabigen Kampagnen, in deren Verlauf eine ganze Reihe unterschiedlicher Techniken genutzt wurde, entwickelt. Diese Verlagerung des Schwerpunkts führte dazu, dass sich die Arbeit archäologischer Geophysik weg vom einfachen Produzieren von Daten hin zum Prospektieren, Analysieren und Interpretieren von Fundstellen auf eine aussagekräftige und ernstzunehmende Weise gewandelt hat. Dass archäologische Stellen, besonders in stark landwirtschaftlich geprägten Regionen oder in schützungswürdigen Arealen, in denen Ausgrabungen jedoch nicht in Frage kommen, adäquat verwaltet werden müssen, um sie sichern zu können, ist schon lange erwiesen.

Eines der Resultate der eben angesprochenen Entwicklung ist, größere Datenmengen auf einmal zu erheben, z.B. durch den Gebrauch von motorbetriebenen oder von Menschenhand gezogenen Wagen oder Schlitten. Im Normalfall werden diese Daten automatisch durch mitgeführte GPS Empfänger georeferenziert. Zusätzlich können manche Vorgehensweisen ohne ein Messgitter durchgeführt werden, was die Dauer dieser Untersuchungen und damit einhergehend die Kosten stark nach unten senkt. Ein begrüßenswertes Resultat neuer, geophysikalischer Strategien ist, dass die Messungen nun beides sein können, nämlich großflächig und genau. Somit ist man um einiges weiter als noch vor fünf Jahren, als sie nur eines waren: großflächig oder genau.

Der Vortrag gibt eine Übersicht über einige moderne technische Entwicklungen. Weiterhin setzt er sich damit auseinander, wie geophysikalische Ergebnisse genutzt werden können, um Archäologie und ihre Quellen einschätzen zu können und sie auf die Belange der Verwaltung abstimmen zu können. Vor allem die Thematik „Geschwindigkeit" gegen „minimale Beeinflussung" wird diskutiert, ebenso wie die Bedeutung eines digitalen Umfelds, in dem die geophysikalischen Daten erhoben werden. All dies wird anhand von aktuellen Beispielen illustriert.

11 | Marine Geophysik: Neue Denkansätze zum Abtasten des Meersbodens

Antony Firth

Der vorliegende Beitrag behandelt den Nutzen von Geophysik und Fernerkundung auf dem Gebiet der Meeresarchäologie unter besonderer Berücksichtigung der entsprechenden Auswertung. Der Gebrauch von Einzel- und Fächerecholoten, Sub Bottom Profiler-Messungen, Side-Scan-Sonaren und Magnetometern werden als geophysikalische Messmethode vorgestellt, ebenso wie das Anfertigen von Kernbohrungen, das Abtasten des Meeresbodens, der Einsatz von (Video-) Kameras, ferngesteuerten Fahrzeugen und Tauchern. Unterschiedliche Motivationen, Fragestellungen, Ausmaß und Art von Arbeitsabläufen werden diskutiert, vor allem im Hinblick auf die Auswirkungen auf methodische Innovationen: Standortbestimmung sowie ereignisbasierte und ereignisorientierte Aufnahmen.

12 | Das English Heritage National Mapping Programme

Pete Horne

Das English National Mapping Programme hat mittlerweile archäologische Luftbilder interpretiert, die über 40% der Fläche Englands abdecken.

Es sind geschätzte 100,000 archäologische Fundstellen aufgenommen worden, die Hälfte davon war vorher nicht bekannt. Die Methode entwickelt sich im selben Rahmen beständig weiter, wie sich Technik und die zu Verfügung stehenden Datenquellen weiterentwickeln. Das grundsätzliche Verfahren des Spezialistenteams, das die Luftbilder interpretiert und alle zur Verfügung stehenden Quellen analysiert, bleibt jedoch dasselbe.

Eine der maßgeblichen Herausforderungen des Langzeitprogramms ist immer wieder die Sicherstellung der Finanzierung, ebenso wie das Aufrechterhalten des Zugangs zu den wachsenden Archiven sowie den Pool an in Luftbildinterpretation gut ausgebildeten Archäologen zu vergrößern. Glücklicherweise gewährleisten die außergewöhnlichen Ergebnisse des NMP das Fortbestehen des Programms und somit seine Schlüsselrolle im Verwalten und Beaufsichtigen von Englands historisch gewachsenem Umfeld.

13 | Die Weiterentwicklung des polnischen Archäologie Registers AZP

Włodek Rączkowski

Viele Male schon wurde das polnische AZP vorgestellt. Es ist eine Basisdatenbank, in der alle wichtigen Informationen zu archäologischen Fundstellen erfasst sind. Da in manchen Regionen Polens diese Erfassung nun abgeschlossen ist, stellt sich die Frage was als nächstes kommt. Hierfür wird momentan das so genannte AZP_2 Projekt geplant. Trotz des frühen Stadiums der Entwicklung steht fest, dass nichtinvasive Methoden einen grundlegenden Teil des Projekts ausmachen werden. Bis jetzt wurden in mehreren Pilotprojekten vier angestrebte Möglichkeiten des Informationsgewinns festgelegt: Geländebegehung (wie im AZP), Luftbildarchäologie, Geophysikalische Prospektion und Sondageschnitte. Diese dadurch gesammelten Informationen müssen im Anschluss in einer relationalen Datenbank zusammengeführt und verwaltet werden. Da sie oft ungleicher Natur sind, bietet sich an dieser Stelle die Möglichkeit, die Gründe hierfür im Vorfeld zu diskutieren.

Das Konzept von AZP_2 bringt mehrere Probleme mit sich, da es momentan noch nicht genug Daten gibt, die herangezogen werden können. Verantwortliche Denkmalpfleger als auch Feldarchäologen müssen das Konzept des AZP_2 verstehen, um daran aktiv mitarbeiten und mithelfen zu können. Hierfür müssen Theorie und Praxis weiterentwickelt werden und alle Beteiligten müssen neue Techniken (inklusive der Anwendung von GIS) und komplexere Vorgänge, verbunden mit dem Verständnis für die erhobenen Daten, lernen.

14 | Soweit der Laser reicht… – Flächenhafte Analyse von LiDAR-Daten als leistungsfähiges Instrument der archäologischen Denkmalpflege in Baden-Württemberg, Deutschland

Jörg Bofinger* und *Ralf Hesse

Seit mehreren Jahren nutzt das Landesamt für Denkmalpflege in Baden-Württemberg LiDAR-Daten für die Entdeckung und Visualisierung von archäologischen Bodendenkmälern. In der Anfangszeit wurden ausgewählte Bodendenkmäler in der archäologischen Landschaft analysiert, beispielsweise die eisenzeitliche Befestigung Heuneburg und Punkte entlang des obergermanisch-raetischen Limes.

Im Gegensatz zu diesen ersten Arbeiten befasst sich ein im Jahr 2009 initiiertes Projekt mit der landesweiten, LiDAR-basierten Erfassung aller als Reliefmerkmale erkennbaren Strukturen, die als archäologische Bodendenkmäler oder als Elemente der historischen Kulturlandschaft interpretiert werden können.

Auf der Grundlage eines vollständigen, hochauflösenden LiDAR-Datensatzes des Landes Baden-Württemberg sollen im Rahmen dieses dreijährigen Projektes alle archäologischen Verdachtsflächen in einem Gebiet von 35,751km² kartiert werden. Neben der Identifizierung einer Vielzahl bisher unbekannter Bodendenkmäler insbesondere in den bewaldeten Teilen der Landschaft sind viele bereits in der archäologischen Datenbank des Landes Baden-Württemberg verzeichnete Objekte hinsichtlich ihrer exakten Lage und Ausdehnung zu überprüfen. Im Gegensatz zur konventionellen visuellen Interpretation von LiDAR-Daten, die meist auf der Basis von Reliefdarstellungen des Digitalen Höhenmodells erfolgt, wurde während der ersten Phase des Projektes eine neue Methodik zur Datenverarbeitung entwickelt. Farbkodierte Karten des resultierenden *Lokalen Reliefmodells* (LRM) können genutzt werden, um kleine topographische Details mit Höhendifferenzen von nur 0,1 oder 0,2m gegenüber ihrer Umgebung deutlich sichtbar zu machen.

Die Bearbeitung von zwei großenteils bewaldeten Testgebieten – dem Schönbuch im Zentrum Baden-Württembergs sowie dem Südschwarzwald – hat das große Potential dieser Methodik für die archäologische Prospektion ausgedehnter Gebiete gezeigt. Bereits während der ersten Monate der LiDAR-basierten Prospektion konnten 1900km² (5,5% der Landesfläche) bearbeitet werden. Das Ergebnis war die Kartierung von 25,600 anthropogenen Strukturen, die als Kulturlandschaftselemente und archäologische Verdachtsflächen interpretiert werden. Im gleichen Gebiet waren bisher 4100 Bodendenkmäler und Fundstellen in der archäologischen Datenbank Baden-Württembergs verzeichnet.

15 | Zwischen den Zeilen – Verbesserung der Methoden zur Erforschung großflächiger, komplexer und überschwemmter Altlandschaften

Simon Fitch, Vincent Gaffney, Benjamin Gearey* und *Eleanor Ramsey

In den vergangenen Jahren ist das archäologische Potential von flachem, küstennahem Festlandboden (sog. Festlandsockel) immer mehr in die Wahrnehmung europäischer Archäologen gerückt. Dies ist nicht auf ein gesteigertes Interesse an „normaler" Marinegeschichte zurückzuführen, sondern vielmehr auf die Entwicklung einer größeren maritimen Archäologie, die zum Ziel hat, Ausschnitte menschlicher Tätigkeiten der riesigen prähistorischen Landschaft, die heute unter dem Meer liegt, zu entdecken und zu erforschen. Bis vor wenigen Jahren, waren diese Gegenden für den Archäologen noch so gut wie unerreichbar. Aber die Entwicklung einer Reihe von Fernerkundungstechniken hat die Erforschung solcher Landschaften nun möglich gemacht. Die dadurch gewonnene Information hat die Forschungsmeinung über das Paläo- und Mesolithikum grundlegend geändert. Der Vortrag zeigt, wie alte seismische Daten genutzt wurden, um diese Paläolandschaft zu kartieren, und diskutiert dabei den Nutzen von 2D und 3D Daten für diese Zwecke. Die Ergebnisse neuer Forschungen in der Nordsee and an der westlichen Küste des Vereinigten Königreichs werden vorgestellt. Hierbei wurde darüber hinaus versucht, welche Erkenntnisse stichprobenartige Untersuchungen des

Meeresbodens, der nicht durch 3D-Technologien erfasst wurde, liefern können.

Der Vortrag stellt zur Diskussion, dass verschiedene Prospektionsmethoden in solchen Untersuchungen angewendet werden müssen. Diese müssen optimiert werden, um die einzelnen wissenschaftlichen Ziele adäquat verfolgen zu können. In Anbetracht der heutigen schlechten finanziellen Lage müssen Archäologen und Denkmalpfleger aus ihren Datensätzen, deren Anfertigung viel Geld gekostet hat, alles herausholen, um das Beste für Forschung und Verwaltung zu erzielen.

16 | Luftbildarchive im Management der archäologischen Denkmalpflege: Das Aerial Reconnaissance Archive (TARA) – eine gemeinsame europäische Quelle

Lesley Ferguson

Das Aerial Reconnaissance Archive (TARA) befindet sich mittlerweile in der Royal Commission on the Ancient and Historic Monuments of Scotland (RCAHMS) in Edinburgh. Hier finden sich außergewöhnliche Aufnahmen von Landschaften und Spuren früherer menschlicher Aktivitäten aus ganz Europa. Die dortigen militärischen Luftaufklärungsbilder umfassen einen Anfertigungszeitraum von den 1940er bis in die 1990er Jahre. Der Vortrag liefert einen kurzen Überblick über die Inhalte von TARA aus archäologischer Sicht und stellt einige Entdeckungen und Verbesserungen seit der Verlagerung zur RCAHMS dar. Auch schaut der Beitrag in die Zukunft und durchleuchtet einige der Möglichkeiten und wesentlichen Partnerschaften für zukünftige Entwicklungen, ebenso die Herausforderungen, denen man ausgesetzt wird, wenn man Archäologen und anderen den Zugang zum Archiv für weitere Forschungen geben möchte.

17 | Fernerkundung und die Verwaltung von (Boden-)Denkmälern – eine zentraleuropäische Perspektive und eine tschechische Fallstudie

Martin Gojda

Neueste Innovationen in der Fernerkundung haben die Möglichkeiten dessen verbessert, was traditionellerweise „Luftbildarchäologie" genannt wird. Diese Veränderungen werden nicht nur in west-, sondern auch in den zentraleuropäischen Ländern, von denen die meisten erst in der zweiten Hälfte der 1990er Jahre Zugang zur Luftbildarchäologie fanden, diskutiert und wahrgenommen. Davor sorgten gesetzliche und administrative Beschränkungen (bezüglich der Fliegerei, des Aufnehmens und Publizierens von Luftbildern etc.) für erhebliche Verzögerungen, wenn dauerhafte und systematische Luftbild-Projekte ins Leben gerufen werden sollten. Trotz allem konnten fernerkundigende Aktivitäten in den letzten zwei Jahrzehnten die Wahrnehmung archäologischer Landschaften und das Verstehen ihrer Siedlungsgeschichte enorm verbessern. Dieser Beitrag zeigt, wie ein laufendes Luftbildarchäologie-Projekt in Böhmen die dortige Forschungsmeinung über frühere Siedlungstätigkeiten und andere dynamische Prozesse verändert hat, und wie wichtig Luftbildfotografie für die Erfassung und Dokumentation tschechischer (Boden)-Denkmäler ist. Ein staatliches gefördertes Pilot-Projekt (2010–2011) testet das Potential von ALS für ein großmaßstabiges Kartieren archäologischer Landschaften und wird den Weg bereiten für eine zukünftige Erfassung der gesamten tschechischen Republik mit LiDAR Daten.

18 | Die Verwaltung archäologischer Denkmäler in Lothringen mithilfe von ALS-Daten

Murielle Georges-Leroy

Der Vortrag stellt zwei Untersuchungsareale in Lothringen vor, die mit LiDAR erfasst wurden, um die dortigen archäologischen Denkmäler adäquat zu verwalten. Das erste Gebiet liegt im *Forêt de Haye* bei Nancy. Dort wurde eine große Anzahl von Denkmälern verschiedener Art und Zeitstellungen aufgenommen. Durch die Aufnahmen war es möglich, Lage und Ausdehnung der Denkmäler zu ermitteln, um sie so vor Zerstörung durch die dortigen Waldarbeiten zu schützen. Das zweite Gebiet wurde im Rahmen des Trassenbaus der neuen Hochgeschwindigkeitsstrecke zwischen Paris und Straßburg untersucht. Hier können deutlich der Nutzen, aber auch die Grenzen dieser Methode als präventive Maßnahme zum Schutz archäologischer Kulturgüter hervorgehoben werden.

19 | Luftbildarchäologie einer unterirdischen Landschaft: Das Tóköz-Projekt

Zoltán Czajlik, *László Rupnik*, *Máté Losonczi* und *Lőrinc Timár*

Das GIS Forschungsinstitut und das Luftbildarchiv des Archäologischen Instituts der Eötvös Loránd Universität wurde im Jahre 1993 gegründet, mit dem Ziel, die archäologischen Notbergungen entlang des M3 Autobahnbaus zu verwalten. Durch eine Kooperation von französischen und ungarischen Archäologen konnte eine Sammlung von mehr als 30,000 Luftbildern aufgebaut werden, die mehr als 1,500 Denkmäler umfasst. Neben der archäologischen Interpretation von Luftbildern und der Erstellung von Fotokarten wurden ebenso gezielt archäologische Felddaten in kleinen, vorher ausgewiesenen Untersuchungsgebieten erhoben. Sie liegen im Nordwesten von Ungarn in der Mikroregion Tóköz. Mit Hilfe der Luftbildarchäologie konnten große oftmals zusammenhängende archäologisch relevante Stellen entdeckt werden, obwohl sie zum Teil unter einer dicken, spätholozänen Schwemmschicht liegen.

20 | Die archäologische Landschaft Nordostislands: die Geister der wikingerzeitlichen Gesellschaft

Árni Einarson und *Oscar Aldred*

Im Zuge eines aktuellen Projekts werden mittelalterliche Hinterlassenschaften in einer Untersuchungsregion von über 3000km^2 im Nordosten Islands kartiert. Genutzt werden u. a. großmaßstabige Luftbilder, darunter über 2000 Schrägaufnahmen, die aus geringer Höhe aufgenommen wurden. Unterstützt durch Feldbegehungen konnte ein ganzes System gut erhaltener mittelalterlicher Hinterlassenschaften aufgedeckt werden. Besonders hervorzuheben ist ein ausgedehntes Netzwerk von Torfmauern, aber auch Bauernhöfe, Kirchen, locker gestreuten Ansammlungen von Gebäuden, heidnische Gräberfelder, Stätten der Torfgewinnung, Kohlemeiler und alte Trampelpfade sind eindeutig erkennbar. Diese mittelalterliche Landschaft gehört hauptsächlich in die Wikingerzeit und ist die Hinterlassenschaft der ersten Besiedlung Islands. Die gute Erhaltung und die Möglichkeiten der zukünftigen Erhaltung sollen im Beitrag diskutiert werden.

21 | Verhaltener Optimismus: Denkmalschutz und Kulturerbemanagement in Slowenien

Gašper Rutar **und** ***Matija Črešnar***

Aufgrund von gesetzlichen Änderungen im Jahr 2008 bestimmten bedeutende Neuerungen die letzten zwei Jahre die Verwaltung von (Boden-)Denkmälern in Slowenien. Einige der bemerkenswertesten Modifikationen betrifft die präventive Archäologie. Es wurde eine neue, hauptsächlich auf Forschung ausgerichtete Körperschaft ins Leben gerufen: das Zentrum für Präventive Archäologie (CPA – *Centre for Preventive Archaeology*). Es ist ein Teil des Instituts zum Schutz des kulturellen Erbes von Slowenien (IPCHS – *Protection of Cultural Heritage of Slovenia*). Seine Aufgaben sind angebunden an die Durchführung archäologischer Untersuchungen im Vorfeld des *National Spatial Plan* (NPS). Die Arbeit und die daraus gewonnenen Ergebnisse der ersten 15 Monate des Bestehens sind ermutigend, lassen aber noch Möglichkeiten zur weiteren Verbesserung zu.

22 | Stätten des Ersten Weltkrieges in Belgien: die Verbindung von historischen Luftaufnahmen und EMI

Birger Stichelbaut, Timothy Saey, Fun Meeuws, Jean Bourgeois **und** ***Marc Van Meirvenne***

Millionen von Luftbildern wurden von allen am 1. Weltkrieg teilnehmenden Nationen gemacht. Diese Aufnahmen sind bemerkenswerte, jedoch lange Zeit unterschätzte Quellen an Information über das damalige Geschehen. Dieser Beitrag beschreibt ein interdisziplinäres Forschungsprojekt, bestehend aus Archivarbeit, Interpretation und detaillierter Kartierung von historischen Luftaufnahmen und die Anwendung von elektromagnetischen Messungen. Diese Techniken werden in zwei Fallstudien (Geluveld & Oostduinkerke) zusammengefasst, die sich mit der Auswertung archäologischer Hinterlassenschaften des 1. Weltkriegs beschäftigen.

23 | Ein Blick aus der Luft in die Vergangenheit – Luftbildarchäologie in Dänemark

Lis Helles Olesen

Für ein vierjähriges Projekt über Luftbildarchäologie wurde dem Holstebro-Museum im Westen Dänemarks ein finanzielles Budget von 65,000 Euro zur Verfügung gestellt. Das Projekt läuft bis 2013 und beinhaltet neun Teilprojekte. Die ersten, vorläufigen Ergebnisse der Entdeckung, Beobachtung und Erfassung archäologischer Denkmäler aus der Luft sollen in diesem Beitrag vorgestellt werden. In 2008 und 2009 konnten mit insgesamt 130 Flugstunden rund 560 Luftbildstellen lokalisiert werden. Der Vortrag gibt einen Überblick über die ersten 100 systematisch und vollständig erfassten Denkmäler: unter anderem befinden sich darunter 33 Stellen mit Hausgrundrissen und Grubenhäusern. Diese geben eine besonders gute Grundlage für einen Abriss der Besiedlungsgeschichte der späten Eisen- und Wikingerzeit (20% aller Luftbildstellen datieren in diese beiden Epochen). Das Resultat einer ersten Aufnahme von 50 weiteren archäologischen Stellen wird ebenfalls präsentiert. Diese vorläufigen Ergebnisse lassen darauf schließen, dass den vorgestellten Teilprojekten ein großes Potential innewohnt.

24 | Wissensbasierte Luftbildinterpretation

Rog Palmer

Die wissensbasierte Bildinterpretation ist ein subjektiver Prozess. Es werden Bilder (Luftbilder, Satellitenbilder oder andere fernerkundliche Aufnahmen) zu Informationen umgewandelt, indem sie ein Auswertungsverfahren durchlaufen, das von einer bestimmten Frage geleitet wird. Der Vortrag setzt sich mit der Art und Weise auseinander, wie wir Bilder anschauen und wie unsere Beweggründe beeinflussen, was wir sehen. Ein Beispiel wird zeigen, wieviel Wert ein Foto haben kann, wenn man eine ganze Reihe von Fragen durchgeht und wieviel Information aus einem einzelnen Luftbild gewonnen werden kann. Es werden bestimmte Arbeitswege und der Umgang mit Luftbildern vorgestellt, und unterschiedliche Level der Interpretation und Kartierung werden aufgezeigt. Der letzte Abschnitt verbindet die Ergebnisse der Luftbildinterpretation mit denen anderer Quellen und verdeutlicht, wie diese uns helfen, die Vergangenheit zu verstehen.

25 | Eine Übung für die Zukunft: Kooperationen in ganz Europa

Chris Musson

Seit den 1990ern wird immer wieder darauf hingewiesen, dass die Möglichkeiten der praktischen und theoretischen Ausbildung in Luftbildarchäologie und anderer Formen der Fernerkundung stark begrenzt sind und infolgedessen ihr Potential innerhalb des Denkmalschutzes und der Landschaftsstudien nicht vollständig ausgeschöpft wird. Im ersten Teil des Beitrages werden einige der Schritte der letzen 15 Jahre in Richtung Verbesserung dieser Situation umrissen. Der zweite Teil beschäftigt sich mit den Zukunftsaussichten einer kürzlich begonnenen internationalen fünfjährigen Zusammenarbeit, gefördert durch das Kulturprogramm der Europäischen Union.